PREALGEBRA

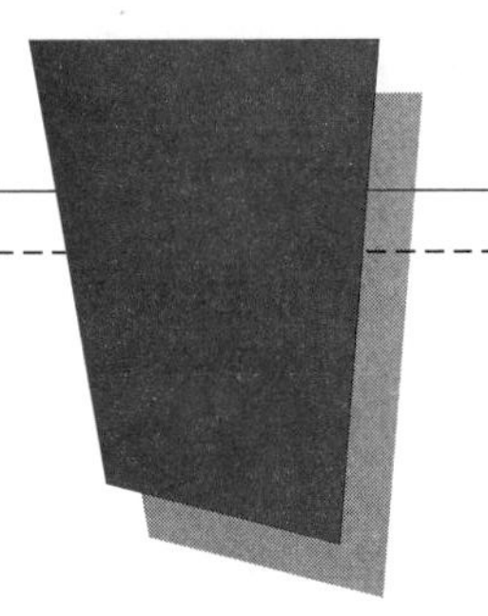

PREALGEBRA

Third Edition

Charles P. McKeague
Cuesta College, *San Luis Obispo*

PWS PUBLISHING COMPANY

I(T)P **An International Thomson Publishing Company**
Boston • Albany • Bonn • Cincinnati • Detroit • London • Madrid • Melbourne • Mexico City • New York • Paris • San Francisco • Singapore • Tokyo • Toronto • Washington

PWS PUBLISHING COMPANY
20 Park Plaza, Boston, MA 02116-4324

Library of Congress Cataloging-in-Publication Data
McKeague, Charles P.
Prealgebra / Charles P. McKeague. — 3rd ed.
p. cm.
Includes index.
ISBN 0-534-94764-6
1. Mathematics. I. Title.
QA39.2.M418 1995 95-35608
513'.14—dc20 CIP

Sponsoring Editor *Susan J. McCulley*
Developmental Editor *Elizabeth R. Deck*
Production Coordinator *Elise S. Kaiser*
Marketing Manager *Marianne C. P. Rutter*
Manufacturing Coordinator *Marcia A. Locke*
Production *Quadrata, Inc.*
Interior Designer *Julia Gecha*
Cover Designer *Elise S. Kaiser*
Interior Illustrator *Academy Artworks, Inc.*
Cover Art *Celia Johnson*
Compositor *Beacon Graphics Corporation*
Cover Printer *New England Book Components, Inc.*
Text Printer and Binder *Courier/Westford*
Printed and bound in the United States of America.
95 96 97 98 99 — 10 9 8 7 6 5 4 3 2 1

 *This book is printed on recycled, acid-free paper.*

For more information, contact:

PWS Publishing Company
20 Park Plaza
Boston, MA 02116

International Thomson Publishing Europe
Berkshire House I68-I73
High Holborn
London WC1V 7AA
England

Thomas Nelson Australia
102 Dodds Street
South Melbourne, 3205
Victoria, Australia

Nelson Canada
1120 Birchmont Road
Scarborough, Ontario
Canada M1K 5G4

International Thomson Editores
Campos Eliseos 385, Piso 7
Col. Polanco
11560 Mexico D.F., Mexico

International Thomson Publishing GmbH
Königswinterer Strasse 418
53227 Bonn, Germany

International Thomson Publishing Asia
221 Henderson Road
#05-10 Henderson Building
Singapore 0315

International Thomson Publishing Japan
Hirakawacho Kyowa Building, 31
2-2-1 Hirakawacho
Chiyoda-ku, Tokyo 102
Japan

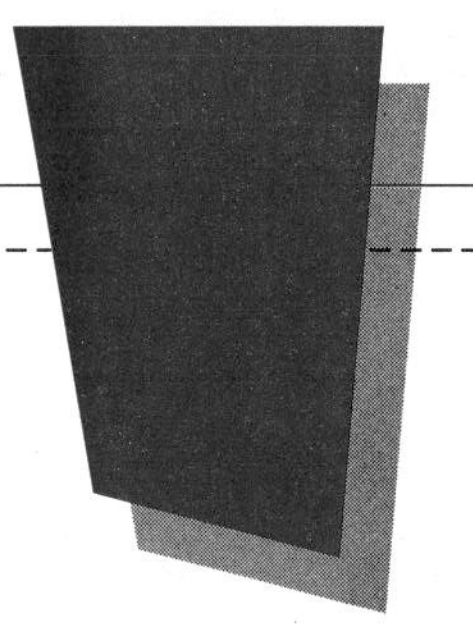

CONTENTS

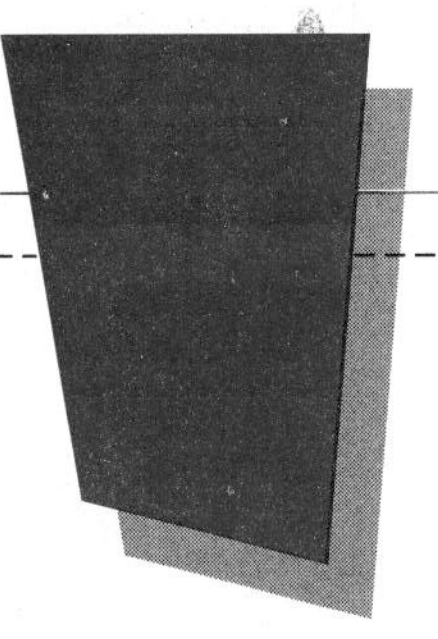

PREFACE TO THE INSTRUCTOR

Prealgebra was originally written to bridge the gap between arithmetic and introductory algebra. The basic concepts of algebra are introduced early in the book and then applied to new topics as they are encountered.

General Overview of the Text

The commutative, associative, and distributive properties are covered in Chapter 1, along with the arithmetic of whole numbers. Chapter 2 gives a thorough coverage of negative numbers, along with an introduction to exponents and polynomials. Fractions (both positive and negative), mixed numbers, and simple algebraic fractions are covered in Chapter 3. Linear equations in one variable are covered in Chapter 4, along with graphing in two dimensions. All the material on positive and negative numbers and linear equations is carried into Chapter 5 on decimals. Chapters 6 and 7 cover ratio and proportion, measurement, and percent.

New for the Third Edition

Chapter Introductions Each chapter opens with a real-world application. Whenever possible, these introductions are expanded in the chapter and then carried through to a topic found later on in the book.

Study Skills The first five chapter introductions contain a list of study skills to help students organize their time efficiently. More detailed than the general study skills listed in the Preface to the Student, these study skills point students in the direction of success. They are intended to benefit students not only in this course but also throughout their college careers.

Blueprint for Problem Solving New to this edition, and introduced in Section 4.5 on page 289, the Blueprint for Problem Solving is a detailed outline of the steps needed to successfully attempt application problems that involve equations in one variable.

Emphasis on Visualization of Topics This edition contains many more diagrams, charts, and graphs than the previous edition. The purpose of this is to give students additional information, in visual form, to help them understand the topics covered.

Calculator Notes Each time a new operation is encountered, there is a note as to how to work that problem on a calculator. Notes are given for both scientific and graphing calculators.

Chapter Tests Each chapter ends with a test that contains a representative sample of the problems covered in the chapter.

Features of the Text

Flexibility In a lecture-format class, each section of the book can be discussed in a 45- or 50-minute class session. A technique that I have used successfully in lecture classes is to have the students work some of the Practice Problems in the margins of the textbook after I have done an example on the board or overhead projector.

In a self-paced class, the Practice Problems in the margins allow the student to become actively involved with the material before starting the Problem Set that follows that section.

Early and Integrated Coverage of Geometry The material on perimeter, area, and volume is introduced in Chapter 1, and then integrated throughout the rest of the book. Here is a list of the sections into which these topics have been integrated:

Section 1.3	Perimeter of a polygon
Section 1.10	Area of a square and rectangle; volume and surface area of a rectangular solid
Section 3.3	Area of a triangle
Section 5.3	Area and circumference of a circle; volume of a right circular cylinder
Section 5.5	Volume of a sphere

Section Organization Each chapter is divided into sections that contain the following items:

1. *Section Openings:* Many sections open with a short application problem that leads into the topic being discussed.
2. *Explanations and Examples:* The explanations are made as simple and intuitive as possible. The examples are chosen to clarify the explanations and preview the Problem Set.
3. *Practice Problems:* In the margin of the text, next to each example, is a Practice Problem with the same number. Each Practice Problem is to be worked after the example with the same number has been studied.
4. *Answers to the Practice Problems:* The answers to the Practice Problems are placed at the bottom of the page on which the problems occur.
5. *Solutions to the Practice Problems:* Solutions to all Practice Problems that require more than one step are provided in the back of the book.

Unit Analysis Conversions within and between units in the U.S. system of measurement and the metric system are accomplished by unit analysis. The advantage to using this method of conversion is that it is used in other disciplines, such as nursing and chemistry. See Chapter 6.

Practice Problems The practice problems, with their answers and solutions, are the key to moving students through the material in this book. In the margin, next to each example, is a practice problem. The practice problems are to be worked by students after they have read through the corresponding example. The *answer* to each practice problem is at the bottom of the page on which the problem occurs. *Solutions* to most practice problems are in the back of the book.

One Step Further Many of the Problem Sets in this book end with a few problems under this heading. These include more challenging problems, those that require written responses, require critical thinking, and extend the topics covered in the sections. In most cases, there are no examples in the text similar to these problems.

Problem Set Organization Following each section is a Problem Set. For this edition, Problem Sets have been expanded to include more work on estimation and reading and using graphs. Problem Sets contain the following items:

1. *Drill:* There are enough drill problems in each Problem Set to ensure that students working the odd-numbered problems become proficient with the material.
2. *Progressive Difficulty:* The problems increase in difficulty as the Problem Set progresses.
3. *Odd–Even Similarities:* Whenever possible, each even-numbered problem is similar to the odd-numbered problem that precedes it. The answers to the odd-numbered problems are listed in the back of the book.
4. *Applications:* Application problems are emphasized throughout the book. The placement of the word problems is such that the student must do more than just read the section title to decide how to set up and solve a word problem. For example, Problem Set 1.7 covers multiplication with whole numbers. The solutions to some of the application problems in the Problem Set require multiplication, but others require addition and subtraction as well.
5. *Calculator Problems:* Most of the Problem Sets contain a set of problems to be worked on a calculator.
6. *Review Problems:* Beginning in Chapter 2, each Problem Set contains a few review problems. Whenever possible, these problems review concepts from preceding sections that are needed for the development of the following section.
7. *One Step Further Problems:* These problems, most of which can be categorized as critical thinking problems, extend the concepts covered in the section.
8. *Answers:* The answers to all odd-numbered problems in the Problem Sets and Chapter Reviews, and the answers to all problems in the Chapter Tests, are given in the back of the book.

Chapter Summaries Each Chapter Summary lists the new properties and definitions found in the chapter. The margins in the Chapter Summaries contain examples that illustrate the topics being reviewed.

Chapter Reviews Following the Chapter Summary is a set of Review Problems that cover all the different types of problems found in the chapter. The Chapter Reviews are longer and more extensive than the Chapter Tests.

1

WHOLE NUMBERS

INTRODUCTION

The table below shows the average amount of caffeine in a number of different beverages. The chart next to the table is a visual presentation of the same information. The relationship between the table and the chart is one of the things we will study in this chapter. The table gives information in numerical form, while the chart gives the same information in a geometrical way. In mathematics it is important to be able to move back and forth between the two forms. Later, in Chapter 4, we will introduce a third form, the algebraic form, in which we summarize relationships with equations.

Drink (6-ounce cup)	Caffeine (in milligrams)
Brewed coffee	100
Instant coffee	70
Tea	50
Cocoa	5
Decaffeinated coffee	4

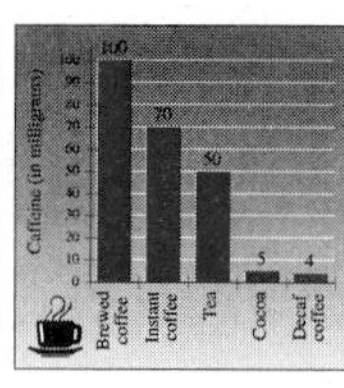

To begin our study of prealgebra, we will develop the rules and properties for adding, subtracting, multiplying, and dividing whole numbers. Once we have covered the four basic operations, we will introduce the rule for order of operations that tells us how to simplify expressions that contain more than one operation. As we progress through the chapter, we will consider ways of estimating answers to mathematics problems, work with averages, and review some of the formulas for perimeter, area, and volume.

For many of the topics we will cover in this course, you will find that understanding the concepts is not always enough to ensure success; some of the material here will have to be memorized. The facts and concepts that need to be memorized are always clearly indicated.

1

Chapter Introductions

A real-world application now opens each chapter, and is often expanded and reiterated in the book.

Emphasis on Visualization of Topics

Clear, informative diagrams, charts, and graphs enliven the new edition and help students visualize the topics they are studying.

Study Skills

New in this edition, this list of detailed study skills specific to the chapter helps students organize their work and become more efficient with their time.

STUDY SKILLS

Some of the students enrolled in my mathematics classes develop difficulties early in the course. Their difficulties are not associated with their ability to learn mathematics; they all have the potential to pass the course. Students who get off to a poor start do so because they have not developed the study skills necessary to be successful in mathematics; they do not put themselves on an effective homework schedule, and when they work problems, they do it their way, not my way. Here is a list of things you can do to begin to develop effective study skills.

1 Put yourself on a schedule. The general rule is that you spend two hours on homework for every hour you are in class. Make a schedule for yourself in which you set aside two hours each day to work on this course. Once you make the schedule, stick to it. Don't just complete your assignments and then stop. Use all the time you have set aside. If you complete the assignment and have time left over, read the next section in the book, and then work more problems. As the course progresses you may find that two hours a day is not enough time to master the material in this course. If it takes you longer than two hours a day to reach your goals for this course, then that's how much time it takes. Trying to get by with less will not work.

2 Find your mistakes and correct them. There is more to studying mathematics than just working problems. You must always check your answers with the answers in the back of the book. When you have made a mistake, find out what it is, and then correct it. Making mistakes is part of the process of learning mathematics. I have never had a successful student who didn't make mistakes—lots of them. Your mistakes are your guides to understanding; look forward to them.

3 Imitate success. Your work should look like the work you see in this book and the work your instructor shows. The steps shown in solving problems in this book were written by someone who has been successful in mathematics. The same is true of your instructor. Your work should imitate the work of people who have been successful in mathematics.

If you are determined to pass this course, make it a point to become organized and effective right from the start, and to show your work in the same way as it is shown in this book.

2

Practice Problems
A hallmark of the McKeague texts, practice problems in the margin are the key to moving students through the material. Answers are at the bottom of the page. Solutions are at the back of the book.

Annotations
Each step of the problem is carefully annotated in color to reinforce for students the logical problem-solving progression.

Calculator Notes
Also new in this edition, calculator notes show students how to work each new problem on both scientific and graphing calculators.

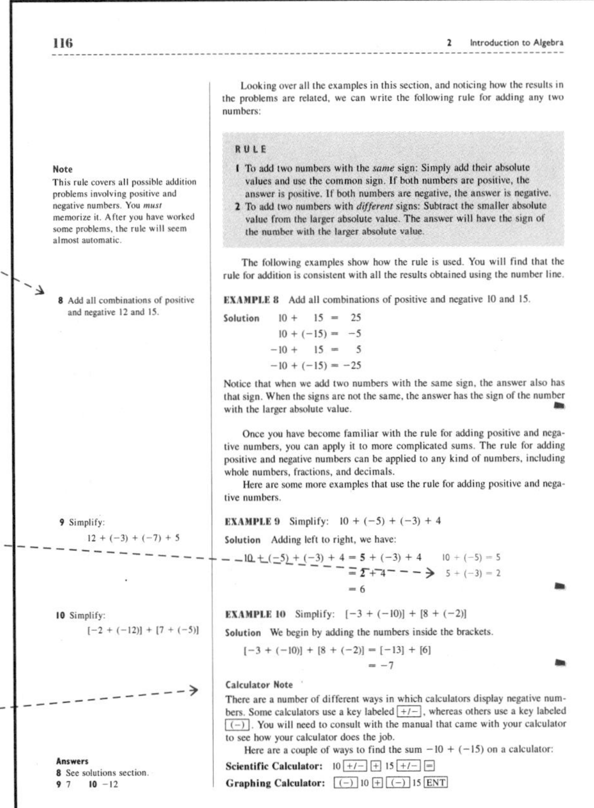

116 2 Introduction to Algebra

Looking over all the examples in this section, and noticing how the results in the problems are related, we can write the following rule for adding any two numbers:

Note
This rule covers all possible addition problems involving positive and negative numbers. You *must* memorize it. After you have worked some problems, the rule will seem almost automatic.

RULE
1 To add two numbers with the *same* sign: Simply add their absolute values and use the common sign. If both numbers are positive, the answer is positive. If both numbers are negative, the answer is negative.
2 To add two numbers with *different* signs: Subtract the smaller absolute value from the larger absolute value. The answer will have the sign of the number with the larger absolute value.

The following examples show how the rule is used. You will find that the rule for addition is consistent with all the results obtained using the number line.

8 Add all combinations of positive and negative 12 and 15.

EXAMPLE 8 Add all combinations of positive and negative 10 and 15.

Solution
$10 + 15 = 25$
$10 + (-15) = -5$
$-10 + 15 = 5$
$-10 + (-15) = -25$

Notice that when we add two numbers with the same sign, the answer also has that sign. When the signs are not the same, the answer has the sign of the number with the larger absolute value.

Once you have become familiar with the rule for adding positive and negative numbers, you can apply it to more complicated sums. The rule for adding positive and negative numbers can be applied to any kind of numbers, including whole numbers, fractions, and decimals.

Here are some more examples that use the rule for adding positive and negative numbers.

9 Simplify: $12 + (-3) + (-7) + 5$

EXAMPLE 9 Simplify: $10 + (-5) + (-3) + 4$

Solution Adding left to right, we have:
$10 + (-5) + (-3) + 4 = 5 + (-3) + 4$ $\quad 10 + (-5) = 5$
$= 2 + 4$ $\quad 5 + (-3) = 2$
$= 6$

10 Simplify: $[-2 + (-12)] + [7 + (-5)]$

EXAMPLE 10 Simplify: $[-3 + (-10)] + [8 + (-2)]$

Solution We begin by adding the numbers inside the brackets.
$[-3 + (-10)] + [8 + (-2)] = [-13] + [6]$
$= -7$

Calculator Note
There are a number of different ways in which calculators display negative numbers. Some calculators use a key labeled [+/−], whereas others use a key labeled [(−)]. You will need to consult with the manual that came with your calculator to see how your calculator does the job.

Here are a couple of ways to find the sum $-10 + (-15)$ on a calculator:
Scientific Calculator: 10 [+/−] [+] 15 [+/−] [=]
Graphing Calculator: [(−)] 10 [+] [(−)] 15 [ENT]

Answers
8 See solutions section.
9 7 **10** −12

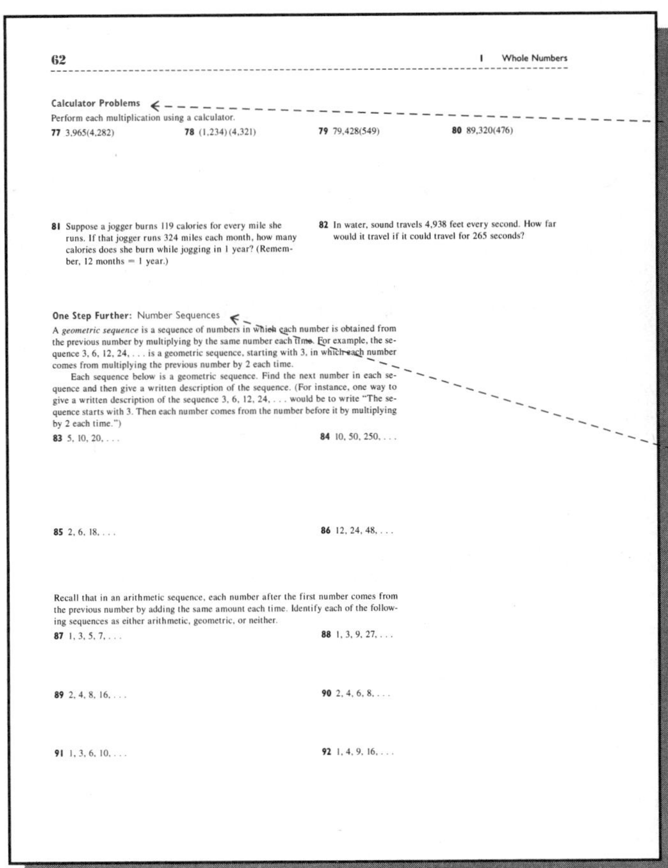

62 1 Whole Numbers

Calculator Problems
Perform each multiplication using a calculator.
77 3,965(4,282) **78** (1,234)(4,321) **79** 79,428(549) **80** 89,320(476)

81 Suppose a jogger burns 119 calories for every mile she runs. If that jogger runs 324 miles each month, how many calories does she burn while jogging in 1 year? (Remember, 12 months = 1 year.)

82 In water, sound travels 4,938 feet every second. How far would it travel if it could travel for 265 seconds?

One Step Further: Number Sequences
A *geometric sequence* is a sequence of numbers in which each number is obtained from the previous number by multiplying by the same number each time. For example, the sequence 3, 6, 12, 24, . . . is a geometric sequence, starting with 3, in which each number comes from multiplying the previous number by 2 each time.

Each sequence below is a geometric sequence. Find the next number in each sequence and then give a written description of the sequence. (For instance, one way to give a written description of the sequence 3, 6, 12, 24, . . . would be to write "The sequence starts with 3. Then each number comes from the number before it by multiplying by 2 each time.")

83 5, 10, 20, . . . **84** 10, 50, 250, . . .

85 2, 6, 18, . . . **86** 12, 24, 48, . . .

Recall that in an arithmetic sequence, each number after the first number comes from the previous number by adding the same amount each time. Identify each of the following sequences as either arithmetic, geometric, or neither.

87 1, 3, 5, 7, . . . **88** 1, 3, 9, 27, . . .

89 2, 4, 8, 16, . . . **90** 2, 4, 6, 8, . . .

91 1, 3, 6, 10, . . . **92** 1, 4, 9, 16, . . .

Calculator Problems
To reinforce the skills in the text's new Calculator Notes, each problem set includes exercises specifically designed for solution with a calculator.

One Step Further
These sections include problems that challenge the student, that require written response or critical thinking, or that extend beyond the topics in the chapter.

Supplements to the Text

Prealgebra, Third Edition, comes with a complete package of supplements to aid both the instructor and the student.

For the Instructor

Annotated Instructor's Edition This is a specially bound version of the complete student text with exercise answers printed adjacent to each exercise and teaching notes written for the instructor by the author.

Instructor's Manual The manual includes tests and even-numbered answers to the text exercises. There are three multiple-choice and two free-response tests for each chapter, as well as two final exams. A complete answer key is provided. All test items are new for this edition.

EXP-Test This is a fast, highly flexible computerized testing system for the IBM-PC and compatibles. It is available for DOS and Windows environments. Instructors can edit and scramble the database of test items or create their own questions. A Macintosh testing system is also available.

For the Student

Videotapes Created and hosted by the author, this set of videotapes is available free upon adoption of the text. Each tape offers one chapter of the text and is broken down into 10–20-minute problem-solving lessons that cover each section of the chapter.

Video Notebook This workbook contains all of the problems that are worked on the videotapes in a format that allows students to work along with the video. It is available for student purchase at a minimal cost.

MathQUEST Tutorial Software This software is available for both Macintosh and Windows environments. It can be used on a network or an individual computer, and has been designed for quick and easy use. Students can find and work the appropriate problems simply by pointing and clicking on the table of contents, which mirrors the text's table of contents. The program features three types of problems: multiple-choice, true/false, and free-response. Fully-worked solution displays of each problem can be accessed at any time during the problem-solving session. MathQUEST also offers hints when students give the wrong answer and includes an optional on-screen calculator. A record-keeping utility lets students track their progress and records their work in class files.

Computerized Interactive Algeblocks This computerized version of the popular manipulative model helps students make the jump from concrete arithmetic reasoning to symbolic algebraic reasoning. The user chooses topics from a menu provided. After a short tutorial, the user manipulates blocks of varying size and color on the screen to set up the "equation." Volume 1 topics include adding variables, simplifying expressions, and solving equations.

Math Facts, Second Edition Created by Theodore Szymanski of Tompkins Cortland Community College, this pocket-sized reference book contains essential arithmetic concepts, formulas, and procedures in a concise and readable format. It is available for student purchase at a minimal cost.

Acknowledgments

A project of this size cannot be completed without help from many people. In particular, Susan McCulley and Elizabeth Deck, my editors at PWS Publishing Company, did an outstanding job of commissioning and summarizing the reviews of the book, and an equally outstanding job handling all of the peripheral items—contract questions, supplement packages, and so on—associated with this project. Elise Kaiser, the production editor on this project, has done an exceptional job of keeping all the pieces together, coordinating the accuracy checking, editing, art rendering, and proofreading for the book, making the corrections I requested, and getting the book published on time. My thanks to all these people. Their help with this project made it much less cumbersome than it could have been.

Thanks also to Diane McKeague for her help with the nutrition label problems, and other health- and fitness-related topics. To Amy McKeague for her encouragement and advice on the new elements of the text. To Patrick McKeague for his help with the answer sections, videotape production, and general advice on application problems. To Geri Davis and Martha Morong of Quadrata, Inc., for their meticulous production work; Kate Pawlik, Chaudra Hallock, and Hali Hallock for their help with the problem checking; Julia Gecha for the wonderful design of the book; and Bob Martin for his assistance with ancillary preparation.

Finally, I am grateful to the following instructors for their suggestions and comments on this revision. Some reviewed the entire manuscript, while others were asked to evaluate the development of specific topics or the overall sequence of topics. My thanks go to the people listed below:

Carla K. Ainsworth
Salt Lake Community College

Sarah Carpenter
Vincennes University

Deann Christianson
University of the Pacific

Anne Harris
Merced College

Jean P. Millen
Dekalb College—Central Campus

Jennie Preston-Sabin
Austin Peay State University

Dr. Leon F. Sagan
Anne Arundel Community College

Karen L. Schwitters
Seminole Community College

Joel Spring
Broward Community College

Charles P. McKeague

PREFACE TO THE STUDENT

Many of the people who enroll in my introductory math classes are apprehensive at first. They think that since they have had a difficult time with mathematics in the past, they are in for a difficult time again. Some of them feel that they have never really understood mathematics and probably never will.

Do you feel like that?

Most people who have a difficult time with mathematics expect something from it that is not there. They believe that since mathematics is a logical subject, they should be able to understand it without any trouble. Mathematics is probably the most logical subject there is, but that doesn't mean it is always easy to understand. Some topics in mathematics must be read and thought about many times before they become understandable.

If you are interested in having a positive experience in this course and have had difficulty with mathematics in the past, then the following list will be of interest to you. The list explains the things you can do to make sure that you are successful in mathematics.

How to Be Successful in Mathematics

1 **If you are in a lecture class, be sure to attend all class sessions on time.** You cannot know exactly what goes on in class unless you are there. Missing class and then expecting to find out what went on from someone else is not the same as being there yourself.

2 **Read the book and work the Practice Problems.** It is best to read the section that will be covered in class beforehand. Reading in advance, even if you do not understand everything you read, is still better than going to class with no idea of what will be discussed. As you read through each section, be sure to work the Practice Problems in the margin of the text. Each Practice Problem is similar to the example with the same number. Look over the example and then try the corresponding Practice Problem. The answers to the Practice Problems are given on the same page as the problems. If you don't get the correct answer, see if you can rework the problem correctly. If you miss it a second time, check your solution with the solution to the Practice Problem in the back of the book.

3 **Work problems every day and check your answers.** This is especially important if you are studying in a self-paced environment. The key to success in mathematics is working problems. The more problems you work,

the better you will become at working them. The answers to the odd-numbered problems are given in the back of the book. When you have finished an assignment, be sure to compare your answers with those in the book. If you have made a mistake, find out what it is, and correct it.

4 **Do it on your own.** Don't be misled into thinking someone else's work is your own. Watching someone else work through a problem is not the same as working the same problem yourself. It is okay to get help when you are stuck; as a matter of fact, it is a good idea. Just be sure that you do the *work* yourself.

5 **Don't expect to understand every new topic the first time you see it.** Sometimes you will understand everything you are doing, and sometimes you won't. That's just the way things are in mathematics. Expecting to understand each topic the first time you see it will only lead to disappointment and frustration. The process of understanding mathematics takes time. It requires that you read the book, work the problems, and get your questions answered.

6 **Review every day.** After you have finished the assigned problems on a certain day, take at least another 15 minutes to go back and review a section you did previously. You can review by working the Practice Problems in the margin or by doing some of the problems in the Problem Set. The more you review, the longer you will retain the material. Also, there are times when material that at first seemed unclear will become understandable when you review it.

7 **Spend as much time as it takes for you to master the material.** There is no set formula for the exact amount of time you will need to spend on the material in this course in order to master it. You will find out quickly—probably on the first test—if you are spending enough time studying. Even if it turns out that you have to spend 2 or 3 hours on each section to master it, then that's how much time you should take. Trying to get by with less time will not work.

8 **Relax.** It's probably not as difficult as you think.

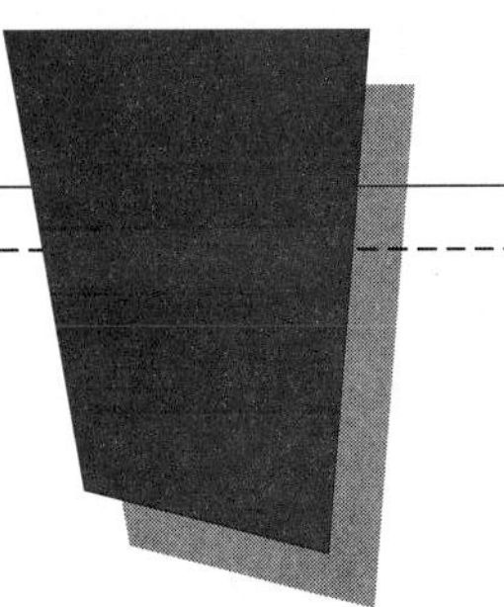

PREALGEBRA

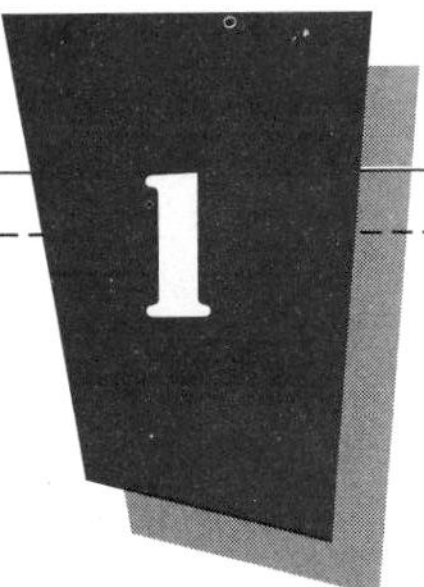

WHOLE NUMBERS

INTRODUCTION

The table below shows the average amount of caffeine in a number of different beverages. The chart next to the table is a visual presentation of the same information. The relationship between the table and the chart is one of the things we will study in this chapter. The table gives information in numerical form, while the chart gives the same information in a geometrical way. In mathematics it is important to be able to move back and forth between the two forms. Later, in Chapter 4, we will introduce a third form, the algebraic form, in which we summarize relationships with equations.

Drink (6-ounce cup)	Caffeine (in milligrams)
Brewed coffee	100
Instant coffee	70
Tea	50
Cocoa	5
Decaffeinated coffee	4

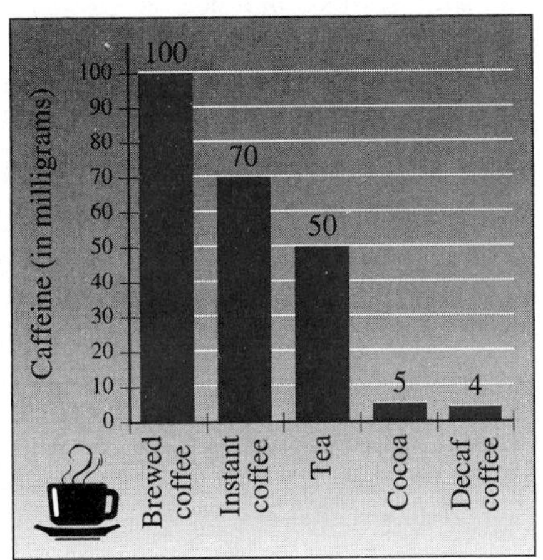

To begin our study of prealgebra, we will develop the rules and properties for adding, subtracting, multiplying, and dividing whole numbers. Once we have covered the four basic operations, we will introduce the rule for order of operations that tells us how to simplify expressions that contain more than one operation. As we progress through the chapter, we will consider ways of estimating answers to mathematics problems, work with averages, and review some of the formulas for perimeter, area, and volume.

For many of the topics we will cover in this course, you will find that understanding the concepts is not always enough to ensure success; some of the material here will have to be memorized. The facts and concepts that need to be memorized are always clearly indicated.

STUDY SKILLS

Some of the students enrolled in my mathematics classes develop difficulties early in the course. Their difficulties are not associated with their ability to learn mathematics; they all have the potential to pass the course. Students who get off to a poor start do so because they have not developed the study skills necessary to be successful in mathematics; they do not put themselves on an effective homework schedule, and when they work problems, they do it their way, not my way. Here is a list of things you can do to begin to develop effective study skills.

1 Put yourself on a schedule. The general rule is that you spend two hours on homework for every hour you are in class. Make a schedule for yourself in which you set aside two hours each day to work on this course. Once you make the schedule, stick to it. Don't just complete your assignments and then stop. Use all the time you have set aside. If you complete the assignment and have time left over, read the next section in the book, and then work more problems. As the course progresses you may find that two hours a day is not enough time to master the material in this course. If it takes you longer than two hours a day to reach your goals for this course, then that's how much time it takes. Trying to get by with less will not work.

2 Find your mistakes and correct them. There is more to studying mathematics than just working problems. You must always check your answers with the answers in the back of the book. When you have made a mistake, find out what it is, and then correct it. Making mistakes is part of the process of learning mathematics. I have never had a successful student who didn't make mistakes—lots of them. Your mistakes are your guides to understanding; look forward to them.

3 Imitate success. Your work should look like the work you see in this book and the work your instructor shows. The steps shown in solving problems in this book were written by someone who has been successful in mathematics. The same is true of your instructor. Your work should imitate the work of people who have been successful in mathematics.

If you are determined to pass this course, make it a point to become organized and effective right from the start, and to show your work in the same way as it is shown in this book.

1.1 Place Value and Names for Numbers

Although there are many different number systems and a variety of symbols used to represent numbers, our number system is based on the number 10 and is therefore called a "base 10" number system. We write all numbers in our number system using the *digits* 0, 1, 2, 3, 4, 5, 6, 7, 8, and 9. The positions of the digits in a number determine the values of the digits. For example, the 5 in the number 251 has a different value from the 5 in the number 542.

The *place values* in our number system are as follows: The first digit on the right is in the *ones column*. The next digit to the left of the ones column is in the *tens column*. The next digit to the left is in the *hundreds column*. For a number like 542, the digit 5 is in the hundreds column, the 4 is in the tens column, and the 2 is in the ones column.

If we keep moving to the left, the columns increase in value. (Actually, each column is 10 times as large as the column on its right. We will say more about this when we cover multiplication.) The following diagram shows the name and value of each of the first seven columns in our number system:

Millions Column	Hundred Thousands Column	Ten Thousands Column	Thousands Column	Hundreds Column	Tens Column	Ones Column
1,000,000	100,000	10,000	1,000	100	10	1

Note

Next to each Example in the text is a Practice Problem with the same number. After you read through an Example, try the Practice Problem next to it. The answers to the Practice Problems are at the bottom of the page. Be sure to check your answers as you work these problems. The worked-out solutions to all Practice Problems with more than one step are given in the back of the book. So if you find a Practice Problem that you cannot work correctly, you can look up the correct solution to that problem in the back of the book.

EXAMPLE 1 Give the place value of each digit in the number 305,964.

Solution Starting with the digit at the right, we have:

4 in the ones column, 6 in the tens column, 9 in the hundreds column, 5 in the thousands column, 0 in the ten thousands column, and 3 in the hundred thousands column.

Practice Problems

1 Give the place value of each digit in the number 46,095.

To find the place values of digits in larger numbers, we can refer to Table 1.

Table 1

Hundred Billions	*Ten Billions*	*Billions*	*Hundred Millions*	*Ten Millions*	*Millions*	*Hundred Thousands*	*Ten Thousands*	*Thousands*	*Hundreds*	*Tens*	*Ones*
100,000,000,000	10,000,000,000	1,000,000,000	100,000,000	10,000,000	1,000,000	100,000	10,000	1,000	100	10	1

Answer

1 5 ones, 9 tens, 0 hundreds, 6 thousands, 4 ten thousands

2 Give the place value of each digit in the number 21,705,328,456.

EXAMPLE 2 Give the place value of each digit in the number 73,890,672,540.

Solution The following diagram shows the place value of each digit.

Ten Billions	*Billions*	*Hundred Millions*	*Ten Millions*	*Millions*	*Hundred Thousands*	*Ten Thousands*	*Thousands*	*Hundreds*	*Tens*	*Ones*
7	3,	8	9	0,	6	7	2,	5	4	0

We can use the idea of place value to write numbers in *expanded form*. For example, the number 542 can be written in expanded form as

$$542 = 500 + 40 + 2$$

because the 5 is in the hundreds column, the 4 is in the tens column, and the 2 is in the ones column.

Here are more examples of numbers written in expanded form.

3 Write 3,972 in expanded form.

EXAMPLE 3 Write 5,478 in expanded form.

Solution $5{,}478 = 5{,}000 + 400 + 70 + 8$

4 Write 271,346 in expanded form.

EXAMPLE 4 Write 354,798 in expanded form.

Solution $354{,}798 = 300{,}000 + 50{,}000 + 4{,}000 + 700 + 90 + 8$

5 Write 71,306 in expanded form.

EXAMPLE 5 Write 56,094 in expanded form.

Solution Notice that there is a 0 in the hundreds column. This means we have 0 hundreds. In expanded form we have

$$56{,}094 = 50{,}000 + 6{,}000 + 90 + 4$$

Note that we don't have to include the 0 hundreds

Answers

2 6 ones, 5 tens, 4 hundreds, 8 thousands, 2 ten thousands, 3 hundred thousands, 5 millions, 0 ten millions, 7 hundred millions, 1 billion, 2 ten billions
3 3,000 + 900 + 70 + 2
4 200,000 + 70,000 + 1,000 + 300 + 40 + 6
5 70,000 + 1,000 + 300 + 6

EXAMPLE 6 Write 5,070,603 in expanded form.

Solution The columns with 0 in them will not appear in the expanded form.

5,070,603 = 5,000,000 + 70,000 + 600 + 3

The idea of place value and expanded form can be used to help write the names for numbers. Naming numbers and writing them in words takes some practice. Let's begin by looking at the names of some two-digit numbers. Table 2 lists a few. Notice that the two-digit numbers that do not end in 0 have two parts. These parts are separated by a hyphen.

Table 2

Number	In English	Number	In English
25	*Twenty-five*	30	*Thirty*
47	*Forty-seven*	62	*Sixty-two*
93	*Ninety-three*	77	*Seventy-seven*
		50	*Fifty*

The following examples give the names for some larger numbers. In each case the names are written according to the place values given in Table 1.

EXAMPLE 7 Write each number in words.

a 452
b 397
c 608

Solution

a Four hundred fifty-two
b Three hundred ninety-seven
c Six hundred eight

EXAMPLE 8 Write each number in words.

a 3,561
b 53,662
c 547,801

Solution

a Three thousand, five hundred sixty-one

↑ Notice how the comma separates the thousands from the hundreds

b Fifty-three thousand, six hundred sixty-two
c Five hundred forty-seven thousand, eight hundred one

6 Write 4,003,560 in expanded form.

7 Write each number in words.

a 724
b 595
c 307

8 Write each number in words.

a 4,758
b 62,779
c 305,440

Answers

6 4,000,000 + 3,000 + 500 + 60
7 a Seven hundred twenty-four
b Five hundred ninety-five
c Three hundred seven
8 a Four thousand, seven hundred fifty-eight
b Sixty-two thousand, seven hundred seventy-nine
c Three hundred five thousand, four hundred forty

9 Write each number in words.

a 707,044,002

b 452,900,008

c 4,008,002,001

EXAMPLE 9 Write each number in words.

a 507,034,005
b 739,600,075
c 5,003,007,006

Solution

a Five hundred seven million, thirty-four thousand, five
b Seven hundred thirty-nine million, six hundred thousand, seventy-five
c Five billion, three million, seven thousand, six

The next examples show how we write a number given in words as a number written with digits.

10 Write six thousand, two hundred twenty-one using digits instead of words.

EXAMPLE 10 Write five thousand, six hundred forty-two using digits instead of words.

Solution *Five Thousand,* *Six Hundred* *Forty-Two*
5, 6 4 2

11 Write each number with digits instead of words.

a Eight million, four thousand, two hundred

b Twenty-five million, forty

c Nine million, four hundred thirty-one

EXAMPLE 11 Write each number with digits instead of words.

a Three million, fifty-one thousand, seven hundred
b Two billion, five
c Seven million, seven hundred seven

Solution

a 3,051,700
b 2,000,000,005
c 7,000,707

In mathematics a collection of numbers is called a *set.* In this chapter we will be working with the set of *whole numbers*, which is defined as follows:

Whole numbers = {0, 1, 2, 3, . . .}

The dots mean "and so on," and the braces { } are used to group the numbers in the set together.

Another way to visualize the whole numbers is with a *number line.* To draw a number line, we simply draw a straight line and mark off equally spaced points along the line, as shown in Figure 1. We label the point at the left with 0, and the rest of the points, in order, with the numbers 1, 2, 3, 4, 5, and so on.

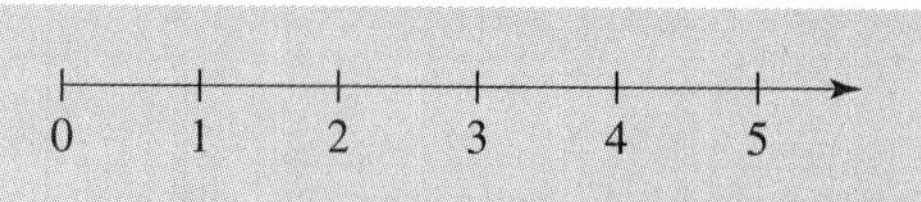

Figure 1

The arrow on the right indicates that the number line can continue in that direction forever. When we refer to numbers in this chapter, we will always be referring to the whole numbers.

Answers

9 a Seven hundred seven million, forty-four thousand, two
b Four hundred fifty-two million, nine hundred thousand, eight
c Four billion, eight million, two thousand, one

10 6,221

11 a 8,004,200
b 25,000,040
c 9,000,431

1.2 Properties and Facts of Addition

Introduction . . . Suppose we put two pieces of wood end to end and measure their combined length with a tape measure. If one of the pieces is 3 feet long and the other is 5 feet long, we could draw a diagram like this:

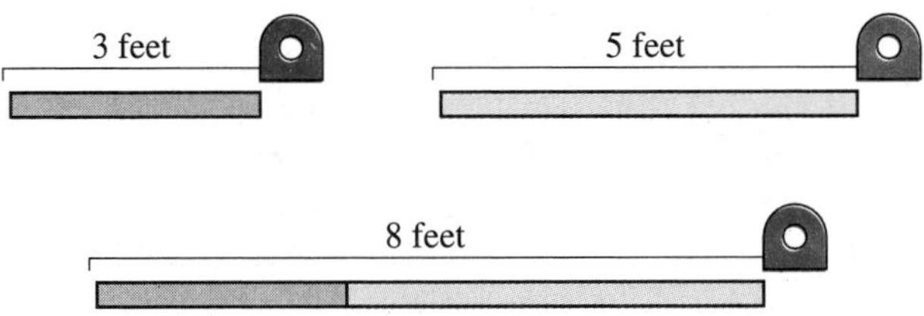

Figure 1

We can summarize all the information above with the following addition problem:

3 feet + 5 feet = 8 feet

Using lengths to visualize addition can be very helpful. In mathematics we generally do so by using the number line. For example, we add 3 and 5 on the number line like this: Start at 0 and move to 3, as shown in Figure 2. From 3, move 5 more units to the right. This brings us to 8. Therefore, $3 + 5 = 8$.

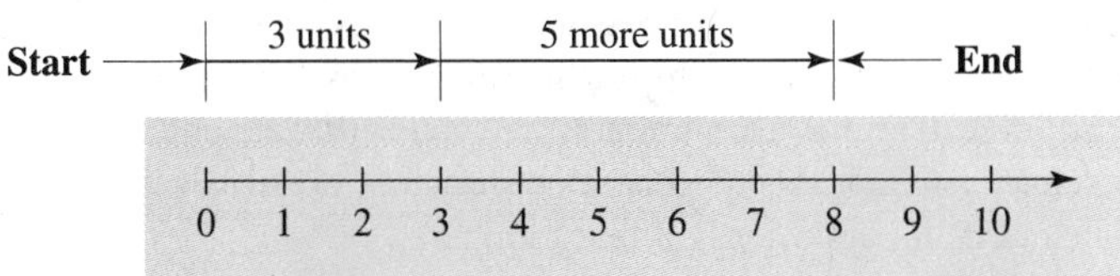

Figure 2

If we do this kind of addition on the number line with all combinations of the numbers 0 through 9, we get the results summarized in Table 1. This addition table shows all the possible ways in which the numbers 0 through 9 can be combined two at a time by using addition.

Table 1 Addition table

+	0	1	2	3	4	5	6	7	8	9
0	0	1	2	3	4	5	6	7	8	9
1	1	2	3	4	5	6	7	8	9	10
2	2	3	4	5	6	7	8	9	10	11
3	3	4	5	6	7	8	9	10	11	12
4	4	5	6	7	8	9	10	11	12	13
5	5	6	7	8	9	10	11	12	13	14
6	6	7	8	9	10	11	12	13	14	15
7	7	8	9	10	11	12	13	14	15	16
8	8	9	10	11	12	13	14	15	16	17
9	9	10	11	12	13	14	15	16	17	18

Note
Table 1 is a summary of the addition facts that you *must* know in order to make a successful start in your study of prealgebra. You *must* know how to add any pair of numbers that come from the list. You *must* be fast and accurate. You don't want to have to think about the answer to $7 + 9$. You should know it's 16. Memorize these facts now. Don't put it off until later.

We read Table 1 in the following manner: Suppose we want to use the table to find the answer to $3 + 5$. We locate the 3 in the column on the left and the 5 in the row at the top. We read *across* from the 3 and *down* from the 5. The entry in the table that is across from 3 and below 5 is 8. To see that $3 + 5 = 8$, we read across from 3 and down from 5.

Variables: An Intuitive Look

When you filled out the application for the school you are attending, there was a space to fill in your first name. "First Name" is a variable quantity, because the value it takes on depends on who is filling out the application. For example, if your first name is Manuel, then the value of "First Name" is Manuel. On the other hand, if your first name is Christa, then the value of "First Name" is Christa.

If we abbreviate "First Name" with FN, "Last Name" with LN, and "Whole Name" with WN, then we take the concept of a variable further and write the relationship between the names this way:

$$\text{FN} + \text{LN} = \text{WN}$$

(We are using the + symbol loosely here to represent writing the names together with a space between them.) The relationship that we have written above holds for all people who have only a first name and a last name. For those people who have a middle name, the relationship between the names is:

$$\text{FN} + \text{MN} + \text{LN} = \text{WN}$$

A similar situation exists in mathematics when we let a letter stand for a number or a group of numbers. For instance, if we say "let a and b represent numbers," then a and b are called *variables* because the values they take on vary. We use the variables a and b in the definition below because we want you to know that the definition is true for all numbers that you will encounter in this book. By using variables, we can write general statements about *all* numbers, rather than specific statements about only a few numbers.

Vocabulary

The word we use to indicate addition is the word *sum*. If we say "the sum of 3 and 5 is 8," what we mean is $3 + 5 = 8$. The word *sum* always indicates addition. We can state this fact in symbols by using the letters a and b to represent numbers.

DEFINITION If a and b are any two numbers, then the **sum** of a and b is $a + b$. To find the sum of two numbers, we add them.

Table 2 gives some word statements and their mathematical equivalents written in symbols.

Note
When mathematics is used to solve everyday problems, the problems are almost always stated in words. The translation of English to symbols is a very important part of mathematics.

Table 2

In English	In Symbols
The sum of 4 and 1	$4 + 1$
4 added to 1	$1 + 4$
8 more than m	$m + 8$
x increased by 5	$x + 5$
The sum of x and y	$x + y$
The sum of 2 and 4 is 6	$2 + 4 = 6$

EXAMPLE 1 Find the sum of 6 and 9.

Solution Since the word *sum* indicates addition, we have $6 + 9 = 15$. We say that the sum of 6 and 9 is 15. If you don't know where we got 15, go back to Table 1 and work on memorizing the addition facts. Both the expressions $6 + 9$ and 15 are the sum of 6 and 9. We could also write this sum in column form as

$$\begin{array}{r} 6 \\ +\ 9 \\ \hline 15 \end{array}$$

Practice Problems

1 Find the sum of 8 and 9.

Properties of Addition

Once we become familiar with the addition table, we may notice some facts about addition that are true regardless of the numbers involved. The first of these facts involves the number 0 (zero).

Notice in Table 1 that whenever we add 0 to a number, the result is the original number. That is:

$0 + 0 = 0$	$5 + 0 = 5$	$0 + 0 = 0$	$0 + 5 = 5$
$1 + 0 = 1$	$6 + 0 = 6$	$0 + 1 = 1$	$0 + 6 = 6$
$2 + 0 = 2$	$7 + 0 = 7$	$0 + 2 = 2$	$0 + 7 = 7$
$3 + 0 = 3$	$8 + 0 = 8$	$0 + 3 = 3$	$0 + 8 = 8$
$4 + 0 = 4$	$9 + 0 = 9$	$0 + 4 = 4$	$0 + 9 = 9$

Since this fact is true no matter what number we add to 0, we call it a property of 0.

Addition Property of 0

If we let a represent any number, then it is always true that

$$a + 0 = a \quad \text{and} \quad 0 + a = a$$

In words: Adding 0 to any number leaves that number unchanged.

A second property we notice by becoming familiar with the addition table is that the order of two numbers in a sum can be changed without changing the result.

$3 + 5 = 8$ and $5 + 3 = 8$

$4 + 9 = 13$ and $9 + 4 = 13$

This fact about addition is true for *all* numbers—not just the numbers listed in the addition table. The order in which you add two numbers doesn't affect the result. We call this fact the *commutative property* of addition, and we write it in symbols as follows.

Commutative Property of Addition

If a and b are any two numbers, then it is always true that

$$a + b = b + a$$

In words: Changing the order of two numbers in a sum doesn't change the result.

Answer

1 17

2 Use the commutative property of addition to rewrite each sum.

a $7 + 9$

b $6 + 3$

c $4 + 0$

d $5 + n$

EXAMPLE 2 Use the commutative property of addition to rewrite each sum.

a $4 + 6$
b $5 + 9$
c $3 + 0$
d $7 + n$

Solution The commutative property of addition indicates that we can change the order of the numbers in a sum without changing the result. Applying this property we have:

a $4 + 6 = 6 + 4$
b $5 + 9 = 9 + 5$
c $3 + 0 = 0 + 3$
d $7 + n = n + 7$

Notice that we did not actually add any of the numbers. The instructions were to use the commutative property, and the commutative property involves only the order of the numbers in a sum.

Note
This discussion is here to show why we write the next property the way we do. Sometimes it is helpful to look ahead to the property itself (in this case, the associative property of addition) to see what it is that is being justified.

The last property of addition we will consider here has to do with sums of more than two numbers. Suppose we want to find the sum of 2 and 3 and then add 4 to it. Since we want to find the sum of 2 and 3 first, and then add 4 to what we get, we use parentheses:

$(2 + 3) + 4$

The parentheses are grouping symbols. They indicate that we are adding the 2 and 3 first.

$(2 + 3) + 4 = 5 + 4$	Add 2 and 3 first
$= 9$	Then add 4 to the result

Now let's find the sum of the same three numbers, but this time we will group the 3 and 4 together.

$2 + (3 + 4) = 2 + 7$	Add 3 and 4 first
$= 9$	Then add 2 to the result

The result in both cases is the same. If we try this with any other numbers, the same thing happens. We call this fact about addition the *associative property* of addition, and we write it in symbols as follows.

Associative Property of Addition
If a, b, and c represent any three numbers, then

$$(a + b) + c = a + (b + c)$$

In words: Changing the grouping of three numbers in a sum doesn't change the result.

Answers
2 a $9 + 7$ **b** $3 + 6$ **c** $0 + 4$ **d** $n + 5$

EXAMPLE 3 Use the associative property of addition to rewrite each sum.

a $(5 + 6) + 7$ **b** $(3 + 9) + 1$
c $6 + (8 + 2)$ **d** $4 + (9 + n)$

Solution The associative property of addition indicates that we are free to regroup the numbers in a sum without changing the result.

a $(5 + 6) + 7 = 5 + (6 + 7)$
b $(3 + 9) + 1 = 3 + (9 + 1)$
c $6 + (8 + 2) = (6 + 8) + 2$
d $4 + (9 + n) = (4 + 9) + n$

3 Use the associative property of addition to rewrite each sum.

a $(3 + 2) + 9$

b $(4 + 10) + 1$

c $5 + (9 + 1)$

d $3 + (8 + n)$

The following examples show how we can use the associative property of addition to simplify expressions that contain both numbers and variables.

In Examples 4 and 5, use the associative property of addition to regroup the numbers and variables so that the expressions can be simplified.

EXAMPLE 4 $(x + 3) + 7 = x + (3 + 7)$ Associative property
$= x + 10$ Addition

EXAMPLE 5 $2 + (4 + y) = (2 + 4) + y$ Associative property
$= 6 + y$ Addition

Use the associative property to simplify.

4 $(x + 5) + 9$

5 $6 + (8 + y)$

The next examples show that it is sometimes necessary to use the commutative property of addition as well as the associative property to simplify expressions.

EXAMPLE 6 $(7 + x) + 2 = (x + 7) + 2$ Commutative property
$= x + (7 + 2)$ Associative property
$= x + 9$ Addition

EXAMPLE 7 $5 + (a + 9) = 5 + (9 + a)$ Commutative property
$= (5 + 9) + a$ Associative property
$= 14 + a$ Addition

Use the commutative property and the associative property to simplify each expression.

6 $(1 + x) + 4$

7 $8 + (a + 4)$

Solving Equations

We can use the addition table to help solve some simple equations. If n is used to represent a number, then the *equation*

$$n + 3 = 5$$

will be true if n is 2. The number 2 is therefore called a *solution* to the equation, since, when we replace n with 2, the equation becomes a true statement:

$$2 + 3 = 5$$

Equations like this are really just puzzles, or questions. When we say, "Solve the equation $n + 3 = 5$," we are asking the question, "What number do we add to 3 to get 5?"

Answers
3 a $3 + (2 + 9)$ **b** $4 + (10 + 1)$
c $(5 + 9) + 1$ **d** $(3 + 8) + n$
4 $x + 14$
5 $14 + y$
6 $x + 5$
7 $12 + a$

8 Use the addition table to find the solution to each equation.

a $n + 9 = 17$

b $n + 2 = 10$

c $8 + n = 9$

d $16 = n + 10$

EXAMPLE 8 Use the addition table to find the solution to each equation.

a $n + 5 = 9$

b $n + 6 = 12$

c $4 + n = 5$

d $13 = n + 8$

Solution We find the solution to each equation by using the addition facts given in Table 1.

a The solution to $n + 5 = 9$ is 4, since $4 + 5 = 9$.

b The solution to $n + 6 = 12$ is 6, since $6 + 6 = 12$.

c The solution to $4 + n = 5$ is 1, since $4 + 1 = 5$.

d The solution to $13 = n + 8$ is 5, since $13 = 5 + 8$.

When we solve equations by reading the equation to ourselves and then stating the solution, as we did with the equation above, we are solving the equation by inspection. The next equations we solve require some simplification before the solution can be found.

9 Solve: $(x + 5) + 4 = 10$

EXAMPLE 9 Solve: $(x + 2) + 3 = 5$

Solution We begin by applying the associative property to the left side of the equation to simplify it. Once the left side has been simplified, we solve the equation by inspection.

$(x + 2) + 3 = 5$	Original equation
$x + (2 + 3) = 5$	Associative property
$x + 5 = 5$	Add 2 and 3
$x = 0$	Solve by inspection

10 Solve: $(a + 4) + 2 = 7 + 9$

EXAMPLE 10 Solve: $(a + 4) + 3 = 2 + 9$

Solution We simplify each side of the equation separately, and then we solve by inspection. Here are the steps:

$(a + 4) + 3 = 2 + 9$	
$a + (4 + 3) = 11$	Simplify each side
$a + 7 = 11$	
$a = 4$	Solve by inspection

Writing about Mathematics

In the *One Step Further* section at the end of the next problem set, you will be asked to give written answers to some questions. Writing about mathematics is a valuable exercise because it requires you to organize your thoughts so that you can communicate what you know. You may think that you have mastered a topic because you can work problems involving that topic. If you write with the intention of explaining and communicating what you know to someone else, you will find that you understand the topic you are writing about even better than you did when you were just working problems.

Here is an example of what one of my students wrote when asked to explain the commutative property of addition:

"The commutative property of addition states that changing the order of the numbers in an addition problem will not change the answer. For example, the commutative property of addition tells us that $3 + 5$ is the same as $5 + 3$, $9 + 7$ is the same as $7 + 9$, and $3 + x$ is the same as $x + 3$."

Answers

8 a 8 b 8 c 1 d 6
9 1
10 10

1.3 Addition with Whole Numbers, and Perimeter

Introduction . . . If a plane flying from Los Angeles to San Francisco leaves Los Angeles with 43 passengers and stops in Bakersfield to pick up 52 more passengers, how many passengers will be on the plane when it reaches San Francisco? To answer this question we add 52 and 43. In this section we will develop a method of adding any two whole numbers, no matter how large or small they are.

EXAMPLE 1 Add: 43 + 52

Solution This type of addition is best done vertically. If we write each number showing the place value of the digits, we have

$$\begin{array}{rl} 43 = 4 \text{ tens} + 3 \text{ ones} & \text{Write each number showing} \\ \underline{+\ 52} = \underline{5 \text{ tens} + 2 \text{ ones}} & \text{the place value of each digit} \\ 9 \text{ tens} + 5 \text{ ones} & \text{Add digits with the same place value to get these numbers} \end{array}$$

We can write the sum 9 tens + 5 ones in standard form as 95. So, to perform addition we add digits with the same place value.

EXAMPLE 2 Add: 165 + 801

Solution Writing the sum vertically and showing the place values of the digits, we have

$$\begin{array}{rl} 165 = & 1 \text{ hundred} + 6 \text{ tens} + 5 \text{ ones} \\ \underline{+\ 801} = & \underline{8 \text{ hundreds} + 0 \text{ tens} + 1 \text{ one}} \\ & 9 \text{ hundreds} + 6 \text{ tens} + 6 \text{ ones} \end{array}$$

This last number written in standard form is 966.

Addition with Carrying

In Examples 1 and 2, the sums of the digits with the same place value were always 9 or less. There are many times when the sum of the digits with the same place value will be a number larger than 9. In these cases we have to do what is called *carrying* in addition. The following examples illustrate this process.

Practice Problems

1 Add: 63 + 25

2 Add: 342 + 605

Note

We are showing the place value of the digits in each number only for the purpose of explanation. The numbers are written like this only to show *why* we add digits with the same place value.

Answers

1 88 2 947

3 Add: 375 + 121 + 473

Note
Look over Example 3 carefully before trying Practice Problem 3. The explanation of why we do carrying in addition is contained in Example 3.

Answer
3 969

EXAMPLE 3 Add: 173 + 224 + 382

Solution We write each number showing the place value of each digit and then add digits with the same place value:

$$\begin{array}{rcl} 173 &=& 1 \text{ hundred} + 7 \text{ tens} + 3 \text{ ones} \\ 224 &=& 2 \text{ hundreds} + 2 \text{ tens} + 4 \text{ ones} \\ \underline{+\ 382} &=& \underline{3 \text{ hundreds} + 8 \text{ tens} + 2 \text{ ones}} \\ && 6 \text{ hundreds} + 17 \text{ tens} + 9 \text{ ones} \end{array}$$

Look at the tens column. We have 17 tens, which is

$$\begin{aligned} 17 \text{ tens} &= 17(10) \\ &= 170 \\ &= 100 + 70 \\ &= 1 \text{ hundred} + 7 \text{ tens} \end{aligned}$$

Since 17 tens is the same as 1 hundred + 7 tens, we carry the 1 hundred to the hundreds column to get

6 hundreds + 1 hundred = 7 hundreds

Our result is

6 hundreds + 17 tens + 9 ones

= 6 hundreds + 1 hundred + 7 tens + 9 ones

= 7 hundreds + 7 tens + 9 ones

We can summarize this whole process by using the shorthand notation for carrying:

$$\begin{array}{r} {}^{1} \\ 173 \\ 224 \\ \underline{+\ 382} \\ 779 \end{array}$$

Step 1 Sum of digits in ones column is 9

Step 2 Sum of digits in tens column is 17. Write the 7 and carry the 1

Step 3 Carry 1 hundred to the hundreds column

Step 4 Sum of digits in hundreds column plus 1 carried from tens column is 7

This shorthand notation for addition is much easier to use than writing out the numbers showing the place values of the digits. We will use the shorthand notation for the rest of the examples in this section. If you have any problems understanding the concept of carrying with addition, you may want to try a few problems the long way. However, the final goal is to be able to find sums quickly and accurately.

EXAMPLE 4 Add: 46,789 + 2,490 + 864

Solution We write the sum vertically—with the digits with the same place value aligned—and then use the shorthand form of addition.

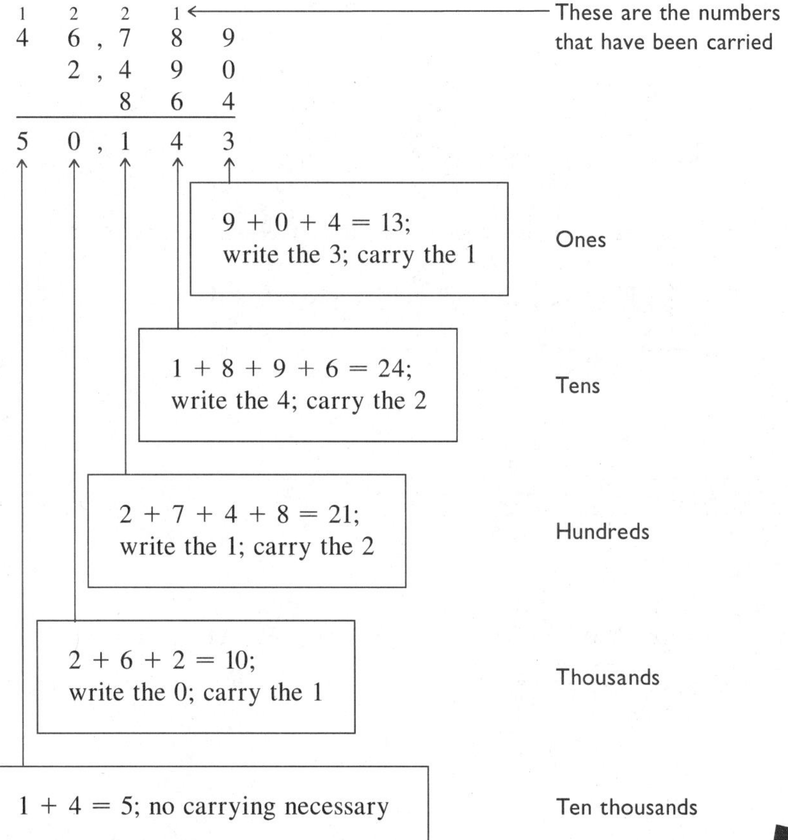

Adding numbers as we are doing here takes some practice. Most people don't make mistakes in carrying. Most mistakes in addition are made in adding the numbers in the columns. That is why it is so important that you are accurate with the basic addition facts given in this chapter.

Facts From Geometry: Perimeter

We end this section with an introduction to perimeter. Let's start with the definition of a *polygon*:

DEFINITION A **polygon** is a closed geometric figure, with at least three sides, in which each side is a straight line segment.

The most common polygons are squares, rectangles, and triangles. Examples of these are shown in Figure 1.

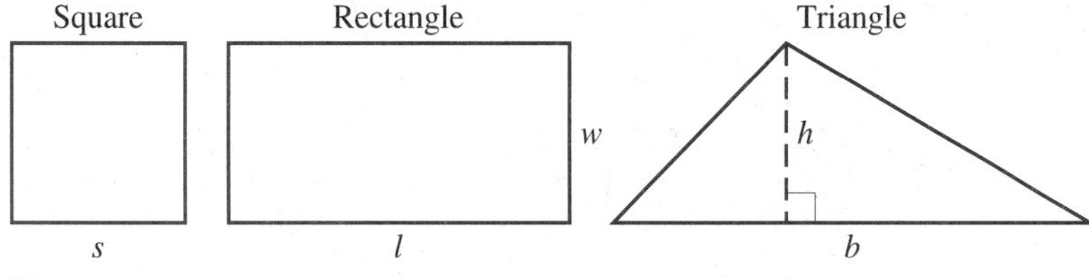

Figure 1

4 Add: 57,904 + 7,193 + 655

Note
In the triangle the small square where the broken line meets the base is the notation we use to show that the two line segments meet at right angles. That is, the height h and the base b are perpendicular to each other; the angle between them is 90°.

Answer
4 65,752

In the square, s is the length of the side, and each side has the same length. In the rectangle, l stands for the length and w stands for the width. The width is usually the lesser of the two. The b and h in the triangle are the base and height, respectively. The height is always perpendicular to the base. That is, the height and base form a 90°, or right, angle where they meet.

> **DEFINITION** The **perimeter** of any polygon is the sum of the lengths of the sides, and it is denoted with the letter P.

To find the perimeter of a polygon we add all the lengths of the sides together.

EXAMPLE 5 Find the perimeter of each geometric figure.

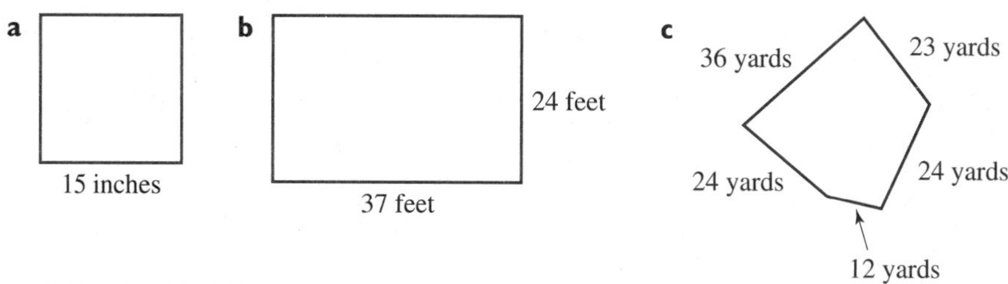

Solution In each case we find the perimeter by adding the lengths of all the sides.

a The figure is a square. Since the length of each side in the square is the same, the perimeter is

$$P = 15 + 15 + 15 + 15 = 60 \text{ inches}$$

b In the rectangle, two of the sides are 24 feet long and the other two are 37 feet long. The perimeter is the sum of the lengths of the sides.

$$P = 24 + 24 + 37 + 37 = 122 \text{ feet}$$

c For this polygon we add the lengths of the sides together. The result is the perimeter.

$$P = 36 + 23 + 24 + 12 + 24 = 119 \text{ yards}$$

Calculator Note

From time to time we will include some notes like this one, which show how a calculator can be used to assist us with some of the calculations in the book. Most calculators on the market today fall into one of two categories: those with algebraic logic and those with function logic. Calculators with algebraic logic have a key with an equals sign on it. Calculators with function logic do not have an equals key. Instead they have a key labeled ENTER or EXE (for execute). Scientific calculators use algebraic logic, and graphing calculators, such as the TI-82 and CASIO 7700, use function logic.

Here are the sequences of keystrokes to use to work the problem shown in Part c of Example 6.

Scientific Calculator: 36 [+] 23 [+] 24 [+] 12 [+] 24 [=]

Graphing Calculator: 36 [+] 23 [+] 24 [+] 12 [+] 24 [ENT]

5 Find the perimeter of each geometric figure.

a

b

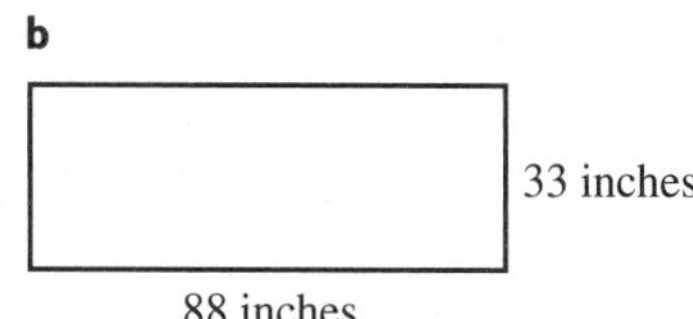

c

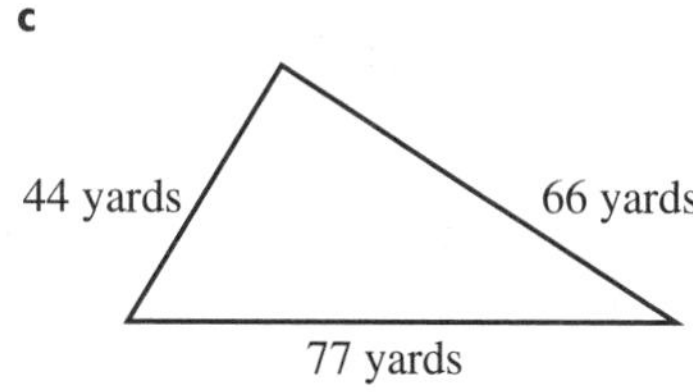

Answers

5 a 28 feet **b** 242 inches **c** 187 yards

1.4 Rounding Numbers, Estimating Answers, and Displaying Information

Introduction . . . Many times when we talk about numbers, it is helpful to use numbers that have been *rounded off*, rather than exact numbers. For example, the city where I live has a population of 35,281. But when I tell people how large the city is, I usually say, "The population is about 35,000." The number 35,000 is the original number rounded to the nearest thousand. The number 35,281 is closer to 35,000 than it is to 36,000, so it is rounded to 35,000. If we wanted to round 35,281 to the nearest hundred, we would get 35,300, because 35,281 is closer to 35,300 than it is to 35,200.

Rounding

The steps used in rounding numbers are given below.

Note
After you have used the steps listed here to work a few problems, you will find that the procedure becomes almost automatic.

Steps for Rounding Whole Numbers

1 Locate the digit just to the right of the place you are to round to.
2 If that digit is less than 5, replace it and all digits to its right with zeros.
3 If that digit is 5 or more, replace it and all digits to its right with zeros, and add 1 to the digit to its left.

You can see from these rules that in order to round a number you must be told what column (or place value) to round to.

EXAMPLE 1 Round 5,382 to the nearest hundred.

Solution The 3 is in the hundreds column. We look at the digit just to its right, which is 8. Since 8 is greater than 5, we add 1 to the 3, and we replace the 8 and 2 with zeros:

5,382 is 5,**400** to the nearest hundred

Greater than 5 (↗ the 8) — Add 1 to get 4 (↑ the 4) — Put zeros here (↖ the 00)

EXAMPLE 2 Round 94 to the nearest ten.

Solution The 9 is in the tens column. To its right is 4. Since 4 is less than 5, we simply replace it with 0:

94 is 9**0** to the nearest ten

Less than 5 (↑ the 4) — Replaced with zero (↑ the 0)

EXAMPLE 3 Round 973 to the nearest hundred.

Solution We have a 9 in the hundreds column. To its right is 7, which is greater than 5. We add 1 to 9 to get 10, and then replace the 7 and 3 with zeros:

973 is 1,0**00** to the nearest hundred

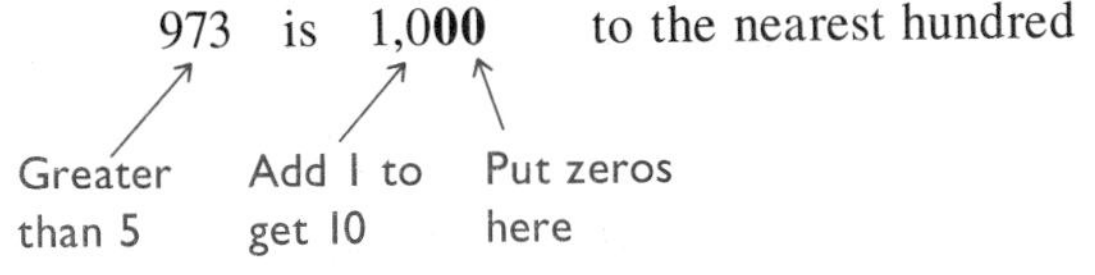

Practice Problems

1 Round 5,742 to the nearest hundred.

2 Round 87 to the nearest ten.

3 Round 980 to the nearest hundred.

Answers
1 5,700 **2** 90 **3** 1,000

4 Round 376,804,909 to the nearest million.

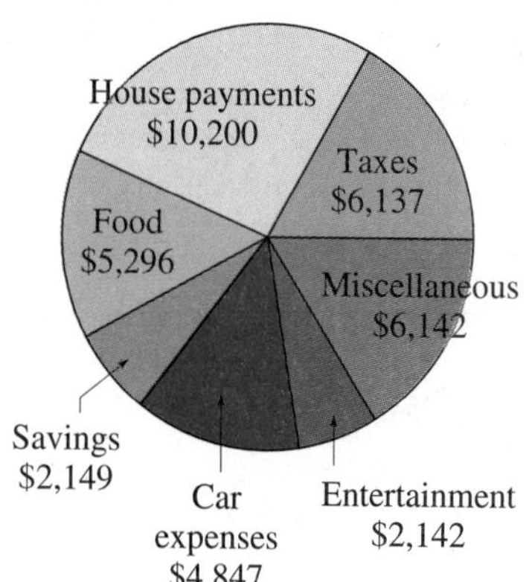

5 Use the pie chart above to answer these questions.

a To the nearest ten dollars, what is the total amount spent on food and car expenses?

b To the nearest hundred dollars, how much is spent on savings and taxes?

c To the nearest thousand dollars, how much of their income is spent on items other than food and entertainment?

Answers
4 377,000,000
5 a $10,140 b $8,300
c $29,000

EXAMPLE 4 Round 47,256,344 to the nearest million.

Solution We have 7 in the millions column. To its right is 2, which is less than 5. We simply replace all the digits to the right of 7 with zeros to get:

47,256,344 is 47,**000,000** to the nearest million

(47,256,344: 2 — Less than 5; 47,000,000: 7 — Leave as is; 000,000 — Replaced with zeros)

Table 1 gives more examples of rounding.

Table 1

Original Number	ROUNDED TO THE NEAREST Ten	Hundred	Thousand
6,914	6,910	6,900	7,000
8,485	8,490	8,500	8,000
5,555	5,560	5,600	6,000
1,234	1,230	1,200	1,000

RULE: Calculating and Rounding If we are doing calculations and are asked to round our answer, we do all our arithmetic first and then round the result. That is, the last step is to round the answer; we don't round the numbers first and then do the arithmetic.

EXAMPLE 5 The pie chart in the margin shows how a family earning $36,913 a year spends their money.

a To the nearest hundred dollars, what is the total amount spent on food and entertainment?

b To the nearest thousand dollars, how much of their income is spent on items other than taxes and savings?

Solution In each case we add the numbers in question and then round the sum to the indicated place.

a We add the amounts spent on food and entertainment and then round that result to the nearest hundred dollars.

Food	$5,296
Entertainment	2,142
Total	$7,438 = $7,400 to the nearest hundred dollars

b We add the numbers for all items except taxes and savings.

House payments	$10,200
Food	5,296
Car expenses	4,847
Entertainment	2,142
Miscellaneous	6,142
Total	$28,627 = $29,000 to the nearest thousand dollars

Estimating

When we *estimate* the answer to a problem, we simplify the problem so that an approximate answer can be found quickly. There are a number of ways of doing this. One common method is to use rounded numbers to simplify the arithmetic necessary to arrive at an approximate answer, as our next example shows.

EXAMPLE 6 Estimate the answer to the following problem by rounding each number to the nearest thousand.

```
   4,872
   1,691
     777
 + 6,124
```

Solution We round each of the four numbers in the sum to the nearest thousand. Then we add the rounded numbers.

4,872	rounds to	5,000
1,691	rounds to	2,000
777	rounds to	1,000
+ 6,124	rounds to	+ 6,000
		14,000

We estimate the answer to this problem to be approximately 14,000. The actual answer, found by adding the original unrounded numbers, is 13,464.

Note The method used in Example 6 above does not conflict with the rule we stated before Example 5. In Example 6 we are asked to *estimate* an answer, so it is okay to round the numbers in the problem before adding them. In Example 5 we are asked for a rounded answer, meaning that we are to find the exact answer to the problem and then round to the indicated place. In this case we must not round the numbers in the problem before adding. Look over the instructions, solutions, and answers to Examples 5 and 6 until you understand the difference between the problems shown there.

6 Estimate the answer by first rounding each number to the nearest thousand.

```
   5,287
   2,561
     888
 + 4,898
```

Displaying Information

In the introduction to this chapter, we gave two representations for the amount of caffeine in five different drinks, one numeric and the other visual. Those two representations are shown below in Table 2 and Figure 1.

Table 2 Caffeine content of hot drinks

Drink (6-ounce cup)	Caffeine (in milligrams)
Brewed coffee	100
Instant coffee	70
Tea	50
Cocoa	5
Decaffeinated coffee	4

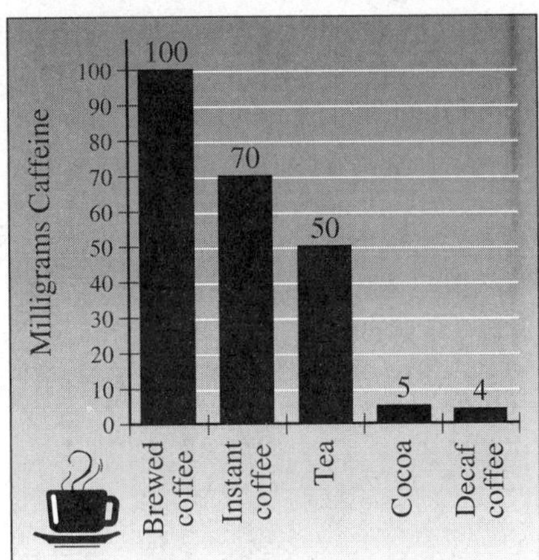

Figure 1

The diagram in Figure 1 is called a *bar chart*, or *histogram*. The horizontal line below which the drinks are listed is called the *horizontal axis*, while the vertical line that is labeled from 0 to 100 is called the *vertical axis*. The key to constructing a readable histogram is in labeling the axes in a simple and straightforward manner, so that the information shown in the histogram is easy to read.

Answer
6 14,000

7 The results of crash tests for two other cars are given below. Extend the horizontal axis in Figure 3 past the Nissan Maxima, so you can add labels for these two cars. Then, for each car below, draw in bars to represent the amount of damage done to each of the two cars.

Honda Accord	\$1,433
Chevrolet Lumina	\$2,629

Answer
7 See solutions section.

EXAMPLE 7 The information in Table 3 was published in 1995 by the Insurance Institute for Highway Safety. It gives the total repair bill from damage done to a car from four separate crashes at 5 miles per hour. Construct a histogram that gives a clear visual summary of the information in the table.

Table 3 Damage at 5 miles per hour

Car	Damage
Chevrolet Cavalier	\$1,795
Mazda Millenia	\$2,031
Toyota Camry LE	\$2,328
Ford Taurus GL	\$2,814
Ford Contour GL	\$3,188
Nissan Maxima	\$3,605

Solution Although there are many ways to construct a histogram of this information, we will construct one in which the information in the first column of the table is given along the horizontal axis and the information in the second column is associated with the vertical axis, as shown in Figure 2 below.

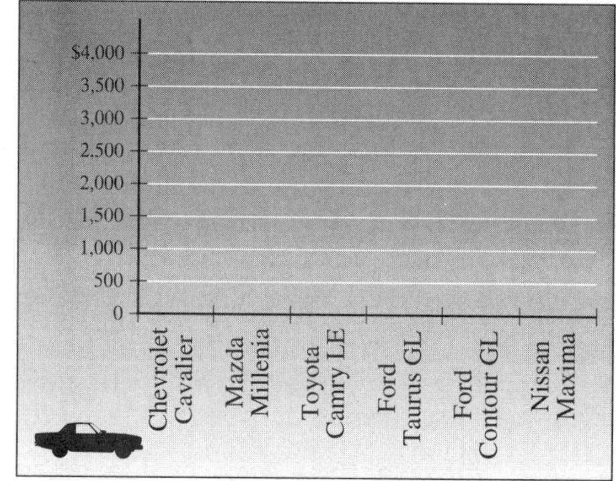

Figure 2

To complete the histogram, we mentally round the numbers in the table to the nearest hundred to better estimate how tall we want the bars. Using the rounded numbers to draw the bars, we have the histogram shown in Figure 3.

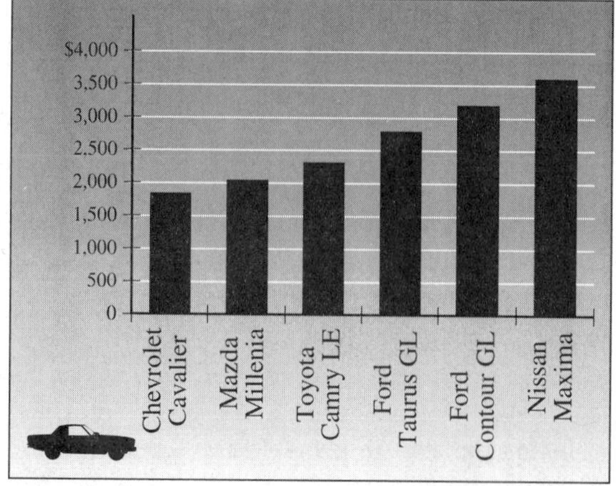

Figure 3

1.5 Subtraction with Whole Numbers

Introduction . . . Suppose your weekly salary is \$324 and you have \$109 withheld from your check for taxes and retirement. If your paycheck is for \$215, is it the correct amount? You can find out by adding your take-home pay and the amount withheld for taxes and retirement: \$215 + \$109 = \$324. Your take-home pay is the correct amount.

A similar problem, and the one your employer would use to find the amount of your take-home pay, is a subtraction problem: \$324 − \$109 = \$215.

Subtraction is the opposite operation of addition. If you understand addition and can work simple addition problems quickly and accurately, then subtraction shouldn't be difficult for you.

Vocabulary

The word **difference** always indicates subtraction. We can state this in symbols by letting the letters a and b represent numbers.

DEFINITION The **difference** of two numbers a and b is

$$a - b$$

Table 1 gives some word statements involving subtraction and their mathematical equivalents written in symbols.

Table 1

In English	In Symbols
The difference of 9 and 1	$9 - 1$
The difference of 1 and 9	$1 - 9$
The difference of m and 4	$m - 4$
The difference of x and y	$x - y$
3 subtracted from 8	$8 - 3$
2 subtracted from t	$t - 2$
The difference of 7 and 4 is 3	$7 - 4 = 3$
The difference of 9 and 3 is 6	$9 - 3 = 6$

The Meaning of Subtraction

When we want to subtract 3 from 8, we write

$8 - 3$, 8 subtract 3, or 8 minus 3

The number we are looking for here is the difference between 8 and 3, or the number we add to 3 to get 8. That is:

$8 - 3 = ?$ is the same as $? + 3 = 8$

In both cases we are looking for the number we add to 3 to get 8. The number we are looking for is 5. We have two ways to write the same statement.

Subtraction		Addition
$8 - 3 = 5$	or	$5 + 3 = 8$

For every subtraction problem, there is an equivalent addition problem. Table 2 lists some examples.

Table 2

Subtraction		Addition
7 − 3 = 4	because	4 + 3 = 7
9 − 7 = 2	because	2 + 7 = 9
10 − 4 = 6	because	6 + 4 = 10
15 − 8 = 7	because	7 + 8 = 15

To subtract numbers with two or more digits, we align the numbers vertically and subtract in columns.

Practice Problems

1 Subtract: 684 − 431

EXAMPLE 1 Subtract: 376 − 241

Solution We write the problem vertically with the place values of the digits showing, and then subtract in columns:

Note

We are showing the place values of the digits to explain subtraction. In actual practice you won't do this.

$$\begin{array}{rcl} 376 & = & 3 \text{ hundreds} + 7 \text{ tens} + 6 \text{ ones} \\ -241 & = & 2 \text{ hundreds} \quad 4 \text{ tens} \quad 1 \text{ one} \\ \hline & & 1 \text{ hundred} + 3 \text{ tens} + 5 \text{ ones} \end{array}$$

← Subtract the bottom number in each column from the number above it

The difference is 1 hundred + 3 tens + 5 ones, which we write in standard form as 135.

2 Subtract 405 from 6,857.

EXAMPLE 2 Subtract 503 from 7,835.

Solution In symbols this statement is equivalent to

7,835 − 503

To subtract we write 503 below 7,835 and then subtract in columns. This time we will not show the place value of each digit but will simply subtract in columns.

$$\begin{array}{r} 7,835 \\ -\ \ 503 \\ \hline 7,332 \end{array}$$

5 − 3 = 2	Ones
3 − 0 = 3	Tens
8 − 5 = 3	Hundreds
7 − 0 = 7	Thousands

Even though no 0 is showing, we can think of it as being there.

Answers

1 253 **2** 6,452

As you can see, subtraction problems like the ones in Examples 1 and 2 are fairly simple. We write the problem vertically, lining up the digits with the same place value, and subtract in columns. We always subtract the bottom number from the top number.

Subtraction with Borrowing

Subtraction must involve *borrowing* when the bottom digit in any column is larger than the digit above it. In one sense borrowing is the reverse of the carrying we did in addition.

EXAMPLE 3 Subtract: 92 − 45

3 Subtract: 63 − 47

Solution We write the problem vertically with the place values of the digits showing:

$$\begin{array}{r} 92 = 9 \text{ tens} + 2 \text{ ones} \\ \underline{-45 = 4 \text{ tens} \quad 5 \text{ ones}} \end{array}$$

Look at the ones column. We cannot subtract immediately because 5 is larger than 2. Instead, we borrow 1 ten from the 9 tens in the tens column. We can rewrite the number 92 as

$$\begin{aligned} & 9 \text{ tens} + 2 \text{ ones} \\ &= 8 \text{ tens} + 1 \text{ ten} + 2 \text{ ones} \\ &= 8 \text{ tens} + 12 \text{ ones} \end{aligned}$$

Note

The discussion here shows why borrowing is necessary and how we go about it. To understand borrowing you should pay close attention to this discussion.

Now we are in a position to subtract.

$$\begin{array}{rcl} 92 = 9 \text{ tens} + 2 \text{ ones} &=& 8 \text{ tens} + 12 \text{ ones} \\ \underline{-45} = \underline{4 \text{ tens} \quad 5 \text{ ones}} &=& \underline{4 \text{ tens} \quad 5 \text{ ones}} \\ && 4 \text{ tens} + \ 7 \text{ ones} \end{array}$$

The result is 4 tens + 7 ones, which can be written in standard form as 47.

Writing the problem out in this way is more trouble than is actually necessary. The shorthand form of the same problem looks like this:

$$\begin{array}{rr} \scriptstyle 8 & \scriptstyle 12 \\ \not 9 & \not 2 \\ -4 & 5 \\ \hline 4 & 7 \end{array}$$

This shows we have borrowed 1 ten to go with the 2 ones

12 − 5 = 7 Ones

8 − 4 = 4 Tens

This shortcut form shows all the necessary work involved in subtraction with borrowing. We will use it from now on.

Answer
3 16

4 Find the difference of 656 and 283.

EXAMPLE 4 Find the difference of 549 and 187.

Solution In symbols the difference of 549 and 187 is written

549 − 187

Writing the problem vertically so that the digits with the same place value are aligned, we have

$$\begin{array}{r} 549 \\ -\ 187 \\ \hline \end{array}$$

The top number in the tens column is smaller than the number below it. This means that we will have to borrow from the next larger column.

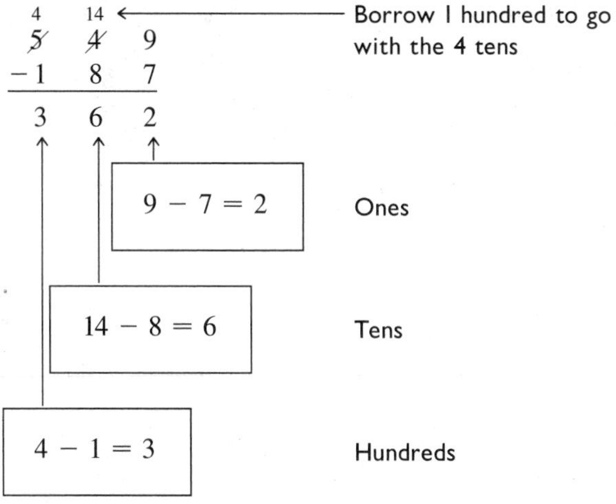

The actual work we did in borrowing looks like this:

5 hundreds + 4 tens + 9 ones

= 4 hundreds + 1 hundred + 4 tens + 9 ones

= 4 hundreds + 14 tens + 9 ones

5 Suppose Jo Ann writes another check for $52. How much does she then have left in her checking account?

EXAMPLE 5 Jo Ann has $742 in her checking account. If she writes a check for $615 to pay the rent, how much is left in her checking account?

Solution To find the amount left in the account after she has written the rent check, we subtract:

$$\begin{array}{r} {\scriptstyle 3\ 12} \\ \$7\not{4}\not{2} \\ -\ 615 \\ \hline \$127 \end{array}$$

She has $127 left in her account after writing a check for the rent.

Calculator Note

Here is how we would work the problem shown in Example 5 on a calculator:

Scientific Calculator: 742 [−] 615 [=]

Graphing Calculator: 742 [−] 615 [ENT]

Estimating

One way to estimate the answer to the problem shown in Example 5 is to round 742 to 700 and 615 to 600 and then subtract 600 from 700 to obtain 100, which is an estimate of the difference. Making a mental estimate in this manner will help you catch some of the errors that will occur if you press the wrong buttons on your calculator.

Answers
4 373 **5** $75

1.6 Properties and Facts of Multiplication

Introduction . . . If Diane makes \$6 an hour for the first 40 hours she works each week and \$9 for every hour after that, how much will she make if she works 45 hours in one week? One way to find the answer to this question is with *multiplication*. We multiply 6 by 40 and 9 by 5 and then add the results. In symbols, the problem looks like this:

$$\begin{aligned} 6(40) + 9(5) &= 240 + 45 \\ &= 285 \end{aligned}$$

She would make \$285 for the week.

The type of multiplication we used in this problem is what we will learn in this section. To begin we can think of multiplication as shorthand for repeated addition. That is, multiplying 3 times 4 can be thought of this way:

$$3 \text{ times } 4 = 4 + 4 + 4 = 12$$

Multiplying 3 times 4 means to add three 4's. In symbols we write 3 times 4 as 3×4, or $3 \cdot 4$.

We can construct a multiplication table using the fact that multiplication can be thought of as repeated addition.

Table 1 Multiplication facts

×	0	1	2	3	4	5	6	7	8	9
0	0	0	0	0	0	0	0	0	0	0
1	0	1	2	3	4	5	6	7	8	9
2	0	2	4	6	8	10	12	14	16	18
3	0	3	6	9	12	15	18	21	24	27
4	0	4	8	12	16	20	24	28	32	36
5	0	5	10	15	20	25	30	35	40	45
6	0	6	12	18	24	30	36	42	48	54
7	0	7	14	21	28	35	42	49	56	63
8	0	8	16	24	32	40	48	56	64	72
9	0	9	18	27	36	45	54	63	72	81

Note
As we have indicated, the numbers in this table were found by repeated addition. To understand multiplication, you must know that 3 times $5 = 5 + 5 + 5$. Once you understand multiplication, the next step is to memorize the multiplication facts contained in Table 1. It is absolutely essential that you become fast and accurate at this kind of multiplication.

Notation

There are many ways to indicate multiplication. All the following statements are equivalent. They all indicate multiplication with the numbers 3 and 4.

$$3 \cdot 4, \quad 3 \times 4, \quad 3(4), \quad (3)4, \quad (3)(4), \quad \begin{array}{r} 4 \\ \times\ 3 \\ \hline \end{array}$$

If one or both of the numbers we are multiplying are represented by letters, we may also use the following notation:

$5n$ means 5 times n
ab means a times b

Note
The kind of notation we will use to indicate multiplication will depend on the situation. For example, when we are solving equations that involve letters, it is not a good idea to indicate multiplication with the symbol ×, since it could be confused with the letter x. The symbol we will use to indicate multiplication most often in this book is the multiplication dot, ·.

Vocabulary

We use the word *product* to indicate multiplication. If we say "The product of 3 and 4 is 12," then we mean

$$3 \cdot 4 = 12$$

Both $3 \cdot 4$ and 12 are called the product of 3 and 4. The 3 and 4 are called *factors*. That is, we say 3 and 4 are factors of 12 since their product is 12.

Table 2 gives some additional translations of statements involving multiplication.

Table 2

In English	In Symbols
The product of 2 and 5	$2 \cdot 5$
The product of 5 and 2	$5 \cdot 2$
The product of 4 and n	$4n$
The product of x and y	xy
The product of 9 and 6 is 54	$9 \cdot 6 = 54$
The product of 2 and 8 is 16	$2 \cdot 8 = 16$

The examples below illustrate the vocabulary associated with multiplication.

Practice Problems

1 Identify the products and factors in the statement

$$6 \cdot 7 = 42$$

2 Identify the products and factors in the statement

$$20 = 2 \cdot 2 \cdot 5$$

EXAMPLE 1 Identify the products and factors in the statement

$$9 \cdot 8 = 72$$

Solution The factors are 9 and 8, and the products are $9 \cdot 8$ and 72.

EXAMPLE 2 Identify the products and factors in the statement

$$30 = 2 \cdot 3 \cdot 5$$

Solution The factors are 2, 3, and 5. The products are $2 \cdot 3 \cdot 5$ and 30.

Properties of Multiplication

You can see from Table 1 that the product of any number and 0 is 0. We state this as a property of 0.

> **Multiplication Property of 0**
>
> If a represents any number, then
>
> $$a \cdot 0 = 0 \quad \text{and} \quad 0 \cdot a = 0$$
>
> *In words:* Multiplication by 0 always results in 0.

We also notice by looking through Table 1 that whenever we multiply a number by 1, the result is the original number.

Answers

1 Factors: 6, 7; products: $6 \cdot 7$ and 42

2 Factors: 2, 2, 5; products: $2 \cdot 2 \cdot 5$ and 20

Multiplication Property of 1

If a represents any number, then

$$a \cdot 1 = a \quad \text{and} \quad 1 \cdot a = a$$

In words: Multiplying any number by 1 leaves that number unchanged.

It is also apparent from Table 1 that multiplication is a commutative operation. For example, the results of multiplying $4 \cdot 5$ and $5 \cdot 4$ are both 20. We can change the order of the factors in a product without changing the result.

Commutative Property of Multiplication

If a and b are any two numbers, then

$$ab = ba$$

In words: The order of the numbers in a product doesn't affect the result.

EXAMPLE 3 Use the commutative property of multiplication to rewrite each of the following products:

a $7 \cdot 9$ **b** $(3)(2)$ **c** $4(6)$ **d** $8 \cdot n$

Solution Applying the commutative property to each expression, we have:

a $7 \cdot 9 = 9 \cdot 7$ **b** $(3)(2) = (2)(3)$
c $4(6) = 6(4)$ **d** $8 \cdot n = n \cdot 8$

And finally, multiplication is also an associative operation.

Associative Property of Multiplication

If a, b, and c represent any three numbers, then

$$(ab)c = a(bc)$$

In words: We can change the grouping of the numbers in a product without changing the result.

EXAMPLE 4 Use the associative property of multiplication to rewrite each of the following products:

a $(2 \cdot 7) \cdot 9$ **b** $(4 \times 5) \times 6$ **c** $3 \cdot (8 \cdot 2)$ **d** $(5 \cdot 7) \cdot n$

Solution Applying the associative property of multiplication, we regroup as follows:

a $(2 \cdot 7) \cdot 9 = 2 \cdot (7 \cdot 9)$ **b** $(4 \times 5) \times 6 = 4 \times (5 \times 6)$
c $3 \cdot (8 \cdot 2) = (3 \cdot 8) \cdot 2$ **d** $(5 \cdot 7) \cdot n = 5 \cdot (7 \cdot n)$

3 Use the commutative property of multiplication to rewrite each of the following products.

a $5 \cdot 8$

b $(4)(6)$

c $7(2)$

d $9n$

4 Use the associative property of multiplication to rewrite each of the following products.

a $(5 \cdot 7) \cdot 4$
b $(3 \times 9) \times 8$
c $4 \cdot (6 \cdot 4)$
d $6 \cdot (9 \cdot n)$

Answers

3 a $8 \cdot 5$ **b** $(6)(4)$ **c** $2(7)$ **d** $n9$
4 a $5 \cdot (7 \cdot 4)$ **b** $3 \times (9 \times 8)$ **c** $(4 \cdot 6) \cdot 4$ **d** $(6 \cdot 9) \cdot n$

Apply the associative property of multiplication and then multiply.

5 $4(7x)$

6 $7(9y)$

7 Use the multiplication table to find the solution to each of the following equations.

a $5 \cdot n = 35$

b $8 \cdot n = 72$

c $49 = 7 \cdot n$

d $27 = 9 \cdot n$

8 Multiply: $5 \cdot 100$

9 Multiply: $4 \cdot 7{,}000$

Answers

5 $28x$ **6** $63y$

7 a 7 **b** 9 **c** 7 **d** 3

8 500 **9** 28,000

We can use the associative property of multiplication to multiply expressions that contain numbers and variables. The next examples illustrate.

EXAMPLE 5 Apply the associative property of multiplication and then multiply.

$3(5x) = (3 \cdot 5)x$ Associative property of multiplication

$= 15x$ Multiply 3 and 5

EXAMPLE 6 $6(8y) = (6 \cdot 8)y$ Associative property of multiplication

$= 48y$ Multiply 6 and 8

Solving Equations

We can use the facts contained in the multiplication table to help solve some equations. If n is used to represent a number, then the equation

$$4 \cdot n = 12$$

is read "4 times n is 12," or "The product of 4 and n is 12." This means that we are looking for the number we multiply by 4 to get 12. The number is 3. Since the equation becomes a true statement if n is 3, we say that 3 is the solution to the equation.

EXAMPLE 7 Use the multiplication table to find the solution to each of the following equations:

a $6 \cdot n = 24$ **b** $4 \cdot n = 36$ **c** $15 = 3 \cdot n$ **d** $21 = 3 \cdot n$

Solution

a The solution to $6 \cdot n = 24$ is 4, since $6 \cdot 4 = 24$.

b The solution to $4 \cdot n = 36$ is 9, since $4 \cdot 9 = 36$.

c The solution to $15 = 3 \cdot n$ is 5, since $15 = 3 \cdot 5$.

d The solution to $21 = 3 \cdot n$ is 7, since $21 = 3 \cdot 7$.

Now let's see what happens when we multiply some familiar whole numbers with multiples of 10.

EXAMPLE 8 Multiply: $3 \cdot 100$

Solution Using the definition of multiplication as repeated addition, we have

$$3 \cdot 100 = 100 + 100 + 100$$
$$= 300$$

Notice that this is the same result we would get if we just multiplied $3 \cdot 1$ and attached two zeros to the result.

EXAMPLE 9 Multiply: $3 \cdot 4{,}000$

Solution Proceeding as we did in Example 8, we have

$$3 \cdot 4{,}000 = 4{,}000 + 4{,}000 + 4{,}000$$
$$= 12{,}000$$

Again if we had multiplied 3 and 4 to get 12, and then attached three zeros on the right, the result would have been the same.

In problems like the ones given in Examples 8 and 9, we can normally leave out the intermediate step that requires showing the repeated addition and just write the result.

1.7 The Distributive Property and Multiplication with Whole Numbers

If a supermarket bought 35 cases of soft drinks, and each case contained 24 bottles, then the number of bottles bought can be found by multiplying 35 by 24. The methods we used to multiply numbers in the previous section will not work well for products like 35(24). To develop a more efficient method of multiplication, we need to use what is called the *distributive property*. To begin consider the following two problems:

Problem 1	Problem 2
$3(4 + 5)$	$3(4) + 3(5)$
$= 3(9)$	$= 12 + 15$
$= 27$	$= 27$

The result in both cases is the same number, 27. This indicates that the original two expressions must have been equal also. That is,

$$3(4 + 5) = 3(4) + 3(5)$$

This is an example of the distributive property. We say that multiplication *distributes* over addition.

$$3(4 + 5) = 3(4) + 3(5)$$

We can write this property in symbols using the letters a, b, and c to represent any three whole numbers.

Distributive Property

If a, b, and c represent any three whole numbers, then

$$a(b + c) = a(b) + a(c)$$

The following examples illustrate how we apply the distributive property.

EXAMPLE 1 $4(5 + 9) = 4(5) + 4(9)$ Distributive property
$= 20 + 36$ Multiplication
$= 56$ Addition

EXAMPLE 2 $7(8 + 6) = 7(8) + 7(6)$ Distributive property
$= 56 + 42$ Multiplication
$= 98$ Addition

The distributive property can also be applied to expressions that contain variables as well as numbers, as shown in Examples 3 and 4.

EXAMPLE 3 $3(x + 5) = 3 \cdot x + 3 \cdot 5$ Distributive property
$= 3x + 15$ Multiply 3 and 5

Practice Problems

Use the distributive property to rewrite each of the following expressions, and then simplify.

1 $3(5 + 7)$

2 $5(6 + 8)$

Apply the distributive property.

3 $4(x + 2)$

Answers

1 36 **2** 70 **3** $4x + 8$

4 $7(a + 9)$

Multiply.

5 $4(7x + 2)$

6 $7(9y + 6)$

7 Multiply: $8(57)$

Answers

4 $7a + 63$ **5** $28x + 8$
6 $63y + 42$ **7** 456

EXAMPLE 4 $9(a + 8) = 9 \cdot a + 9 \cdot 8$ Distributive property
$= 9a + 72$ Multiply 9 and 8

The next examples show how we use the distributive property and then the associative property of multiplication to multiply.

EXAMPLE 5 $5(3x + 2) = 5(3x) + 5 \cdot 2$ Distributive property
$= (5 \cdot 3)x + 5 \cdot 2$ Associative property
$= 15x + 10$ Multiply

EXAMPLE 6 $6(8y + 3) = 6(8y) + 6 \cdot 3$ Distributive property
$= (6 \cdot 8)y + 6 \cdot 3$ Associative property
$= 48y + 18$

Multiplication with Whole Numbers

We can use the distributive property and information from our discussion on multiplication in Section 1.6 to explain how we do multiplication with whole numbers that have two or more digits.

Suppose we want to find the product 7(65). By writing 65 as 60 + 5 and applying the distributive property, we have:

$7(65) = 7(60 + 5)$ $65 = 60 + 5$
$= 7(60) + 7(5)$ Distributive property
$= 420 + 35$ Multiplication
$= 455$ Addition

We can write the same problem vertically like this:

$$\begin{array}{r} 60 + 5 \\ \times \quad 7 \\ \hline 35 \\ +\ 420 \\ \hline 455 \end{array} \qquad \begin{array}{l} \\ \\ \leftarrow 7(5) = 35 \\ \leftarrow 7(60) = 420 \\ \\ \end{array}$$

This saves some space in writing. But notice that we can cut down on the amount of writing even more if we write the problem this way:

$$\begin{array}{r} {}^{3} \\ 65 \\ \times\ 7 \\ \hline 455 \end{array}$$

Step 1 7(5) = 35; carry the 3 to the tens column, and then write the 5 in the ones column

Step 2 7(6) = 42; add the 3 we carried to 42 to get 45

This shortcut notation takes some practice. We will do two more problems showing both the long method and the shortcut method.

EXAMPLE 7 Multiply: 9(43)

Solution We show both methods. The shortcut method is the one you want to become proficient with. The long method is shown so you can see the correlation between the two.

Long Method

$$43 = 40 + 3 \longrightarrow \begin{array}{r} 40 + 3 \\ \times \quad 9 \\ \hline 27 \\ +360 \\ \hline 387 \end{array} \begin{array}{l} \\ \\ \longleftarrow 9(3) = 27 \\ \longleftarrow 9(40) = 360 \\ \longleftarrow 360 + 27 = 387 \end{array}$$

Shortcut Method

$$\begin{array}{r} {}^{2} \\ 43 \\ \times\ 9 \\ \hline 387 \end{array}$$

Step 2 9(4) = 36; add the 2 we carried to 36 to get 38

Step 1 9(3) = 27; carry the 2 to the tens column, and then write the 7 in the ones column

EXAMPLE 8 Multiply: 8(487)

Solution Again we show both methods. This time we will leave out the explanations. Study these solutions by comparing them with the solutions in Example 7 until you understand the relationship between the two methods.

Long Method

$$\begin{array}{r} 400 + 80 + 7 \\ \times \qquad\qquad 8 \\ \hline 56 \\ 640 \\ +3{,}200 \\ \hline 3{,}896 \end{array}$$

Shortcut Method

$$\begin{array}{r} {}^{6\,5} \\ 487 \\ \times\ \ 8 \\ \hline 3{,}896 \end{array}$$

8 Multiply: 6(572)

EXAMPLE 9 Multiply: 52(37)

Solution This is the same as 52(30 + 7) or

52(30) + 52(7)

We can find each of these products by using the shortcut method:

$$\begin{array}{r} 52 \\ \times\ 30 \\ \hline 1{,}560 \end{array} \qquad \begin{array}{r} {}^{1} \\ 52 \\ \times\ 7 \\ \hline 364 \end{array}$$

The sum of these two numbers is 1,560 + 364 = 1,924. Here is a summary of what we have so far:

52(37) = 52(30 + 7)	37 = 30 + 7
= 52(30) + 52(7)	Distributive property
= 1,560 + 364	Multiplication
= 1,924	Addition

The shortcut form for this problem is

$$\begin{array}{r} 52 \\ \times 37 \\ \hline 364 \\ +1{,}560 \\ \hline 1{,}924 \end{array} \begin{array}{l} \\ \\ \longleftarrow 7(52) = 364 \\ \longleftarrow 30(52) = 1{,}560 \\ \\ \end{array}$$

9 Multiply: 45(62)

Note
This discussion is to show why we multiply the way we do. You should go over it in detail, so you will understand the reasons behind the process of multiplication. Besides being able to do multiplication, you should understand it.

Answers
8 3,432 **9** 2,790

In this case we have not shown any of the numbers we carried simply because it becomes very messy. Notice that since multiplication is a commutative operation, we can get the same result by reversing the order of the factors.

$$
\begin{array}{r l}
37 & \\
\times 52 & \\
\hline
74 \leftarrow & 2(37) = 74 \\
+1,850 \leftarrow & 50(37) = 1,850 \\
\hline
1,924 &
\end{array}
$$

10 Multiply: 356(641)

EXAMPLE 10 Multiply: 279(428)

Solution

$$
\begin{array}{r l}
279 & \\
\times 428 & \\
\hline
2,232 \leftarrow & 8(279) = 2,232 \\
5,580 \leftarrow & 20(279) = 5,580 \\
+111,600 \leftarrow & 400(279) = 111,600 \\
\hline
119,412 &
\end{array}
$$

If we reverse the order of the factors, we again arrive at the same result. (This just demonstrates that multiplication is commutative, and gives us another example to look at.)

$$
\begin{array}{r l}
428 & \\
\times 279 & \\
\hline
3,852 \leftarrow & 9(428) = 3,852 \\
29,960 \leftarrow & 70(428) = 29,960 \\
+85,600 \leftarrow & 200(428) = 85,600 \\
\hline
119,412 &
\end{array}
$$

Calculator Note

Here is how we would work the problem shown in Example 10 on a calculator:

Scientific Calculator: 279 [×] 428 [=]

Graphing Calculator: 279 [×] 428 [ENT]

Estimating

One way to estimate the answer to the problem shown in Example 10 is to round each number to the nearest hundred, and then multiply the rounded numbers. Doing so would give us this:

$$300(400) = 120,000$$

Our estimate of the answer is 120,000, which is close to the actual answer, 119,412. Making estimates is important when we are using calculators; having an estimate of the answer will keep us from making major errors in multiplication.

11 If each tablet of vitamin C contains 550 milligrams of vitamin C, what is the total number of milligrams of vitamin C in a bottle that contains 365 tablets?

EXAMPLE 11 A supermarket orders 35 cases of a certain soft drink. If each case contains 24 bottles of the drink, how many bottles were ordered?

Solution We have 35 cases and each case has 24 bottles. The total number of bottles is the product of 35 and 24, which is 35(24):

$$
\begin{array}{r l}
24 & \\
\times 35 & \\
\hline
120 \leftarrow & 5(24) = 120 \\
+720 \leftarrow & 30(24) = 720 \\
\hline
840 &
\end{array}
$$

There is a total of 840 bottles of the soft drink.

Answers
10 228,196
11 200,750 milligrams

EXAMPLE 12 Shirley earns \$12 an hour for the first 40 hours she works each week. If she has \$109 deducted from her weekly check for taxes and retirement, how much money will she take home if she works 38 hours this week?

Solution To find the amount of money she earned for the week, we multiply 12 and 38. From that total we subtract 109. The result is her take-home pay. Without showing all the work involved in the calculations, here is the solution:

38(\$12) = \$456	Her total weekly earnings
\$456 − \$109 = \$347	Her take-home pay

12 If Shirley works 36 hours the next week and has the same amount deducted from her check for taxes and retirement, how much will she take home?

EXAMPLE 13 In 1993, the government standardized the way in which nutrition information is presented on the labels of most packaged food products. Figure 1 below shows one of these standardized food labels. It is from a package of Fritos Corn Chips that I ate the day I was writing this example. Approximately how many chips are in the bag, and what is the total number of calories consumed if all the chips in the bag are eaten?

Nutrition Facts
Serving Size 1 oz. (28g/About 32 chips)
Servings Per Container: 3

Amount Per Serving	
Calories 160	Calories from Fat 90
	% Daily Value*
Total Fat 10 g	16%
Saturated Fat 1.5g	8%
Cholesterol 0mg	0%
Sodium 160mg	7%
Total Carbohydrate 15g	5%
Dietary Fiber 1g	4%
Sugars 0g	
Protein 2g	
Vitamin A 0% •	Vitamin C 0%
Calcium 2% •	Iron 0%

*Percent Daily Values are based on a 2,000 calorie diet.

Figure 1

Solution Reading toward the top of the label, we see that there are about 32 chips in one serving, and approximately 3 servings in the bag. Therefore, the total number of chips in the bag is

3(32) = 96 chips

This is an approximate number since each serving is approximately 32 chips. Reading further we find that each serving contains 160 calories. Therefore, the total number of calories consumed by eating all the chips in the bag is

3(160) = 480 calories

As we progress through the book, we will study more of the information in nutrition labels.

Note
The letter g that is shown after some of the numbers in the nutrition label in Figure 1 stands for grams, a unit used to measure weight. The unit mg stands for milligrams, another, smaller unit of weight. We will have more to say about these units later in the book.

13 The amounts given in the middle of the nutrition label in Figure 1 are for one serving of chips. If all the chips in the bag are eaten, how much fat has been consumed? How much sodium?

Answers
12 \$323
13 30 g of fat, 480 mg of sodium

14 If a 150-pound person bowls for 3 hours, will he or she burn all the calories consumed by eating two bags of the chips mentioned in Example 13?

EXAMPLE 14 The table below was first introduced in Section 1.4. Suppose a 150-pound person goes bowling for 2 hours after having eaten the bag of chips mentioned in Example 13. Will he or she burn all the calories consumed from the chips?

Activity	Calories Burned in 1 Hour by a 150-Pound Person
Bicycling	374
Bowling	265
Handball	680
Jazzercize	340
Jogging	680
Skiing	544

Solution Each hour of bowling burns 265 calories. If the person bowls for 2 hours, a total of

$$2(265) = 530 \text{ calories}$$

will have been burned. Since the bag of chips contained only 480 calories, all of them have been burned with 2 hours of bowling.

Answer
14 No

1.8 Division with Whole Numbers

Introduction . . . Darlene is planning a party and would like to serve 8-ounce glasses of soda. The glasses will be filled from 32-ounce bottles of soda. In order to know how many bottles of soda to buy, she needs to find out how many of the 8-ounce glasses can be filled by one of the 32-ounce bottles. One way to solve this problem is with division: dividing 32 by 8. A diagram of the problem is shown in Figure 1.

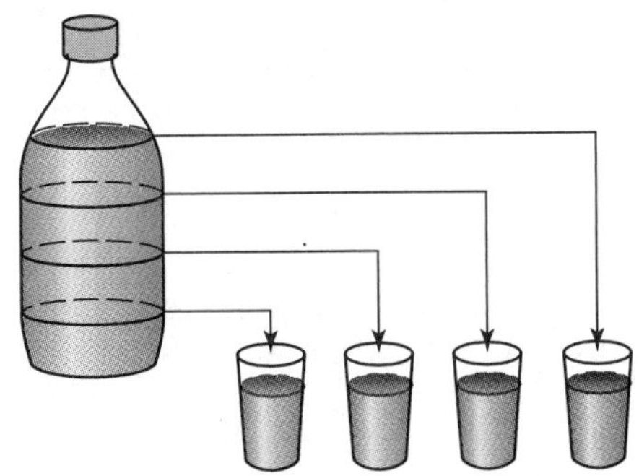

Figure 1

As a division problem:	As a multiplication problem:
$32 \div 8 = 4$	$4 \cdot 8 = 32$

Notation

As was the case with multiplication, there are many ways to write division statements. All the following statements are equivalent. They all mean 10 divided by 5.

$$10 \div 5, \quad \frac{10}{5}, \quad 10/5, \quad 5\overline{)10}, \quad \tfrac{10}{5}$$

The kind of notation we use to write division problems will depend on the situation. We will use the notation $5\overline{)10}$ mostly with the longer division problems found in this chapter. The notation $\tfrac{10}{5}$ will be used in the chapter on fractions and in later chapters. The horizontal line used with the notation $\tfrac{10}{5}$ is called the *fraction bar.*

Vocabulary

The word *quotient* is used to indicate division. If we say "The quotient of 10 and 5 is 2," then we mean

$$10 \div 5 = 2 \quad \text{or} \quad \frac{10}{5} = 2$$

The 10 is called the *dividend* and the 5 is called the *divisor.* All the expressions, $10 \div 5$, $\tfrac{10}{5}$, and 2, are called the *quotient* of 10 and 5.

Some additional examples of division statements are given in Table 1.

Table 1

In English	In Symbols
The quotient of 15 and 3	$15 \div 3$, or $\frac{15}{3}$, or $15/3$
The quotient of 3 and 15	$3 \div 15$, or $\frac{3}{15}$, or $3/15$
The quotient of 8 and n	$8 \div n$, or $\frac{8}{n}$, or $8/n$
x divided by 2	$x \div 2$, or $\frac{x}{2}$, or $x/2$
The quotient of 21 and 3 is 7	$21 \div 3 = 7$, or $\frac{21}{3} = 7$

The Meaning of Division

One way to arrive at an answer to a division problem is by thinking in terms of multiplication. For example, if we want to find the quotient of 32 and 8, we may ask, "What do we multiply by 8 to get 32?"

$$32 \div 8 = ? \quad \text{means} \quad 8 \cdot ? = 32$$

Since we know from our work with multiplication that $8 \cdot 4 = 32$, it must be true that

$$32 \div 8 = 4$$

Table 2 lists some additional examples.

Table 2

Division		Multiplication
$18 \div 6 = 3$	because	$6 \cdot 3 = 18$
$32 \div 8 = 4$	because	$8 \cdot 4 = 32$
$10 \div 2 = 5$	because	$2 \cdot 5 = 10$
$72 \div 9 = 8$	because	$9 \cdot 8 = 72$

Another way to think of division is in terms of repeated subtraction. When we say "54 divided by 6 is 9," we also mean that if we subtract 9 sixes from 54, the result will be 0.

Division by One-Digit Numbers

Consider the following division problem:

$$465 \div 5$$

We can think of this problem as asking the question, "How many fives can we subtract from 465?" To answer the question we begin subtracting multiples of 5. One way to organize this process is shown below:

```
   90 ←— We first guess that there are at least 90 fives in 465
5)465
 −450 ←— 90(5) = 450
   15 ←— 15 is left after we subtract 90 fives from 465
```

What we have done so far is subtract 90 fives from 465 and found that 15 is still left. Since there are 3 fives in 15, we continue the process.

$$\begin{array}{rl} 3 & \longleftarrow \text{There are 3 fives in 15} \\ 90 & \\ 5\overline{)465} & \\ -450 & \\ \hline 15 & \\ -15 & \longleftarrow 3 \cdot 5 = 15 \\ \hline 0 & \longleftarrow \text{The difference is 0} \end{array}$$

The total number of fives we have subtracted from 465 is

$90 + 3 = 93$ The number of fives subtracted from 465

We now summarize the results of our work.

$465 \div 5 = 93$ which we could check with multiplication $\longrightarrow$

$$\begin{array}{r} {}^{1} \\ 93 \\ \times\ 5 \\ \hline 465 \end{array}$$

Notation

The division problem just shown can be shortened by eliminating the subtraction signs, eliminating the zeros in each estimate, and eliminating some of the numbers that are repeated in the problem.

The shorthand form for this problem

$$\begin{array}{r} 3 \\ 90 \\ 5\overline{)465} \\ 450 \\ \hline 15 \\ 15 \\ \hline 0 \end{array}$$

looks like this.

$$\begin{array}{r} 93 \\ 5\overline{)465} \\ 45\downarrow \\ \hline 15 \\ 15 \\ \hline 0 \end{array}$$

The arrow indicates that we bring down the 5 after we subtract.

The problem shown above on the right is the shortcut form of what is called *long division*. Here is an example showing this shortcut form of long division from start to finish.

EXAMPLE 1 Divide: $595 \div 7$

Solution Since $7(8) = 56$, our first estimate of the number of sevens that can be subtracted from 595 is 80:

$$\begin{array}{rl} 8 & \longleftarrow \text{The 8 is placed above the tens column} \\ 7\overline{)595} & \quad\ \ \text{so we know our first estimate is 80} \\ 56\downarrow & \longleftarrow 8(7) = 56 \\ \hline 35 & \longleftarrow 59 - 56 = 3\text{; then bring down the 5} \end{array}$$

Since $7(5) = 35$, we have

$$\begin{array}{rl} 85 & \longleftarrow \text{There are 5 sevens in 35} \\ 7\overline{)595} & \\ 56\downarrow & \\ \hline 35 & \\ 35 & \longleftarrow 5(7) = 35 \\ \hline 0 & \longleftarrow 35 - 35 = 0 \end{array}$$

Our result is $595 \div 7 = 85$, which we can check with multiplication:

$$\begin{array}{r} {}^{3} \\ 85 \\ \times\ 7 \\ \hline 595 \end{array}$$

Practice Problems

1 Divide: $288 \div 4$

Answer

1 72

Division by Two-Digit Numbers

2 Divide: 6,792 ÷ 24

EXAMPLE 2 Divide: 9,380 ÷ 35

Solution In this case our divisor, 35, is a two-digit number. The process of division is the same. We still want to find the number of thirty-fives we can subtract from 9,380.

```
     2   ← The 2 is placed above the hundreds column
35)9,380
   7 0↓  ← 2(35) = 70
   2 38  ← 93 − 70 = 23; then bring down the 8
```

We can make a few preliminary calculations to help estimate how many thirty-fives are in 238:

$5 \times 35 = 175 \qquad 6 \times 35 = 210 \qquad 7 \times 35 = 245$

Since 210 is the closest to 238 without being larger than 238, we use 6 as our next estimate:

```
     26  ← 6 in the tens column means this estimate is 60
35)9,380
   7 0
   2 38
   2 10↓ ← 6(35) = 210
     280 ← 238 − 210 = 28; bring down the 0
```

Since 35(8) = 280, we have

```
     268
35)9,380
   7 0
   2 38
   2 10
     280
     280 ← 8(35) = 280
       0 ← 280 − 280 = 0
```

We can check our result with multiplication:

```
  268
×  35
1,340
8,040
9,380
```

Division with Remainders

Suppose Darlene were planning to use 6-ounce glasses instead of 8-ounce glasses for her party. To see how many glasses she could fill from the 32-ounce bottle, she would divide 32 by 6. If she did so, she would find that she could fill 5 glasses, but after doing so she would have 2 ounces of soda left in the bottle. A diagram of this problem is shown in Figure 2.

Answer
2 283

32-ounce bottle 6-ounce glasses

2 ounces left in bottle

30 ounces total

Figure 2

Writing the results in the diagram as a division problem looks like this:

```
                5 <— Quotient
Divisor —> 6)32 <— Dividend
              30
               2 <— Remainder
```

EXAMPLE 3 Divide: 1,690 ÷ 67

Solution Dividing as we have previously, we get

```
      25
 67)1,690
    1 34↓
     350
     335
      15 <— 15 is left over
```

We have 15 left, and since 15 is less than 67, no more sixty-sevens can be subtracted. In a situation like this we call 15 the *remainder* and write

These indicate that the remainder is 15

```
    25 R 15              25 15/67
 67)1,690       or   67)1,690
    1 34                 1 34
     350                  350
     335                  335
      15                   15
```

Both forms of notation shown above indicate that 15 is the remainder. The notation R 15 is the notation we will use in this chapter. The notation $\frac{15}{67}$ will be useful in Chapter 3.

To check a problem like this, we multiply the divisor and the quotient as usual, and then add the remainder to this result:

```
   67
  ×25
  335
1,340
1,675 <— Product of divisor and quotient
```

$$1{,}675 + 15 = 1{,}690$$

Remainder ↑   ↖ Dividend

3 Divide: 1,883 ÷ 27

Calculator Note

Here is how we would work the problem shown in Example 3 on a calculator:

Scientific Calculator:

1690 [÷] 67 [=]

Graphing Calculator:

1690 [÷] 67 [ENT]

In both cases the calculator will display 25.223881 (give or take a few digits at the end), which gives the remainder in decimal form. We will discuss decimals later in the book.

Answer

3 69 R 20, or $69\frac{20}{27}$

Estimating

So far in this chapter we have done all our estimating by using the rule for rounding numbers. There are times, however, when we round for convenience instead of using the rules for rounding. For example, if we want to estimate the answer to the division problem in Example 3 (1,690 ÷ 67) by rounding each of the numbers in the problem, we would round 1,690 to 1,700 and 67 to 70, and then make our estimate from

$$1{,}700 \div 70$$

which is still a difficult problem to do mentally. A more convenient way to find an estimate in this case is to round 67 to 70, and then find a number close to 1,690 that is easily divisible by 70. The number 1,400 is close to 1,690, and also easily divisible by 70. Our estimate is now based on the problem

$$1{,}400 \div 70$$

which can be calculated mentally as 20.

4 A family spends $1,872 on a 12-day vacation. How much did they spend each day?

Note

To estimate the answer to Example 4 quickly, we can replace 11,520 with 12,000 and mentally calculate

$$12{,}000 \div 12$$

which gives us an estimate of 1,000. Our actual answer, 960, is close enough to our estimate to convince us that we have not made a major error in our calculation.

EXAMPLE 4 A family has an annual income of $11,520. How much is their monthly income?

Solution Since there are 12 months in a year and the yearly (annual) income is $11,520, we want to know what $11,520 divided into 12 equal parts is. Therefore we have

```
      960
12)11,520
   10 8
    72
    72
     00
```

Since 11,520 ÷ 12 = 960, the monthly income for this family is $960.

Division by Zero

We cannot divide by 0. That is, we cannot use 0 as a divisor in any division problem. Here's why.

Suppose there was an answer to the problem

$$\frac{8}{0} = ?$$

That would mean that

$$0 \cdot ? = 8$$

But we already know that multiplication by 0 always produces 0. There is no number we can use for the ? to make a true statement out of

$$0 \cdot ? = 8$$

Since this was equivalent to the original division problem

$$\frac{8}{0} = ?$$

we have no number to associate with the expression $\frac{8}{0}$. It is undefined.

RULE Division by 0 is undefined. Any expression with a divisor of 0 is undefined. We cannot divide by 0.

Answer

4 $156

1.9 Exponents and Order of Operations

Exponents are a shorthand way of writing repeated multiplication. In the expression 2^3, 2 is called the *base* and 3 is called the *exponent*. The expression 2^3 is read "2 to the third power," or "2 cubed." The exponent 3 tells us to use the base 2 as a multiplication factor three times.

$2^3 = 2 \cdot 2 \cdot 2$ 2 is used as a factor three times

We can simplify the expression by multiplication:

$$\begin{aligned} 2^3 &= 2 \cdot 2 \cdot 2 \\ &= 4 \cdot 2 \\ &= 8 \end{aligned}$$

The expression 2^3 is equal to the number 8. We can summarize this discussion with the following definition:

DEFINITION An **exponent** is a number that indicates how many times the base is to be used as a factor. Exponents indicate repeated multiplication.

For example, in the expression 5^2, 5 is the base and 2 is the exponent. The meaning of the expression is

$$\begin{aligned} 5^2 &= 5 \cdot 5 \quad \text{5 is used as a factor two times} \\ &= 25 \end{aligned}$$

The expression 5^2 is read "5 to the second power," or "5 squared."

Here are some more examples.

EXAMPLE 1 3^2 The base is 3, and the exponent is 2. The expression is read "3 to the second power," or "3 squared."

EXAMPLE 2 3^3 The base is 3, and the exponent is 3. The expression is read "3 to the third power," or "3 cubed."

EXAMPLE 3 2^4 The base is 2, and the exponent is 4. The expression is read "2 to the fourth power."

As you can see from these examples, a base raised to the second power is also said to be *squared*, and a base raised to the third power is also said to be *cubed*. These are the only two exponents (^{2} and ^{3}) that have special names. All other exponents are referred to only as "fourth powers," "fifth powers," "sixth powers," and so on.

The next examples show how we can simplify expressions involving exponents by using repeated multiplication.

EXAMPLE 4 $3^2 = 3 \cdot 3 = 9$

EXAMPLE 5 $4^2 = 4 \cdot 4 = 16$

EXAMPLE 6 $3^3 = 3 \cdot 3 \cdot 3 = 9 \cdot 3 = 27$

Practice Problems

For each expression, name the base and the exponent, and write the expression in words.

1 5^2

2 2^3

3 1^4

Simplify each of the following by using repeated multiplication.

4 5^2

5 9^2

6 2^3

Answers

1–3 See solutions section.
4 25 **5** 81 **6** 8

7 1^4

8 2^5

EXAMPLE 7 $3^4 = 3 \cdot 3 \cdot 3 \cdot 3 = 9 \cdot 9 = 81$

EXAMPLE 8 $2^4 = 2 \cdot 2 \cdot 2 \cdot 2 = 4 \cdot 4 = 16$

Calculator Note

Here is how we use a calculator to evaluate exponents, as we did in Example 8:

Scientific Calculator: 2 [x^y] 4 [=]

Graphing Calculator: 2 [^] 4 [ENT] or 2 [x^y] 4 [ENT]
(depending on the calculator)

Finally, we should consider what happens when the numbers 0 and 1 are used as exponents. First of all, any number raised to the first power is itself. That is, if we let the letter a represent any number, then

$$a^1 = a$$

To take care of the cases when 0 is used as an exponent, we must use the following definition:

> **DEFINITION** Any number other than 0 raised to the 0 power is 1. That is, if a represents any nonzero number, then it is always true that
>
> $$a^0 = 1$$

Simplify each of the following expressions.

9 7^1

10 4^1

11 9^0

12 1^0

EXAMPLE 9 $5^1 = 5$
EXAMPLE 10 $9^1 = 9$
Any number raised to the first power is itself.

EXAMPLE 11 $4^0 = 1$
EXAMPLE 12 $8^0 = 1$
Any nonzero number raised to the 0 power is 1.

Order of Operations

The symbols we use to specify operations, $+$, $-$, $\cdot$, and $\div$, along with the symbols we use for grouping, ( ) and [ ], serve the same purpose in mathematics as punctuation marks in English. They may be called the punctuation marks of mathematics.

Consider the following sentence:

Bob said John is tall.

It can have two different meanings, depending on how we punctuate it:

1 "Bob," said John, "is tall."
2 Bob said, "John is tall."

Without the punctuation marks we don't know which meaning the sentence has.

Now, consider the following mathematical expression:

$$4 + 5 \cdot 2$$

It can have two different meanings, depending on the order in which we perform the operations. If we add 4 and 5 first, and then multiply by 2, we get 18. On the other hand, if we multiply 5 and 2 first, and then add 4, we get 14. There seem to be two different answers. In mathematics we want to avoid situations in which two different results are possible. Therefore we follow the rule for order of operations stated on the following page.

Answers
7 1 **8** 32 **9** 7 **10** 4
11 1 **12** 1

RULE: Order of Operations When evaluating mathematical expressions, we will perform the operations in the following order:

1. If the expression contains grouping symbols, such as parentheses ( ), brackets [ ], or a fraction bar, then we perform the operations inside the grouping symbols, or above and below the fraction bar, first.
2. Then we evaluate, or simplify, any numbers with exponents.
3. Then we do all multiplications and divisions in order, starting at the left and moving right.
4. Finally, we do all additions and subtractions, from left to right.

According to our rule, the expression $4 + 5 \cdot 2$ would have to be evaluated by multiplying 5 and 2 first, and then adding 4. The correct answer—and the only answer—to this problem is 14.

$$
\begin{aligned}
4 + 5 \cdot 2 &= 4 + 10 && \text{Multiply first} \\
&= 14 && \text{Then add}
\end{aligned}
$$

Here are some more examples that illustrate how we apply the rule for order of operations to simplify (or evaluate) expressions.

EXAMPLE 13 Simplify: $4 \cdot 8 - 2 \cdot 6$

Solution We multiply first and then subtract:

$$
\begin{aligned}
4 \cdot 8 - 2 \cdot 6 &= 32 - 12 && \text{Multiply first} \\
&= 20 && \text{Then subtract}
\end{aligned}
$$

13 Simplify: $5 \cdot 7 - 3 \cdot 6$

EXAMPLE 14 Simplify: $5 + 2(7 - 1)$

Solution According to the rule for the order of operations, we must do what is inside the parentheses first:

$$
\begin{aligned}
5 + 2(7 - 1) &= 5 + 2(6) && \text{Inside parentheses first} \\
&= 5 + 12 && \text{Then multiply} \\
&= 17 && \text{Then add}
\end{aligned}
$$

14 Simplify: $7 + 3(6 + 4)$

EXAMPLE 15 Simplify: $9 \cdot 2^3 + 36 \div 3^2 - 8$

$$
\begin{aligned}
9 \cdot 2^3 + 36 \div 3^2 - 8 &= 9 \cdot 8 + 36 \div 9 - 8 && \text{Exponents first} \\
&= 72 + 4 - 8 && \text{Then multiply and divide, left to right} \\
&= 76 - 8 && \text{Add and subtract,} \\
&= 68 && \text{left to right}
\end{aligned}
$$

15 Simplify: $6 \cdot 3^2 + 64 \div 2^4 - 2$

Calculator Note

Here is how we use a calculator to work the problem shown in Example 14:

Scientific Calculator: 5 [+] 2 [×] [(] 7 [−] 1 [)] [=]

Graphing Calculator: 5 [+] 2 [(] 7 [−] 1 [)] [ENT]

Example 15 on a calculator looks like this:

Scientific Calculator: 9 [×] 2 [x^y] 3 [+] 36 [÷] 3 [x^y] 2 [−] 8 [=]

Graphing Calculator: 9 [×] 2 [^] 3 [+] 36 [÷] 3 [^] 2 [−] 8 [ENT]

Answers

13 17 **14** 37 **15** 56

16 Simplify:

$5 + 3[24 - 5(6 - 2)]$

EXAMPLE 16 Simplify: $3 + 2[10 - 3(5 - 2)]$

Solution The brackets, [ ], are used in the same way as parentheses. In a case like this we move to the innermost grouping symbols first and begin simplifying:

$$\begin{aligned} 3 + 2[10 - 3(5 - 2)] &= 3 + 2[10 - 3(3)] \\ &= 3 + 2[10 - 9] \\ &= 3 + 2[1] \\ &= 3 + 2 \\ &= 5 \end{aligned}$$

Examples 14–16 all involve expressions that contain more than one operation. We want to be able to translate phrases written in English, that involve more than one operation, into symbols. Table 1 lists some English expressions and their corresponding mathematical expressions written in symbols.

Table 1

In English	Mathematical Equivalent
5 times the sum of 3 and 8	$5(3 + 8)$
Twice the difference of 4 and 3	$2(4 - 3)$
6 added to 7 times the sum of 5 and 6	$6 + 7(5 + 6)$
The sum of 4 times 5 and 8 times 9	$4 \cdot 5 + 8 \cdot 9$
3 subtracted from the quotient of 10 and 2	$10 \div 2 - 3$

Averages

We conclude this section with a look at *averages*. The average of a set of numbers is a number that indicates where the middle, or center, of the set of numbers is. We define average as follows:

DEFINITION To find the **average** of a set of numbers, we add the numbers and then divide their sum by the number of numbers we added.

17 A woman traveled the following distances on a 4-day business trip: 187 miles, 273 miles, 173 miles, and 227 miles. What was the average distance the woman traveled each day?

EXAMPLE 17 A man has had the following yearly salaries over the past 3 years: \$19,500, \$20,700, and \$21,900. What is his average income over this 3-year period?

Solution To find the man's average salary over the 3-year period, we add all three salaries and divide the sum by 3. The sum of the three salaries is

$$\begin{array}{r} \$19{,}500 \\ 20{,}700 \\ \underline{21{,}900} \\ \$62{,}100 \end{array}$$

This sum divided by 3 is

$\$62{,}100 \div 3 = \$20{,}700$ ← Average

In this case the average happens to be one of the original numbers.

Answers

16 17 **17** 215 miles

1.10 Area and Volume

The *area* of a flat object is a measure of the amount of surface the object has. The rectangle in Figure 1 below has an area of 6 square inches, because that is the number of squares (each of which is 1 inch long and 1 inch wide) it takes to cover the rectangle.

one square inch	one square inch	one square inch	2 inches
one square inch	one square inch	one square inch	
	3 inches		

Figure 1 A rectangle with an area of 6 square inches

It is no coincidence that the area of the rectangle in Figure 1 and the product of the length and the width are the same number: We can calculate the area of the rectangle in Figure 1 by simply multiplying the length and the width together:

$$\begin{aligned}\text{Area} &= (\text{length}) \cdot (\text{width}) \\ &= (3 \text{ inches}) \cdot (2 \text{ inches}) \\ &= (3 \cdot 2) \cdot (\text{inches} \cdot \text{inches}) \\ &= 6 \text{ square inches}\end{aligned}$$

The unit *square inches* can be abbreviated as *sq. in.* or in^2.

Formulas for Area

Figure 2 shows three common geometric figures along with the formulas for their areas. The only figure we have not seen before is the parallelogram.

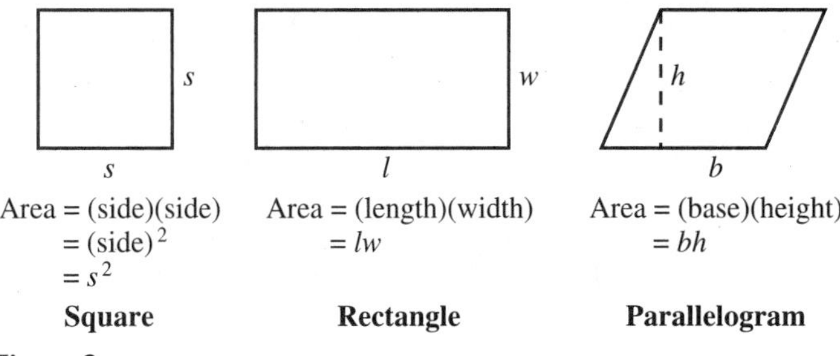

Figure 2

Practice Problems

1 Find the total area enclosed by the figure.

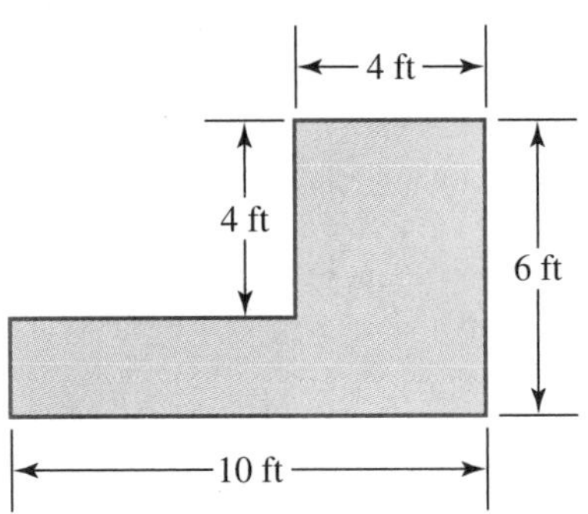

EXAMPLE 1 Find the total area enclosed by Figure 3 below.

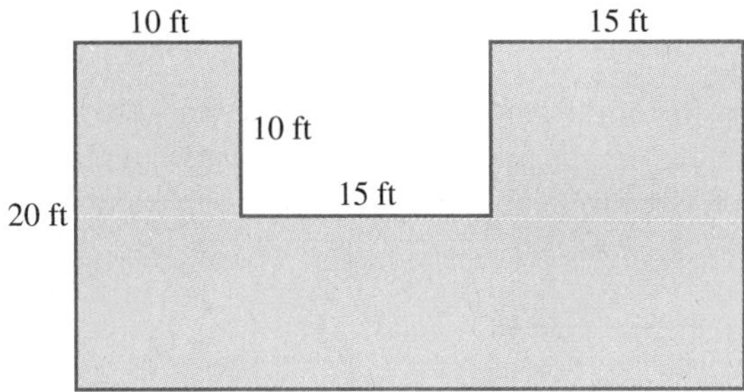

Figure 3

Solution We begin by drawing in two additional lines (shown as broken lines in Figure 4) so that the original figure is now composed of three individual figures.

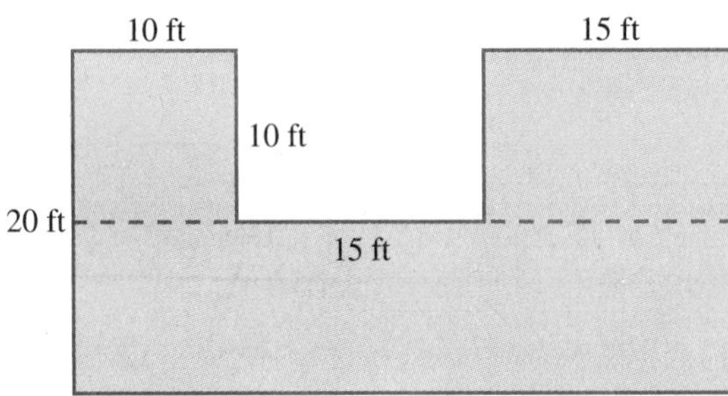

Figure 4

Next we fill in the missing dimensions on the three individual figures (Figure 5). Then we number them for reference.

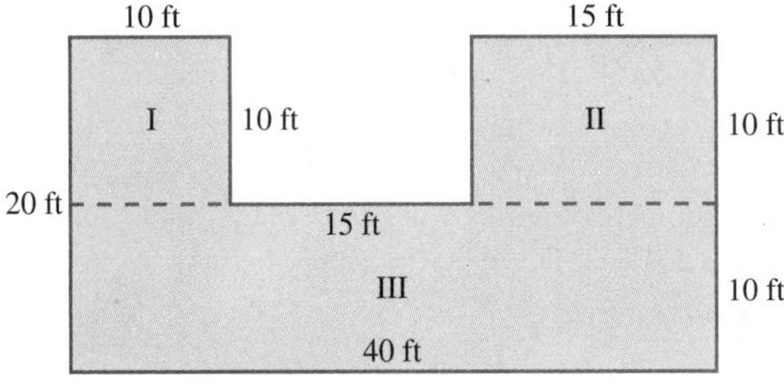

Figure 5

Finally, we calculate the area of the original figure by adding the areas of the three individual figures:

$$\begin{aligned} \text{Area} &= \text{Area I} + \text{Area II} + \text{Area III} \\ &= 10 \cdot 10 + 15 \cdot 10 + 40 \cdot 10 \\ &= 100 + 150 + 400 \\ &= 650 \text{ square feet} \end{aligned}$$

Answer

1 36 square feet

EXAMPLE 2 How many square inches are contained in a square foot?

Solution To solve this problem, we draw a square and label each side with 1 foot = 12 inches, as in Figure 6.

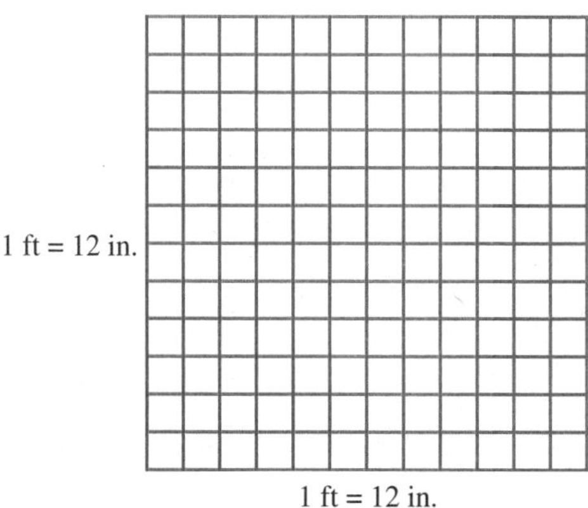

Figure 6 1 square foot

To find the number of square inches in Figure 6, we can either count them to get 144 square inches, or we can calculate it directly by using the formula for the area of a square:

$$\begin{aligned} A &= (1 \text{ foot}) \cdot (1 \text{ foot}) \\ &= (12 \text{ inches}) \cdot (12 \text{ inches}) \\ &= (12 \cdot 12) \text{ square inches} \\ &= 144 \text{ square inches} \end{aligned}$$

In addition to the relationship between square inches and square feet as given in Example 2, we would like to know the relationship between other measures of area. The table below gives some relationships between the most common units of area measure. Some of these will be familiar to you, while others may not.

Table 1 Units of area in the U.S. system

The Relationship between	Is
square inches (in^2) and square feet (ft^2)	$144\ in^2 = 1\ ft^2$
square yards (yd^2) and square feet (ft^2)	$9\ ft^2 = 1\ yd^2$
acres and square feet	$1 \text{ acre} = 43{,}560\ ft^2$
acres and square miles (mi^2)	$640 \text{ acres} = 1\ mi^2$

2 Find the number of square feet in 1 square yard.

Answer
2 9 square feet

3 One square has a side 4 feet long, while a second square has a side 3 times that, or 12 feet long. How many times larger than the first square is the area of the second square?

EXAMPLE 3 For the two squares shown in Figure 7, a side in the larger square is 3 times as large as a side in the smaller square. Give a similar comparison of the areas of the two squares.

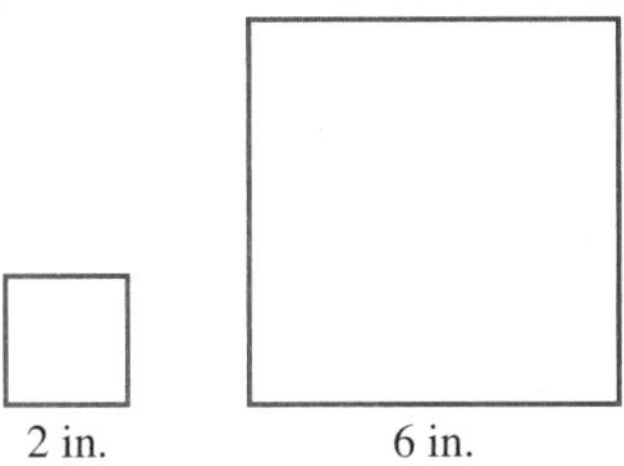

Figure 7 Two squares

Solution First we find the area of each square, as shown in Figure 8.

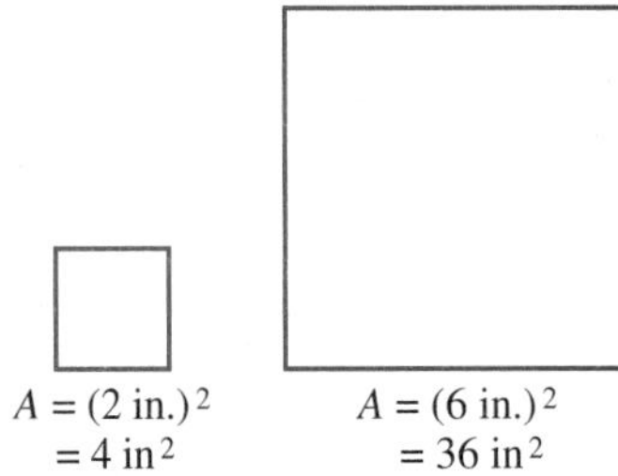

Figure 8 Two squares with areas calculated

The two areas are 4 square inches and 36 square inches. Since we multiply 4 by 9 to get 36, we say the area of the larger square is 9 times the area of the smaller square.

Volume

Next, we move up one dimension and consider what is called *volume*. Volume is the measure of the space enclosed by a solid. For instance, if each edge of a cube is 3 feet long, as shown in Figure 9, then we can think of the cube as being made up of a number of smaller cubes, each of which is 1 foot long, 1 foot wide, and 1 foot high. Each of these smaller cubes is called a cubic foot. To count the number of them in the larger cube, think of the large cube as having three layers. You can see that the top layer contains 9 cubic feet. Since there are three layers, the total number of cubic feet in the large cube is $9 \cdot 3 = 27$.

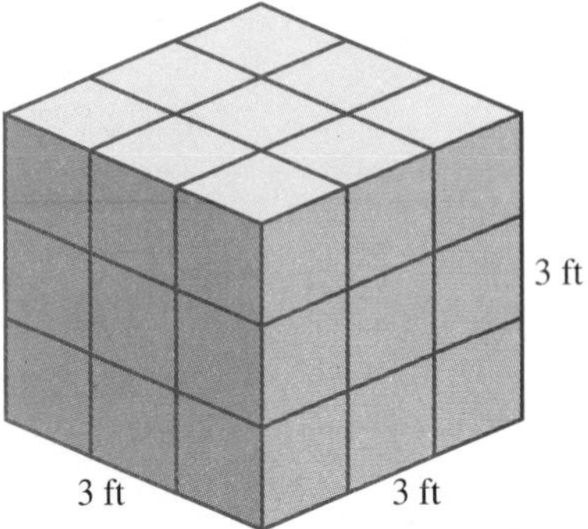

Figure 9 A cube in which each edge is 3 feet long

Answer
3 9 times as large

On the other hand, if we multiply the length, the width, and the height of the cube, we have the same result:

$$\begin{aligned}\text{Volume} &= (3 \text{ feet})(3 \text{ feet})(3 \text{ feet}) \\ &= (3 \cdot 3 \cdot 3)(\text{feet} \cdot \text{feet} \cdot \text{feet}) \\ &= 27 \text{ cubic feet}\end{aligned}$$

For the present we will confine our discussion of volume to volumes of *rectangular solids.* Rectangular solids are the three-dimensional equivalents of rectangles: Opposite sides are parallel and any two sides that meet, meet at right angles. A rectangular solid is shown in Figure 10, along with the formula used to calculate its volume.

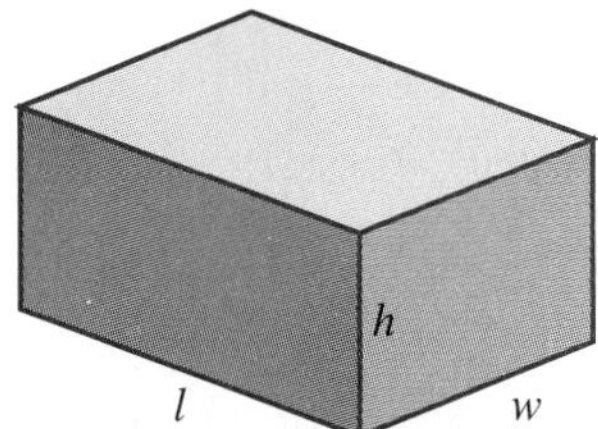

Volume = (length)(width)(height)
$V = lwh$

Figure 10 A rectangular solid

EXAMPLE 4 Find the volume of a rectangular solid with length 15 inches, width 3 inches, and height 5 inches.

Solution To find the volume we apply the formula shown in Figure 10 above:

$$\begin{aligned}V &= l \cdot w \cdot h \\ &= (15 \text{ in.}) \cdot (3 \text{ in.}) \cdot (5 \text{ in.}) \\ &= 225 \text{ in}^3\end{aligned}$$

Note in Example 4 that we used in^3 to represent cubic inches. Another abbreviation for cubic inches is cu in.

Table 2 Units of volume in the U.S. system

The Relationship between	Is
cubic inches (in^3) and cubic feet (ft^3)	$1 \text{ ft}^3 = 1{,}728 \text{ in}^3$
cubic feet and cubic yards (yd^3)	$1 \text{ yd}^3 = 27 \text{ ft}^3$
fluid ounces (fl oz) and pints (pt)	1 pt = 16 fl oz
pints and quarts (qt)	1 qt = 2 pt
quarts and gallons (gal)	1 gal = 4 qt

4 A home has a dining room that is 12 feet wide and 15 feet long. If the ceiling is 8 feet high, find the volume of the dining room.

Answer
4 1,440 cubic feet

Surface Area

Figure 11 shows a closed box with length l, width w, and height h. The surfaces of the box are labeled as sides, top, bottom, front, and back.

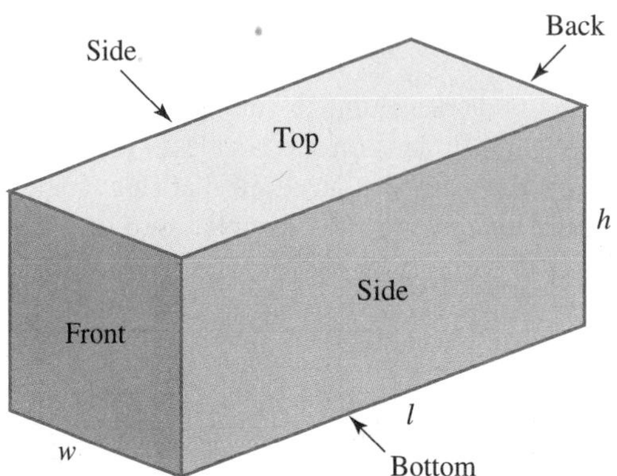

Figure 11 A box with dimensions l, w, and h

To find the *surface area* of the box, we add the areas of each of the six surfaces that are labeled in Figure 11.

$$\begin{aligned}\text{Surface area} &= \text{side} + \text{side} + \text{front} + \text{back} + \text{top} + \text{bottom}\\ S &= l \cdot h + l \cdot h + h \cdot w + h \cdot w + l \cdot w + l \cdot w\\ &= 2lh + 2hw + 2lw\end{aligned}$$

5 A family is painting a dining room that is 12 feet wide and 15 feet long.

a If the ceiling is 8 feet high, find the surface area of the walls and the ceiling, but not the floor.

b If a gallon of paint will cover 400 square feet, how many gallons should they buy to paint the walls and the ceiling?

EXAMPLE 5 Find the surface area of the box shown in Figure 12.

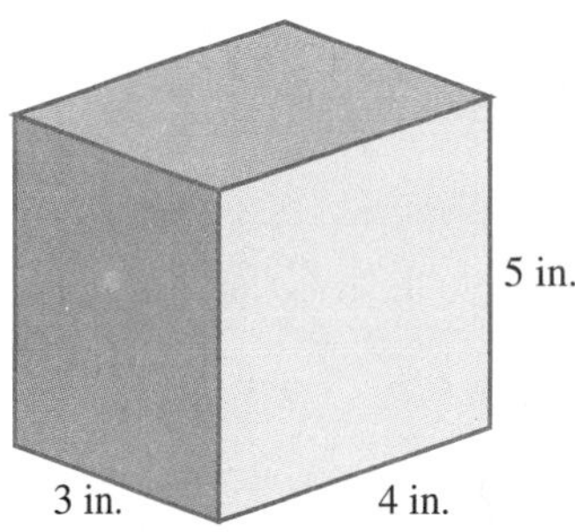

Figure 12 A box 4 inches long, 3 inches wide, and 5 inches high

Solution To find the surface area we find the area of each surface individually, and then we add them together:

$$\begin{aligned}\text{Surface area} &= 2(3 \text{ in.} \cdot 4 \text{ in.}) + 2(3 \text{ in.} \cdot 5 \text{ in.}) + 2(4 \text{ in.} \cdot 5 \text{ in.})\\ &= 24 \text{ in}^2 + 30 \text{ in}^2 + 40 \text{ in}^2\\ &= 94 \text{ in}^2\end{aligned}$$

The total surface area is 94 square inches. If we calculate the volume enclosed by the box, it is $V = 3 \text{ in.} \cdot 4 \text{ in.} \cdot 5 \text{ in.} = 60 \text{ in}^3$. The surface area measures how much material it takes to make the box, whereas the volume measures how much space the box will hold.

Answer

5 a 612 square feet

b 2 gallons will cover everything, with some paint left over.

CHAPTER 1 SUMMARY

Examples
The margins of the chapter summaries will be used for examples of the topics being reviewed, whenever it is convenient.

The numbers in brackets indicate the sections in which the topics were discussed.

Place Values for Decimal Numbers [1.1]
The place values for the digits of any base 10 number are as follows:

Billions			Millions			Thousands					
Hundred Billions	Ten Billions	Billions	Hundred Millions	Ten Millions	Millions	Hundred Thousands	Ten Thousands	Thousands	Hundreds	Tens	Ones

1 The number 42,103,045 written in words is "forty-two million, one hundred three thousand, forty-five."

The number 5,745 written in expanded form is

5,000 + 700 + 40 + 5

Vocabulary Associated with Addition, Subtraction, Multiplication, and Division [1.2, 1.5, 1.6, 1.8]

The word *sum* indicates addition.

The word *difference* indicates subtraction.

The word *product* indicates multiplication.

The word *quotient* indicates division.

2 The sum of 5 and 2 is $5 + 2$.

The difference of 5 and 2 is $5 - 2$.

The product of 5 and 2 is $5 \cdot 2$.

The quotient of 10 and 2 is $10 \div 2$.

Properties of Addition and Multiplication [1.2, 1.6, 1.7]
If a, b, and c represent any three numbers, then the properties of addition and multiplication used most often are:

Commutative property of addition: $a + b = b + a$

Commutative property of multiplication: $a \cdot b = b \cdot a$

Associative property of addition: $(a + b) + c = a + (b + c)$

Associative property of multiplication: $(a \cdot b) \cdot c = a \cdot (b \cdot c)$

Distributive property: $a(b + c) = a(b) + a(c)$

3 $3 + 2 = 2 + 3$

$3 \cdot 2 = 2 \cdot 3$

$(x + 3) + 5 = x + (3 + 5)$

$(4 \cdot 5) \cdot 6 = 4 \cdot (5 \cdot 6)$

$3(4 + 7) = 3(4) + 3(7)$

Perimeter of a Polygon [1.3]
The *perimeter* of any polygon is the sum of the sides, and it is denoted with the letter P.

4 The perimeter of the rectangle below is

$$P = 37 + 37 + 24 + 24$$
$$= 122 \text{ feet}$$

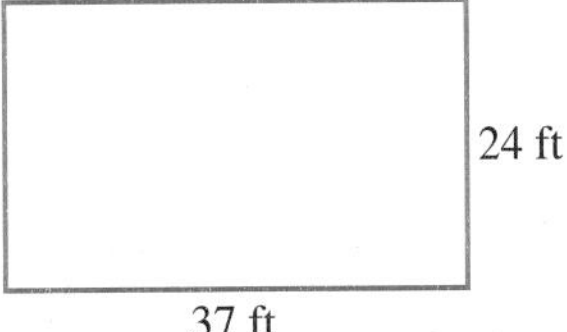

Steps for Rounding Whole Numbers [1.4]

1. Locate the digit just to the right of the place you are to round to.
2. If that digit is less than 5, replace it and all digits to its right with zeros.
3. If that digit is 5 or more, replace it and all digits to its right with zeros, and add 1 to the digit to its left.

5 5,482 to the nearest ten is 5,480
5,482 to the nearest hundred is 5,500
5,482 to the nearest thousand is 5,000

6 $4 + 6(8 - 2) = 4 + 6(6)$ Inside parentheses first
$= 4 + 36$ Then multiply
$= 40$ Then add

Order of Operations [1.9]

To simplify a mathematical expression:

1 We simplify the expression inside the grouping symbols first. Grouping symbols are parentheses ( ), brackets [ ], or a fraction bar.
2 Then we evaluate any numbers with exponents.
3 We then perform all multiplications and divisions in order, starting at the left and moving right.
4 Finally, we do all the additions and subtractions, from left to right.

7 The average of 4, 7, 9 and 12 is

$(4 + 7 + 9 + 12) \div 4 = 32 \div 4 = 8$

Average [1.9]

The *average* for a set of numbers is their sum divided by the number of numbers in the set.

8 Each expression below is undefined.

$5 \div 0 \quad \frac{7}{0} \quad 4/0$

Division by 0 (Zero) [1.8]

Division by 0 is undefined. We cannot use 0 as a divisor in any division problem.

9 $2^3 = 2 \cdot 2 \cdot 2 = 8$
$5^0 = 1$

Exponents [1.9]

In the expression 2^3, 2 is the *base* and 3 is the *exponent*. An exponent is a shorthand notation for repeated multiplication. The exponent 0 is a special exponent. Any nonzero number to the 0 power is 1.

Formulas for Area [1.10]

Below are three common geometric figures, along with the formulas for their areas.

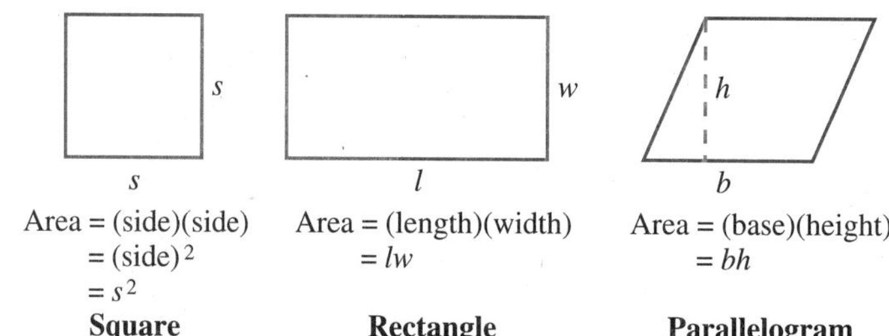

Formulas for Volume and Surface Area [1.10]

Below is a rectangular solid, along with the formulas for its volume and surface area.

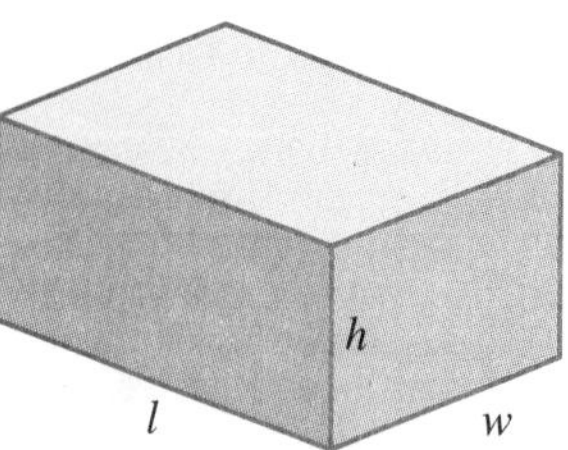

Volume = lwh
Surface area = $2lh + 2hw + 2lw$

COMMON MISTAKE

A common mistake is to write numbers in the wrong order when translating subtraction or division statements from words to symbols.

The difference of 5 and 2 is $5 - 2$, *not* $2 - 5$.
The quotient of 12 and 4 is $12 \div 4$, *not* $4 \div 12$.

CHAPTER 1 REVIEW

The numbers in brackets indicate the sections in which problems of a similar type can be found.

1 One of the largest fish ever caught was found off the coast of Guam in 1969. It was a Pacific blue marlin and weighed 1,153 pounds. Write 1,153 in words. **[1.1]**

2 In 1978 more people attended professional baseball games at Dodger Stadium than any other ball park. The attendance that year for the Dodgers was 3,347,776. Write 3,347,776 in words. **[1.1]**

For Problems 3 and 4, write each number with digits instead of words. **[1.1]**

3 Five million, two hundred forty-five thousand, six hundred fifty-two

4 Twelve million, twelve thousand, twelve

5 In 1978 the Oakland A's had the lowest attendance in the league. They drew only 527,007 people to their home games. Write 527,007 in expanded form. **[1.1]**

6 In 1977 there were more marriages in the state of Texas than in any other state. During that year, there were 161,331 marriages in Texas. Write 161,331 in expanded form. **[1.1]**

Identify each of the statements in Problems 7–14 as an example of one of the following properties. **[1.2, 1.6]**

a Addition property of 0
b Multiplication property of 0
c Multiplication property of 1
d Commutative property of addition
e Commutative property of multiplication
f Associative property of addition
g Associative property of multiplication

7 $5 + 7 = 7 + 5$

8 $(4 + 3) + 2 = 4 + (3 + 2)$

9 $6 \cdot 1 = 6$

10 $8 + 0 = 8$

11 $5 \cdot 0 = 0$

12 $4 \cdot 6 = 6 \cdot 4$

13 $5 \cdot (3 \cdot 2) = (5 \cdot 3) \cdot 2$

14 $(6 + 2) + 3 = (2 + 6) + 3$

Find each of the following sums. (Add.) **[1.3]**

15 $\begin{array}{r} 498 \\ +\ 251 \\ \hline \end{array}$

16 $\begin{array}{r} 784 \\ +\ 598 \\ \hline \end{array}$

17 $\begin{array}{r} 7{,}384 \\ 251 \\ +\ \ 637 \\ \hline \end{array}$

18 $\begin{array}{r} 4{,}901 \\ 648 \\ +\ 3{,}592 \\ \hline \end{array}$

Find each of the following differences. (Subtract.) **[1.5]**

19 $\begin{array}{r} 789 \\ -\,475 \\ \hline \end{array}$

20 $\begin{array}{r} 792 \\ -\,178 \\ \hline \end{array}$

21 $\begin{array}{r} 5{,}908 \\ -\,2{,}759 \\ \hline \end{array}$

22 $\begin{array}{r} 3{,}527 \\ -\,1{,}789 \\ \hline \end{array}$

Find each of the following products. (Multiply.) **[1.7]**

23 8(73)

24 7(984)

25 63(59)

26 49(876)

Find each of the following quotients. (Divide.) **[1.8]**

27 $692 \div 4$

28 $1{,}020 \div 15$

29 $36\overline{)15{,}408}$

30 $286\overline{)21{,}736}$

Round the number 3,781,092 to the nearest: **[1.4]**

31 Ten

32 Hundred

33 Hundred thousand

34 Million

Use the rule for the order of operations to simplify each expression as much as possible. **[1.9]**

35 $4 + 3 \cdot 5^2$

36 $7(9)^2 - 6(4)^3$

37 $3(2 + 8 \cdot 9)$

38 $7 - 2(6 - 4)$

39 A first-year math student had grades of 79, 64, 78, and 95 on the first four tests. What is the student's average test grade? **[1.9]**

40 If a person has scores of 205, 222, 174, 236, 185, and 214 for six games of bowling, what is the average score for the six games? **[1.9]**

Write an expression using symbols that is equivalent to each of the following expressions; then simplify. **[1.9]**

41 3 times the sum of 4 and 6

42 9 times the difference of 5 and 3

43 Twice the difference of 17 and 5

44 The product of 5 with the sum of 8 and 2

45 A person has a monthly income of $1,783 and monthly expenses of $1,295. What is the difference between the monthly income and the expenses?

46 A rancher bought 395 sheep and then sold 197 of them. How many were left?

47 An airplane manufacturer bought 28 crates of airplane parts. If each crate contained 75 parts, how many parts were purchased?

48 A bottle contains 150 vitamin C tablets. If each tablet contains 250 milligrams of vitamin C, how many milligrams of vitamin C are in the bottle?

49 Each month a family budgets $1,150 for rent, $625 for food, and $257 for entertainment. What is the sum of these numbers?

50 If a person wrote 23 checks in January, 37 checks in February, 40 checks in March, and 27 checks in April, what is the total number of checks written in the 4-month period?

51 A person has a yearly income of $23,256. What is the person's monthly income?

52 It takes a jogger 126 minutes to run 14 miles. At that rate, how long does it take the jogger to run 1 mile?

53 Jeff makes $6 an hour for the first 40 hours he works in a week and $9 an hour for every hour after that. Each week he has $102 deducted from his check for income taxes and retirement. If he works 45 hours in one week, how much is his take-home pay?

54 Barbara earns $8 an hour for the first 40 hours she works in a week and $12 an hour for every hour after that. Each week she has $123 deducted from her check for income taxes and retirement. What is her take-home pay for a week in which she works 50 hours?

55 Find the perimeter and the area of the parallelogram below. **[1.3, 1.10]**

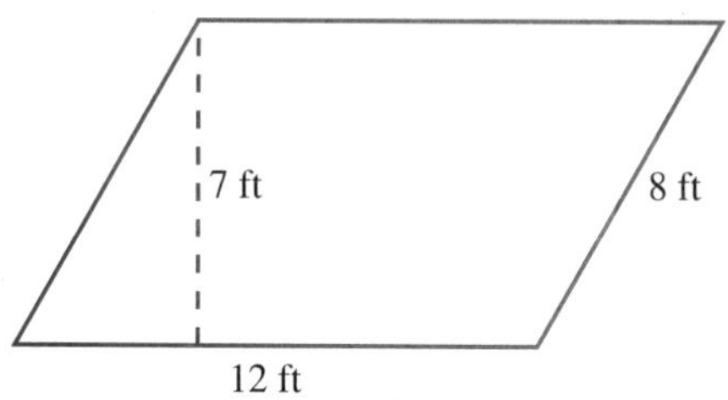

56 Find the volume and the surface area of the rectangular solid shown below. **[1.10]**

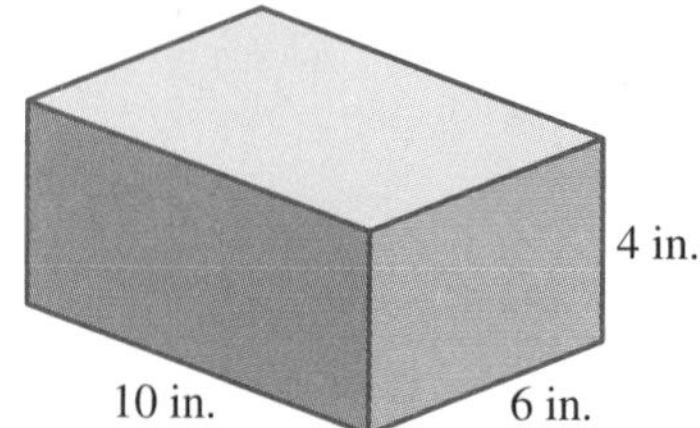

The tables below are similar to two of the tables we have worked with in this chapter. Use the information in the tables to work Problems 57–64. **[1.7]**

Number of calories in fast food

Food	Calories
McDonald's Hamburger	270
Burger King Hamburger	260
Jack in the Box Hamburger	280
McDonald's Big Mac	510
Burger King Whopper	630
Jack in the Box Colossus Burger	940
Roy Rogers Roast Beef Sandwich	260
McDonald's Chicken McNuggets (6)	300
Taco Bell Chicken Burrito	345
McDonald's French Fries (large)	450
Burger King BK Broiler	540
Burger King Chicken Sandwich	700

Number of calories burned in 30 minutes

	130-pound person	170-pound person
Indoor Activities		
Vacuuming	100	130
Mopping floors	105	135
Shopping for food	110	135
Ironing clothes	115	145
Outdoor Activities		
Chopping wood	150	200
Ice skating	170	235
Cross-country skiing	250	330
Shoveling snow	260	350

57 How many calories do you consume if you eat one large order of McDonald's French Fries and 2 Big Macs?

58 How many calories do you consume if you eat one order of Chicken McNuggets and a McDonald's Hamburger?

59 How many more calories are in one Colossus Burger than in two Taco Bell Chicken Burritos?

60 What is the difference in calories between a Whopper and a BK Broiler?

61 If you weigh 170 pounds and ice skate for 1 hour, will you burn all the calories consumed by eating one Whopper?

62 If you weigh 130 pounds and go cross-country skiing for 1 hour, will you burn all the calories consumed by eating one large order of McDonald's French Fries?

63 Suppose you eat a Big Mac and a large order of fries for lunch. If you weigh 130 pounds, what combination of 30-minute activities could you do to burn all the calories you consumed at lunch?

64 Suppose you weigh 170 pounds and you eat two Taco Bell Chicken Burritos for lunch. What combination of 30-minute activities could you do to burn all the calories in the burritos?

CHAPTER 1 TEST

1 Write the number 20,347 in words.

2 Write the number two million, forty-five thousand, six with digits instead of words.

3 Write the number 123,407 in expanded form.

Identify each of the statements in Problems 4–7 as an example of one of the following properties.

a Addition property of 0
b Multiplication property of 0
c Multiplication property of 1
d Commutative property of addition
e Commutative property of multiplication
f Associative property of addition
g Associative property of multiplication

4 $(5 + 6) + 3 = 5 + (6 + 3)$

5 $7 \cdot 1 = 7$

6 $9 + 0 = 9$

7 $5 \cdot 6 = 6 \cdot 5$

Find each of the following sums. (Add.)

8 $\begin{array}{r} 135 \\ +741 \\ \hline \end{array}$

9 $\begin{array}{r} 5{,}401 \\ 329 \\ +10{,}653 \\ \hline \end{array}$

Find each of the following differences. (Subtract.)

10 $\begin{array}{r} 937 \\ -413 \\ \hline \end{array}$

11 $\begin{array}{r} 7{,}052 \\ -3{,}967 \\ \hline \end{array}$

Find each of the following products. (Multiply.)

12 9(186)

13 62(359)

Find each of the following quotients. (Divide.)

14 $1{,}105 \div 13$

15 $583\overline{)12{,}243}$

16 Round the number 516,249 to the nearest ten thousand.

Use the rule for the order of operations to simplify each expression as much as possible.

17 $8(5)^2 - 7(3)^3$

18 $8 - 2(5 - 3)$

19 If a person has scores of 143, 187, 150, and 176 for four games of bowling, what is the average score for the four games?

Write an expression using symbols that is equivalent to each of the following expressions; then simplify.

20 Twice the sum of 11 and 7

21 The quotient of 20 and 5 increased by 9

22 A person has a yearly income of $18,324. What is the person's monthly income?

23 Karen earns $7 an hour for the first 40 hours she works in a week and $10 an hour for every hour after that. Each week she has $115 deducted from her check for income taxes and retirement. If she works 47 hours in one week, how much is her take-home pay?

24 Find the perimeter and the area of the rectangle below.

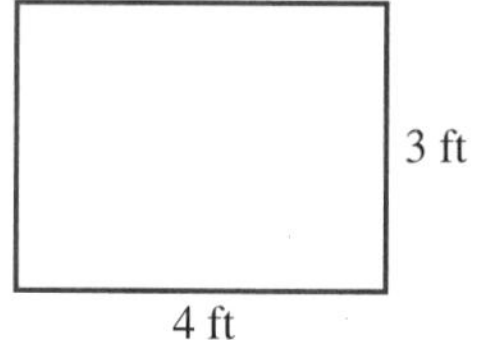

25 Find the volume and the surface area of the rectangular solid shown below.

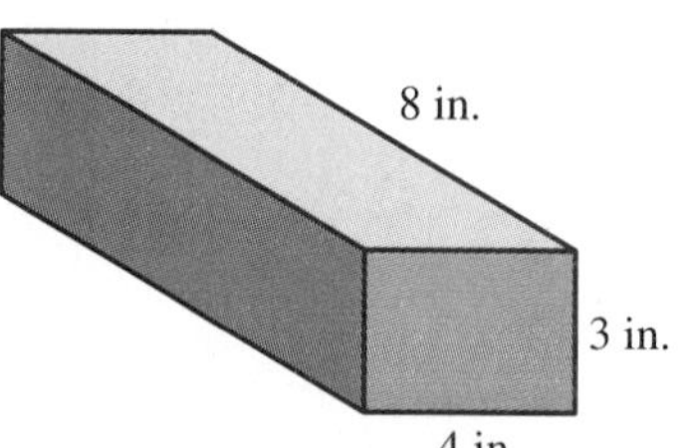

2

INTRODUCTION TO ALGEBRA

INTRODUCTION

The table below gives the record low temperature, in degrees Fahrenheit, for each month of the year in the city of Jackson, Wyoming. The two diagrams are drawn from the information in the table. Figure 1 is a bar chart or histogram similar to the ones we studied in Chapter 1. Figure 2 is called a line graph. Line graphs are one of the items that we will study in this chapter.

Month	Temperature
January	−50°F
February	−44°F
March	−32°F
April	−5°F
May	12°F
June	19°F
July	24°F
August	18°F
September	14°F
October	2°F
November	−27°F
December	−49°F

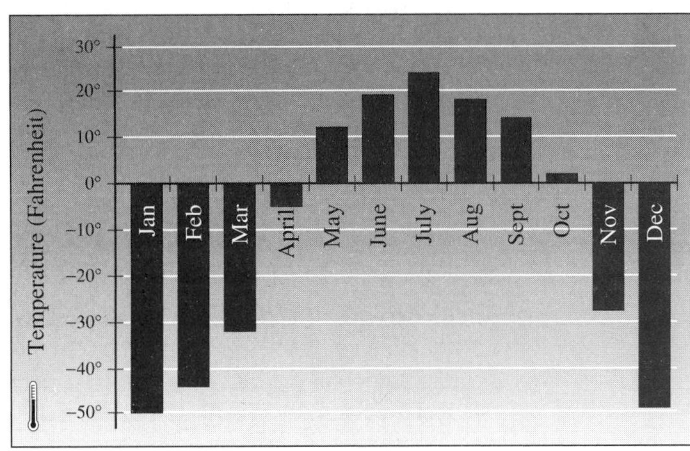

Figure 1

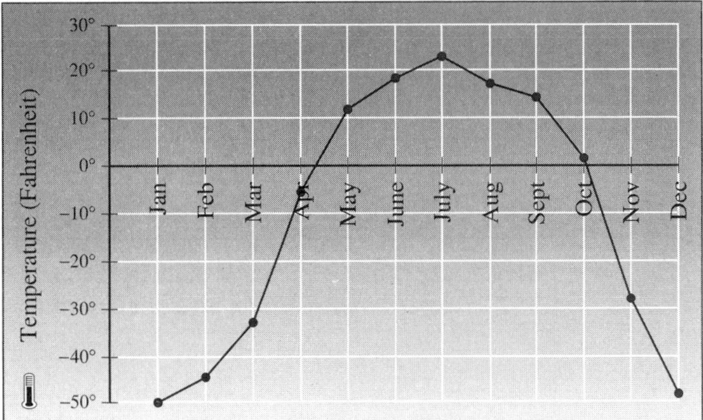

Figure 2

As you can see, the temperatures in the table and the two figures that are below zero are given as negative numbers. Negative numbers are the topic we will spend the most time with in this chapter. To give you an idea of the type of question we will ask in this chapter, can you find the difference in record low temperatures between September and November from the table or figures above? The table gives the record low temperature for September as 14°F and the record low for November as −27°F. Since the word *difference* is associated with subtraction, our question is answered with the following problem:

$$14 - (-27)$$

In order to work problems such as these, we need to develop rules for addition, subtraction, multiplication, and division with negative numbers. And as we said previously, that is our main goal for this chapter.

To be successful in this chapter, you need a working knowledge of the commutative, associative, and distributive properties from Chapter 1, and, of course, you must be able to add, subtract, multiply, and divide whole numbers.

STUDY SKILLS

If you have successfully completed Chapter 1, then you have made a good start at developing the study skills necessary to succeed in all math classes. Here is the list of study skills for this chapter. Some are a continuation of the skills from Chapter 1, while others are new to this chapter.

1 Continue to set and keep a schedule. Sometimes I find that students who do well in Chapter 1 become overconfident. They begin to put in less time with their homework. Don't make that mistake. Keep to the same schedule.

2 Increase effectiveness. You want to become more and more effective with the time you spend on your homework. You want to increase the amount of learning you obtain in the time you have set aside. Increase those activities that you feel are the most beneficial and decrease those that have not given you the results you wanted. As time goes by you will get more out of the time you spend studying mathematics. But you must make an effort to have that happen by constantly evaluating the effectiveness of what you are doing.

3 List difficult problems. Begin to make lists of problems that give you the most difficulty—those that you are repeatedly making mistakes with.

4 Look for difficulties caused by your intuition. You will probably find, as you progress through this chapter, that some of the problems we work will be contrary to your own intuition. For example, many people will answer 5, if they are asked for the difference between 8 and −3. As you will see when we get to Section 2.3, this difference is actually 11. However, even though you may understand the rules we will introduce for subtraction, you may still have a tendency to use your intuition, rather than the rules we develop, to answer questions. The solution to this is simply to notice when your intuition is getting in the way, and make it a point to work problems the way you see them done in the book and the way your instructor does them. The more problems you work correctly, the more closely your intuition will align with the mathematics you see here.

Many of my students keep a notebook that contains everything that they need for the course: class notes, homework, quizzes, tests, and other projects. A three-ring binder with tabs is ideal. Your main goal with the notebook is to organize it so that you can easily get to any item you need.

2.1 Positive and Negative Numbers

Introduction . . . Suppose you have a balance of \$20 in your checkbook and then write a check for \$30. You are now overdrawn by \$10. How will you write this new balance? One way is with a negative number. You could write the balance as −\$10, which is a negative number. Negative numbers can be used to describe other situations as well—for instance, temperature below zero and distance below sea level.

To see the relationship between negative and positive numbers, we can extend the number line as shown in Figure 1. We first draw a straight line and label a convenient point with 0. This is called the *origin,* and it is usually in the middle of the line. We then label positive numbers to the right (as we have done previously), and negative numbers to the left.

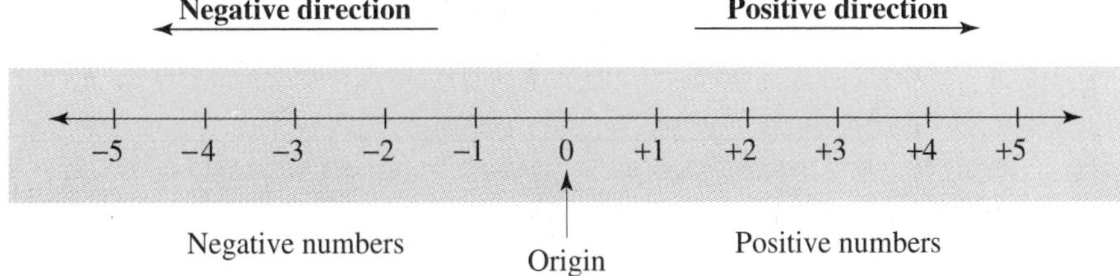

Figure 1

The numbers increase going from left to right. If we move to the right, we are moving in the positive direction. If we move to the left, we are moving in the negative direction. *Any number to the left of another number is considered to be smaller than the number to its right.*

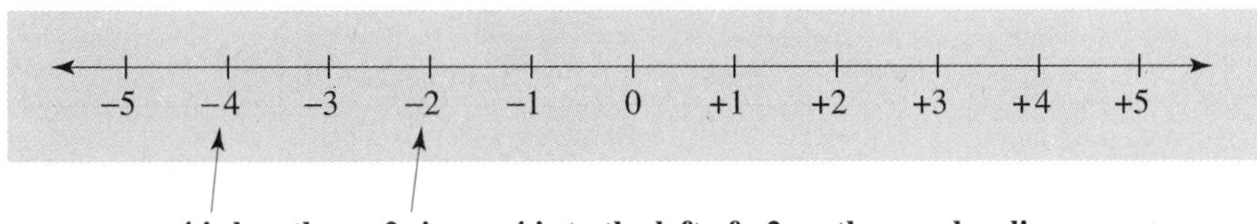

Figure 2

We see from the line that every negative number is less than every positive number.

In algebra we can use inequality symbols when comparing numbers.

> **Notation**
>
> If a and b are any two numbers on the number line, then
>
> $a < b$ is read "a is less than b."
>
> $a > b$ is read "a is greater than b."

As you can see, the inequality symbols always point to the smaller of the two numbers being compared. Here are some examples that illustrate how we use the inequality symbols.

EXAMPLE 1 $3 < 5$ is read "3 is less than 5." Note that it would also be correct to write $5 > 3$. Both statements, "3 is less than 5" and "5 is greater than 3," have the same meaning. The inequality symbols always point to the smaller number.

Practice Problems

Write each statement in words.

1 $2 < 8$

Answer

1 2 is less than 8.

2 $5 > 10$ (Is this a true statement?)

3 $-4 < 4$

4 $-7 < -2$

Give the absolute value of each of the following.

5 $|6|$

6 $|-5|$

7 $|-8|$

8 Give the opposite of each of the following numbers: 8, 10, 0, −4.

Answers
2 5 is greater than 10. No
3 −4 is less than 4.
4 −7 is less than −2.
5 6 **6** 5 **7** 8
8 −8, −10, 0, 4

EXAMPLE 2 $0 > 100$ is a false statement, because 0 is less than 100, not greater than 100. To write a true inequality statement using the numbers 0 and 100, we would have to write either $0 < 100$ or $100 > 0$.

EXAMPLE 3 $-3 < 5$ is a true statement, because -3 is to the left of 5 on the number line, and, therefore, it must be less than 5. Another statement that means the same thing is $5 > -3$.

EXAMPLE 4 $-5 < -2$ is a true statement, because -5 is to the left of -2 on the number line, meaning that -5 is less than -2. Both statements $-5 < -2$ and $-2 > -5$ have the same meaning; they both say that -5 is a smaller number than -2.

It is sometimes convenient to talk about only the numerical part of a number and disregard the sign (+ or −) in front of it. The following definition gives us a way of doing this:

DEFINITION The **absolute value** of a number is its distance from 0 on the number line. We denote the absolute value of a number with vertical lines. For example, the absolute value of -3 is written $|-3|$.

The absolute value of a number is never negative because it is a distance, and a distance is always measured in positive units (unless it happens to be 0).

Here are some examples of absolute value problems.

EXAMPLE 5 $|5| = 5$ The number 5 is 5 units from 0.

EXAMPLE 6 $|-3| = 3$ The number -3 is 3 units from 0.

EXAMPLE 7 $|-7| = 7$ The number -7 is 7 units from 0.

DEFINITION Two numbers that are the same distance from 0 but in opposite directions from 0 are called **opposites.***

EXAMPLE 8 Give the opposite of each of the following numbers:

5, 7, 1, −5, −8

Solution The opposite of 5 is −5.

The opposite of 7 is −7.

The opposite of 1 is −1.

The opposite of −5 is −(−5), or 5.

The opposite of −8 is −(−8), or 8.

We see from this example that the opposite of every positive number is a negative number, and, likewise, the opposite of every negative number is a

*In some books opposites are called *additive inverses.*

positive number. The last two parts of Example 8 illustrate the following property:

> **Property**
> If a represents any positive number, then it is always true that
> $-(-a) = a$

In other words, this property states that the opposite of a negative number is a positive number.

It should be evident now that the symbols + and − can be used to indicate several different ideas in mathematics. In the past we have used them to indicate addition and subtraction. They can also be used to indicate the direction a number is from 0 on the number line. For instance, the number +3 (read "positive 3") is the number that is 3 units from 0 in the positive direction. On the other hand, the number −3 (read "negative 3") is the number that is 3 units from 0 in the negative direction. The symbol − can also be used to indicate the opposite of a number, as in $-(-2) = 2$. The interpretation of the symbols + and − depends on the situation in which they are used. For example:

$3 + 5$	The + sign indicates addition.
$7 - 2$	The − sign indicates subtraction.
-7	The − sign is read "negative" 7.
$-(-5)$	The first − sign is read "the opposite of." The second − sign is read "negative" 5.

This may seem confusing at first, but as you work through the problems in this chapter you will get used to the different interpretations of the symbols + and −.

We should mention here that the set of whole numbers along with their opposites forms the set of *integers.* That is:

$$\text{Integers} = \{\ldots, -3, -2, -1, 0, 1, 2, 3, \ldots\}$$

Displaying Information

In the introduction to this chapter, we showed a table of temperatures in which the temperatures below zero were represented by negative numbers. The table and Figure 1 from that introduction are repeated here for reference.

Table 1 Record low temperatures for Jackson, Wyoming

Month	Temperature
January	−50°F
February	−44°F
March	−32°F
April	−5°F
May	12°F
June	19°F
July	24°F
August	18°F
September	14°F
October	2°F
November	−27°F
December	−49°F

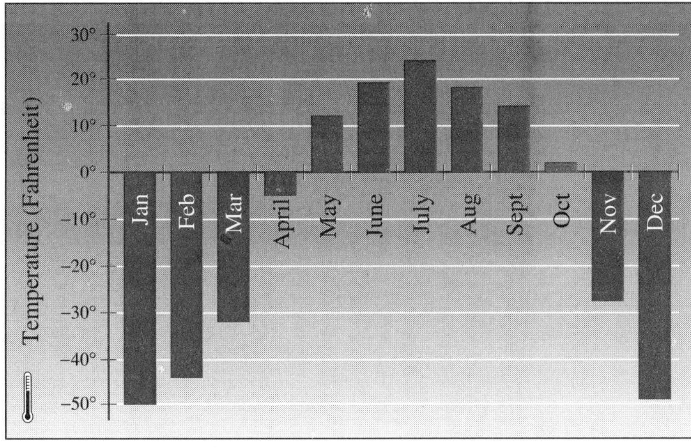

Figure 3 A histogram of Table 1

The only difference between the histogram in Figure 3 and those we encountered in Chapter 1 is the use of negative numbers. The vertical axis now resembles a number line that extends into the negative numbers. The horizontal axis extends out from the vertical axis at 0, the number that separates the positive numbers from the negative numbers.

Scatter Diagrams and Line Graphs

Figure 4 shows another way to visualize the information in Table 1. It is known as a *scatter diagram.* Dots are used instead of the bars shown in Figure 3; they represent the record low temperature for each of the 12 months of the year. If we connect the dots in Figure 4 with straight lines, we produce the diagram in Figure 5, which is known as a *line graph.*

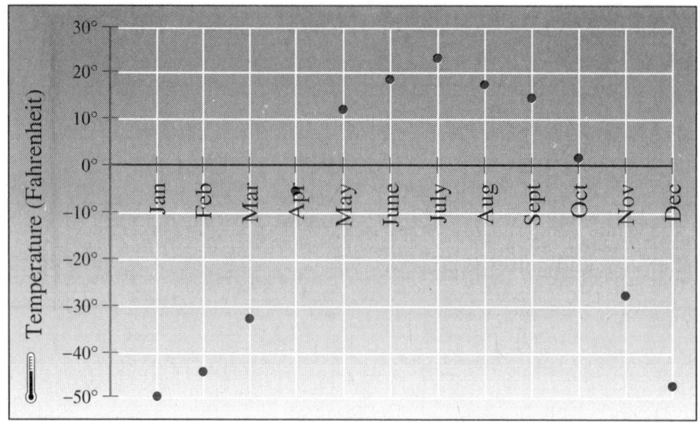

Figure 4 A scatter diagram of Table 1

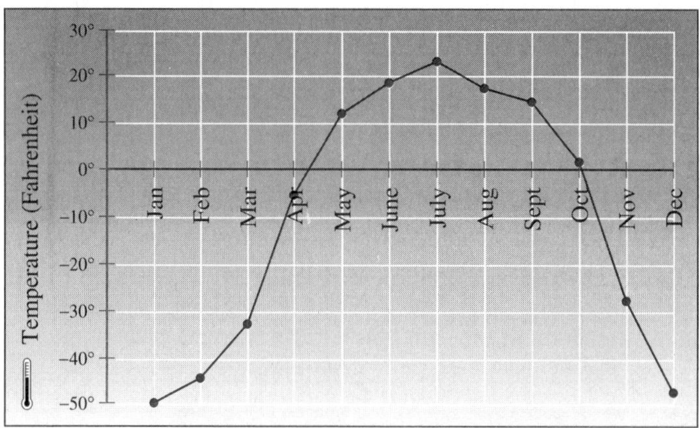

Figure 5 A line graph of Table 1

90 Use Figure 8 below to draw a histogram of the information in Table 4. Use Figure 9 to construct a scatter diagram of the same information. Then connect the dots in the scatter diagram to obtain a line graph of that same information. (Again, we have used the numbers 1 through 12 to represent the months January through December.)

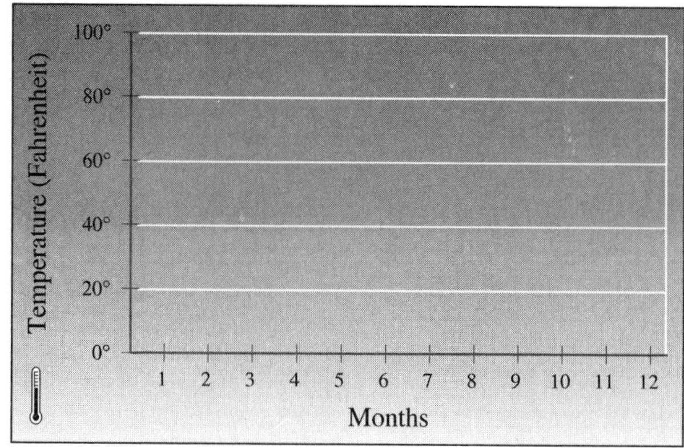

Figure 8 A histogram of Table 4

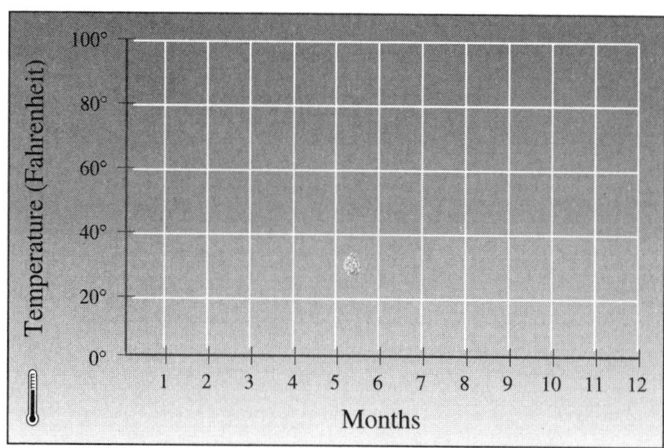

Figure 9 A scatter diagram, then line graph of Table 4

Calculator Problems

Display each of the following numbers on your calculator.

91 -18 **92** -6 **93** -144 **94** -5

Perform each of the following additions on your calculator.

95 $-18 + 18$ **96** $-6 + 6$ **97** $-144 + 144$ **98** $-5 + 5$

Review Problems

From here on, each problem set will contain some review problems. In mathematics it is very important to review the material you have studied. The more you do so, the better you will understand the topics we cover and the longer you will remember them.

The problems below review the basic facts and properties of addition with whole numbers from Chapter 1.

Complete each statement using the commutative property of addition.

99 $3 + 5 =$ **100** $9 + x =$

Complete each statement using the associative property of addition.

101 $7 + (2 + 6) =$

102 $(x + 3) + 5 =$

Write each of the following in symbols.

103 The sum of x and 4

104 The sum of x and 4 is 9.

105 5 more than y

106 x increased by 8

Add.

107 $276 + 32 + 4{,}005$

108 $17 + 3 + 152 + 1{,}200$

109 $(43{,}567 + 76{,}543) + 11{,}111$

110 $43{,}567 + (76{,}543 + 11{,}111)$

One Step Further: Extending the Concepts

111 There are two numbers that are 5 units from 2 on the number line. One of them is 7. What is the other one?

112 There are two numbers that are 5 units from -2 on the number line. One of them is 3. What is the other one?

113 In your own words and in complete sentences, explain what the opposite of a number is.

114 In your own words and in complete sentences, explain what the absolute value of a number is.

115 The expression $-(-3)$ is read "the opposite of negative 3," and it simplifies to just 3. Give a similar written description of the expression $-|-3|$, and then simplify it.

116 Give written descriptions of the expressions $-(-4)$ and $-|-4|$, and then simplify each of them.

2.2 Addition with Negative Numbers

Introduction . . . Suppose you are in Las Vegas playing blackjack and you lose \$3 on the first hand and then you lose \$5 on the next hand. If you represent winning with positive numbers and losing with negative numbers, how will you represent the results from your first two hands? Since you lost \$3 and \$5 for a total of \$8, one way to represent the situation is with addition of negative numbers:

$$(-\$3) + (-\$5) = -\$8$$

From this example we see that the sum of two negative numbers is a negative number. To generalize addition of positive and negative numbers, we can use our number line.

We can think of each number on the number line as having two characteristics: (1) a *distance* from 0 (absolute value) and (2) a *direction* from 0 (positive or negative). The distance from 0 is represented by the numerical part of the number (like the 5 in the number -5), and its direction is represented by the + or − sign in front of the number.

We can visualize addition of numbers on the number line by thinking in terms of distance and direction from 0. Let's begin with a simple problem we know the answer to. We interpret the sum $3 + 5$ on the number line as follows:

1 The first number is 3, which tells us "start at the origin and move 3 units in the positive direction."

2 The + sign is read "and then move."

3 The 5 means "5 units in the positive direction."

Note
This method of adding numbers may seem a little complicated at first, but it will allow us to add numbers we couldn't otherwise add.

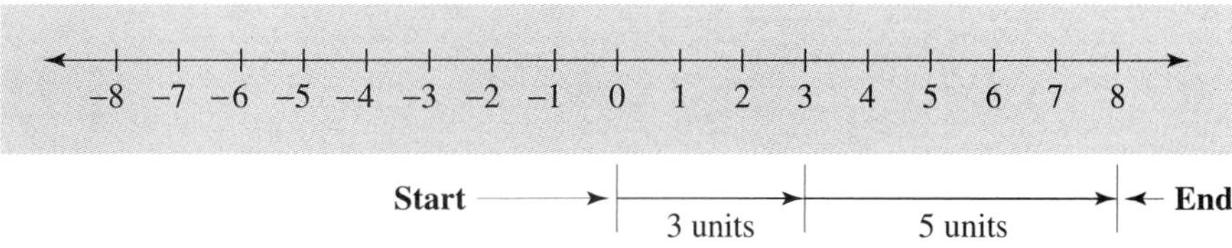

Figure 1

Figure 1 shows these steps. To summarize, $3 + 5$ means to start at the origin (0), move 3 units in the *positive* direction, and then move 5 units in the *positive* direction. We end up at 8, which is the sum we are looking for: $3 + 5 = 8$.

Practice Problems
1 Add: $2 + (-5)$

EXAMPLE 1 Add $3 + (-5)$ using the number line.

Solution We start at the origin, move 3 units in the positive direction, and then move 5 units in the negative direction, as shown in Figure 2. The last arrow ends at -2, which must be the sum of 3 and -5. That is:

$$3 + (-5) = -2$$

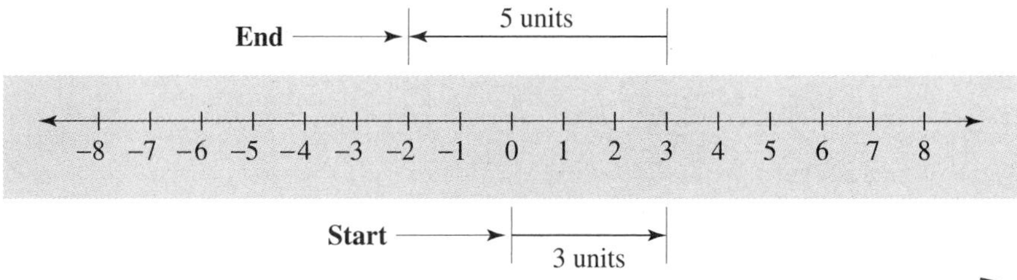

Figure 2

Answer
1 -3

2 Add: $-2 + 5$

EXAMPLE 2 Add $-3 + 5$ using the number line.

Solution We start at the origin, move 3 units in the negative direction, and then move 5 units in the positive direction, as shown in Figure 3. We end up at 2, which is the sum of -3 and 5. That is:

$$-3 + 5 = 2$$

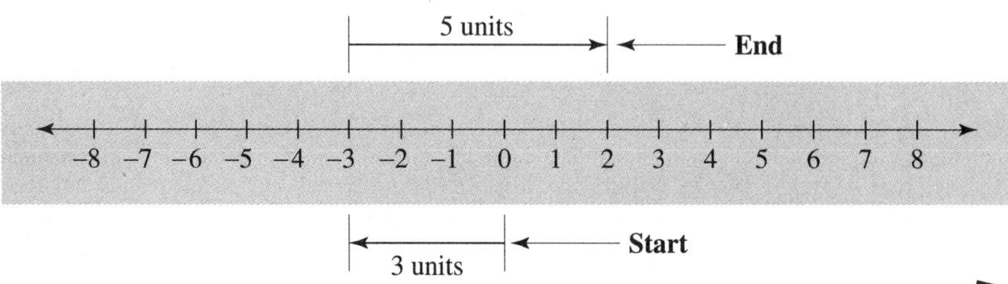

Figure 3

3 Add: $-2 + (-5)$

EXAMPLE 3 Add $-3 + (-5)$ using the number line.

Solution We start at the origin, move 3 units in the negative direction, and then move 5 more units in the negative direction. This is shown on the number line in Figure 4. As you can see, the last arrow ends at -8. We must conclude that the sum of -3 and -5 is -8. That is:

$$-3 + (-5) = -8$$

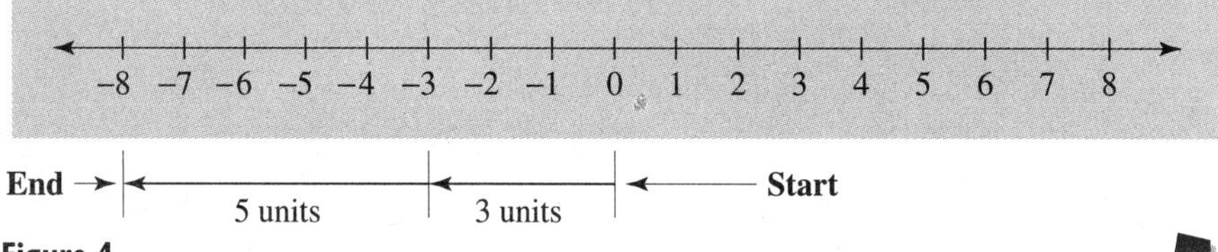

End 5 units 3 units Start

Figure 4

Adding numbers on the number line as we have done in these first three examples gives us a way of visualizing addition of positive and negative numbers. We eventually want to be able to write a rule for addition of positive and negative numbers that doesn't involve the number line. The number line is a way of justifying the rule we will eventually write. Here is a summary of the results we have so far:

$$3 + 5 = 8 \qquad -3 + 5 = 2$$
$$3 + (-5) = -2 \qquad -3 + (-5) = -8$$

Examine these results to see if you notice any pattern in the answers.

Answers
2 3 **3** -7

EXAMPLE 4 $4 + 7 = 11$

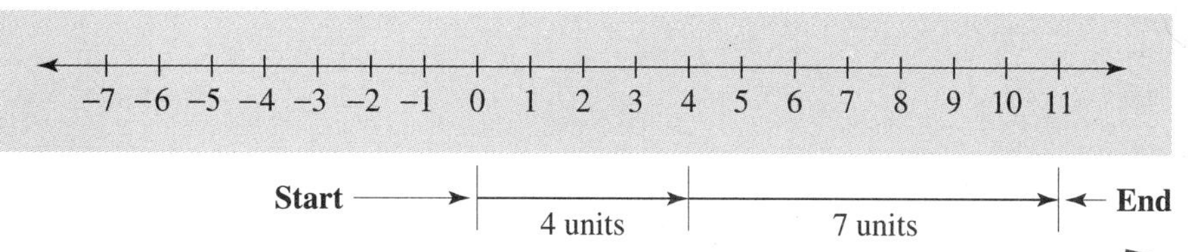

4 Add: $2 + 6$

EXAMPLE 5 $4 + (-7) = -3$

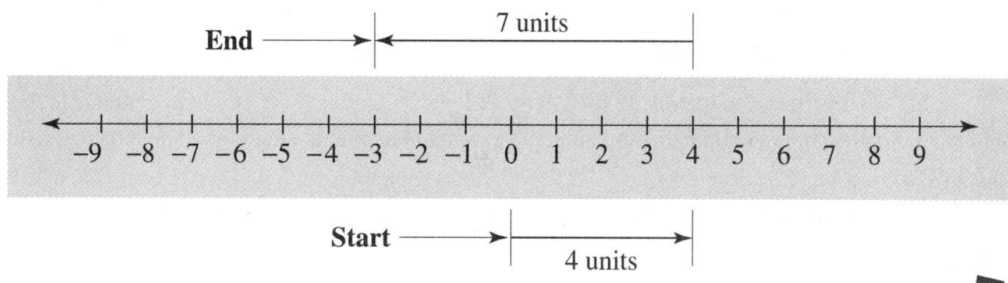

5 Add: $2 + (-6)$

EXAMPLE 6 $-4 + 7 = 3$

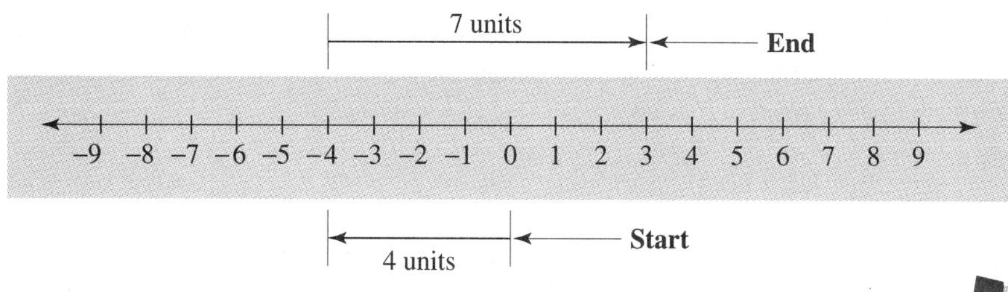

6 Add: $-2 + 6$

EXAMPLE 7 $-4 + (-7) = -11$

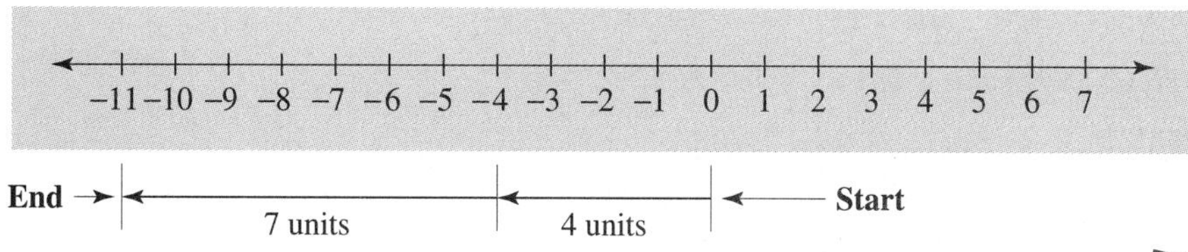

7 Add: $-2 + (-6)$

A summary of the results of these last four examples looks like this:

$$
\begin{aligned}
4 + 7 &= 11 \\
4 + (-7) &= -3 \\
-4 + 7 &= 3 \\
-4 + (-7) &= -11
\end{aligned}
$$

Answers
4 8 **5** −4 **6** 4 **7** −8

Looking over all the examples in this section, and noticing how the results in the problems are related, we can write the following rule for adding any two numbers:

Note
This rule covers all possible addition problems involving positive and negative numbers. You *must* memorize it. After you have worked some problems, the rule will seem almost automatic.

RULE

1 To add two numbers with the *same* sign: Simply add their absolute values and use the common sign. If both numbers are positive, the answer is positive. If both numbers are negative, the answer is negative.
2 To add two numbers with *different* signs: Subtract the smaller absolute value from the larger absolute value. The answer will have the sign of the number with the larger absolute value.

The following examples show how the rule is used. You will find that the rule for addition is consistent with all the results obtained using the number line.

8 Add all combinations of positive and negative 12 and 15.

EXAMPLE 8 Add all combinations of positive and negative 10 and 15.

Solution

$$10 + 15 = 25$$
$$10 + (-15) = -5$$
$$-10 + 15 = 5$$
$$-10 + (-15) = -25$$

Notice that when we add two numbers with the same sign, the answer also has that sign. When the signs are not the same, the answer has the sign of the number with the larger absolute value.

Once you have become familiar with the rule for adding positive and negative numbers, you can apply it to more complicated sums. The rule for adding positive and negative numbers can be applied to any kind of numbers, including whole numbers, fractions, and decimals.

Here are some more examples that use the rule for adding positive and negative numbers.

9 Simplify:

$$12 + (-3) + (-7) + 5$$

EXAMPLE 9 Simplify: $10 + (-5) + (-3) + 4$

Solution Adding left to right, we have:

$$10 + (-5) + (-3) + 4 = \mathbf{5} + (-3) + 4 \qquad 10 + (-5) = 5$$
$$= \mathbf{2} + 4 \qquad 5 + (-3) = 2$$
$$= 6$$

10 Simplify:

$$[-2 + (-12)] + [7 + (-5)]$$

EXAMPLE 10 Simplify: $[-3 + (-10)] + [8 + (-2)]$

Solution We begin by adding the numbers inside the brackets.

$$[-3 + (-10)] + [8 + (-2)] = [-13] + [6]$$
$$= -7$$

Calculator Note

There are a number of different ways in which calculators display negative numbers. Some calculators use a key labeled [+/−], whereas others use a key labeled [(−)]. You will need to consult with the manual that came with your calculator to see how your calculator does the job.

Here are a couple of ways to find the sum $-10 + (-15)$ on a calculator:

Scientific Calculator: 10 [+/−] [+] 15 [+/−] [=]

Graphing Calculator: [(−)] 10 [+] [(−)] 15 [ENT]

Answers
8 See solutions section.
9 7 10 −12

2.3 Subtraction with Negative Numbers

Introduction . . . At the beginning of this chapter we asked how we would represent the final balance in a checkbook if the original balance was \$20 and we wrote a check for \$30. We decided that the final balance would be −\$10. We can summarize the whole situation with subtraction:

$$\$20 - \$30 = -\$10$$

From this we see that subtracting 30 from 20 gives us −10. Another example that gives the same answer but involves addition is this:

$$20 + (-30) = -10$$

From the two examples above, we find that subtracting 30 gives the same result as adding −30. We use this kind of reasoning to give a definition for subtraction that will allow us to use the rules we developed for addition to do our subtraction problems. Here is that definition:

DEFINITION: Subtraction If a and b represent any two numbers, then it is always true that

$$a - b = a + (-b)$$

To subtract b → Add its opposite, $-b$

In words: Subtracting a number is equivalent to adding its opposite.

Note
This definition of subtraction may seem a little strange at first. In Example 1 you will notice that using the definition gives us the same results we are used to getting with subtraction. As we progress further into the section, we will use the definition to subtract numbers we haven't been able to subtract before.

Let's see if this definition conflicts with what we already know to be true about subtraction.

EXAMPLE 1 Subtract: $5 - 2$

Solution From previous experience we know that

$$5 - 2 = 3$$

We can get the same answer by using the definition we just gave for subtraction. Instead of subtracting 2, we can add its opposite, −2. Here is how it looks:

$5 - 2 = 5 + (-2)$ Change subtraction to addition of the opposite

$= 3$ Apply the rule for addition of positive and negative numbers

The result is the same whether we use our previous knowledge of subtraction or the new definition. The new definition is essential when the problems begin to get more complicated.

Practice Problems

1 Subtract: $7 - 3$

Answer

1 4

EXAMPLE 2 Subtract: $-7 - 2$

Solution We have never subtracted a positive number from a negative number before. We must apply our definition of subtraction:

$$-7 - 2 = -7 + (-2)$$ Instead of subtracting 2, we add its opposite, -2

$$= -9$$ Apply the rule for addition

EXAMPLE 3 Subtract: $-10 - 5$

Solution We apply the definition of subtraction (if you don't know the definition of subtraction yet, go back and read it) and add as usual.

$$-10 - 5 = -10 + (-5)$$ Definition of subtraction

$$= -15$$ Addition

EXAMPLE 4 Subtract: $12 - (-6)$

Solution The first $-$ sign is read "subtract," and the second one is read "negative." The problem in words is "12 subtract negative 6." We can use the definition of subtraction to change this to the addition of positive 6:

$$12 - (-6) = 12 + 6$$ Subtracting -6 is equivalent to adding $+6$

$$= 18$$ Addition

EXAMPLE 5 Subtract: $-20 - (-30)$

Solution Instead of subtracting -30, we can use the definition of subtraction to write the problem again as the addition of 30:

$$-20 - (-30) = -20 + 30$$ Definition of subtraction

$$= 10$$ Addition

Examples 1–5 illustrate all the possible combinations of subtraction with positive and negative numbers. There are no new rules for subtraction. We apply the definition to change each subtraction problem into an equivalent addition problem. The rule for addition can then be used to obtain the correct answer.

EXAMPLE 6 The following table shows the relationship between subtraction and addition:

Subtraction	Addition of the Opposite	Answer
$7 - 9$	$7 + (-9)$	-2
$-7 - 9$	$-7 + (-9)$	-16
$7 - (-9)$	$7 + 9$	16
$-7 - (-9)$	$-7 + 9$	2
$15 - 10$	$15 + (-10)$	5
$-15 - 10$	$-15 + (-10)$	-25
$15 - (-10)$	$15 + 10$	25
$-15 - (-10)$	$-15 + 10$	-5

2 Subtract: $-7 - 3$

Note
A real-life analogy to Example 2 would be: "If the temperature were 7° below 0 and then it dropped another 2°, what would the temperature be then?"

3 Subtract: $-8 - 6$

4 Subtract: $10 - (-6)$

5 Subtract: $-10 - (-15)$

Note
Examples 4 and 5 may give results you are not used to getting. But you must realize that the results are correct. That is, $12 - (-6)$ is 18, and $-20 - (-30)$ is 10. If you think these results should be different, then you are not thinking of subtraction correctly.

6 Subtract each of the following.

a $8 - 5$

b $-8 - 5$

c $8 - (-5)$

d $-8 - (-5)$

e $12 - 10$

f $-12 - 10$

g $12 - (-10)$

h $-12 - (-10)$

Answers
2 -10 **3** -14 **4** 16 **5** 5
6 a 3 **b** -13 **c** 13 **d** -3
e 2 **f** -22 **g** 22 **h** -2

EXAMPLE 7 Combine: $-3 + 6 - 2$

Solution The first step is to change subtraction to addition of the opposite. After that has been done we add left to right.

$$-3 + 6 - 2 = -3 + 6 + (-2)$$ Subtracting 2 is equivalent to adding -2

$$= 3 + (-2)$$ Add left to right

$$= 1$$

7 Combine: $-4 + 6 - 7$

EXAMPLE 8 Combine: $10 - (-4) - 8$

Solution Changing subtraction to addition of the opposite, we have

$$10 - (-4) - 8 = 10 + 4 + (-8)$$
$$= 14 + (-8)$$
$$= 6$$

8 Combine: $15 - (-5) - 8$

EXAMPLE 9 Subtract 3 from -5.

Solution Subtracting 3 is equivalent to adding -3:

$$-5 - 3 = -5 + (-3) = -8$$

Subtracting 3 from -5 gives us -8.

9 Subtract 2 from -8.

EXAMPLE 10 Subtract -4 from 9.

Solution Subtracting -4 is the same as adding $+4$:

$$9 - (-4) = 9 + 4 = 13$$

Subtracting -4 from 9 gives us 13.

10 Subtract -5 from 7.

EXAMPLE 11 Find the difference of 8 and -2.

Solution Subtracting -2 from 8, we have

$$8 - (-2) = 8 + 2 = 10$$

The difference of 8 and -2 is 10.

11 Find the difference of 7 and -3.

EXAMPLE 12 Find the difference of -7 and -4.

Solution Subtracting -4 from -7 looks like this:

$$-7 - (-4) = -7 + 4 = -3$$

The difference of -7 and -4 is -3.

12 Find the difference of -8 and -6.

Calculator Note

Here is how we work the subtraction problem shown in Example 12 on a calculator:

Scientific Calculator: 7 [+/−] [−] 4 [+/−] [=]

Graphing Calculator: [(−)] 7 [−] [(−)] 4 [ENT]

Answers

7 -5 **8** 12 **9** -10
10 12 **11** 10 **12** -2

13 Suppose the temperature is 42°F at take-off and then drops to -42°F when the plane reaches its cruising altitude. Find the difference in temperature at take-off and at 28,000 feet.

EXAMPLE 13 Many of the planes used by the United States during World War II were not pressurized or sealed from outside air. As a result the temperature inside these planes was the same as the surrounding air temperature outside. Suppose the temperature inside a B-17 Flying Fortress is 50°F at take-off and then drops to -30°F when the plane reaches its cruising altitude of 28,000 feet. Find the difference in temperature inside this plane at take-off and at 28,000 feet.

Solution The temperature at take-off is 50°F, whereas the temperature at 28,000 feet is -30°F. To find the difference we subtract, with the numbers in the same order as they are given in the problem:

$$50 - (-30) = 50 + 30 = 80$$

The difference in temperature is 80°F. ■

Subtraction and Taking Away

For some people studying algebra for the first time, subtraction of positive and negative numbers can be a problem. These people may believe that the answer to $-5 - 9$ should be -4 or 4, not -14. If this is happening to you, you are probably thinking of subtraction in terms of taking one number away from another. Thinking of subtraction in this way works well with positive numbers if you always subtract the smaller number from the larger. In algebra, however, we encounter many situations other than this. The definition of subtraction, that $a - b = a + (-b)$, clearly indicates the correct way to use subtraction. That is, when working subtraction problems, you should think "addition of the opposite," not "take one number away from another." To be successful in algebra, you need to apply these properties and definitions exactly as they are presented here.

Answer
13 84°F

2.4 Multiplication with Negative Numbers

Introduction . . . Suppose you own three shares of a stock and the price per share drops \$5. How much money have you lost? The answer is \$15. Since it is a loss, we can express it as −\$15. The multiplication problem below can be used to describe the relationship among the numbers.

3 shares → each loses \$5 → for a total of −\$15

$$3(-5) = -15$$

From this we conclude that it is reasonable to say that the product of a positive number and a negative number is a negative number.

In order to generalize multiplication with negative numbers, recall that we first defined multiplication by whole numbers to be repeated addition. That is:

$$3 \cdot 5 = 5 + 5 + 5$$

Multiplication ↑ Repeated addition ↑

This concept is very helpful when it comes to developing the rule for multiplication problems that involve negative numbers. For the first example we look at what happens when we multiply a negative number by a positive number.

Practice Problems

1 Multiply: $2(-6)$

EXAMPLE 1 Multiply: $3(-5)$

Solution Writing this product as repeated addition, we have

$$\begin{aligned} 3(-5) &= (-5) + (-5) + (-5) \\ &= -10 + (-5) \\ &= -15 \end{aligned}$$

The result, −15, is obtained by adding the three negative 5's.

2 Multiply: $-2(6)$

EXAMPLE 2 Multiply: $-3(5)$

Solution In order to write this multiplication problem in terms of repeated addition, we will have to reverse the order of the two numbers. This is easily done since multiplication is a commutative operation.

$-3(5) = 5(-3)$	Commutative property
$= (-3) + (-3) + (-3) + (-3) + (-3)$	Repeated addition
$= -15$	Addition

The product of −3 and 5 is −15.

3 Multiply: $-2(-6)$

EXAMPLE 3 Multiply: $-3(-5)$

Solution It is impossible to write this product in terms of repeated addition. We will find the answer to $-3(-5)$ by solving a different problem. Look at the following problem:

$$-3[5 + (-5)] = -3[0] = 0$$

Answers
1 −12 **2** −12 **3** 12

Note
The discussion here explains why $-3(-5) = 15$. We want to be able to justify everything we do in mathematics. The discussion tells *why* $-3(-5) = 15$.

The result is 0, since multiplying by 0 always produces 0. Now we can work the same problem another way and in the process find the answer to $-3(-5)$. Applying the distributive property to the same expression, we have

$$-3[5 + (-5)] = -3(5) + (-3)(-5) \qquad \text{Distributive property}$$
$$= -15 + (?) \qquad -3(5) = -15$$

The question mark must be +15, because we already know that the answer to the problem is 0, and +15 is the only number we can add to −15 to get 0. So, our problem is solved:

$$-3(-5) = +15$$

Table 1 gives a summary of what we have done so far in this section.

Table 1

Original Numbers Have	For Example	The Answer Is
Same signs	$3(5) = 15$	Positive
Different signs	$-3(5) = -15$	Negative
Different signs	$3(-5) = -15$	Negative
Same signs	$-3(-5) = 15$	Positive

From the examples we have done so far in this section and their summaries in Table 1, we write the following rule for multiplication of positive and negative numbers:

RULE To multiply any two numbers, we multiply their absolute values.

1 The answer is *positive* if both the original numbers have the same sign. That is, the product of two numbers with the same sign is positive.
2 The answer is *negative* if the original two numbers have different signs. The product of two numbers with different signs is negative.

This rule should be memorized. By the time you have finished reading this section and working the problems at the end of the section, you should be fast and accurate at multiplication with positive and negative numbers.

Multiply.
4 $3(2)$
5 $-3(-2)$
6 $3(-2)$
7 $-3(2)$

EXAMPLE 4 $2(4) = 8$ Like signs; positive answer

EXAMPLE 5 $-2(-4) = 8$ Like signs; positive answer

EXAMPLE 6 $2(-4) = -8$ Unlike signs; negative answer

EXAMPLE 7 $-2(4) = -8$ Unlike signs; negative answer

Answers
4 6 5 6 6 −6 7 −6

EXAMPLE 8 $7(-6) = -42$ Unlike signs; negative answer

EXAMPLE 9 $-5(-8) = 40$ Like signs; positive answer

EXAMPLE 10
$$\begin{aligned} -3(2)(-5) &= -6(-5) && \text{Multiply } -3 \text{ and } 2 \text{ to get } -6 \\ &= 30 \end{aligned}$$

EXAMPLE 11
$$\begin{aligned} (-5)^3 &= -5(-5)(-5) && \text{Expand exponent 3} \\ &= 25(-5) && \text{Multiply } -5 \text{ and } -5 \text{ to get } 25 \\ &= -125 \end{aligned}$$

EXAMPLE 12 Simplify: $-6[3 + (-5)]$

Solution We begin inside the brackets and work our way out:

$$\begin{aligned} -6[3 + (-5)] &= -6[-2] \\ &= 12 \end{aligned}$$

EXAMPLE 13 Simplify: $-4 + 5(-6 + 2)$

Solution Simplifying inside the parentheses first, we have

$$\begin{aligned} -4 + 5(-6 + 2) &= -4 + 5(-4) \\ &= -4 + (-20) \\ &= -24 \end{aligned}$$

EXAMPLE 14 Simplify: $-2(7) + 3(-6)$

Solution Multiplying left to right before we add gives us

$$\begin{aligned} -2(7) + 3(-6) &= -14 + (-18) \\ &= -32 \end{aligned}$$

EXAMPLE 15 Simplify: $-3(2 - 9) + 4(-7 - 2)$

Solution We begin by subtracting inside the parentheses:

$$\begin{aligned} -3(2 - 9) + 4(-7 - 2) &= -3(-7) + 4(-9) \\ &= 21 + (-36) \\ &= -15 \end{aligned}$$

EXAMPLE 16 Simplify: $(-3 - 7)(2 - 6)$

Solution Again, we begin by simplifying inside the parentheses:

$$\begin{aligned} (-3 - 7)(2 - 6) &= (-10)(-4) \\ &= 40 \end{aligned}$$

Calculator Note

Here is how we work the problem shown in Example 16 on a calc he [×] key on the first line may, or may not, be necessary. Try your c vithout it and see.)

Scientific Calculator: [(] 3 [+/−] [−] 7 [)] [×] [(] 2 [−] 6 [)]

Graphing Calculator: [(] [(−)] 3 [−] 7 [)] [(] 2 [−] 6 [)] [EN

8 $8(-9)$

9 $-6(-4)$

10 $-5(2)(-4$

11 $(-5)^2$

12 Sim $-2[5 + (-8)]$

1? ify: $-3 + 4(-7 + 3)$

Simplify: $-3(5) + 4(-4)$

15 Simplify:
$-2(3 - 5) - 7(-2 - 4)$

16 Simplify: $(-6 - 1)(4 - 9)$

Answers

8 −72 **9** 24 **10** 40
11 25 **12** 6 **13** −19
14 −31 **15** 46 **16** 35

2.5 Division with Negative Numbers

Introduction . . . If you and three friends bought equal shares of an investment for a total of \$10,000, and then sold it later for only \$9,000, how much did each person lose? Since the total loss can be represented by −\$1,000 and there are four people with equal shares, each person's loss can be found by division:

$$(-\$1{,}000) \div 4 = -\$250$$

From this example it seems reasonable to assume that a negative number divided by a positive number will give a negative answer.

To cover all the possible situations we can encounter with division of negative numbers, we use the relationship between multiplication and division. If we let n be the answer to the problem $12 \div (-2)$, then we know that

$$12 \div (-2) = n \quad \text{and} \quad -2(n) = 12$$

From our work with multiplication, we know that n must be -6 in the multiplication problem above, because -6 is the only number we can multiply -2 by to get 12. Because of the relationship between the two problems above, it must be true that 12 divided by -2 is -6.

The following pairs of problems show more quotients of positive and negative numbers. In each case the multiplication problem on the right justifies the answer to the division problem on the left.

$$6 \div 3 = 2 \quad \text{because} \quad 3(2) = 6$$
$$6 \div (-3) = -2 \quad \text{because} \quad -3(-2) = 6$$
$$-6 \div 3 = -2 \quad \text{because} \quad 3(-2) = -6$$
$$-6 \div (-3) = 2 \quad \text{because} \quad -3(2) = -6$$

The results given above can be used to write the rule for division with negative numbers.

RULE To divide two numbers, we divide their absolute values.

1. The answer is *positive* if both the original numbers have the same sign. That is, the quotient of two numbers with the same signs is positive.
2. The answer is *negative* if the original two numbers have different signs. That is, the quotient of two numbers with different signs is negative.

EXAMPLE 1 $-12 \div 4 = -3$ Unlike signs; negative answer

EXAMPLE 2 $12 \div (-4) = -3$ Unlike signs; negative answer

EXAMPLE 3 $-12 \div (-4) = 3$ Like signs, positive answer

EXAMPLE 4 $\dfrac{15}{-5} = -3$ Unlike signs; negative answer

Practice Problems

Divide.

1. $-8 \div 2$
2. $8 \div (-2)$
3. $-8 \div (-2)$
4. $\dfrac{20}{-5}$

Answers

1. −4 2. −4 3. 4 4. −4

5 $\dfrac{-30}{-5}$

EXAMPLE 5 $\dfrac{-20}{-4} = 5$ Like signs; positive answer

From the examples we have done so far, we can make the following generalization about quotients that contain negative signs:

> If a and b are numbers and b is not equal to 0, then
>
> $$\frac{-a}{b} = \frac{a}{-b} = -\frac{a}{b} \quad \text{and} \quad \frac{-a}{-b} = \frac{a}{b}$$

The last examples in this section involve more than one operation. We use the rules developed previously in this chapter and the rule for order of operations to simplify each.

6 Simplify: $\dfrac{8(-5)}{-4}$

EXAMPLE 6 Simplify: $\dfrac{6(-3)}{-2}$

Solution We begin by multiplying 6 and −3:

$$\frac{6(-3)}{-2} = \frac{-18}{-2} \quad \text{Multiplication; } 6(-3) = -18$$

$$= 9 \quad \text{Like signs; positive answer}$$

7 Simplify: $\dfrac{-20 + 6(-2)}{7 - 11}$

EXAMPLE 7 Simplify: $\dfrac{-15 + 5(-4)}{12 - 17}$

Solution Simplifying above and below the fraction bar, we have

$$\frac{-15 + 5(-4)}{12 - 17} = \frac{-15 + (-20)}{-5}$$

$$= \frac{-35}{-5}$$

$$= 7$$

8 Simplify:

$-3(4^2) + 10 \div (-5)$

EXAMPLE 8 Simplify: $-4(10^2) + 20 \div (-4)$

Solution Applying the rule for order of operations, we have

$$-4(10^2) + 20 \div (-4) = -4(100) + 20 \div (-4) \quad \text{Exponents first}$$

$$= -400 + (-5) \quad \text{Multiply and divide}$$

$$= -405 \quad \text{Add}$$

9 Simplify:

$-80 \div 2 \div 10$

EXAMPLE 9 Simplify: $-80 \div 10 \div 2$

Solution In a situation like this, the rule for order of operations states that we are to divide left to right.

$$-80 \div 10 \div 2 = -8 \div 2 \quad \text{Divide } -80 \text{ by } 10$$

$$= -4$$

Answers
5 6 **6** 10 **7** 8 **8** −50
9 −4

2.6 Simplifying Algebraic Expressions

In this section we want to combine some of the work we have done in this chapter and in Chapter 1 to simplify expressions containing variables—that is, algebraic expressions.

To begin let's review how we use the associative properties for addition and multiplication to simplify expressions.

Consider the expression $4(5x)$. We can apply the associative property of multiplication to this expression to change the grouping so that the 4 and the 5 are grouped together, instead of the 5 and the x. Here's how it looks:

$4(5x) = (4 \cdot 5)x$ Associative property

$= 20x$ Multiplication: $4 \cdot 5 = 20$

We have simplified the expression to $20x$, which in most cases in algebra will be easier to work with than the original expression.

Here are some more examples.

Practice Problems

Multiply.

1 $5(7a)$

2 $-3(9x)$

3 $5(-8y)$

EXAMPLE 1 $7(3a) = (7 \cdot 3)a$ Associative property

$= 21a$ 7 times 3 is 21

EXAMPLE 2 $-2(5x) = (-2 \cdot 5)x$ Associative property

$= -10x$ The product of -2 and 5 is -10

EXAMPLE 3 $3(-4y) = [3(-4)]y$ Associative property

$= -12y$ 3 times -4 is -12

We can use the associative property of addition to simplify expressions also.

Simplify.

4 $6 + (9 + x)$

5 $(3x + 7) + 4$

EXAMPLE 4 $3 + (8 + x) = (3 + 8) + x$ Associative property

$= 11 + x$ The sum of 3 and 8 is 11

EXAMPLE 5 $(2x + 5) + 10 = 2x + (5 + 10)$ Associative property

$= 2x + 15$ Addition

In Chapter 1 we introduced the distributive property. In symbols it looks like this:

$$a(b + c) = ab + ac$$

Since subtraction is defined as addition of the opposite, the distributive property holds for subtraction as well as addition. That is,

$$a(b - c) = ab - ac$$

We say that multiplication distributes over addition and subtraction. Here are some examples that review how the distributive property is applied to expressions that contain variables.

Apply the distributive property.

6 $6(x + 4)$

EXAMPLE 6 $4(x + 5) = 4(x) + 4(5)$ Distributive property

$= 4x + 20$ Multiplication

Answers

1 $35a$ 2 $-27x$ 3 $-40y$

4 $15 + x$ 5 $3x + 11$

6 $6x + 24$

7 $7(a - 5)$

EXAMPLE 7 $2(a - 3) = 2(a) - 2(3)$ Distributive property
$= 2a - 6$ Multiplication

In Examples 1–3 we simplified expressions such as $4(5x)$ by using the associative property. Here are some examples that use a combination of the associative property and the distributive property.

8 $6(4x + 5)$

EXAMPLE 8 $4(5x + 3) = 4(5x) + 4(3)$ Distributive property
$= (4 \cdot 5)x + 4(3)$ Associative property
$= 20x + 12$ Multiplication

9 $3(8a - 4)$

EXAMPLE 9 $7(3a - 6) = 7(3a) - 7(6)$ Distributive property
$= 21a - 42$ Associative property and multiplication

10 $8(3x + 4y)$

EXAMPLE 10 $5(2x + 3y) = 5(2x) + 5(3y)$ Distributive property
$= 10x + 15y$ Associative property and multiplication

We can also use the distributive property to simplify expressions like $4x + 3x$. Since multiplication is a commutative operation, we can also rewrite the distributive property like this:

$$b \cdot a + c \cdot a = (b + c)a$$

Applying the distributive property in this form to the expression $4x + 3x$, we have

$4x + 3x = (4 + 3)x$ Distributive property
$= 7x$ Addition

Similar Terms

Expressions like $4x$ and $3x$ are called *similar terms* because the variable parts are the same. Some other examples of similar terms are $5y$ and $-6y$, and the terms $7a$, $-13a$, $\frac{3}{4}a$. To simplify an algebraic expression (an expression that involves both numbers and variables), we combine similar terms by applying the distributive property. Table 1 shows several pairs of similar terms and how they can be combined using the distributive property.

Table 1

Original Expression		Apply Distributive Property		Simplified Expression
$4x + 3x$	=	$(4 + 3)x$	=	$7x$
$7a + a$	=	$(7 + 1)a$	=	$8a$
$-5x + 7x$	=	$(-5 + 7)x$	=	$2x$
$8y - y$	=	$(8 - 1)y$	=	$7y$
$-4a - 2a$	=	$(-4 - 2)a$	=	$-6a$
$3x - 7x$	=	$(3 - 7)x$	=	$-4x$

Answers
7 $7a - 35$ 8 $24x + 30$
9 $24a - 12$ 10 $24x + 32y$

As you can see from the table, the distributive property can be applied to any combination of positive and negative terms so long as they are similar terms.

Exponents and Variables

Next, we want to simplify products like $2x(3x)$. Since the problem involves only multiplication, the commutative property tells us we can change the order of the numbers and variables. Likewise, the associative property tells us we can change the grouping of the numbers and variables. We use the two properties together to group the numbers separately from the variables.

$$2x(3x) = (2 \cdot 3)(x \cdot x) \quad \text{Commutative and associative properties}$$
$$= 6x^2 \quad 2 \cdot 3 = 6 \text{ and } x \cdot x = x^2$$

Note that the expression $6x^2$ means $6 \cdot x \cdot x$, not $6x \cdot 6x$. The exponent 2 in the expression $6x^2$ is associated only with the number directly to its left, unless parentheses are used. That is:

$$6x^2 = 6 \cdot x \cdot x \quad \text{and} \quad (6x)^2 = (6x)(6x)$$

EXAMPLE 11 Expand and simplify: $(5a)^3$

Solution We begin by writing the expression as $(5a)(5a)(5a)$:

$$(5a)^3 = (5a)(5a)(5a) \quad \text{Definition of exponents}$$
$$= (5 \cdot 5 \cdot 5)(a \cdot a \cdot a) \quad \text{Commutative and associative properties}$$
$$= 125a^3 \quad 5 \cdot 5 \cdot 5 = 125;\ a \cdot a \cdot a = a^3$$

11 Expand and simplify: $(2a)^3$

EXAMPLE 12 Expand and simplify: $(8xy)^2$

Solution Proceeding as we did above, we have

$$(8xy)^2 = (8xy)(8xy) \quad \text{Definition of exponents}$$
$$= (8 \cdot 8)(x \cdot x)(y \cdot y) \quad \text{Commutative and associative properties}$$
$$= 64x^2y^2 \quad 8 \cdot 8 = 64;\ x \cdot x = x^2;\ y \cdot y = y^2$$

12 Expand and simplify: $(7xy)^2$

EXAMPLE 13 Simplify: $(7x)^2(8xy)^2$

Solution We begin by applying the definition of exponents:

$$(7x)^2(8xy)^2 = (7x)(7x)(8xy)(8xy)$$
$$= (7 \cdot 7 \cdot 8 \cdot 8)(x \cdot x \cdot x \cdot x)(y \cdot y) \quad \text{Commutative and associative properties}$$
$$= 3{,}136x^4y^2$$

13 Simplify: $(3x)^2(7xy)^2$

EXAMPLE 14 Simplify: $(2x)^3(4x)^2$

Solution Proceeding as we did above, we have

$$(2x)^3(4x)^2 = (2x)(2x)(2x)(4x)(4x)$$
$$= (2 \cdot 2 \cdot 2 \cdot 4 \cdot 4)(x \cdot x \cdot x \cdot x \cdot x)$$
$$= 128x^5$$

14 Simplify: $(5x)^3(2x)^2$

Answers
11 $8a^3$ **12** $49x^2y^2$
13 $441x^4y^2$ **14** $500x^5$

2.7 Multiplication Properties of Exponents

In this section we will extend our work with exponents to develop some general properties of exponents that we can use to simplify expressions. Each of the properties we will develop has its foundation in the definition for exponents.

Practice Problems

1 Multiply: $x^3 \cdot x^5$

EXAMPLE 1 Multiply: $x^2 \cdot x^4$

Solution We write x^2 as $x \cdot x$ and x^4 as $x \cdot x \cdot x \cdot x$ and then use the definition of exponents to write the answer with just one exponent.

$$\begin{aligned} x^2 \cdot x^4 &= (x \cdot x)(x \cdot x \cdot x \cdot x) && \text{Expand exponents} \\ &= x \cdot x \cdot x \cdot x \cdot x \cdot x \\ &= x^6 && \text{Write with a single exponent} \end{aligned}$$

Notice that the two exponents in the original problem given in Example 1 add up to the exponent in the answer: $2 + 4 = 6$. We use this result as justification for writing our first property of exponents.

> **Property 1 for Exponents**
>
> If r and s are any two whole numbers and a is an integer, then
>
> $$a^r \cdot a^s = a^{r+s}$$
>
> *In words:* To multiply two expressions with the same base, add exponents and use the common base.

2 Multiply: $4x^2 \cdot 6x^5$

EXAMPLE 2 Multiply: $5x^3 \cdot 3x^2$

Solution We apply the commutative and associative properties so that the numbers 5 and 3 are together, and the x's are also. Then we use Property 1 for exponents to add exponents.

$$\begin{aligned} 5x^3 \cdot 3x^2 &= (5 \cdot 3)(x^3 \cdot x^2) && \text{Commutative and associative properties} \\ &= (5 \cdot 3)(x^{3+2}) && \text{Property 1 for exponents} \\ &= 15x^5 && \text{Multiply 3 and 5; add 3 and 2} \end{aligned}$$

3 Multiply: $2x^5 \cdot 3x^4 \cdot 4x^3$

EXAMPLE 3 Multiply: $4x^3 \cdot 3x^4 \cdot 5x^2$

Solution Here we have the product of three expressions with exponents. Using the same steps shown in Example 2, we have

$$\begin{aligned} 4x^3 \cdot 3x^4 \cdot 5x^2 &= (4 \cdot 3 \cdot 5)(x^3 \cdot x^4 \cdot x^2) \\ &= (4 \cdot 3 \cdot 5)(x^{3+4+2}) \\ &= 60x^9 \end{aligned}$$

Answers

1 x^8 **2** $24x^7$ **3** $24x^{12}$

4 Multiply: $(10xy^2)(9x^3y^5)$

EXAMPLE 4 Multiply: $(2x^2y^3)(7x^4y)$

Solution This time we have two different variables to work with. We use the commutative and associative properties to group the numbers together, the x's together, and the y's together.

$$\begin{aligned}(2x^2y^3)(7x^4y) &= (2 \cdot 7)(x^2 \cdot x^4)(y^3 \cdot y)\\ &= (2 \cdot 7)(x^{2+4})(y^{3+1})\\ &= 14x^6y^4\end{aligned}$$

Another common expression in algebra is that of a power raised to another power—for example, $(x^2)^3$, $(10^4)^5$, and $(y^6)^7$. To see how these expressions can be written with a single exponent, we return to the definition of exponents.

$$\begin{aligned}(x^2)^3 &= x^2 \cdot x^2 \cdot x^2 && \text{Definition of exponents}\\ &= x^{2+2+2} && \text{Property 1 for exponents}\\ &= x^6 && \text{Add exponents}\end{aligned}$$

As you can see, to obtain the exponent in the answer, we multiply the exponents in the original problem. The result leads us to our second property of exponents.

Property 2 for Exponents

If r and s are any two whole numbers and a is an integer, then

$$(a^r)^s = a^{r \cdot s}$$

In words: A power raised to another power is the base raised to the product of the powers.

5 Simplify the expression: $(2^2)^3$

EXAMPLE 5 Simplify the expression $(2^3)^2$.

Solution We apply Property 2 by multiplying the exponents; then we simplify if we can.

$$\begin{aligned}(2^3)^2 &= 2^{3 \cdot 2} && \text{Property 2 for exponents}\\ &= 2^6 && \text{Multiply exponents}\\ &= 64 && \text{Multiplication}\end{aligned}$$

6 Simplify: $(x^3)^4 \cdot (x^5)^2$

EXAMPLE 6 Simplify: $(x^5)^6 \cdot (x^4)^7$

Solution In this case we apply Property 2 for exponents first to write each expression with a single exponent. Then we apply Property 1 to simplify further.

$$\begin{aligned}(x^5)^6 \cdot (x^4)^7 &= x^{5 \cdot 6} \cdot x^{4 \cdot 7} && \text{Property 2 for exponents}\\ &= x^{30} \cdot x^{28} && \text{Multiply exponents}\\ &= x^{30+28} && \text{Property 1 for exponents}\\ &= x^{58} && \text{Add exponents}\end{aligned}$$

Answers

4 $90x^4y^7$ 5 64 6 x^{22}

The third, and final, property of exponents for this section covers the situation that occurs when we have the product of more than one number or variable, all raised to a power. Expressions such as $(4x)^2$, $(2ab)^3$, and $(5x^2y^3)^5$ all fall into this category.

$(4x)^2 = (4x)(4x)$ Definition of exponents

$= (4 \cdot 4)(x \cdot x)$ Commutative and associative properties

$= 4^2 \cdot x^2$ Definition of exponents

As you can see, both the numbers, 4 and x, contained within the parentheses end up raised to the second power.

Property 3 for Exponents

If r is a whole number and a and b are integers, then

$$(a \cdot b)^r = a^r \cdot b^r$$

In words: A power of a product is the product of the powers.

EXAMPLE 7 Simplify: $(3xy)^4$

Solution Applying our new property, we have

$(3xy)^4 = 3^4 \cdot x^4 \cdot y^4$ Property 3 for exponents

$= 81x^4y^4$ Simplify

7 Simplify: $(5xy)^3$

EXAMPLE 8 Simplify: $(2x^3y^2)^5$

Solution In this case we will have to apply Property 3 first and then Property 2, in order to simplify the expression.

$(2x^3y^2)^5 = 2^5(x^3)^5(y^2)^5$ Property 3 for exponents

$= 32x^{15}y^{10}$ Property 2 for exponents

8 Simplify: $(4x^5y^2)^3$

EXAMPLE 9 Simplify: $(4x^3y)^2(5x^2y^4)^3$

Solution This simplification will require all three properties of exponents, as well as the commutative and associative properties of multiplication. Here are all the steps:

$(4x^3y)^2(5x^2y^4)^3 = 4^2(x^3)^2y^2 \cdot 5^3(x^2)^3(y^4)^3$ Property 3 for exponents

$= 16x^6y^2 \cdot 125x^6y^{12}$ Property 2 for exponents

$= (16 \cdot 125)(x^6x^6)(y^2y^{12})$ Commutative and associative properties

$= 2{,}000x^{12}y^{14}$ Property 1 for exponents

9 Simplify: $(2x^3y^4)^3(3xy^5)^2$

Answers
7 $125x^3y^3$ **8** $64x^{15}y^6$
9 $72x^{11}y^{22}$

2.8 Adding and Subtracting Polynomials

Previously we wrote numbers in expanded form to show the place value of each of the digits. For example, the number 456 in expanded form looks like this:

$$456 = 4 \cdot 100 + 5 \cdot 10 + 6 \cdot 1$$

If we replace 100 with 10^2, we have

$$456 = 4 \cdot 10^2 + 5 \cdot 10 + 6 \cdot 1$$

If we replace the 10's with x's, we get what is called a *polynomial.* It looks like this:

$$4x^2 + 5x + 6$$

Polynomials are to algebra what whole numbers written in expanded form are to arithmetic. As in other expressions in algebra, we can use any variable we choose. Here are some other examples of polynomials:

$$5a + 3 \qquad y^2 - 2y + 4 \qquad x^3 - 2x^2 + 5x - 1$$

Adding Polynomials

When we add two whole numbers, we add in columns. That is, if we add 234 and 345, we write one number under the other and add the numbers in the ones column, then the numbers in the tens column, and finally the numbers in the hundreds column. Here is how it looks:

$$\begin{array}{rr} 234 & \quad 2 \cdot 10^2 + 3 \cdot 10 + 4 \\ \underline{345} & \quad \underline{3 \cdot 10^2 + 4 \cdot 10 + 5} \\ 579 & \quad 5 \cdot 10^2 + 7 \cdot 10 + 9 \end{array}$$

We add polynomials in the same manner. If we want to add $2x^2 + 3x + 4$ and $3x^2 + 4x + 5$, we write one polynomial under the other, and then add in columns:

$$\begin{array}{r} 2x^2 + 3x + 4 \\ \underline{3x^2 + 4x + 5} \\ 5x^2 + 7x + 9 \end{array}$$

The sum of the two polynomials is the polynomial $5x^2 + 7x + 9$. We add only the digits. Notice that the variable parts (the x's) stay the same, just as the powers of 10 did when we added 234 and 345. The reason we add the numbers, while the variable parts of each term stay the same, can be explained with the commutative, associative, and distributive properties. Here is the same problem again, but this time we show the properties in use:

$$\begin{aligned} &(2x^2 + 3x + 4) + (3x^2 + 4x + 5) \\ &\quad = (2x^2 + 3x^2) + (3x + 4x) + (4 + 5) && \text{Commutative and associative properties} \\ &\quad = (2 + 3)x^2 + (3 + 4)x + (4 + 5) && \text{Distributive property} \\ &\quad = 5x^2 + 7x + 9 && \text{Addition} \end{aligned}$$

Practice Problems

1 Add $5x^2 - 3x + 2$ and $2x^2 + 10x - 9$.

EXAMPLE 1 Add $3x^2 - 2x + 1$ and $5x^2 + 3x - 4$.

Solution We will work the problem two ways. First, we write one polynomial under the other, and add in columns:

$$\begin{array}{r} 3x^2 - 2x + 1 \\ \underline{5x^2 + 3x - 4} \\ 8x^2 + 1x - 3 \end{array}$$

The sum of the two polynomials is $8x^2 + x - 3$.

Next we add horizontally, showing the commutative and associative properties in the first step, then the distributive property in the second step:

$$\begin{aligned} (3x^2 - 2x + 1) + (5x^2 + 3x - 4) \\ &= (3x^2 + 5x^2) + (-2x + 3x) + (1 - 4) \\ &= (3 + 5)x^2 + (-2 + 3)x + (1 - 4) \\ &= 8x^2 + 1x + (-3) \\ &= 8x^2 + x - 3 \end{aligned}$$

2 Add $3y^2 + 9y - 5$ and $6y^2 - 4$.

EXAMPLE 2 Add $6y^2 + 4y - 3$ and $2y^2 - 4$.

Solution We write one polynomial under the other, so that the terms with y^2 line up, and the terms without any y's line up:

$$\begin{array}{r} 6y^2 + 4y - 3 \\ \underline{2y^2 \qquad\; - 4} \\ 8y^2 + 4y - 7 \end{array}$$

The same problem, written horizontally, looks like this:

$$\begin{aligned} (6y^2 + 4y - 3) + (2y^2 - 4) &= (6y^2 + 2y^2) + 4y + (-3 - 4) \\ &= (6 + 2)y^2 + 4y + (-3 - 4) \\ &= 8y^2 + 4y + (-7) \\ &= 8y^2 + 4y - 7 \end{aligned}$$

3 Add $8x^3 + 4x^2 + 3x + 2$ and $4x^2 + 5x + 6$.

EXAMPLE 3 Add $2x^3 + 5x^2 + 3x + 4$ and $3x^2 + 2x + 1$.

Solution Showing only the vertical method, we line up terms with the same variable part and add:

$$\begin{array}{r} 2x^3 + 5x^2 + 3x + 4 \\ \underline{3x^2 + 2x + 1} \\ 2x^3 + 8x^2 + 5x + 5 \end{array}$$

Subtracting Polynomials

If there is a negative sign directly preceding the parentheses surrounding a polynomial, we may remove the parentheses and preceding negative sign by applying the distributive property. For example, to remove the parentheses from the expression

$$-(3x + 4)$$

we think of the negative sign as representing -1. Doing so allows us to apply the distributive property:

$$\begin{aligned} -(3x + 4) &= -1(3x + 4) \\ &= -1(3x) + (-1)(4) && \text{Distributive property} \\ &= -3x + (-4) && \text{Multiply} \\ &= -3x - 4 && \text{Simplify} \end{aligned}$$

Answers

1 $7x^2 + 7x - 7$ 2 $9y^2 + 9y - 9$
3 $8x^3 + 8x^2 + 8x + 8$

As a result, the sign of each term inside the parentheses changes. Without showing all the steps involved in removing the parentheses, here are some more examples:

$$-(2x - 8) = -2x + 8$$
$$-(x^2 + 2x + 3) = -x^2 - 2x - 3$$
$$-(-4x^2 + 5x - 7) = 4x^2 - 5x + 7$$
$$-(3y^3 - 6y^2 + 7y - 3) = -3y^3 + 6y^2 - 7y + 3$$

In each case we remove the parentheses and preceding negative sign by changing the sign of each term that is found within the parentheses.

EXAMPLE 4 Subtract: $(6x^2 - 3x + 5) - (3x^2 + 2x - 3)$

Solution Since subtraction is addition of the opposite, we simply change the sign of each term of the second polynomial and then add. With subtraction, there is less chance of making mistakes if we add horizontally, rather than in columns.

$$(6x^2 - 3x + 5) - (3x^2 + 2x - 3) \quad \text{Subtraction}$$
$$= 6x^2 - 3x + 5 - 3x^2 - 2x + 3 \quad \text{Addition of the opposite}$$
$$= (6x^2 - 3x^2) + (-3x - 2x) + (5 + 3)$$
$$= (6 - 3)x^2 + (-3 - 2)x + (5 + 3) \quad \text{Distributive property}$$
$$= 3x^2 - 5x + 8$$

EXAMPLE 5 Subtract: $(2y^3 - 3y^2 - 4y - 2) - (3y^3 - 6y^2 + 7y - 3)$

Solution Again, to subtract, we add the opposite of the polynomial that follows the subtraction sign. That is, we change the sign of each term in the second polynomial; then we combine similar terms.

$$(2y^3 - 3y^2 - 4y - 2) - (3y^3 - 6y^2 + 7y - 3) \quad \text{Subtraction}$$
$$= 2y^3 - 3y^2 - 4y - 2 - 3y^3 + 6y^2 - 7y + 3$$
$$= -y^3 + 3y^2 - 11y + 1$$

EXAMPLE 6 Subtract $4x^2 - 3x + 1$ from $-3x^2 + 5x - 2$.

Solution We must supply our own subtraction sign and write the two polynomials in the correct order.

$$(-3x^2 + 5x - 2) - (4x^2 - 3x + 1)$$
$$= -3x^2 + 5x - 2 - 4x^2 + 3x - 1$$
$$= -7x^2 + 8x - 3$$

Finding the Value of a Polynomial

The last topic we want to consider in this section is finding the value of a polynomial for a given value of the variable.

EXAMPLE 7 Find the value of $4x^2 - 7x + 2$ when $x = -2$.

Solution When $x = -2$, the polynomial $4x^2 - 7x + 2$ becomes

$$4(-2)^2 - 7(-2) + 2 = 4(4) + 14 + 2$$
$$= 16 + 14 + 2$$
$$= 32$$

4 Subtract:
$(5x^2 - 2x + 7) - (4x^2 + 8x - 4)$

5 Subtract: $(3y^3 - 2y^2 + 7y - 6) - (8y^3 - 6y^2 + 4y - 8)$

Note
Examples 5 and 6 show the minimum number of steps needed to subtract two polynomials.

6 Subtract $6x^2 - 2x + 5$ from $-2x^2 + 5x - 1$.

7 Find the value of $5x^2 - 3x + 8$ when $x = -3$.

Answers
4 $x^2 - 10x + 11$
5 $-5y^3 + 4y^2 + 3y + 2$
6 $-8x^2 + 7x - 6$ 7 62

Multiplying Polynomials: An Introduction

Recall that the distributive property allows us to multiply across parentheses when a sum or difference is enclosed within the parentheses. That is:

$$a(b + c) = a \cdot b + a \cdot c$$

We can use the distributive property to multiply polynomials.

Practice Problems

1 Multiply: $x^3(x^5 + x^7)$

EXAMPLE 1 Multiply: $x^2(x^3 + x^4)$

Solution Applying the distributive property, we have

$$\begin{aligned} \mathbf{x^2}(x^3 + x^4) &= \mathbf{x^2} \cdot x^3 + \mathbf{x^2} \cdot x^4 \quad \text{Distributive property} \\ &= x^5 + x^6 \end{aligned}$$

The distributive property works for multiplication from the right as well as the left. That is, we can also write the distributive property this way:

$$(b + c)a = b \cdot a + c \cdot a$$

2 Multiply: $(x^5 + x^7)x^3$

EXAMPLE 2 Multiply: $(x^3 + x^4)x^2$

Solution Since multiplication is a commutative operation, we should expect to obtain the same answer as in Example 1 above.

$$\begin{aligned} (x^3 + x^4)\mathbf{x^2} &= x^3 \cdot \mathbf{x^2} + x^4 \cdot \mathbf{x^2} \\ &= x^5 + x^6 \end{aligned}$$

3 Multiply: $5x^2(6x^3 - 4)$

EXAMPLE 3 Multiply: $4x^3(6x^2 - 8)$

Solution The distributive property allows us to multiply $4x^3$ by both $6x^2$ and 8:

$$\begin{aligned} \mathbf{4x^3}(6x^2 - 8) &= \mathbf{4x^3} \cdot 6x^2 - \mathbf{4x^3} \cdot 8 \\ &= (4 \cdot 6)(x^3 \cdot x^2) - (4 \cdot 8)x^3 \\ &= 24x^5 - 32x^3 \end{aligned}$$

4 Multiply: $5a^3b^5(2a^2 + 7b^2)$

EXAMPLE 4 Multiply: $2a^4b^2(3a^3 + 4b^5)$

Solution Applying the distributive property as we did in the previous two examples, we have

$$\begin{aligned} \mathbf{2a^4b^2}(3a^3 + 4b^5) &= \mathbf{2a^4b^2} \cdot 3a^3 + \mathbf{2a^4b^2} \cdot 4b^5 \\ &= (2 \cdot 3)(a^4 \cdot a^3)b^2 + (2 \cdot 4)(a^4)(b^2 \cdot b^5) \\ &= 6a^7b^2 + 8a^4b^7 \end{aligned}$$

Multiplying Binomials

Polynomials with exactly two terms are called *binomials*. We multiply binomials by applying the distributive property.

5 Multiply: $(x + 2)(x + 6)$

EXAMPLE 5 Multiply: $(x + 3)(x + 5)$

Solution We can think of the first binomial, $x + 3$, as a single number. (Remember, for any value of x, $x + 3$ will be just a number.) We apply the distributive property by multiplying $x + 3$ times both x and 5.

$$(x + 3)(x + 5) = (x + 3) \cdot x + (x + 3) \cdot 5$$

Answers

1 $x^8 + x^{10}$ **2** $x^8 + x^{10}$
3 $30x^5 - 20x^2$
4 $10a^5b^5 + 35a^3b^7$
5 $x^2 + 8x + 12$

Next, we apply the distributive property again to multiply x times both x and 3, and 5 times both x and 3.

$$= x \cdot x + 3 \cdot x + x \cdot 5 + 3 \cdot 5$$
$$= x^2 + 3x + 5x + 15$$

The last thing to do is to combine the similar terms $3x$ and $5x$ to get $8x$. (Remember, this is also an application of the distributive property.)

$$= x^2 + 8x + 15$$

6 Multiply: $(x - 2)(x + 6)$

EXAMPLE 6 Multiply: $(x - 3)(x + 5)$

Solution The only difference between the binomials in this example and those in Example 5 is the subtraction sign in $x - 3$. The steps in multiplying are exactly the same.

$(x - 3)(x + 5) = (x - 3) \cdot x + (x - 3) \cdot 5$	Multiply $x - 3$ times both x and 5
$= x \cdot x - 3 \cdot x + x \cdot 5 - 3 \cdot 5$	Distributive property two more times
$= x^2 - 3x + 5x - 15$	Simplify each term
$= x^2 + 2x - 15$	$-3x + 5x = 2x$

7 Multiply: $(3x - 2)(5x + 4)$

EXAMPLE 7 Multiply: $(2x - 3)(4x + 7)$

Solution Using the same steps shown in Examples 5 and 6, we have

$$(2x - 3)(4x + 7) = (2x - 3) \cdot 4x + (2x - 3) \cdot 7$$
$$= 2x \cdot 4x - 3 \cdot 4x + 2x \cdot 7 - 3 \cdot 7$$
$$= 8x^2 - 12x + 14x - 21$$
$$= 8x^2 + 2x - 21$$

Our next two examples show how we raise binomials to the second power.

8 Expand and multiply: $(x + 3)^2$

EXAMPLE 8 Expand and multiply: $(x + 5)^2$

Solution We use the definition of exponents to write $(x + 5)^2$ as $(x + 5)(x + 5)$. Then we multiply as we did in the previous examples.

$(x + 5)^2 = (x + 5)(x + 5)$	Definition of exponents
$= (x + 5) \cdot x + (x + 5) \cdot 5$	Distributive property
$= x \cdot x + 5 \cdot x + x \cdot 5 + 5 \cdot 5$	Distributive property
$= x^2 + 5x + 5x + 25$	Simplify each term
$= x^2 + 10x + 25$	$5x + 5x = 10x$

9 Expand and multiply: $(3x - 5)^2$

EXAMPLE 9 Expand and multiply: $(3x - 7)^2$

Solution We know that $(3x - 7)^2 = (3x - 7)(3x - 7)$. It will be easier to apply the distributive property to this last expression if we think of the second $3x - 7$ as $3x + (-7)$. In doing so we will also be less likely to make a mistake in our signs. (Try the problem without changing subtraction to addition of the opposite and see how your answer compares to the answer in this example.)

$$(3x - 7)(3x - 7) = (3x - 7)[3x + (-7)]$$
$$= (3x - 7) \cdot 3x + (3x - 7)(-7)$$
$$= 3x \cdot 3x - 7 \cdot 3x + 3x(-7) - 7(-7)$$
$$= 9x^2 - 21x - 21x + 49$$
$$= 9x^2 - 42x + 49$$

Answers
6 $x^2 + 4x - 12$
7 $15x^2 + 2x - 8$
8 $x^2 + 6x + 9$
9 $9x^2 - 30x + 25$

CHAPTER 2 SUMMARY

Examples

Absolute Value [2.1]

The absolute value of a number is its distance from 0 on the number line. It is the numerical part of a number. The absolute value of a number is never negative.

1 $|3| = 3$ and $|-3| = 3$

Opposites [2.1]

Two numbers are called opposites if they are the same distance from 0 on the number line but in opposite directions from 0. The opposite of a positive number is a negative number, and the opposite of a negative number is a positive number.

2 $-(5) = -5$ and $-(-5) = 5$

Addition of Positive and Negative Numbers [2.2]

1 To add two numbers with *the same sign:* Simply add absolute values and use the common sign. If both numbers are positive, the answer is positive. If both numbers are negative, the answer is negative.

2 To add two numbers with *different signs:* Subtract the smaller absolute value from the larger absolute value. The answer has the same sign as the number with the larger absolute value.

3 $3 + 5 = 8$
$-3 + (-5) = -8$

$5 + (-3) = 2$
$-5 + 3 = -2$

Subtraction [2.3]

Subtracting a number is equivalent to adding its opposite. If a and b represent numbers, then subtraction is defined in terms of addition as follows:

$$a - b = a + (-b)$$

↑ Subtraction ↑ Addition of the opposite

4 $3 - 5 = 3 + (-5) = -2$
$-3 - 5 = -3 + (-5) = -8$
$3 - (-5) = 3 + 5 = 8$
$-3 - (-5) = -3 + 5 = 2$

Multiplication with Positive and Negative Numbers [2.4]

To multiply two numbers multiply their absolute values.

1 The answer is *positive* if both numbers have the same sign.

2 The answer is *negative* if the numbers have different signs.

5 $3(5) = 15$
$3(-5) = -15$
$-3(5) = -15$
$-3(-5) = 15$

Division [2.5]

The rule for assigning the correct sign to the answer in a division problem is the same as the rule for multiplication. That is, like signs give a positive answer, and unlike signs give a negative answer.

6 $\frac{12}{4} = 3$
$\frac{-12}{4} = -3$
$\frac{12}{-4} = -3$
$\frac{-12}{-4} = 3$

Property 1 for Exponents [2.7]

If r and s are any two whole numbers and a is an integer, then

$$a^r \cdot a^s = a^{r+s}$$

In words: To multiply two expressions with the same base, add exponents and use the common base.

7 $x^3 \cdot x^4 \cdot x^2 = x^{3+4+2}$
$= x^9$

Property 2 for Exponents [2.7]

If r and s are any two whole numbers and a is an integer, then

$$(a^r)^s = a^{r \cdot s}$$

In words: A power raised to another power is the base raised to the product of the powers.

8 $(2^3)^2 = 2^{3 \cdot 2} = 2^6$

9 $(3xy)^4 = 3^4 \cdot x^4 \cdot y^4$
$= 81x^4y^4$

Property 3 for Exponents [2.7]

If r is a whole number and a and b are integers, then

$$(a \cdot b)^r = a^r \cdot b^r$$

In words: A power of a product is the product of the powers.

10
$$\begin{array}{r} 2x^2 + 3x + 4 \\ \underline{3x^2 + 4x + 5} \\ 5x^2 + 7x + 9 \end{array}$$

Adding Polynomials [2.8]

We add polynomials by writing one polynomial under the other, and then adding in columns.

11 $(6x^2 - 3x + 5) - (3x^2 + 2x - 3)$
$= 6x^2 - 3x + 5 - 3x^2 - 2x + 3$
$= (6x^2 - 3x^2) + (-3x - 2x) + (5 + 3)$
$= (6 - 3)x^2 + (-3 - 2)x + (5 + 3)$
$= 3x^2 - 5x + 8$

Subtracting Polynomials [2.8]

Since subtraction is addition of the opposite, we simply change the sign of each term of the second polynomial and then add. With subtraction, there is less chance of making mistakes if we add horizontally, rather than in columns.

12 $\mathbf{x}^2(x^3 + x^4) = \mathbf{x}^2 \cdot x^3 + \mathbf{x}^2 \cdot x^4$
$= x^5 + x^6$

$(\mathbf{2x - 3})(4x + 7)$
$= (\mathbf{2x - 3}) \cdot 4x + (\mathbf{2x - 3}) \cdot 7$
$= 2x \cdot 4x - 3 \cdot 4x + 2x \cdot 7 - 3 \cdot 7$
$= 8x^2 - 12x + 14x - 21$
$= 8x^2 + 2x - 21$

Multiplying Polynomials [2.9]

We use the distributive property to multiply polynomials.

CHAPTER 2 REVIEW

Give the opposite of each number. **[2.1]**

1 17

2 −32

Place an inequality symbol (< or >) between each pair of numbers so that the resulting statement is true. **[2.1]**

3 6 $\quad$ −6

4 −8 $\quad$ −3

5 $|-3|$ $\quad$ 2

6 $|-4|$ $\quad$ $|6|$

Simplify each expression. **[2.1]**

7 $-(-4)$

8 $|-6|$

Perform the indicated operations. **[2.2, 2.3, 2.4, 2.5]**

9 $5 + (-7)$

10 $-3 + 8$

11 $-345 + (-626)$

12 $7 - 9 - 4 - 6$

13 $-7 - 5 - 2 - 3$

14 $4 - (-3)$

15 $5(-4)$

16 $-4(-3)$

17 $\frac{48}{-16}$

18 $\frac{-20}{5}$

19 $\frac{-14}{-7}$

Simplify the following expressions as much as possible. **[2.2, 2.3, 2.4, 2.5]**

20 $(-6)^2$

21 $(-2)^3$

22 $7 + 4(6 - 9)$

23 $(-3)(-4) + 2(-5)$

24 $(7 - 3)(7 - 9)$

25 $3(-6) + 8(2 - 5)$

26 $\frac{8 - 4}{-8 + 4}$

27 $\frac{-4 + 2(-5)}{6 - 4}$

28 $\frac{8(-2) + 5(-4)}{12 - 3}$

29 $\frac{-2(5) + 4(-3)}{10 - 8}$

30 Give the sum of −19 and −23.

31 Give the sum of −78 and −51.

32 Find the difference of −6 and 5.

33 Subtract −8 from −10.

34 What is the product of −9 and 3?

35 What is −3 times the sum of −9 and −4?

36 Divide the product of 8 and −4 by −16.

37 Give the quotient of −38 and 2.

Indicate whether each statement is *True* or *False*.

38 $\frac{-10}{-5} = -2$

39 $10 - (-5) = 15$

40 $2(-3) = -3 + (-3)$

41 $-6 - (-2) = -8$

42 $3 - 5 = 5 - 3$

43 $3 - 5 = -5 + 3$

44 A gambler wins \$58 Saturday night and then loses \$86 on Sunday. Use positive and negative numbers to describe this situation. Then give the gambler's net loss or gain as a positive or negative number.

45 Name two numbers that are 7 units from -8 on the number line.

46 On Wednesday, the temperature reaches a high of 17° above 0 and a low of 7° below 0. What is the difference between the high and low temperatures for Wednesday?

47 If the difference between two numbers is -3, and one of the numbers is 5, what is the other number?

Use the associative properties to simplify each expression. **[2.6]**

48 $(3x + 4) + 8$

49 $8(3x)$

50 $-3(7a)$

51 $6(-5y)$

Apply the distributive property and then simplify if possible. **[2.6]**

52 $4(x + 3)$

53 $2(x - 5)$

54 $7(3y - 8)$

55 $3(2a + 5b)$

Expand and simplify. **[2.6]**

56 $(4x)(5x)$

57 $(5x)^3$

58 $(3x)^3(4x)^2$

Use the properties of exponents to simplify each expression. **[2.7]**

59 $x^3 \cdot x^5$

60 $(x^3)^6$

61 $(3xy)^4$

62 $4x^2 \cdot 7x^7$

63 $(5x^6y^2)^3$

64 $(2x^4y^5)^3 \cdot (4x^5y^8)^2$

65 Add $4x^2 + 4x - 5$ and $8x^2 - 7x + 2$. **[2.8]**

66 Add $6x^3 - 2x^2 + 5x - 2$ and $9x^3 + 4x - 3$. **[2.8]**

67 Subtract $5x + 3$ from $8x - 9$. **[2.8]**

68 Subtract $6x^2 - 3x - 2$ from $10x^2 + 3x - 6$. **[2.8]**

Multiply. **[2.9]**

69 $x^2(x^3 + x^5)$

70 $4a^4(2a^3 - 5)$

71 $(x + 2)(x + 7)$

72 $(3x + 5)(2x - 7)$

Expand and multiply. **[2.9]**

73 $(x + 5)^2$

74 $(4x - 3)^2$

CHAPTER 2 TEST

Give the opposite of each number.

1 14

2 -5

Place an inequality symbol ($<$ or $>$) between each pair of numbers so that the resulting statement is true.

3 $-1 \quad -4$

4 $|-4| \quad |2|$

Simplify each expression.

5 $-(-7)$

6 $-|-2|$

Perform the indicated operations.

7 $8 + (-17)$

8 $-4 - 2$

9 $-9 + (-12)$

10 $-65 - (-29)$

11 $(-6)(-7)$

12 $-3(-18)$

13 $\dfrac{-80}{16}$

14 $\dfrac{-35}{-7}$

Simplify the following expressions as much as possible.

15 $(-3)^2$

16 $(-2)^3$

17 $(-7)(3) + (-2)(-5)$

18 $(8 - 5)(6 - 11)$

19 $\dfrac{-5 + 3(-3)}{5 - 7}$

20 $\dfrac{-3(2) + 5(-2)}{7 - 3}$

21 Give the sum of -15 and -46.

22 Subtract -5 from -12.

23 What is the product of -8 and -3?

24 Give the quotient of 45 and -9.

25 What is -4 times the sum of -1 and -6?

26 A gambler loses \$100 Saturday night and wins \$65 on Sunday. Give the gambler's net loss or gain as a positive or negative number.

27 On Friday, the temperature reaches a high of 21° above 0 and a low of 4° below 0. What is the difference between the high and low temperatures for Friday?

28 Use the properties of exponents to simplify $6x^3 \cdot 5x^4$.

29 Multiply $2a^2(4a^3 - 6)$.

30 Multiply $(2x + 4)(3x - 5)$.

31 Expand and multiply $(x + 3)^2$.

32 Add $2x^2 + 5x - 7$ and $3x^2 - 8x + 2$.

33 Subtract $4x + 2$ from $7x - 6$.

FRACTIONS AND MIXED NUMBERS

INTRODUCTION

The figure below shows one of the nutrition labels we worked with in Chapter 1. It is from a can of Italian tomatoes. Notice toward the top of the label, the number of servings in the can is $3\frac{1}{2}$. The number $3\frac{1}{2}$ is called a *mixed number.* If we want to know how many calories are in the whole can of tomatoes, we must be able to multiply $3\frac{1}{2}$ by 25 (the number of calories per serving). Multiplication with mixed numbers is one of the topics we will cover in this chapter.

Canned Italian tomatoes

Nutrition Facts
Serving Size 1/2 cup (121g)
Servings Per Container: about 3 1/2

Amount Per Serving	
Calories 25	Calories from Fat 0
	% Daily Value*
Total Fat 0g	0%
Saturated Fat 0g	0%
Cholesterol 0mg	0%
Sodium 300mg	12%
Potassium 145mg	4%
Total Carbohydrate 4g	2%
Dietary Fiber 1g	4%
Sugars 4g	
Protein 1g	
Vitamin A 20% •	Vitamin C 15%
Calcium 4% •	Iron 15%

*Percent Daily Values are based on a 2,000 calorie diet. Your daily values may be higher or lower depending on your calorie needs.

We begin this chapter with the definition of a fraction and the properties that we can apply to fractions. Then we discuss how to reduce fractions to lowest terms. The remainder of the chapter is spent learning how to add, subtract, multiply, and divide fractions and mixed numbers. In addition, the chapter has been written so that, in the process of learning mathematical skills with fractions and mixed numbers, you will also sharpen your "common sense" with fractions.

STUDY SKILLS

The study skills for this chapter are concerned with getting ready to take an exam.

1 Get ready to take an exam. Try to arrange your daily study habits so that you have very little studying to do the night before your next exam. The next two goals will help you achieve this goal.

2 Review with the exam in mind. Review material that will be covered on the next exam every day. Your review should consist of working problems. Preferably, the problems you work should be problems from your list of difficult problems.

3 Continue to list difficult problems. This study skill was started in the previous chapter. You should continue to list and rework the problems that give you the most difficulty. It is this list that you will use to study for the next exam. Your goal is to go into the next exam knowing that you can successfully work any problem from your list of difficult problems.

4 Pay attention to instructions. Taking a test is not like doing homework. On a test, the problems will be varied. When you do your homework, you usually work a number of similar problems. I have some students who do very well on their homework but become confused when they see the same problems on a test. The reason for their confusion is that they have not paid attention to the instructions on their homework. If a test problem asks for the *average* of some numbers, then you must know what the word average means. Likewise, if a test problem asks you to find a *sum* and then to *round* your answer to the nearest hundred, then you must know that the word sum indicates addition, and, after you have added, you must round your answer as indicated.

If you train yourself to pay attention to the instructions that accompany a problem as you work through the assigned problems, you will not find yourself confused about what to do with a problem when you see it on a test.

3.1 The Meaning and Properties of Fractions

Introduction . . . In the United States, a 1-foot ruler is broken down into 12 equal parts, each of which is called an inch. Therefore an inch is said to be $\frac{1}{12}$, one twelfth, of a foot. In the metric system, a meter stick is one meter long and is divided into 100 equal parts, each of which is called a centimeter. A centimeter, then, is $\frac{1}{100}$, one hundredth, of a meter. When we refer to time, one hour is divided into 60 equal parts called minutes. Therefore one minute is $\frac{1}{60}$, one sixtieth, of an hour. In general a fraction like $\frac{1}{60}$ tells us two things: 60 is the number of equal parts that together make a whole, and 1 is the number of those equal parts that we have. The fraction $\frac{3}{4}$ gives us 3 of 4 equal parts; that is, the quantity in question has been divided into 4 equal parts and we are interested in 3 of them.

Figure 1 below shows a rectangle that has been divided into equal parts, four different ways. The shaded area for each rectangle is $\frac{1}{2}$ the total area.

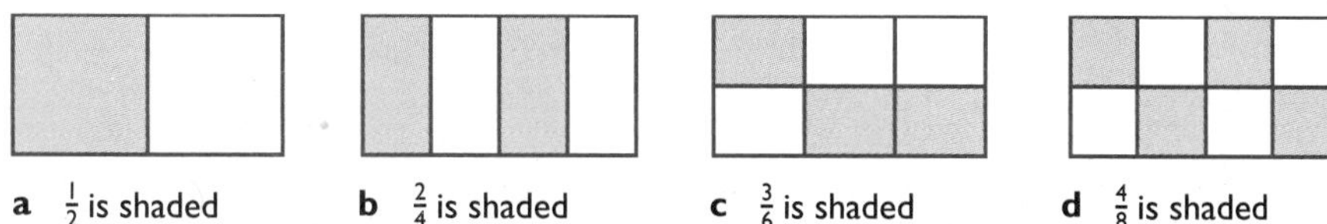

a $\frac{1}{2}$ is shaded **b** $\frac{2}{4}$ is shaded **c** $\frac{3}{6}$ is shaded **d** $\frac{4}{8}$ is shaded

Figure 1 Rectangles divided into equal parts

Now that we have an intuitive idea of the meaning of fractions, here are the more formal definitions and vocabulary associated with fractions.

> **DEFINITION** A **fraction** is any number that can be put in the form $\dfrac{a}{b}$ (also sometimes written a/b), where a and b are integers and b cannot be 0.

Some examples of fractions are:

$\dfrac{1}{2}$	$\dfrac{3}{4}$	$\dfrac{7}{8}$	$\dfrac{9}{5}$
One-half	*Three-fourths*	*Seven-eighths*	*Nine-fifths*

> **DEFINITION** For the fraction $\dfrac{a}{b}$, a and b are called the **terms** of the fraction. More specifically, a is called the **numerator** and b is called the **denominator**.

EXAMPLE 1 The terms of the fraction $\frac{3}{4}$ are 3 and 4. The 3 is called the numerator, and the 4 is called the denominator.

EXAMPLE 2 The numerator of the fraction $\dfrac{a}{5}$ is a. The denominator is 5. Both a and 5 are called terms.

EXAMPLE 3 The number 7 may also be put in fraction form, since it can be written as $\frac{7}{1}$. In this case, 7 is the numerator and 1 is the denominator.

Practice Problems

1 Name the terms of the fraction $\frac{5}{6}$. Which is the numerator and which is the denominator?

2 Name the numerator and the denominator of the fraction $\dfrac{x}{3}$.

3 Why is the number 9 considered to be a fraction?

Answers

1 Terms: 5 and 6; numerator: 5; denominator: 6

2 Numerator: x; denominator: 3

3 Because it can be written $\frac{9}{1}$

DEFINITION A **proper fraction** is a fraction in which the numerator is less than the denominator. If the numerator is greater than or equal to the denominator, the fraction is called an **improper fraction**.

4 Which of the following are proper fractions?

$$\frac{1}{6} \quad \frac{2}{3} \quad \frac{8}{5}$$

5 Which of the following are improper fractions?

$$\frac{5}{9} \quad \frac{6}{5} \quad \frac{4}{3} \quad 7$$

EXAMPLE 4 The fractions $\frac{3}{4}$, $\frac{1}{8}$, and $\frac{9}{10}$ are all proper fractions, since in each case the numerator is less than the denominator.

EXAMPLE 5 The numbers $\frac{9}{5}$, $\frac{10}{10}$, and 6 are all improper fractions, since in each case the numerator is greater than or equal to the denominator. (Remember that 6 can be written as $\frac{6}{1}$, in which case 6 is the numerator and 1 is the denominator.)

We can give meaning to the fraction $\frac{2}{3}$ by using a number line. If we take that part of the number line from 0 to 1 and divide it into *three equal parts*, we say that we have divided it into *thirds* (see Figure 2). Each of the three segments is $\frac{1}{3}$ (one third) of the whole segment from 0 to 1.

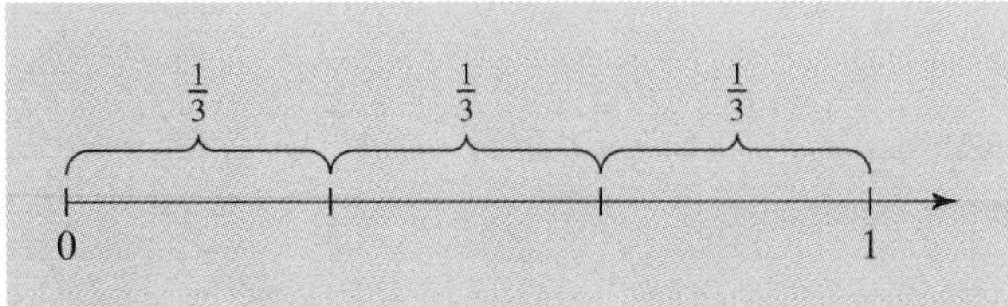

Figure 2

Note
There are many ways to give meaning to fractions like $\frac{2}{3}$ other than by using the number line. One popular way is to think of cutting a pie into three equal pieces, as shown below. If you take two of the pieces, you have taken $\frac{2}{3}$ of the pie.

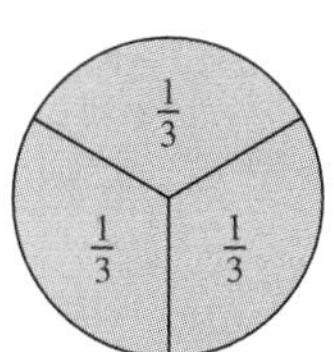

Two of these smaller segments together are $\frac{2}{3}$ (two thirds) of the whole segment. And three of them would be $\frac{3}{3}$ (three thirds), or the whole segment, as indicated in Figure 3.

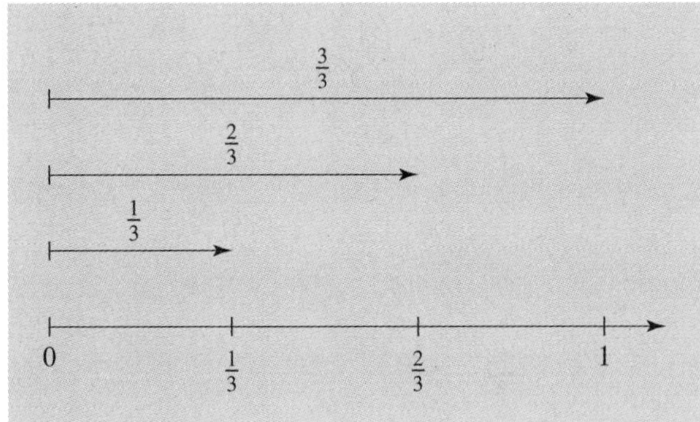

Figure 3

Let's do the same thing again with six equal divisions of the segment from 0 to 1 (see Figure 4). In this case we say that each of the smaller segments has a length of $\frac{1}{6}$ (one sixth).

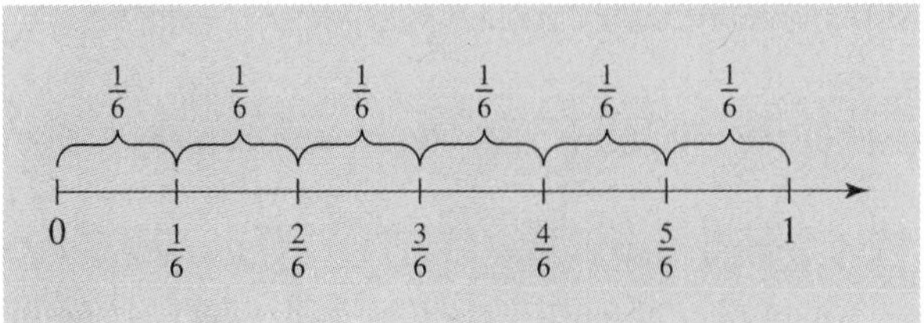

Figure 4

Answers
4 $\frac{1}{6}, \frac{2}{3}$ **5** $\frac{6}{5}, \frac{4}{3}, 7$

The same point that we labeled with $\frac{1}{3}$ in Figure 3 is now labeled with $\frac{2}{6}$. Likewise, the point that we labeled earlier with $\frac{2}{3}$ is now labeled $\frac{4}{6}$. It must be true then that

$$\frac{2}{6} = \frac{1}{3} \quad \text{and} \quad \frac{4}{6} = \frac{2}{3}$$

Actually, there are many fractions that name the same point as $\frac{1}{3}$. If we were to divide the segment between 0 and 1 into twelve equal parts, four of these twelve equal parts $\left(\frac{4}{12}\right)$ would be the same as $\frac{2}{6}$ or $\frac{1}{3}$. That is,

$$\frac{4}{12} = \frac{2}{6} = \frac{1}{3}$$

Although these three fractions look different, each names the same point on the number line, as shown in Figure 5. All three fractions have the same *value*, because they all represent the same number.

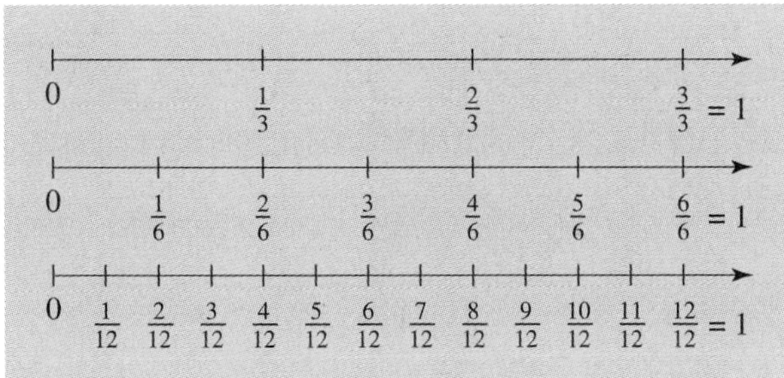

Figure 5

DEFINITION Fractions that represent the same number are said to be **equivalent**. Equivalent fractions may look different, but they must have the same value.

It is apparent that every fraction has many different representations, each of which is equivalent to the original fraction. The next two properties give us a way of changing the terms of a fraction without changing its value.

Property 1 for Fractions

If a, b, and c are integers and b and c are not 0, then it is always true that

$$\frac{a}{b} = \frac{a \cdot c}{b \cdot c}$$

In words: If the numerator and the denominator of a fraction are multiplied by the same nonzero number, the resulting fraction is equivalent to the original fraction.

EXAMPLE 6 Write $\frac{3}{4}$ as an equivalent fraction with denominator 20.

Solution The denominator of the original fraction is 4. The fraction we are trying to find must have a denominator of 20. We know that if we multiply 4 by 5,

6 Write $\frac{2}{3}$ as an equivalent fraction with denominator 12.

Answer

6 $\frac{8}{12}$

we get 20. Property 1 indicates that we are free to multiply the denominator by 5 so long as we do the same to the numerator.

$$\frac{3}{4} = \frac{3 \cdot \mathbf{5}}{4 \cdot \mathbf{5}} = \frac{15}{20}$$

The fraction $\frac{15}{20}$ is equivalent to the fraction $\frac{3}{4}$.

7 Write $\frac{2}{3}$ as an equivalent fraction with denominator $12x$.

EXAMPLE 7 Write $\frac{3}{4}$ as an equivalent fraction with denominator $12x$.

Solution If we multiply 4 by $3x$, we will have $12x$:

$$\frac{3}{4} = \frac{3 \cdot \mathbf{3x}}{4 \cdot \mathbf{3x}} = \frac{9x}{12x}$$

Note

When we encounter fractions that have variables in their denominator, like $\frac{9x}{12x}$, we will assume that those variables cannot be 0, since division by 0 is undefined.

Property 2 for Fractions

If a, b, and c are integers and b and c are not 0, then it is always true that

$$\frac{a}{b} = \frac{a \div c}{b \div c}$$

In words: If the numerator and the denominator of a fraction are divided by the same nonzero number, the resulting fraction is equivalent to the original fraction.

8 Write $\frac{15}{20}$ as an equivalent fraction with denominator 4.

EXAMPLE 8 Write $\frac{10}{12}$ as an equivalent fraction with denominator 6.

Solution If we divide the original denominator 12 by 2, we obtain 6. Property 2 indicates that if we divide both the numerator and the denominator by 2, the resulting fraction will be equal to the original fraction:

$$\frac{10}{12} = \frac{10 \div \mathbf{2}}{12 \div \mathbf{2}} = \frac{5}{6}$$

The Number 1 and Fractions

There are two situations involving fractions and the number 1 that occur frequently in mathematics. The first is when the denominator of a fraction is 1. In this case, if we let a represent any number, then

$$\frac{a}{1} = a \quad \text{for any number } a$$

The second situation occurs when the numerator and the denominator of a fraction are the same nonzero number:

$$\frac{a}{a} = 1 \quad \text{for any nonzero number } a$$

9 Simplify.

a $\frac{18}{1}$ **b** $\frac{18}{18}$

c $\frac{36}{18}$ **d** $\frac{72}{18}$

EXAMPLE 9 Simplify each expression.

a $\frac{24}{1}$ **b** $\frac{24}{24}$ **c** $\frac{48}{24}$ **d** $\frac{72}{24}$

Solution In each case we divide the numerator by the denominator:

a $\frac{24}{1} = 24$ **b** $\frac{24}{24} = 1$ **c** $\frac{48}{24} = 2$ **d** $\frac{72}{24} = 3$

Answers

7 $\frac{8x}{12x}$ **8** $\frac{3}{4}$

9 a 18 **b** 1 **c** 2 **d** 4

Prime Numbers, Factors, and Reducing to Lowest Terms

Introduction . . . Suppose you and a friend decide to split a medium-sized pizza for lunch. When the pizza is delivered you find that it has been cut into eight equal pieces. If you eat four pieces, you have eaten $\frac{4}{8}$ of the pizza, but you also know that you have eaten $\frac{1}{2}$ of the pizza. The fraction $\frac{4}{8}$ is equivalent to the fraction $\frac{1}{2}$; that is, they both have the same value. The mathematical process we use to rewrite $\frac{4}{8}$ as $\frac{1}{2}$ is called *reducing to lowest terms.* Before we look at that process, we need to define some new terms. Here is our first one.

DEFINITION A **prime number** is any whole number greater than 1 that has exactly two divisors—itself and 1. (A number is a divisor of another number if it divides it without a remainder.)

Prime numbers = {2, 3, 5, 7, 11, 13, 17, 19, 23, 29, 31, 37, . . . }

The list goes on indefinitely. Each number in the list has exactly two distinct divisors—itself and 1.

DEFINITION Any whole number greater than 1 that is not a prime number is called a **composite number**. A composite number always has at least one divisor other than itself and 1.

EXAMPLE 1 Identify each of the numbers below as either a prime number or a composite number. For those that are composite, give two divisors other than the number itself or 1.

a 43 **b** 12

Solution

a 43 is a prime number, since the only numbers that divide it without a remainder are 43 and 1.

b 12 is a composite number, since it can be written as $12 = 4 \cdot 3$, which means that 4 and 3 are divisors of 12. (These are not the only divisors of 12; other divisors are 1, 2, 6, and 12.)

Practice Problems

1 Which of the numbers below are prime numbers, and which are composite? For those that are composite, give two divisors other than the number itself and 1.

37, 39, 51, 59

Note

You may have already noticed that the word *divisor* as we are using it here means the same as the word *factor.* A divisor and a factor of a number are the same thing. A number can't be a divisor of another number without also being a factor of it.

Every composite number can be written as the product of prime factors. Let's look at the composite number 108. We know we can write 108 as $2 \cdot 54$. The number 2 is a prime number, but 54 is not prime. Since 54 can be written as $2 \cdot 27$, we have

$$\begin{aligned} 108 &= 2 \cdot 54 \\ &= 2 \cdot 2 \cdot 27 \end{aligned}$$

Now the number 27 can be written as $3 \cdot 9$ or $3 \cdot 3 \cdot 3$ (since $9 = 3 \cdot 3$), so

$$108 = 2 \cdot 54$$
$$108 = 2 \cdot 2 \cdot 27$$
$$108 = 2 \cdot 2 \cdot 3 \cdot 9$$
$$108 = 2 \cdot 2 \cdot 3 \cdot 3 \cdot 3$$

Answer

1 See solutions section.

This last line is the number 108 written as the product of prime factors. We can use exponents to simplify the last line:

$$108 = 2^2 \cdot 3^3$$

This process works by writing the original composite number as the product of any two of its factors, and then writing any factor that is not prime as the product of any two of its factors. The process is continued until all factors are prime numbers.

2 Factor 90 into a product of prime factors.

EXAMPLE 2 Factor 60 into a product of prime factors.

Solution We begin by writing 60 as $6 \cdot 10$ and continue factoring until all factors are prime numbers:

$$\begin{aligned} 60 &= 6 \cdot 10 \\ &= 2 \cdot 3 \cdot 2 \cdot 5 \\ &= 2^2 \cdot 3 \cdot 5 \end{aligned}$$

Notice that if we had started by writing 60 as $3 \cdot 20$, we would have achieved the same result:

$$\begin{aligned} 60 &= 3 \cdot 20 \\ &= 3 \cdot 2 \cdot 10 \\ &= 3 \cdot 2 \cdot 2 \cdot 5 \\ &= 2^2 \cdot 3 \cdot 5 \end{aligned}$$

There are some "tricks" to finding the divisors of a number. For instance, if a number ends in 0 or 5, then it is divisible by 5. If a number ends in an even number (0, 2, 4, 6, or 8), then it is divisible by 2. A number is divisible by 3 if the sum of its digits is divisible by 3. For example, 921 is divisible by 3 because the sum of its digits is $9 + 2 + 1 = 12$, which is divisible by 3.

We can use the method of factoring numbers into prime factors to help reduce fractions to lowest terms. Here is the definition for lowest terms.

DEFINITION A fraction is said to be in **lowest terms** if the numerator and the denominator have no factors in common other than the number 1.

3 Which of the following fractions are in lowest terms?

$$\frac{1}{6}, \frac{2}{8}, \frac{15}{25}, \frac{9}{13}$$

EXAMPLE 3 The fractions $\frac{1}{2}, \frac{1}{3}, \frac{2}{3}, \frac{1}{4}, \frac{3}{4}, \frac{1}{5}, \frac{2}{5}, \frac{3}{5}$, and $\frac{4}{5}$ are all in lowest terms, since in each case the numerator and the denominator have no factors other than 1 in common. That is, in each fraction, no number other than 1 divides both the numerator and the denominator exactly (without a remainder).

4 Reduce $\frac{12}{18}$ to lowest terms by dividing the numerator and the denominator by 6.

EXAMPLE 4 The fraction $\frac{6}{8}$ is not written in lowest terms, since the numerator and the denominator are both divisible by 2. To write $\frac{6}{8}$ in lowest terms, we apply Property 2 from Section 3.1 and divide both the numerator and the denominator by 2:

$$\frac{6}{8} = \frac{6 \div 2}{8 \div 2} = \frac{3}{4}$$

The fraction $\frac{3}{4}$ is in lowest terms, since 3 and 4 have no factors in common except the number 1.

Reducing a fraction to lowest terms is simply a matter of dividing the numerator and the denominator by all the factors they have in common. We know

Answers

2 $2 \cdot 3^2 \cdot 5$ **3** $\frac{1}{6}, \frac{9}{13}$ **4** $\frac{2}{3}$

from Property 2 of Section 3.1 that this will produce an equivalent fraction. All we need to do is recognize what factors the numerator and the denominator have in common and then divide the numerator and the denominator by these common factors. One method of finding the factors that are common to the numerator and the denominator is to factor each of them into prime factors.

EXAMPLE 5 Reduce the fraction $\frac{12}{15}$ to lowest terms by first factoring the numerator and the denominator into prime factors and then dividing both the numerator and the denominator by the factor they have in common.

Solution The numerator and the denominator factor as follows:

$$12 = 2 \cdot 2 \cdot 3 \quad \text{and} \quad 15 = 3 \cdot 5$$

The factor they have in common is 3. Property 2 tells us that we can divide both terms of a fraction by 3 to produce an equivalent fraction. So

$$\frac{12}{15} = \frac{2 \cdot 2 \cdot 3}{3 \cdot 5} \qquad \text{Factor the numerator and the denominator completely}$$

$$= \frac{2 \cdot 2 \cdot 3 \div \mathbf{3}}{3 \cdot 5 \div \mathbf{3}} \qquad \text{Divide by 3}$$

$$= \frac{2 \cdot 2}{5} = \frac{4}{5}$$

The fraction $\frac{4}{5}$ is equivalent to $\frac{12}{15}$ and is in lowest terms, since the numerator and the denominator have no factors other than 1 in common.

We can shorten the work involved in reducing fractions to lowest terms by using a slash to indicate division. For example, we can write the above problem this way:

$$\frac{12}{15} = \frac{2 \cdot 2 \cdot \cancel{3}}{\cancel{3} \cdot 5} = \frac{4}{5}$$

So long as we understand that the slashes through the 3's indicate that we have divided both the numerator and the denominator by 3, we can use this notation.

EXAMPLE 6 Reduce $\frac{20}{24}$ to lowest terms.

Solution Factoring 20 and 24 completely and then dividing out both the factors they have in common gives us

$$\frac{20}{24} = \frac{\cancel{2} \cdot \cancel{2} \cdot 5}{\cancel{2} \cdot \cancel{2} \cdot 2 \cdot 3} = \frac{5}{6}$$

Note The slashes in Example 6 indicate that we have divided both the numerator and the denominator by $2 \cdot 2$, which is equal to 4. With some fractions it is apparent at the start what number divides the numerator and the denominator. For instance, you may have recognized that both 20 and 24 in Example 6 are divisible by 4. We can divide both terms by 4 without factoring first, just as we did in Section 3.1. Property 2 guarantees that dividing both terms of a fraction by 4 will produce an equivalent fraction:

$$\frac{20}{24} = \frac{20 \div \mathbf{4}}{24 \div \mathbf{4}} = \frac{5}{6}$$

5 Reduce the fraction $\frac{15}{20}$ to lowest terms by first factoring the numerator and the denominator into prime factors and then dividing out the factors they have in common.

6 Reduce $\frac{30}{35}$ to lowest terms.

Answers

5 $\frac{3}{4}$ **6** $\frac{6}{7}$

7 Reduce $\frac{8}{72}$ to lowest terms.

EXAMPLE 7 Reduce $\frac{6}{42}$ to lowest terms.

Solution We begin by factoring both terms. We then divide through by any factors common to both terms:

$$\frac{6}{42} = \frac{\cancel{2} \cdot \cancel{3}}{\cancel{2} \cdot \cancel{3} \cdot 7} = \frac{1}{7}$$

We must be careful in a problem like this to remember that the slashes indicate division. They are used to indicate that we have divided both the numerator and the denominator by $2 \cdot 3 = 6$. The result of dividing the numerator 6 by $2 \cdot 3$ is 1. It is a very common mistake to call the numerator 0 instead of 1, or to leave the numerator out of the answer.

Reduce each fraction to lowest terms.

8 $\frac{5}{50}$

9 $\frac{120}{25}$

In Examples 8 and 9, reduce each fraction to lowest terms.

EXAMPLE 8

$$\frac{4}{40} = \frac{\cancel{2} \cdot \cancel{2} \cdot 1}{\cancel{2} \cdot \cancel{2} \cdot 2 \cdot 5} = \frac{1}{10}$$

EXAMPLE 9

$$\frac{105}{30} = \frac{\cancel{3} \cdot \cancel{5} \cdot 7}{2 \cdot \cancel{3} \cdot \cancel{5}} = \frac{7}{2}$$

If a fraction contains variables (letters) in its numerator and/or denominator, we treat the variables in the same way we treat the other factors in the numerator or denominator. If a variable factor is common to the numerator and the denominator, we can reduce to lowest terms by dividing the numerator and the denominator by that variable. (Remember also that any variable that appears in a denominator is assumed to be nonzero.)

10 $\frac{54x}{90xy}$

EXAMPLE 10 Reduce $\frac{36xy}{120x}$ to lowest terms.

Solution First we factor 36 and 120 completely. Then we divide out all the factors that are common to the numerator and the denominator, including any variable factors.

$$\frac{36xy}{120x} = \frac{\cancel{2} \cdot \cancel{2} \cdot \cancel{3} \cdot 3 \cdot \cancel{x} \cdot y}{\cancel{2} \cdot \cancel{2} \cdot 2 \cdot \cancel{3} \cdot 5 \cdot \cancel{x}} = \frac{3y}{10}$$

11 $\frac{306a^2}{228a}$

EXAMPLE 11 Reduce $\frac{204a^2}{342a}$ to lowest terms.

Solution We factor both the numerator and the denominator completely and then divide out any factors common to both.

$$\frac{204a^2}{342a} = \frac{\cancel{2} \cdot 2 \cdot \cancel{3} \cdot 17 \cdot \cancel{a} \cdot a}{\cancel{2} \cdot \cancel{3} \cdot 3 \cdot 19 \cdot \cancel{a}} = \frac{34a}{57}$$

Answers

7 $\frac{1}{9}$ 8 $\frac{1}{10}$ 9 $\frac{24}{5}$ 10 $\frac{3}{5y}$ 11 $\frac{51a}{38}$

3.3 Multiplication with Fractions, and the Area of a Triangle

Multiplication with Fractions

Introduction . . . A recipe calls for $\frac{3}{4}$ cup of flour. If you are making only $\frac{1}{2}$ the recipe, how much flour do you use? This question can be answered by multiplying $\frac{1}{2}$ and $\frac{3}{4}$. But, if you imagine taking half of each of the fourths, you can also see that the answer is $\frac{3}{8}$. That is, $\frac{1}{2}$ of $\frac{3}{4}$ is $\frac{3}{8}$. Here is the problem written in symbols:

$$\frac{1}{2} \cdot \frac{3}{4} = \frac{3}{8}$$

As you can see from this example, to multiply two fractions, we multiply the numerators and then multiply the denominators. We begin this section with the rule for multiplication of fractions.

RULE The product of two fractions is the fraction whose numerator is the product of the two numerators, and whose denominator is the product of the two denominators. We can write this rule in symbols as follows:

If a, b, c, and d represent any numbers, and b and d are not zero, then

$$\frac{a}{b} \cdot \frac{c}{d} = \frac{a \cdot c}{b \cdot d}$$

EXAMPLE 1 Multiply: $\frac{3}{5} \cdot \frac{2}{7}$

Solution Using our rule for multiplication, we multiply the numerators and multiply the denominators:

$$\frac{3}{5} \cdot \frac{2}{7} = \frac{3 \cdot 2}{5 \cdot 7} = \frac{6}{35}$$

The product of $\frac{3}{5}$ and $\frac{2}{7}$ is the fraction $\frac{6}{35}$. The numerator 6 is the product of 3 and 2, and the denominator 35 is the product of 5 and 7.

EXAMPLE 2 Multiply: $-\frac{3}{8} \cdot 5$

Solution The number 5 can be written as $\frac{5}{1}$. That is, 5 can be considered a fraction with numerator 5 and denominator 1. Writing 5 this way enables us to apply the rule for multiplying fractions. Remember also that the product of two numbers with different signs is negative:

$$-\frac{3}{8} \cdot 5 = -\frac{3}{8} \cdot \frac{5}{1}$$

$$= -\frac{3 \cdot 5}{8 \cdot 1} \quad \text{Unlike signs give a negative answer}$$

$$= -\frac{15}{8}$$

Practice Problems

1 Multiply: $\frac{2}{3} \cdot \frac{5}{9}$

2 Multiply: $-\frac{2}{5} \cdot 7$

Answers

1 $\frac{10}{27}$ **2** $-\frac{14}{5}$

3 Multiply: $\frac{1}{3}\left(\frac{4}{5} \cdot \frac{1}{3}\right)$

EXAMPLE 3 Multiply: $\frac{1}{2}\left(\frac{3}{4} \cdot \frac{1}{5}\right)$

Solution We find the product inside the parentheses first, and then multiply the result by $\frac{1}{2}$:

$$\begin{aligned} \frac{1}{2}\left(\frac{3}{4} \cdot \frac{1}{5}\right) &= \frac{1}{2}\left(\frac{3 \cdot 1}{4 \cdot 5}\right) \\ &= \frac{1}{2}\left(\frac{3}{20}\right) \\ &= \frac{1 \cdot 3}{2 \cdot 20} = \frac{3}{40} \end{aligned}$$

The properties of multiplication that we developed in Chapter 1 for whole numbers apply to fractions as well. That is, if a, b, and c are fractions, then

$a \cdot b = b \cdot a$	Multiplication with fractions is commutative
$a \cdot (b \cdot c) = (a \cdot b) \cdot c$	Multiplication with fractions is associative

To demonstrate the associative property for fractions, let's do Example 3 again, but this time we will apply the associative property first:

$$\begin{aligned} \frac{1}{2}\left(\frac{3}{4} \cdot \frac{1}{5}\right) &= \left(\frac{1}{2} \cdot \frac{3}{4}\right) \cdot \frac{1}{5} && \text{Associative property} \\ &= \left(\frac{1 \cdot 3}{2 \cdot 4}\right) \cdot \frac{1}{5} \\ &= \left(\frac{3}{8}\right) \cdot \frac{1}{5} \\ &= \frac{3 \cdot 1}{8 \cdot 5} = \frac{3}{40} \end{aligned}$$

The result is identical to that of Example 3.

Here is another example that involves the associative property. Problems like this will be useful in the next chapter when we solve equations.

4 Multiply: $\frac{1}{4}(4y)$

EXAMPLE 4 Multiply: $\frac{1}{3}(3x)$

Solution Remember that $3x$ means 3 times x. The 3 and the x are not combined. If we apply the associative property so that we can group the $\frac{1}{3}$ and 3 together, we will be able to multiply them:

$$\begin{aligned} \frac{1}{3}(3x) &= \left(\frac{1}{3} \cdot 3\right)x && \text{Associative property} \\ &= 1x && \text{The product of } \frac{1}{3} \text{ and 3 is 1} \\ &= x && \text{The product of 1 and } x \text{ is } x \end{aligned}$$

Answers

3 $\frac{4}{45}$ **4** y

The answers to all the examples so far in this section have been in lowest terms. Let's see what happens when we multiply two fractions to get a product that is not in lowest terms.

EXAMPLE 5 Multiply: $\frac{15}{8} \cdot \frac{4}{9}$

Solution Multiplying the numerators and multiplying the denominators, we have

$$\frac{15}{8} \cdot \frac{4}{9} = \frac{15 \cdot 4}{8 \cdot 9} = \frac{60}{72}$$

The product is $\frac{60}{72}$, which can be reduced to lowest terms by factoring 60 and 72 and then dividing out any factors they have in common:

$$\frac{60}{72} = \frac{\cancel{2} \cdot \cancel{2} \cdot \cancel{3} \cdot 5}{\cancel{2} \cdot \cancel{2} \cdot 2 \cdot \cancel{3} \cdot 3} = \frac{5}{6}$$

We can actually save ourselves some time by factoring before we multiply. Here's how it is done:

$$\frac{15}{8} \cdot \frac{4}{9} = \frac{15 \cdot 4}{8 \cdot 9} = \frac{(3 \cdot 5) \cdot (2 \cdot 2)}{(2 \cdot 2 \cdot 2) \cdot (3 \cdot 3)} = \frac{\cancel{3} \cdot 5 \cdot \cancel{2} \cdot \cancel{2}}{\cancel{2} \cdot \cancel{2} \cdot 2 \cdot \cancel{3} \cdot 3} = \frac{5}{6}$$

The result is the same in both cases. Reducing to lowest terms before we actually multiply takes less time. Here are some additional examples.

EXAMPLE 6 $-\frac{9}{2} \cdot \left(-\frac{8}{18}\right) = \frac{9 \cdot 8}{2 \cdot 18}$ Like signs give a positive product

$$= \frac{(3 \cdot 3) \cdot (2 \cdot 2 \cdot 2)}{2 \cdot (2 \cdot 3 \cdot 3)} = \frac{\cancel{3} \cdot \cancel{3} \cdot \cancel{2} \cdot \cancel{2} \cdot 2}{\cancel{2} \cdot \cancel{2} \cdot \cancel{3} \cdot \cancel{3}} = \frac{2}{1} = 2$$

5 Multiply: $\frac{12}{25} \cdot \frac{5}{6}$

6 Find the product: $\frac{8}{3} \cdot \frac{9}{24}$

Note
Although $\frac{2}{1}$ is in lowest terms, it is still simpler to write the answer as just 2. We will always do this when the denominator is the number 1.

Answers
5 $\frac{2}{5}$ **6** 1

Find each product.

7 $\dfrac{yz^2}{x} \cdot \dfrac{x^3}{yz}$

Note

Remember, we are assuming that any variables that appear in a denominator are not 0.

EXAMPLE 7 $\dfrac{x^2y}{z} \cdot \dfrac{z^3}{xy} = \dfrac{x^2y \cdot z^3}{z \cdot xy}$

$$= \frac{(x \cdot x \cdot y)(z \cdot z \cdot z)}{z \cdot x \cdot y}$$

$$= \frac{\not{x} \cdot x \cdot \not{y} \cdot \not{z} \cdot z \cdot z}{\not{z} \cdot \not{x} \cdot \not{y}}$$

$$= \frac{x \cdot z \cdot z}{1}$$

$$= xz^2$$

8 $\dfrac{3}{4} \cdot \dfrac{8}{3} \cdot \dfrac{1}{6}$

EXAMPLE 8 $\dfrac{2}{3} \cdot \dfrac{6}{5} \cdot \dfrac{5}{8} = \dfrac{2 \cdot 6 \cdot 5}{3 \cdot 5 \cdot 8}$

$$= \frac{2 \cdot (2 \cdot 3) \cdot 5}{3 \cdot 5 \cdot (2 \cdot 2 \cdot 2)}$$

$$= \frac{\not{2} \cdot \not{2} \cdot \not{3} \cdot \not{5}}{\not{3} \cdot \not{5} \cdot \not{2} \cdot \not{2} \cdot 2}$$

$$= \frac{1}{2}$$

In Chapter 1 we did some work with exponents. We can extend our work with exponents to include fractions, as the following examples indicate.

Apply the definition of exponents and then multiply.

9 $\left(\dfrac{2}{3}\right)^2$

EXAMPLE 9 $\left(-\dfrac{3}{4}\right)^2 = -\dfrac{3}{4}\left(-\dfrac{3}{4}\right)$

$$= \frac{3 \cdot 3}{4 \cdot 4} \quad \text{Like signs give a positive product}$$

$$= \frac{9}{16}$$

10 $\left(\dfrac{3}{4}\right)^2 \cdot \dfrac{1}{2}$

EXAMPLE 10 $\left(\dfrac{5}{6}\right)^2 \cdot \dfrac{1}{2} = \dfrac{5}{6} \cdot \dfrac{5}{6} \cdot \dfrac{1}{2}$

$$= \frac{5 \cdot 5 \cdot 1}{6 \cdot 6 \cdot 2}$$

$$= \frac{25}{72}$$

The word *of* used in connection with fractions indicates multiplication. If we want to find $\frac{1}{2}$ of $\frac{2}{3}$, then what we do is multiply $\frac{1}{2}$ and $\frac{2}{3}$.

Answers

7 x^2z **8** $\dfrac{1}{3}$ **9** $\dfrac{4}{9}$ **10** $\dfrac{9}{32}$

EXAMPLE 11 Find $\frac{1}{2}$ of $\frac{2}{3}$.

Solution Knowing the word *of* as used here indicates multiplication, we have

$$\frac{1}{2} \text{ of } \frac{2}{3} = \frac{1}{2} \cdot \frac{2}{3}$$

$$= \frac{1 \cdot \cancel{2}}{\cancel{2} \cdot 3} = \frac{1}{3}$$

This seems to make sense. Logically, $\frac{1}{2}$ of $\frac{2}{3}$ should be $\frac{1}{3}$, as Figure 1 shows.

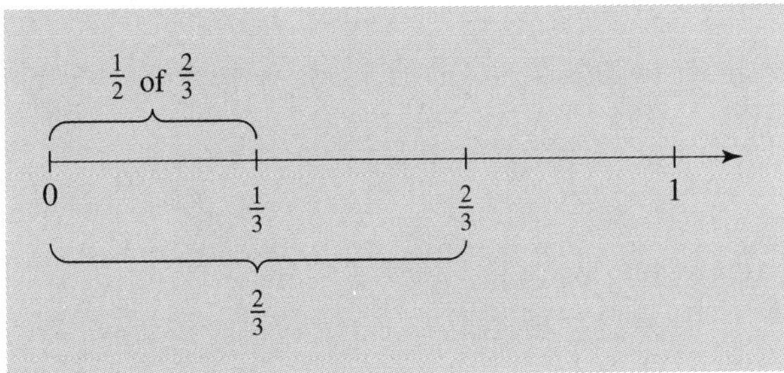

Figure 1

11 Find $\frac{2}{3}$ of $\frac{1}{2}$.

EXAMPLE 12 What is $\frac{3}{4}$ of -12?

Solution Again, *of* means multiply.

$$\frac{3}{4} \text{ of } -12 = \frac{3}{4}(-12)$$

$$= \frac{3}{4}\left(-\frac{12}{1}\right)$$

$$= -\frac{3 \cdot 12}{4 \cdot 1} \quad \text{Unlike signs give a negative product}$$

$$= -\frac{3 \cdot \cancel{2} \cdot \cancel{2} \cdot 3}{\cancel{2} \cdot \cancel{2} \cdot 1}$$

$$= -\frac{9}{1} = -9$$

12 What is $\frac{2}{3}$ of -12?

Note on Shortcuts

As you become familiar with multiplying fractions, you may notice shortcuts that reduce the number of steps in the problems. It's okay to use these shortcuts if you understand why they work and are consistently getting correct answers. If you are using shortcuts and not consistently getting correct answers, then go back to showing all the work until you completely understand the process.

The Area of a Triangle

The formula for the area of a triangle is one application of multiplication with fractions. Figure 2 shows a triangle with base b and height h. Below the triangle is the formula for its area. As you can see, it is a product containing the fraction $\frac{1}{2}$.

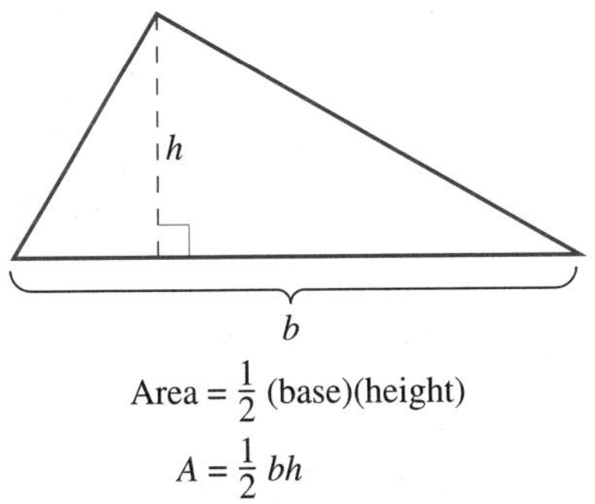

Figure 2 The area of a triangle

Answers

11 $\frac{1}{3}$ **12** -8

13 Find the area of the triangle below.

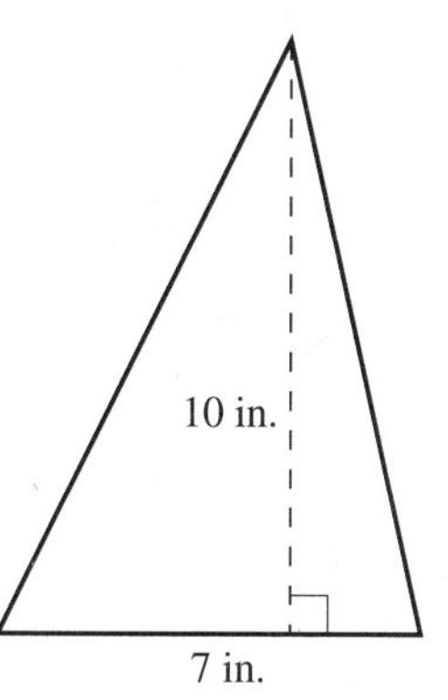

14 Find the total area enclosed by the figure.

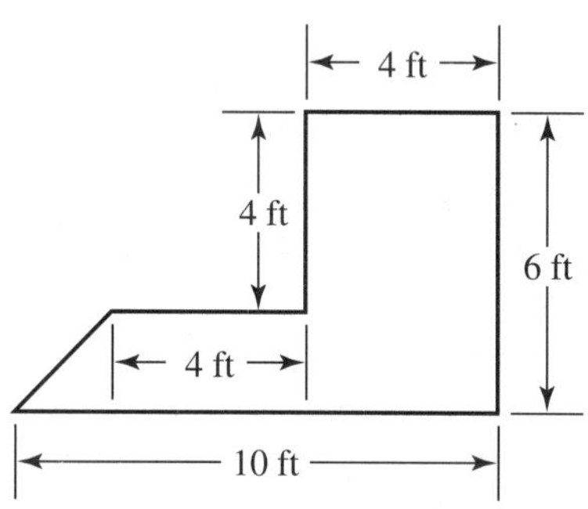

Answers
13 35 in^2 **14** 34 ft^2

EXAMPLE 13 Find the area of the triangle in Figure 3.

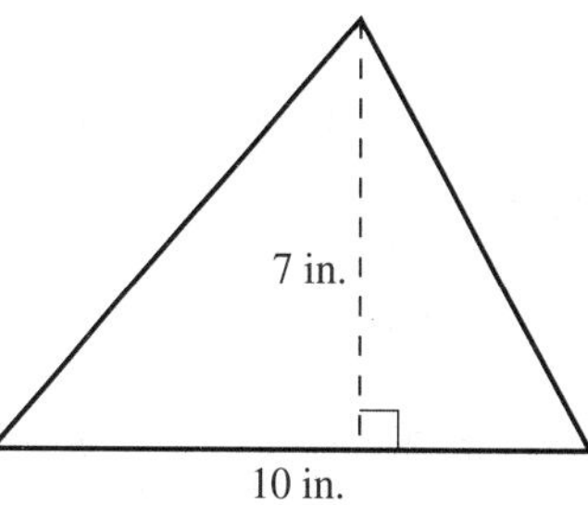

Figure 3 A triangle with base 10 inches and height 7 inches

Solution Applying the formula for the area of a triangle, we have

$$A = \frac{1}{2}bh = \frac{1}{2} \cdot 10 \cdot 7 = 5 \cdot 7 = 35 \text{ in}^2$$

EXAMPLE 14 Find the area of the figure in Figure 4.

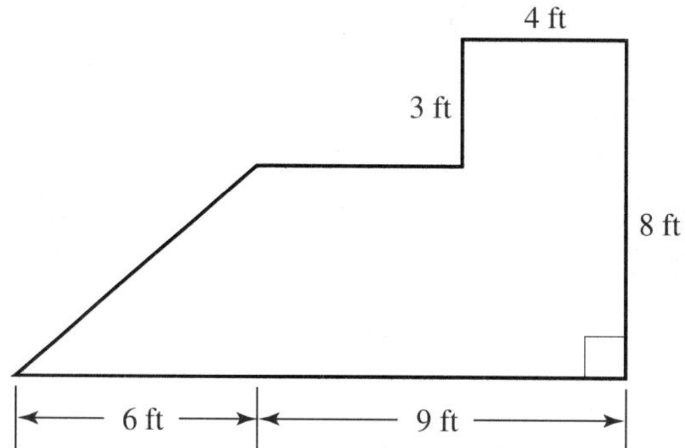

Figure 4

Solution We divide the figure into three parts and then find the area of each part (see Figure 5). The area of the whole figure is the sum of the areas of its parts.

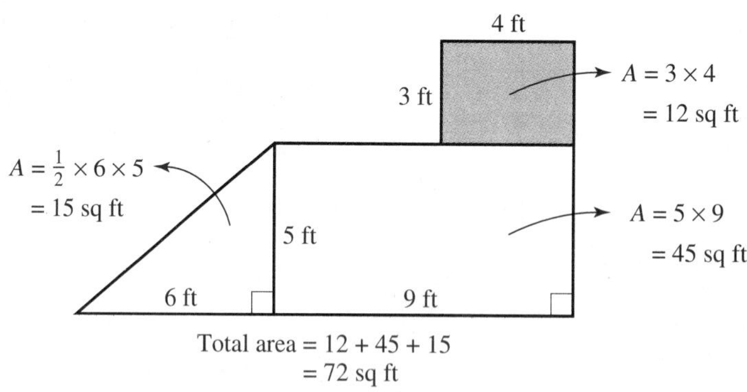

Figure 5

78 a The base and the height of the triangle shown below are $\frac{1}{4}$ inch each. Draw three more triangles on the grid. The base and the height in the first one will be $\frac{1}{2}$ inch each, the base and the height for the second one will be $\frac{3}{4}$ inch each, and the base and the height of the third one will be 1 inch each.

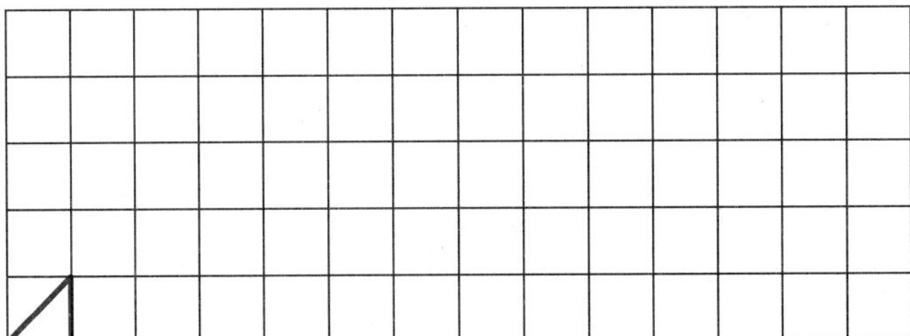

b Use the triangles you have drawn above to complete the following table.

Table 3 Areas of triangles

Length of Base and Height (in inches)	Area (in square inches)
$\frac{1}{4}$	
$\frac{1}{2}$	
$\frac{3}{4}$	
1	

c Use Figure 8 to construct a line graph from the information in Table 3.

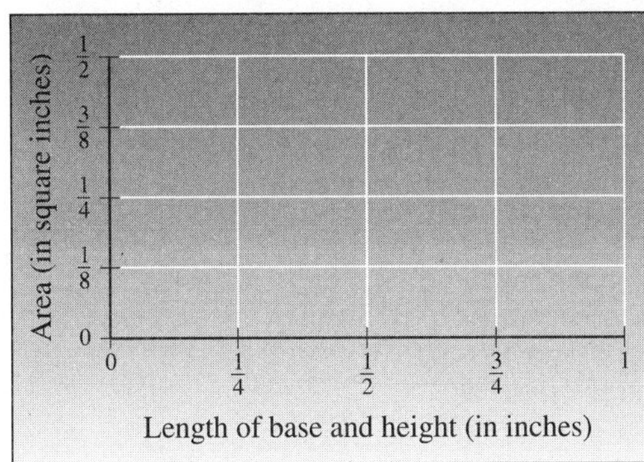

Figure 8 A line graph of Table 3

Estimating

For each problem below, mentally estimate which of the numbers 0, 1, 2, or 3 is closest to the answer. Make your estimate without using pencil and paper or a calculator.

79 $\frac{11}{5} \cdot \frac{19}{20}$ **80** $\frac{3}{5} \cdot \frac{1}{20}$ **81** $\frac{16}{5} \cdot \frac{23}{24}$ **82** $\frac{9}{8} \cdot \frac{31}{32}$

83 $\frac{1}{8} \cdot \frac{15}{32}$ **84** $\frac{7}{8} \cdot \frac{17}{16}$

Calculator Problems

Use a calculator to find the following products. Do *not* reduce your answers to lowest terms.

85 $\frac{31}{32} \cdot \frac{41}{42} \cdot \frac{43}{44}$ **86** $\frac{16}{17} \cdot \frac{18}{19} \cdot \frac{20}{21}$ **87** $\frac{93}{97} \cdot 18 \cdot \frac{46}{53}$ **88** $\frac{32}{79} \cdot 25 \cdot \frac{87}{91}$

Review Problems

In the next section we will do division with fractions. As you already know, division and multiplication are closely related. These review problems are intended to let you see more of the relationship between multiplication and division.

Perform the indicated operations.

89 $6 \div 2$ **90** $6 \cdot \frac{1}{2}$ **91** $8 \div 4$ **92** $8 \cdot \frac{1}{4}$

93 $15 \div 3$ **94** $15 \cdot \frac{1}{3}$ **95** $18 \div 6$ **96** $18 \cdot \frac{1}{6}$

One Step Further: Number Sequences

Recall that a geometric sequence is a sequence in which each term comes from the previous term by multiplying by the same number each time. For example, the sequence $1, \frac{1}{2}, \frac{1}{4}, \frac{1}{8}, \ldots$ is a geometric sequence in which each term is found by multiplying the previous term by $\frac{1}{2}$. By observing this fact, we know that the next term in the sequence will be $\frac{1}{8} \cdot \frac{1}{2} = \frac{1}{16}$.

Find the next number in each of the geometric sequences below.

97 $1, \frac{1}{3}, \frac{1}{9}, \ldots$ **98** $1, \frac{1}{4}, \frac{1}{16}, \ldots$ **99** $\frac{3}{2}, 1, \frac{2}{3}, \frac{4}{9}, \ldots$ **100** $\frac{2}{3}, 1, \frac{3}{2}, \frac{9}{4}, \ldots$

3.4 Division with Fractions

Introduction . . . A few years ago our 4-H club was making blankets to keep their lambs clean at the County Fair. Each blanket required $\frac{3}{4}$ yard of material. We had 9 yards of material left over from the year before. To see how many blankets we could make, we divided 9 by $\frac{3}{4}$. The result was 12, meaning that we could make 12 lamb blankets for the fair.

Before we define division with fractions, we must first introduce the idea of *reciprocals*. Look at the following multiplication problems:

$$\frac{3}{4} \cdot \frac{4}{3} = \frac{12}{12} = 1 \qquad \frac{7}{8} \cdot \frac{8}{7} = \frac{56}{56} = 1$$

In each case the product is 1. Whenever the product of two numbers is 1, we say the two numbers are *reciprocals*.

DEFINITION Two numbers whose product is 1 are said to be **reciprocals**. In symbols, the reciprocal of $\frac{a}{b}$ is $\frac{b}{a}$, because

$$\frac{a}{b} \cdot \frac{b}{a} = \frac{a \cdot b}{b \cdot a} = \frac{a \cdot b}{a \cdot b} = 1 \qquad (a \neq 0,\ b \neq 0)$$

Every number has a reciprocal except 0. The reason 0 does not have a reciprocal is because the product of *any* number with 0 is 0. It can never be 1. Reciprocals of whole numbers are fractions with 1 as the numerators. For example, the reciprocal of 5 is $\frac{1}{5}$, since

$$5 \cdot \frac{1}{5} = \frac{5}{1} \cdot \frac{1}{5} = \frac{5}{5} = 1$$

Table 1 lists some numbers and their reciprocals.

Table 1

Number	Reciprocal	Reason
$\frac{3}{4}$	$\frac{4}{3}$	Because $\frac{3}{4} \cdot \frac{4}{3} = \frac{12}{12} = 1$
$-\frac{9}{5}$	$-\frac{5}{9}$	Because $-\frac{9}{5}\left(-\frac{5}{9}\right) = \frac{45}{45} = 1$
$\frac{1}{3}$	3	Because $\frac{1}{3} \cdot 3 = \frac{1}{3} \cdot \frac{3}{1} = \frac{3}{3} = 1$
-7	$-\frac{1}{7}$	Because $-7\left(-\frac{1}{7}\right) = -\frac{7}{1}\left(-\frac{1}{7}\right) = \frac{7}{7} = 1$

Division with fractions is accomplished by using reciprocals. More specifically, we can define division by a fraction to be the same as multiplication by its reciprocal. Here is the precise definition:

Note
Defining division to be the same as multiplication by the reciprocal does make sense. If we divide 6 by 2, we get 3. On the other hand, if we multiply 6 by $\frac{1}{2}$ (the reciprocal of 2), we also get 3. Whether we divide by 2 or multiply by $\frac{1}{2}$, we get the same result.

DEFINITION If a, b, c, and d are integers and b, c, and d are all not equal to 0, then

$$\frac{a}{b} \div \frac{c}{d} = \frac{a}{b} \cdot \frac{d}{c}$$

This defi' ,n states that dividing by the fraction $\frac{c}{d}$ is exactly the same as multiplyin' its reciprocal $\frac{d}{c}$. Since we developed the rule for multiplying fraction' Section 3.3, we do not need a new rule for division. We simply *replace t' ivisor by its reciprocal* and multiply. Here are some examples to illustrate ' procedure.

Practice Problems

1 Divide: $\frac{1}{3} \div \frac{1}{6}$

2 Divi $\frac{5}{9} \div \frac{10}{3}$

3 Divide: $-\frac{3}{4} \div 3$

EX /PLE 1 Divide: $\frac{1}{2} \div \frac{1}{4}$

' ution The divisor is $\frac{1}{4}$, and its reciprocal is $\frac{4}{1}$. Applying the definition of divi- n for fractions, we have

$$\begin{aligned}
\frac{1}{2} \div \frac{1}{4} &= \frac{1}{2} \cdot \frac{4}{1} \\
&= \frac{1 \cdot 4}{2 \cdot 1} \\
&= \frac{1 \cdot \not{2} \cdot 2}{\not{2} \cdot 1} \\
&= \frac{2}{1} \\
&= 2
\end{aligned}$$

The quotient of $\frac{1}{2}$ and $\frac{1}{4}$ is 2. Or, $\frac{1}{4}$ "goes into" $\frac{1}{2}$ two times. Logically, our definition for division of fractions seems to be giving us answers that are consistent with what we know about fractions from previous experience. Since 2 times $\frac{1}{4}$ is $\frac{2}{4}$ or $\frac{1}{2}$, it seems logical that $\frac{1}{2}$ divided by $\frac{1}{4}$ should be 2.

EXAMPLE 2 Divide: $\frac{3}{8} \div \frac{9}{4}$

Solution Dividing by $\frac{9}{4}$ is the same as multiplying by its reciprocal, which is $\frac{4}{9}$:

$$\begin{aligned}
\frac{3}{8} \div \frac{9}{4} &= \frac{3}{8} \cdot \frac{4}{9} \\
&= \frac{\not{3} \cdot \not{2} \cdot \not{2}}{\not{2} \cdot \not{2} \cdot 2 \cdot \not{3} \cdot 3} \\
&= \frac{1}{6}
\end{aligned}$$

The quotient of $\frac{3}{8}$ and $\frac{9}{4}$ is $\frac{1}{6}$.

EXAMPLE 3 Divide: $-\frac{2}{3} \div 2$

Solution The reciprocal of 2 is $\frac{1}{2}$. Applying the definition for division of fractions, we have

$$\begin{aligned}
-\frac{2}{3} \div 2 &= -\frac{2}{3} \cdot \frac{1}{2} \\
&= -\frac{\not{2} \cdot 1}{3 \cdot \not{2}} \\
&= -\frac{1}{3}
\end{aligned}$$

Answers

1 2 **2** $\frac{1}{6}$ **3** $-\frac{1}{4}$

EXAMPLE 4 Divide: $2 \div \left(-\frac{1}{3}\right)$

Solution We replace $-\frac{1}{3}$ by its reciprocal, which is -3, and multiply:

$$2 \div \left(-\frac{1}{3}\right) = 2(-3) = -6$$

Here are some further examples of division with fractions. Notice in each case that the first step is the only new part of the process.

EXAMPLE 5 $$\frac{4}{27} \div \frac{16}{9} = \frac{4}{27} \cdot \frac{9}{16} = \frac{\not{4} \cdot \not{9}}{3 \cdot \not{9} \cdot \not{4} \cdot 4} = \frac{1}{12}$$

In this example we did not factor the numerator and the denominator completely in order to reduce to lowest terms because, as you have probably already noticed, it is not necessary to do so. We need to factor only enough to show what numbers are common to the numerator and the denominator. If we factored completely in the second step, it would look like this:

$$= \frac{\not{2} \cdot \not{2} \cdot \not{3} \cdot \not{3}}{\not{3} \cdot \not{3} \cdot 3 \cdot \not{2} \cdot \not{2} \cdot 2 \cdot 2} = \frac{1}{12}$$

The result is the same in both cases. From now on we will factor numerators and denominators only enough to show the factors we are dividing out.

EXAMPLE 6 $$\frac{16}{35} \div 8 = \frac{16}{35} \cdot \frac{1}{8} = \frac{2 \cdot \not{8} \cdot 1}{35 \cdot \not{8}} = \frac{2}{35}$$

EXAMPLE 7 $$-27 \div \left(-\frac{3}{2}\right) = -27\left(-\frac{2}{3}\right) = \frac{\not{3} \cdot 9 \cdot 2}{\not{3}} = 18$$

Like signs give a positive answer

4 Divide: $-4 \div \left(-\frac{1}{5}\right)$

Find each quotient.

5 $\frac{5}{32} \div \frac{10}{42}$

6 $\frac{12}{25} \div 6$

7 $-12 \div \left(-\frac{4}{3}\right)$

Answers

4 20 **5** $\frac{21}{32}$ **6** $\frac{2}{25}$ **7** 9

8 $\frac{x^3}{y} \div \frac{x^2}{y^2}$

EXAMPLE 8 $\frac{x^2}{y} \div \frac{x}{y^3} = \frac{x^2}{y} \cdot \frac{y^3}{x}$

$$= \frac{\not{x} \cdot x \cdot \not{y} \cdot y \cdot y}{\not{y} \cdot \not{x}}$$

$$= \frac{x \cdot y \cdot y}{1}$$

$$= xy^2$$

The next two examples combine what we have learned about division of fractions with the rule for order of operations.

9 The quotient of $\frac{5}{4}$ and $\frac{1}{8}$ is increased by 8. What number results?

EXAMPLE 9 The quotient of $\frac{8}{3}$ and $\frac{1}{6}$ is increased by 5. What number results?

Solution Translating to symbols, we have

$$\frac{8}{3} \div \frac{1}{6} + 5 = \frac{8}{3} \cdot \frac{6}{1} + 5$$

$$= 16 + 5$$

$$= 21$$

10 Simplify:

$$18 \div \left(\frac{3}{5}\right)^2 + 48 \div \left(\frac{2}{5}\right)^2$$

EXAMPLE 10 Simplify: $32 \div \left(\frac{4}{3}\right)^2 + 75 \div \left(\frac{5}{2}\right)^2$

Solution According to the rule for order of operations, we must first evaluate the numbers with exponents, then we divide, and finally we add.

$$32 \div \left(\frac{4}{3}\right)^2 + 75 \div \left(\frac{5}{2}\right)^2 = 32 \div \frac{16}{9} + 75 \div \frac{25}{4}$$

$$= 32 \cdot \frac{9}{16} + 75 \cdot \frac{4}{25}$$

$$= 18 + 12$$

$$= 30$$

11 How many blankets can the 4-H Club make with 12 yards of material?

EXAMPLE 11 A 4-H Club is making blankets to keep their lambs clean at the County Fair. If each blanket requires $\frac{3}{4}$ yard of material, how many blankets can they make from 9 yards of material?

Solution To answer this question we must divide 9 by $\frac{3}{4}$:

$$9 \div \frac{3}{4} = 9 \cdot \frac{4}{3}$$

$$= 3 \cdot 4$$

$$= 12$$

They can make 12 blankets from the 9 yards of material.

Answers

8 xy **9** 18 **10** 350 **11** 16

3.5 Addition and Subtraction with Fractions

Adding and subtracting fractions is actually just another application of the distributive property. The distributive property looks like this:

$$a(b + c) = a(b) + a(c)$$

where a, b, and c may be whole numbers or fractions. We will want to apply this property to expressions like

$$\frac{2}{7} + \frac{3}{7}$$

But before we do, we must make one additional observation about fractions.

The fraction $\frac{2}{7}$ can be written as $2 \cdot \frac{1}{7}$, since

$$2 \cdot \frac{1}{7} = \frac{2}{1} \cdot \frac{1}{7} = \frac{2}{7}$$

Likewise, the fraction $\frac{3}{7}$ can be written as $3 \cdot \frac{1}{7}$, since

$$3 \cdot \frac{1}{7} = \frac{3}{1} \cdot \frac{1}{7} = \frac{3}{7}$$

In general, we can say that the fraction $\frac{a}{b}$ can always be written as $a \cdot \frac{1}{b}$, because

$$a \cdot \frac{1}{b} = \frac{a}{1} \cdot \frac{1}{b} = \frac{a}{b}$$

To add the fractions $\frac{2}{7}$ and $\frac{3}{7}$, we simply rewrite each of them as we have done above and apply the distributive property. Here is how it works:

$$\frac{2}{7} + \frac{3}{7} = 2 \cdot \frac{1}{7} + 3 \cdot \frac{1}{7}$$ Rewrite each fraction

$$= (2 + 3) \cdot \frac{1}{7}$$ Apply the distributive property

$$= 5 \cdot \frac{1}{7}$$ Add 2 and 3 to get 5

$$= \frac{5}{7}$$ Rewrite $5 \cdot \frac{1}{7}$ as $\frac{5}{7}$

The fraction $\frac{5}{7}$ is the sum of $\frac{2}{7}$ and $\frac{3}{7}$. The steps above show why we add numerators *but do not add denominators*. Using this example as justification, we can write a rule for adding two fractions that have the same denominator.

RULE To add two fractions that have the same denominator, we add their numerators to get the numerator of the answer. The denominator in the answer is the same denominator as in the original fractions.

What we have here is the sum of the numerators placed over the *common denominator.* In symbols we have the following:

Note

Most people who have done any work with adding fractions know that you add fractions that have the same denominator by adding their numerators, but not their denominators. However, most people don't know why this works. The reason why we add numerators but not denominators is because of the distributive property. And that is what the discussion at the left is all about. If you really want to understand addition of fractions, pay close attention to this discussion.

If a, b, and c are integers and c is not equal to 0, then

$$\frac{a}{c} + \frac{b}{c} = \frac{a+b}{c}$$

This rule holds for subtraction as well. That is,

$$\frac{a}{c} - \frac{b}{c} = \frac{a-b}{c}$$

In Examples 1–4, find the sum or difference. (Add or subtract as indicated.) Reduce all answers to lowest terms. (Assume all variables represent nonzero numbers.)

EXAMPLE 1 $\frac{3}{8} + \frac{1}{8} = \frac{3+1}{8}$ Add numerators; keep the same denominator

$= \frac{4}{8}$ The sum of 3 and 1 is 4

$= \frac{1}{2}$ Reduce to lowest terms

EXAMPLE 2 $\frac{a+5}{8} - \frac{3}{8} = \frac{a+5-3}{8}$ Combine numerators; keep the same denominator

$= \frac{a+2}{8}$

EXAMPLE 3 $\frac{9}{x} - \frac{3}{x} = \frac{9-3}{x}$ Subtract numerators; keep the same denominator

$= \frac{6}{x}$ The difference of 9 and 3 is 6

EXAMPLE 4 $\frac{3}{7} + \frac{2}{7} - \frac{9}{7} = \frac{3+2-9}{7}$

$= \frac{-4}{7}$

$= -\frac{4}{7}$ Unlike signs give a negative answer

As Examples 1–4 indicate, addition and subtraction are simple, straightforward processes when all the fractions have the same denominator. We will now turn our attention to the process of adding fractions that have different denominators. In order to get started, we need the following definition:

DEFINITION The **least common denominator** (LCD) for a set of denominators is the smallest number that is exactly divisible by each denominator. (Note that, in some books, the least common denominator is also called the **least common multiple**.)

In other words, all the denominators of the fractions involved in a problem must divide into the least common denominator exactly. That is, they divide it without leaving a remainder.

Practice Problems

Find the sum or difference. Reduce all answers to lowest terms.

1 $\frac{3}{10} + \frac{1}{10}$

2 $\frac{a-5}{12} + \frac{3}{12}$

3 $\frac{8}{x} - \frac{5}{x}$

4 $\frac{5}{9} - \frac{8}{9} + \frac{5}{9}$

Answers

1 $\frac{2}{5}$ 2 $\frac{a-2}{12}$ 3 $\frac{3}{x}$ 4 $\frac{2}{9}$

EXAMPLE 5 Find the LCD for the fractions $\frac{5}{12}$ and $\frac{7}{18}$.

Solution The least common denominator for the denominators 12 and 18 must be the smallest number divisible by both 12 and 18. We can factor 12 and 18 completely and then build the LCD from these factors. Factoring 12 and 18 completely gives us

$$12 = 2 \cdot 2 \cdot 3 \qquad 18 = 2 \cdot 3 \cdot 3$$

Now, if 12 is going to divide the LCD exactly, then the LCD must have factors of $2 \cdot 2 \cdot 3$. If 18 is to divide it exactly, it must have factors of $2 \cdot 3 \cdot 3$. We don't need to repeat the factors that 12 and 18 have in common:

12 divides the LCD

$$\left.\begin{array}{l} 12 = 2 \cdot 2 \cdot 3 \\ 18 = 2 \cdot 3 \cdot 3 \end{array}\right\} \quad \text{LCD} = 2 \cdot 2 \cdot 3 \cdot 3 = 36$$

18 divides the LCD

The LCD for 12 and 18 is 36. It is the smallest number that is divisible by both 12 and 18; 12 divides it exactly three times, and 18 divides it exactly two times.

5 Find the LCD for the fractions $\frac{5}{18}$ and $\frac{3}{14}$.

Note
The ability to find least common denominators is very important in mathematics. The discussion here is a detailed explanation of how to do it.

We can use the results of Example 5 to find the sum of the fractions $\frac{5}{12}$ and $\frac{7}{18}$.

EXAMPLE 6 Add: $\frac{5}{12} + \frac{7}{18}$

Solution We can add fractions only when they have the same denominators. In Example 5, we found the LCD for $\frac{5}{12}$ and $\frac{7}{18}$ to be 36. We change $\frac{5}{12}$ and $\frac{7}{18}$ to equivalent fractions that have 36 for a denominator by applying Property 1 for fractions:

$$\frac{5}{12} = \frac{5 \cdot \mathbf{3}}{12 \cdot \mathbf{3}} = \frac{15}{36}$$

$$\frac{7}{18} = \frac{7 \cdot \mathbf{2}}{18 \cdot \mathbf{2}} = \frac{14}{36}$$

The fraction $\frac{15}{36}$ is equivalent to $\frac{5}{12}$, since it was obtained by multiplying both the numerator and the denominator by 3. Likewise, $\frac{14}{36}$ is equivalent to $\frac{7}{18}$, since it was obtained by multiplying the numerator and the denominator by 2. All we have left to do is to add numerators:

$$\frac{15}{36} + \frac{14}{36} = \frac{29}{36}$$

The sum of $\frac{5}{12}$ and $\frac{7}{18}$ is the fraction $\frac{29}{36}$. Let's write the complete problem again step by step.

$$\frac{5}{12} + \frac{7}{18} = \frac{5 \cdot \mathbf{3}}{12 \cdot \mathbf{3}} + \frac{7 + \mathbf{2}}{18 \cdot \mathbf{2}}$$ Rewrite each fraction as an equivalent fraction with denominator 36

$$= \frac{15}{36} + \frac{14}{36}$$

$$= \frac{29}{36}$$ Add numerators; keep the common denominator

6 Add: $\frac{5}{18} + \frac{3}{14}$

Answers
5 126 6 $\frac{31}{63}$

7 Find the LCD for $\frac{2}{9}$ and $\frac{4}{15}$.

8 Add: $\frac{2}{9} + \frac{4}{15}$

9 Subtract: $\frac{8}{25} - \frac{3}{20}$

Answers

7 45 8 $\frac{22}{45}$ 9 $\frac{17}{100}$

EXAMPLE 7 Find the LCD for $\dfrac{3}{4}$ and $\dfrac{1}{6}$.

Solution We factor 4 and 6 into products of prime factors and build the LCD from these factors:

$$\left.\begin{array}{l} 4 = 2 \cdot 2 \\ 6 = 2 \cdot 3 \end{array}\right\} \quad \text{LCD} = 2 \cdot 2 \cdot 3 = 12$$

The LCD is 12. Both denominators divide it exactly; 4 divides 12 exactly 3 times, and 6 divides 12 exactly 2 times.

EXAMPLE 8 Add: $\dfrac{3}{4} + \dfrac{1}{6}$

Solution In Example 7, we found that the LCD for these two fractions is 12. We begin by changing $\frac{3}{4}$ and $\frac{1}{6}$ to equivalent fractions with denominator 12:

$$\frac{3}{4} = \frac{3 \cdot \mathbf{3}}{4 \cdot \mathbf{3}} = \frac{9}{12}$$

$$\frac{1}{6} = \frac{1 \cdot \mathbf{2}}{6 \cdot \mathbf{2}} = \frac{2}{12}$$

The fraction $\frac{9}{12}$ is equal to the fraction $\frac{3}{4}$, since it was obtained by multiplying the numerator and the denominator of $\frac{3}{4}$ by 3. Likewise, $\frac{2}{12}$ is equivalent to $\frac{1}{6}$, since it was obtained by multiplying the numerator and the denominator of $\frac{1}{6}$ by 2. To complete the problem we add numerators:

$$\frac{9}{12} + \frac{2}{12} = \frac{11}{12}$$

The sum of $\frac{3}{4}$ and $\frac{1}{6}$ is $\frac{11}{12}$. Here is how the complete problem looks:

$$\frac{3}{4} + \frac{1}{6} = \frac{3 \cdot \mathbf{3}}{4 \cdot \mathbf{3}} + \frac{1 \cdot \mathbf{2}}{6 \cdot \mathbf{2}}$$ Rewrite each fraction as an equivalent fraction with denominator 12

$$= \frac{9}{12} + \frac{2}{12}$$

$$= \frac{11}{12}$$ Add numerators; keep the same denominator

EXAMPLE 9 Subtract: $\dfrac{7}{15} - \dfrac{3}{10}$

Solution Let's factor 15 and 10 completely and use these factors to build the LCD:

$$\left.\begin{array}{l} 15 = 3 \cdot 5 \\ 10 = 2 \cdot 5 \end{array}\right\} \quad \text{LCD} = 2 \cdot 3 \cdot 5 = 30$$

15 divides the LCD

10 divides the LCD

Changing to equivalent fractions and subtracting, we have

$$\frac{7}{15} - \frac{3}{10} = \frac{7 \cdot \mathbf{2}}{15 \cdot \mathbf{2}} - \frac{3 \cdot \mathbf{3}}{10 \cdot \mathbf{3}}$$ Rewrite as equivalent fractions with the LCD for the denominator

$$= \frac{14}{30} - \frac{9}{30}$$

$$= \frac{5}{30}$$ Subtract numerators; keep the LCD

$$= \frac{1}{6}$$ Reduce to lowest terms

As a summary of what we have done so far, and as a guide to working other problems, we now list the steps involved in adding and subtracting fractions with different denominators.

To Add or Subtract Any Two Fractions

Step 1 Factor each denominator completely and use the factors to build the LCD. (Remember, the LCD is the smallest number divisible by each of the denominators in the problem.)

Step 2 Rewrite each fraction as an equivalent fraction that has the LCD for its denominator. This is done by multiplying both the numerator and the denominator of the fraction in question by the appropriate whole number.

Step 3 Add or subtract the numerators of the fractions produced in Step 2. This is the numerator of the sum or difference. The denominator of the sum or difference is the LCD.

Step 4 Reduce the fraction produced in Step 3 to lowest terms if it is not already in lowest terms.

The idea behind adding or subtracting fractions is really very simple. We can only add or subtract fractions that have the same denominators. If the fractions we are trying to add or subtract do not have the same denominators, we rewrite each of them as an equivalent fraction with the LCD for a denominator.

Here are some additional examples of sums and differences of fractions.

EXAMPLE 10 Subtract: $\frac{x}{5} - \frac{1}{6}$

10 Subtract: $\frac{x}{4} - \frac{1}{5}$

Solution The LCD for 5 and 6 is their product, 30. We begin by rewriting each fraction with this common denominator:

$$\frac{x}{5} - \frac{1}{6} = \frac{x \cdot \mathbf{6}}{5 \cdot \mathbf{6}} - \frac{1 \cdot \mathbf{5}}{6 \cdot \mathbf{5}}$$

$$= \frac{6x}{30} - \frac{5}{30}$$

$$= \frac{6x - 5}{30}$$

Answer

10 $\frac{5x - 4}{20}$

11 Add: $\frac{1}{9} + \frac{1}{4} + \frac{1}{6}$

EXAMPLE 11 Add: $\frac{1}{6} + \frac{1}{8} + \frac{1}{4}$

Solution We begin by factoring the denominators completely and building the LCD from the factors that result:

$$\left.\begin{aligned} 6 &= 2 \cdot 3 \\ 8 &= 2 \cdot 2 \cdot 2 \\ 4 &= 2 \cdot 2 \end{aligned}\right\} \quad \text{LCD} = 2 \cdot 2 \cdot 2 \cdot 3 = 24$$

8 divides the LCD

4 divides the LCD

6 divides the LCD

We then change to equivalent fractions and add as usual:

$$\begin{aligned} \frac{1}{6} + \frac{1}{8} + \frac{1}{4} &= \frac{1 \cdot \mathbf{4}}{6 \cdot \mathbf{4}} + \frac{1 \cdot \mathbf{3}}{8 \cdot \mathbf{3}} + \frac{1 \cdot \mathbf{6}}{4 \cdot \mathbf{6}} \\ &= \frac{4}{24} + \frac{3}{24} + \frac{6}{24} \\ &= \frac{13}{24} \end{aligned}$$

12 Subtract: $2 - \frac{3}{4}$

EXAMPLE 12 Subtract: $3 - \frac{5}{6}$

Solution The denominators are 1 $\left(\text{because } 3 = \frac{3}{1}\right)$ and 6. The smallest number divisible by both 1 and 6 is 6.

$$\begin{aligned} 3 - \frac{5}{6} &= \frac{3}{1} - \frac{5}{6} \\ &= \frac{3 \cdot \mathbf{6}}{1 \cdot \mathbf{6}} - \frac{5}{6} \\ &= \frac{18}{6} - \frac{5}{6} \\ &= \frac{13}{6} \end{aligned}$$

13 Add: $\frac{5}{x} + \frac{2}{3}$

Note

In Example 13, it is understood that x cannot be 0. Do you know why?

EXAMPLE 13 Add: $\frac{4}{x} + \frac{2}{3}$

Solution The LCD for x and 3 is $3x$. We multiply the numerator and the denominator of the first fraction by 3, and the numerator and the denominator of the second fraction by x, to get two fractions with the same denominator. We then add the numerators:

$$\begin{aligned} \frac{4}{x} + \frac{2}{3} &= \frac{4 \cdot \mathbf{3}}{x \cdot \mathbf{3}} + \frac{2 \cdot \boldsymbol{x}}{3 \cdot \boldsymbol{x}} && \text{Change to equivalent fractions} \\ &= \frac{12}{3x} + \frac{2x}{3x} \\ &= \frac{12 + 2x}{3x} && \text{Add the numerators} \end{aligned}$$

Answers

11 $\frac{19}{36}$ **12** $\frac{5}{4}$ **13** $\frac{15 + 2x}{3x}$

3.6 Mixed-Number Notation

Introduction . . . If you are interested in the stock market, you know that stock prices are given in eighths. For example, on the day I am writing this introduction, one share of Intel Corporation is selling at \$$73\frac{5}{8}$, or seventy-three and five-eighths dollars. The number $73\frac{5}{8}$ is called a *mixed number.* It is the sum of a whole number and a proper fraction. With mixed-number notation, we leave out the addition sign.

Notation

A number such as $5\frac{3}{4}$ is called a *mixed number* and is equal to $5 + \frac{3}{4}$. It is simply the sum of the whole number 5 and the proper fraction $\frac{3}{4}$, written without a + sign. Here are some further examples:

$$2\tfrac{1}{8} = 2 + \frac{1}{8}, \qquad 6\tfrac{5}{9} = 6 + \frac{5}{9}, \qquad 11\tfrac{2}{3} = 11 + \frac{2}{3}$$

The notation used in writing mixed numbers (writing the whole number and the proper fraction next to each other) must always be interpreted as addition. It is a mistake to read $5\frac{3}{4}$ as meaning 5 times $\frac{3}{4}$. If we want to indicate multiplication, we must use parentheses or a multiplication symbol. That is:

$5\frac{3}{4}$ is not the same as $5\left(\frac{3}{4}\right)$

$5\frac{3}{4}$ is not the same as $5 \cdot \frac{3}{4}$

This implies addition (arrows to $5\frac{3}{4}$). These imply multiplication (arrows to $5\left(\frac{3}{4}\right)$ and $5 \cdot \frac{3}{4}$).

Changing Mixed Numbers to Improper Fractions

To change a mixed number to an improper fraction, we write the mixed number with the + sign showing and then add the two numbers, as we did earlier.

EXAMPLE 1 Change $2\frac{3}{4}$ to an improper fraction.

Solution

$2\frac{3}{4} = 2 + \frac{3}{4}$	Write the mixed number as a sum
$= \frac{2}{1} + \frac{3}{4}$	Show that the denominator of 2 is 1
$= \frac{\mathbf{4} \cdot 2}{\mathbf{4} \cdot 1} + \frac{3}{4}$	Multiply the numerator and the denominator of $\frac{2}{1}$ by 4 so both fractions will have the same denominator
$= \frac{8}{4} + \frac{3}{4}$	
$= \frac{11}{4}$	Add the numerators; keep the common denominator

The mixed number $2\frac{3}{4}$ is equal to the improper fraction $\frac{11}{4}$. Figure 1 further illustrates the equivalence of $2\frac{3}{4}$ and $\frac{11}{4}$.

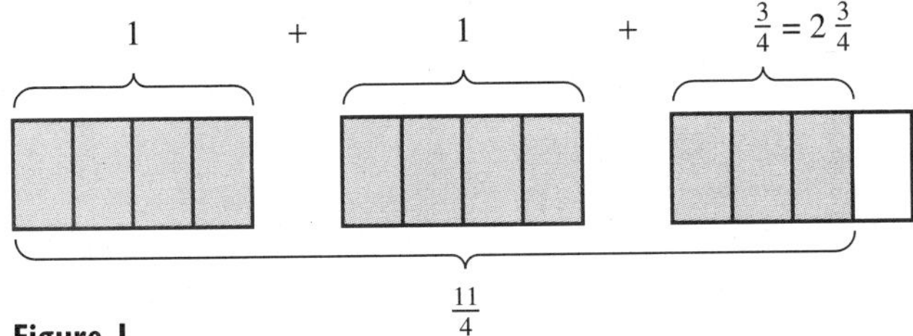

Figure 1

Practice Problems

1 Change $5\frac{2}{3}$ to an improper fraction.

Answer

1 $\frac{17}{3}$

2 Change $3\frac{1}{6}$ to an improper fraction.

EXAMPLE 2 Change $2\frac{1}{8}$ to an improper fraction.

Solution $2\frac{1}{8} = 2 + \frac{1}{8}$ Write as addition

$= \frac{2}{1} + \frac{1}{8}$ Write the whole number 2 as a fraction

$= \frac{\mathbf{8} \cdot 2}{\mathbf{8} \cdot 1} + \frac{1}{8}$ Change $\frac{2}{1}$ to a fraction with denominator 8

$= \frac{16}{8} + \frac{1}{8}$

$= \frac{17}{8}$ Add the numerators

If we look closely at Examples 1 and 2, we can see a shortcut that will let us change a mixed number to an improper fraction without so many steps. Here is the shortcut:

Shortcut: To change a mixed number to an improper fraction, simply multiply the denominator of the fraction part of the mixed number by the whole number and add the result to the numerator of the fraction. The result is the numerator of the improper fraction we are looking for. The denominator is the same as the original denominator.

3 Use the shortcut to change $5\frac{2}{3}$ to an improper fraction.

EXAMPLE 3 Use the shortcut to change $5\frac{3}{4}$ to an improper fraction.

Solution

1 First, we multiply 4×5 to get 20.

2 Next, we add 20 to 3 to get 23.

3 The improper fraction equal to $5\frac{3}{4}$ is $\frac{23}{4}$.

Here is a diagram showing what we have done:

Step 1 Multiply $4 \times 5 = 20$.

$5\frac{3}{4}$

Step 2 Add $20 + 3 = 23$.

Mathematically, our shortcut is written like this:

$$5\frac{3}{4} = \frac{(4 \cdot 5) + 3}{4} = \frac{20 + 3}{4} = \frac{23}{4}$$

The result will always have the same denominator as the original mixed number

The shortcut shown in Example 3 works because the whole-number part of a mixed number can always be written with a denominator of 1. Therefore, the LCD for a whole number and fraction will always be the denominator of the fraction. That is why we multiply the whole number by the denominator of the fraction:

$$5\frac{3}{4} = 5 + \frac{3}{4} = \frac{5}{1} + \frac{3}{4} = \frac{\mathbf{4} \cdot 5}{\mathbf{4} \cdot 1} + \frac{3}{4} = \frac{4 \cdot 5 + 3}{4} = \frac{23}{4}$$

Answers

2 $\frac{19}{6}$ **3** $\frac{17}{3}$

EXAMPLE 4 Change $6\frac{5}{9}$ to an improper fraction.

Solution Using the first method, we have

$$6\tfrac{5}{9} = 6 + \frac{5}{9} = \frac{6}{1} + \frac{5}{9} = \frac{\mathbf{9 \cdot 6}}{\mathbf{9 \cdot 1}} + \frac{5}{9} = \frac{54}{9} + \frac{5}{9} = \frac{59}{9}$$

Using the shortcut method, we have

$$6\tfrac{5}{9} = \frac{(9 \cdot 6) + 5}{9} = \frac{54 + 5}{9} = \frac{59}{9}$$

4 Change $6\frac{4}{9}$ to an improper fraction.

Calculator Note

The sequence of keys to press on a calculator to obtain the numerator in Example 4 looks like this:

9 [×] 6 [+] 5 [=]

Changing Improper Fractions to Mixed Numbers

To change an improper fraction to a mixed number, we divide the numerator by the denominator. The result is used to write the mixed number.

EXAMPLE 5 Change $\frac{11}{4}$ to a mixed number.

Solution Dividing 11 by 4 gives us

$$\begin{array}{r} 2 \\ 4\overline{)11} \\ \underline{8} \\ 3 \end{array}$$

We see that 4 goes into 11 two times with 3 for a remainder. We write this result as

$$\frac{11}{4} = 2 + \frac{3}{4} = 2\tfrac{3}{4}$$

The improper fraction $\frac{11}{4}$ is equivalent to the mixed number $2\frac{3}{4}$.

5 Change $\frac{11}{3}$ to a mixed number.

Note

This division process shows us how many ones are in $\frac{11}{4}$ and, when the ones are taken out, how many fourths are left.

Here are some further examples that illustrate the same procedures.

EXAMPLE 6 $\frac{10}{3}$:

$$\begin{array}{r} \mathbf{3} \\ 3\overline{)10} \\ \underline{9} \\ \mathbf{1} \end{array} \quad \text{so} \quad \frac{10}{3} = \mathbf{3} + \frac{\mathbf{1}}{3} = 3\tfrac{1}{3}$$

EXAMPLE 7 $\frac{208}{24}$:

$$\begin{array}{r} \mathbf{8} \\ 24\overline{)208} \\ \underline{192} \\ \mathbf{16} \end{array} \quad \text{so} \quad \frac{208}{24} = \mathbf{8} + \frac{\mathbf{16}}{24} = 8 + \frac{2}{3} = 8\tfrac{2}{3}$$

(Reduce to lowest terms: $\frac{16}{24} \to \frac{2}{3}$)

Change each improper fraction to a mixed number.

6 $\frac{14}{5}$

7 $\frac{207}{26}$

Notice that in Example 7 we had to reduce the fraction part of the mixed number to lowest terms to get the final result. We could have done this to begin with and the result would be the same:

$$\frac{208}{24} = \frac{26}{3}: \quad \begin{array}{r} 8 \\ 3\overline{)26} \\ \underline{24} \\ 2 \end{array} \quad \text{so} \quad \frac{26}{3} = 8\tfrac{2}{3}$$

(Reduce: $\frac{208}{24} \to \frac{26}{3}$)

Answers

4 $\frac{58}{9}$ 5 $3\frac{2}{3}$ 6 $2\frac{4}{5}$ 7 $7\frac{25}{26}$

In the first part of this section, we changed mixed numbers to improper fractions. For example we changed $2\frac{3}{4}$ to an improper fraction by adding 2 and $\frac{3}{4}$ (see Example 1). An extension of this concept to algebra would be to add 2 and $\frac{3}{x}$.

8 Add: $3 + \frac{2}{x}$

EXAMPLE 8 Add: $2 + \frac{3}{x}$

Solution We can write 2 as $\frac{2}{1}$:

$$2 + \frac{3}{x} = \frac{2}{1} + \frac{3}{x}$$

Now, the LCD for 1 and x is $1x$ or just x. To change to equivalent fractions, we multiply the numerator and the denominator of $\frac{2}{1}$ by x.

$$\frac{2}{1} + \frac{3}{x} = \frac{2 \cdot \boldsymbol{x}}{1 \cdot \boldsymbol{x}} + \frac{3}{x} \qquad \text{The LCD is } x$$

$$= \frac{2x}{x} + \frac{3}{x}$$

$$= \frac{2x + 3}{x} \qquad \text{Add numerators}$$

9 Subtract: $x - \frac{2}{3}$

EXAMPLE 9 Subtract: $x - \frac{3}{4}$

Solution This time we write x as $\frac{x}{1}$. The LCD for 1 and 4 is 4.

$$x - \frac{3}{4} = \frac{x}{1} - \frac{3}{4} \qquad \text{Write } x \text{ as } \frac{x}{1}$$

$$= \frac{\mathbf{4} \cdot x}{\mathbf{4} \cdot 1} - \frac{3}{4} \qquad \text{The LCD is 4}$$

$$= \frac{4x}{4} - \frac{3}{4}$$

$$= \frac{4x - 3}{4} \qquad \text{Subtract numerators}$$

As a final note in this section, we should mention that negative mixed numbers are thought of this way:

$$-3\tfrac{2}{5} = -3 - \frac{2}{5}$$

Both the whole-number part and the fraction part of the mixed number are negative when the mixed number is negative. It would be a mistake to think of mixed numbers this way:

$$-3\tfrac{2}{5} = -3 + \frac{2}{5} \qquad \text{Mistake}$$

Because negative mixed numbers can be difficult to work with in algebra, you are more likely to see improper fractions than mixed numbers if you go on to take an algebra class.

Answers

8 $\frac{3x + 2}{x}$ **9** $\frac{3x - 2}{3}$

3.7 Multiplication and Division with Mixed Numbers

Introduction . . . A recipe for making 36 chocolate chip cookies calls for $1\frac{1}{2}$ cups of flour. If you are making 48 cookies, how many cups of flour should you use? The answer to this question can be found in the problem

$$\frac{48}{36} \cdot 1\frac{1}{2}$$

which requires that we multiply a fraction and a mixed number.

The procedures for multiplying and dividing mixed numbers are the same as those we used in Sections 3.3 and 3.4 to multiply and divide fractions. The only additional work involved is in changing the mixed numbers to improper fractions before we actually multiply or divide.

Practice Problems

1 Multiply: $2\frac{3}{4} \cdot 4\frac{1}{3}$

EXAMPLE 1 Multiply: $2\frac{3}{4} \cdot 3\frac{1}{5}$

Solution We begin by changing each mixed number to an improper fraction:

$$2\frac{3}{4} = \frac{11}{4} \quad \text{and} \quad 3\frac{1}{5} = \frac{16}{5}$$

Using the resulting improper fractions, we multiply as usual. (That is, we multiply numerators and multiply denominators.)

$$\begin{aligned} \frac{11}{4} \cdot \frac{16}{5} &= \frac{11 \cdot 16}{4 \cdot 5} \\ &= \frac{11 \cdot \cancel{4} \cdot 4}{\cancel{4} \cdot 5} \\ &= \frac{44}{5} \quad \text{or} \quad 8\frac{4}{5} \end{aligned}$$

2 Multiply: $2 \cdot 3\frac{5}{8}$

EXAMPLE 2 Multiply: $3 \cdot 4\frac{5}{8}$

Solution Writing each number as an improper fraction, we have

$$3 = \frac{3}{1} \quad \text{and} \quad 4\frac{5}{8} = \frac{37}{8}$$

The complete problem looks like this:

$$\begin{aligned} 3 \cdot 4\frac{5}{8} &= \frac{3}{1} \cdot \frac{37}{8} && \text{Change to improper fractions} \\ &= \frac{111}{8} && \text{Multiply numerators and multiply denominators} \\ &= 13\frac{7}{8} && \text{Write the answer as a mixed number} \end{aligned}$$

Note

As you can see, once you have changed each mixed number to an improper fraction, you multiply the resulting fractions the same way you did in Section 3.3.

Answers

1 $11\frac{11}{12}$ **2** $7\frac{1}{4}$

Dividing mixed numbers also requires that we change all mixed numbers to improper fractions before we actually do the division.

3 Divide: $1\frac{3}{5} \div 3\frac{2}{5}$

EXAMPLE 3 Divide: $1\frac{3}{5} \div 2\frac{4}{5}$

Solution We begin by rewriting each mixed number as an improper fraction:

$$1\frac{3}{5} = \frac{8}{5} \quad \text{and} \quad 2\frac{4}{5} = \frac{14}{5}$$

We then divide by the same method we used in Section 3.4. Remember? We multiply by the reciprocal of the divisor. Here is the complete problem:

$$\begin{aligned}
1\frac{3}{5} \div 2\frac{4}{5} &= \frac{8}{5} \div \frac{14}{5} && \text{Change to improper fractions} \\
&= \frac{8}{5} \cdot \frac{5}{14} && \text{To divide by } \tfrac{14}{5}\text{, multiply by } \tfrac{5}{14} \\
&= \frac{8 \cdot 5}{5 \cdot 14} && \text{Multiply numerators and multiply denominators} \\
&= \frac{4 \cdot \cancel{2} \cdot \cancel{5}}{\cancel{5} \cdot \cancel{2} \cdot 7} && \text{Divide out factors common to the numerator and denominator} \\
&= \frac{4}{7} && \text{Answer in lowest terms}
\end{aligned}$$

4 Divide: $4\frac{5}{8} \div 2$

EXAMPLE 4 Divide: $5\frac{9}{10} \div 2$

Solution We change to improper fractions and proceed as usual:

$$\begin{aligned}
5\frac{9}{10} \div 2 &= \frac{59}{10} \div \frac{2}{1} && \text{Write each number as an improper fraction} \\
&= \frac{59}{10} \cdot \frac{1}{2} && \text{Write division as multiplication by the reciprocal} \\
&= \frac{59}{20} && \text{Multiply numerators and multiply denominators} \\
&= 2\frac{19}{20} && \text{Change to a mixed number}
\end{aligned}$$

Answers

3 $\frac{8}{17}$ **4** $2\frac{5}{16}$

3.8 Addition and Subtraction with Mixed Numbers

Introduction . . . In March 1995, rumors that Michael Jordan would return to basketball sent stock prices for the companies whose products he endorses higher. The price of one share of General Mills, the maker of Wheaties, which Michael Jordan endorses, went from $\$60\frac{1}{2}$ to $\$63\frac{3}{8}$. To find the increase in the price of this stock, we must be able to subtract mixed numbers.

The notation we use for mixed numbers is especially useful for addition and subtraction. When adding and subtracting mixed numbers, we will assume you recall how to go about finding a least common denominator (LCD). (If you don't remember, then review Section 3.5.)

EXAMPLE 1 Add: $3\frac{2}{3} + 4\frac{1}{5}$

Practice Problems

1 Add: $3\frac{2}{3} + 2\frac{1}{4}$

Solution We begin by writing each mixed number showing the + sign. We then apply the commutative and associative properties to rearrange the order and grouping:

$$3\tfrac{2}{3} + 4\tfrac{1}{5} = 3 + \frac{2}{3} + 4 + \frac{1}{5}$$ Expand each number to show the + sign

$$= 3 + 4 + \frac{2}{3} + \frac{1}{5}$$ Commutative property

$$= (3 + 4) + \left(\frac{2}{3} + \frac{1}{5}\right)$$ Associative property

$$= 7 + \left(\frac{\mathbf{5} \cdot 2}{\mathbf{5} \cdot 3} + \frac{\mathbf{3} \cdot 1}{\mathbf{3} \cdot 5}\right)$$ Add 3 + 4 = 7; then multiply to get the LCD

$$= 7 + \left(\frac{10}{15} + \frac{3}{15}\right)$$ Write each fraction with the LCD

$$= 7 + \frac{13}{15}$$ Add the numerators

$$= 7\tfrac{13}{15}$$ Write the answer in mixed-number notation

As you can see, we obtain our result by adding the whole-number parts (3 + 4 = 7) and the fraction parts $\left(\frac{2}{3} + \frac{1}{5} = \frac{13}{15}\right)$ of each mixed number. Knowing this, we can save ourselves some writing by doing the same problem in columns:

$$\begin{array}{rcccl} 3\frac{2}{3} & = & 3\frac{2\cdot 5}{3\cdot 5} & = & 3\frac{10}{15} \\ +4\frac{1}{5} & = & 4\frac{1\cdot 3}{5\cdot 3} & = & 4\frac{3}{15} \\ \hline & & & & 7\frac{13}{15} \end{array}$$

Add whole numbers. Then add fractions.

Write each fraction with LCD 15

Note

You should try both methods given in Example 1 on Practice Problem 1.

The second method shown above requires less writing and lends itself to mixed-number notation. We will use this method for the rest of this section.

Answer

1 $5\frac{11}{12}$

2 Add: $5\frac{3}{4} + 6\frac{4}{5}$

EXAMPLE 2 Add: $5\frac{3}{4} + 9\frac{5}{6}$

Solution The LCD for 4 and 6 is 12. Writing the mixed numbers in a column and then adding looks like this:

$$\begin{array}{rcrcr} 5\frac{3}{4} & = & 5\frac{3\cdot 3}{4\cdot 3} & = & 5\frac{9}{12} \\ +9\frac{5}{6} & = & 9\frac{5\cdot 2}{6\cdot 2} & = & 9\frac{10}{12} \\ \hline & & & & 14\frac{19}{12} \end{array}$$

Note
Once you see how to change from a whole number and an improper fraction to a whole number and a proper fraction, you will be able to do this step without showing any work.

The fraction part of the answer is an improper fraction. We rewrite it as a whole number and a proper fraction:

$$\begin{aligned} 14\tfrac{19}{12} &= 14 + \frac{19}{12} && \text{Write the mixed number with a + sign} \\ &= 14 + 1\tfrac{7}{12} && \text{Write } \tfrac{19}{12} \text{ as a mixed number} \\ &= 15\tfrac{7}{12} && \text{Add 14 and 1} \end{aligned}$$

3 Add: $6\frac{3}{4} + 2\frac{7}{8}$

EXAMPLE 3 Add: $5\frac{2}{3} + 6\frac{8}{9}$

Solution

$$\begin{array}{rcrcr} 5\frac{2}{3} & = & 5\frac{2\cdot 3}{3\cdot 3} & = & 5\frac{6}{9} \\ +6\frac{8}{9} & = & 6\frac{8}{9} & = & 6\frac{8}{9} \\ \hline & & & & 11\frac{14}{9} = 12\frac{5}{9} \end{array}$$

The last step involves writing $\frac{14}{9}$ as $1\frac{5}{9}$ and then adding 11 and 1 to get 12.

4 Add: $2\frac{1}{3} + 1\frac{1}{4} + 3\frac{11}{12}$

EXAMPLE 4 Add: $3\frac{1}{4} + 2\frac{3}{5} + 1\frac{9}{10}$

Solution The LCD is 20. We rewrite each fraction as an equivalent fraction with denominator 20 and add:

$$\begin{array}{rcrcr} 3\frac{1}{4} & = & 3\frac{1\cdot 5}{4\cdot 5} & = & 3\frac{5}{20} \\ 2\frac{3}{5} & = & 2\frac{3\cdot 4}{5\cdot 4} & = & 2\frac{12}{20} \\ +1\frac{9}{10} & = & 1\frac{9\cdot 2}{10\cdot 2} & = & 1\frac{18}{20} \\ \hline & & & & 6\frac{35}{20} = 7\frac{15}{20} = 7\frac{3}{4} \end{array}$$

Reduce to lowest terms

$$\frac{35}{20} = 1\frac{15}{20}$$

Change to a mixed number

We should note here that we could have worked each of the first four examples in this section by first changing each mixed number to an improper fraction and then adding as we did in Section 3.5. To illustrate, if we were to work Example 4 this way, it would look like this:

$$\begin{aligned} 3\tfrac{1}{4} + 2\tfrac{3}{5} + 1\tfrac{9}{10} &= \frac{13}{4} + \frac{13}{5} + \frac{19}{10} && \text{Change to improper fractions} \\ &= \frac{13\cdot \mathbf{5}}{4\cdot \mathbf{5}} + \frac{13\cdot \mathbf{4}}{5\cdot \mathbf{4}} + \frac{19\cdot \mathbf{2}}{10\cdot \mathbf{2}} && \text{LCD is 20} \\ &= \frac{65}{20} + \frac{52}{20} + \frac{38}{20} && \text{Equivalent fractions} \\ &= \frac{155}{20} && \text{Add numerators} \\ &= 7\tfrac{15}{20} = 7\tfrac{3}{4} && \text{Change to a mixed number and reduce} \end{aligned}$$

Answers
2 $12\frac{11}{20}$ **3** $9\frac{5}{8}$ **4** $7\frac{1}{2}$

As you can see, the result is the same as the result we obtained in Example 4.

There are advantages to both methods. The method just shown works well when the whole-number parts of the mixed numbers are small. The vertical method shown in Examples 1–4 works well when the whole-number parts of the mixed numbers are large.

Subtraction with mixed numbers is very similar to addition with mixed numbers.

EXAMPLE 5 Subtract: $3\frac{9}{10} - 1\frac{3}{10}$

Solution Since the denominators are the same, we simply subtract the whole numbers and subtract the fractions:

$$\begin{array}{r} 3\frac{9}{10} \\ -\,1\frac{3}{10} \\ \hline 2\frac{6}{10} = 2\frac{3}{5} \end{array}$$

Reduce to lowest terms

5 Subtract: $4\frac{7}{8} - 1\frac{5}{8}$

EXAMPLE 6 Subtract: $12\frac{7}{10} - 8\frac{3}{5}$

Solution The common denominator is 10. We must rewrite $\frac{3}{5}$ as an equivalent fraction with denominator 10:

$$\begin{array}{rcrcr} 12\frac{7}{10} & = & 12\frac{7}{10} & = & 12\frac{7}{10} \\ -\;8\frac{3}{5} & = & -\;8\frac{3\cdot 2}{5\cdot 2} & = & -\;8\frac{6}{10} \\ \hline & & & & 4\frac{1}{10} \end{array}$$

6 Subtract: $12\frac{7}{10} - 7\frac{2}{5}$

When the fraction we are subtracting is larger than the fraction we are subtracting it from, it is necessary to borrow during subtraction of mixed numbers. This is the same idea we used when borrowing was necessary in subtraction of whole numbers.

EXAMPLE 7 Subtract: $10 - 5\frac{2}{7}$

Solution In order to have a fraction from which to subtract $\frac{2}{7}$, we borrow 1 from 10 and rewrite the 1 we borrow as $\frac{7}{7}$. The process looks like this:

$$\begin{array}{rcr} 10 & = & 9\frac{7}{7} \\ -\;5\frac{2}{7} & = & -5\frac{2}{7} \\ \hline & & 4\frac{5}{7} \end{array}$$

← We rewrite 10 as $9 + 1$, which is $9 + \frac{7}{7} = 9\frac{7}{7}$. Then we can subtract as usual.

7 Subtract: $10 - 5\frac{4}{7}$

Note

Convince yourself that 10 is the same as $9\frac{7}{7}$. The reason we choose to write the 1 we borrowed as $\frac{7}{7}$ is that the fraction we eventually subtracted from $\frac{7}{7}$ was $\frac{2}{7}$. Both fractions must have the same denominator, 7, so that we can subtract.

EXAMPLE 8 Subtract: $8\frac{1}{4} - 3\frac{3}{4}$

Solution Since $\frac{3}{4}$ is larger than $\frac{1}{4}$, we again need to borrow 1 from the whole number. The 1 that we borrow from the 8 is rewritten as $\frac{4}{4}$, since 4 is the denominator of both fractions:

$$\begin{array}{rcr} 8\frac{1}{4} & = & 7\frac{5}{4} \\ -\,3\frac{3}{4} & = & -\,3\frac{3}{4} \\ \hline & & 4\frac{2}{4} = 4\frac{1}{2} \end{array}$$

← Borrow 1 in the form $\frac{4}{4}$; then $\frac{4}{4} + \frac{1}{4} = \frac{5}{4}$

Reduce to lowest terms

8 Subtract: $6\frac{1}{3} - 2\frac{2}{3}$

Answers

5 $3\frac{1}{4}$ **6** $5\frac{3}{10}$ **7** $4\frac{3}{7}$ **8** $3\frac{2}{3}$

9 Subtract: $6\frac{3}{4} - 2\frac{5}{6}$

EXAMPLE 9 Subtract: $4\frac{3}{4} - 1\frac{5}{6}$

Solution This is about as complicated as it gets with subtraction of mixed numbers. We begin by rewriting each fraction with the common denominator 12:

$$\begin{array}{rcrcr} 4\frac{3}{4} & = & 4\frac{3\cdot 3}{4\cdot 3} & = & 4\frac{9}{12} \\ -1\frac{5}{6} & = & -1\frac{5\cdot 2}{6\cdot 2} & = & -1\frac{10}{12} \end{array}$$

Since $\frac{10}{12}$ is larger than $\frac{9}{12}$, we must borrow 1 from 4 in the form $\frac{12}{12}$ before we subtract:

$$\begin{array}{rcr} 4\frac{9}{12} & = & 3\frac{21}{12} \\ -1\frac{10}{12} & = & -1\frac{10}{12} \\ \hline & & 2\frac{11}{12} \end{array} \longleftarrow 4 = 3 + 1 = 3 + \frac{12}{12}, \text{ so}$$

$$\begin{aligned} 4\tfrac{9}{12} &= \left(3 + \tfrac{12}{12}\right) + \tfrac{9}{12} \\ &= 3 + \left(\tfrac{12}{12} + \tfrac{9}{12}\right) \\ &= 3 + \tfrac{21}{12} \\ &= 3\tfrac{21}{12} \end{aligned}$$

10 Subtract: $17\frac{1}{8} - 12\frac{4}{5}$

EXAMPLE 10 Subtract: $15\frac{1}{10} - 11\frac{2}{3}$

Solution The LCD is 30.

$$\begin{array}{rcrcr} 15\frac{1}{10} & = & 15\frac{1\cdot 3}{10\cdot 3} & = & 15\frac{3}{30} \\ -11\frac{2}{3} & = & -11\frac{2\cdot 10}{3\cdot 10} & = & -11\frac{20}{30} \end{array}$$

Since $\frac{20}{30}$ is larger than $\frac{3}{30}$, we must borrow 1 from 15:

$$\begin{array}{rcr} 15\frac{3}{30} & = & 14\frac{33}{30} \\ -11\frac{20}{30} & = & -11\frac{20}{30} \\ \hline & & 3\frac{13}{30} \end{array}$$

Borrow 1 from 15 in the form $\frac{30}{30}$. Then add $\frac{30}{30}$ to $\frac{3}{30}$ to get $\frac{33}{30}$.

11 Using the table in Example 11, how much money would you lose if you bought 10,000 shares of the stock on Tuesday and sold them all on Wednesday?

EXAMPLE 11 The table below shows the selling price of one share of HBJ stock during each trading day of one week in 1990.

M	T	W	T	F
$2\frac{3}{8}$	$2\frac{1}{8}$	$1\frac{15}{16}$	2	$2\frac{3}{4}$

If you bought 1,000 shares on Monday and sold them all on Friday, how much money would you make?

Solution The simplest way to work this problem is to find the difference in the price of the stock on Friday and Monday, and then multiply that difference by 1,000:

$$1{,}000\left(2\tfrac{3}{4} - 2\tfrac{3}{8}\right) = 1{,}000\left(\frac{3}{8}\right) = \frac{3{,}000}{8} = \$375$$

Note that, when you buy and sell stock like this, you must also pay a commission to a stockbroker when you buy the stock and when you sell it. So, your actual profit would be \$375, less what you pay in commission. Also, you will have to pay income tax on your profit, which will reduce it even further.

Answers
9 $3\frac{11}{12}$ **10** $4\frac{13}{40}$
11 You would lose \$1,875.

3.9 Combinations of Operations, and Complex Fractions

The problems in this section all involve combinations of the four basic operations we've already discussed—addition, subtraction, multiplication, and division—involving mixed numbers and fractions.

EXAMPLE 1 Simplify the expression: $5 + \left(2\frac{1}{2}\right)\left(3\frac{2}{3}\right)$

Solution The rule for order of operations indicates that we should multiply $2\frac{1}{2}$ times $3\frac{2}{3}$ and then add 5 to the result:

$$5 + \left(2\tfrac{1}{2}\right)\left(3\tfrac{2}{3}\right) = 5 + \left(\frac{5}{2}\right)\left(\frac{11}{3}\right)$$ Change the mixed numbers to improper fractions

$$= 5 + \frac{55}{6}$$ Multiply the improper fractions

$$= \frac{30}{6} + \frac{55}{6}$$ Write 5 as $\frac{30}{6}$ so both numbers have the same denominator

$$= \frac{85}{6}$$ Add fractions by adding their numerators

$$= 14\tfrac{1}{6}$$ Write the answer as a mixed number

EXAMPLE 2 Simplify: $\left(\frac{3}{4} + \frac{5}{8}\right)\left(2\frac{3}{8} + 1\frac{1}{4}\right)$

Solution We begin by combining the numbers inside the parentheses:

$$\frac{3}{4} + \frac{5}{8} = \frac{3 \cdot \mathbf{2}}{4 \cdot \mathbf{2}} + \frac{5}{8}$$ Write each fraction as an equivalent fraction with LCD = 8

$$= \frac{6}{8} + \frac{5}{8}$$

$$= \frac{11}{8}$$

$$\begin{array}{r} 2\frac{3}{8} \\ +1\frac{1}{4} \\ \hline \end{array} = \begin{array}{r} 2\frac{3}{8} \\ +1\frac{1\cdot 2}{4\cdot 2} \\ \hline \end{array} = \begin{array}{r} 2\frac{3}{8} \\ +1\frac{2}{8} \\ \hline 3\frac{5}{8} \end{array}$$

Now that we have combined the expressions inside the parentheses, we can complete the problem by multiplying the results:

$$\left(\frac{3}{4} + \frac{5}{8}\right)\left(2\tfrac{3}{8} + 1\tfrac{1}{4}\right) = \left(\frac{11}{8}\right)\left(3\tfrac{5}{8}\right)$$

$$= \frac{11}{8} \cdot \frac{29}{8}$$ Change $3\frac{5}{8}$ to an improper fraction

$$= \frac{319}{64}$$ Multiply fractions

$$= 4\tfrac{63}{64}$$ Write the answer as a mixed number

Practice Problems

1 Simplify the expression:
$4 + \left(1\frac{1}{2}\right)\left(2\frac{3}{4}\right)$

2 Simplify:
$\left(\frac{2}{3} + \frac{1}{6}\right)\left(2\frac{5}{6} + 1\frac{1}{3}\right)$

Answers
1 $8\frac{1}{8}$ **2** $3\frac{17}{36}$

3 Simplify: $\frac{3}{7} + \frac{1}{3}\left(1\frac{1}{2} + 4\frac{1}{2}\right)^2$

EXAMPLE 3 Simplify: $\frac{3}{5} + \frac{1}{2}\left(3\frac{2}{3} + 4\frac{1}{3}\right)^2$

Solution We begin by combining the expressions inside the parentheses:

$$\frac{3}{5} + \frac{1}{2}\left(3\frac{2}{3} + 4\frac{1}{3}\right)^2 = \frac{3}{5} + \frac{1}{2}(8)^2 \quad \text{The sum inside the parentheses is 8}$$

$$= \frac{3}{5} + \frac{1}{2}(64) \quad \text{The square of 8 is 64}$$

$$= \frac{3}{5} + 32 \quad \frac{1}{2} \text{ of 64 is 32}$$

$$= 32\frac{3}{5} \quad \text{The result is a mixed number}$$

Complex Fractions

DEFINITION A **complex fraction** is a fraction in which the numerator and/or the denominator are themselves fractions or combinations of fractions.

Each of the following is a complex fraction:

$$\frac{\frac{3}{4}}{\frac{5}{6}}, \quad \frac{3 + \frac{1}{2}}{2 - \frac{3}{4}}, \quad \frac{\frac{1}{2} + \frac{2}{3}}{\frac{3}{4} - \frac{1}{6}}$$

Simplify each complex fraction as much as possible.

4 $\dfrac{\frac{2}{3}}{\frac{5}{9}}$

EXAMPLE 4 Simplify: $\dfrac{\frac{3}{4}}{\frac{5}{6}}$

Solution This is actually the same as the problem $\frac{3}{4} \div \frac{5}{6}$, because the bar between $\frac{3}{4}$ and $\frac{5}{6}$ indicates division. Therefore, it must be true that

$$\frac{\frac{3}{4}}{\frac{5}{6}} = \frac{3}{4} \div \frac{5}{6}$$

$$= \frac{3}{4} \cdot \frac{6}{5}$$

$$= \frac{18}{20}$$

$$= \frac{9}{10}$$

Answers
3 $12\frac{3}{7}$ **4** $1\frac{1}{5}$

EXAMPLE 5 Simplify: $\dfrac{\dfrac{1}{2}+\dfrac{2}{3}}{\dfrac{3}{4}-\dfrac{1}{6}}$

Solution Let's decide to call the numerator of this complex fraction the *top* of the fraction and its denominator the *bottom* of the complex fraction. It will be less confusing if we name them this way. The LCD for all the denominators on the top and bottom is 12, so we can multiply the top and bottom of this complex fraction by 12 and be sure all the denominators will divide it exactly. This will leave us with only whole numbers on the top and bottom:

$$\dfrac{\dfrac{1}{2}+\dfrac{2}{3}}{\dfrac{3}{4}-\dfrac{1}{6}} = \dfrac{\mathbf{12}\left(\dfrac{1}{2}+\dfrac{2}{3}\right)}{\mathbf{12}\left(\dfrac{3}{4}-\dfrac{1}{6}\right)}$$ Multiply the top and bottom by the LCD

$$= \dfrac{\mathbf{12}\cdot\dfrac{1}{2}+\mathbf{12}\cdot\dfrac{2}{3}}{\mathbf{12}\cdot\dfrac{3}{4}-\mathbf{12}\cdot\dfrac{1}{6}}$$ Distributive property

$$= \dfrac{6+8}{9-2}$$ Multiply each fraction by 12

$$= \dfrac{14}{7}$$ Add on top and subtract on bottom

$$= 2$$ Reduce to lowest terms

The problem can be worked in another way also. We can simplify the top and bottom of the complex fraction separately. Simplifying the top, we have

$$\frac{1}{2}+\frac{2}{3}=\frac{1\cdot\mathbf{3}}{2\cdot\mathbf{3}}+\frac{2\cdot\mathbf{2}}{3\cdot\mathbf{2}}=\frac{3}{6}+\frac{4}{6}=\frac{7}{6}$$

Simplifying the bottom, we have

$$\frac{3}{4}-\frac{1}{6}=\frac{3\cdot\mathbf{3}}{4\cdot\mathbf{3}}-\frac{1\cdot\mathbf{2}}{6\cdot\mathbf{2}}=\frac{9}{12}-\frac{2}{12}=\frac{7}{12}$$

We now write the original complex fraction again using the simplified expressions for the top and bottom. Then we proceed as we did in Example 4.

$$\dfrac{\dfrac{1}{2}+\dfrac{2}{3}}{\dfrac{3}{4}-\dfrac{1}{6}} = \dfrac{\dfrac{7}{6}}{\dfrac{7}{12}}$$

$$= \frac{7}{6}\div\frac{7}{12}$$ The divisor is $\frac{7}{12}$

$$= \frac{7}{6}\cdot\frac{12}{7}$$ Replace $\frac{7}{12}$ by its reciprocal and multiply

$$= \frac{\not{7}\cdot 2\cdot \not{6}}{\not{6}\cdot\not{7}}$$ Divide out common factors

$$= 2$$

In both cases the result is the same. We can use either method to simplify complex fractions.

5 $\dfrac{\dfrac{1}{2}+\dfrac{3}{4}}{\dfrac{2}{3}-\dfrac{1}{4}}$

Note
We are going to simplify this complex fraction by two different methods. This is the first method.

Note
The fraction bar that separates the numerator of the complex fraction from its denominator works like parentheses. If we were to rewrite this problem without it, we would write it like this:

$$\left(\frac{1}{2}+\frac{2}{3}\right)\div\left(\frac{3}{4}-\frac{1}{6}\right)$$

That is why we simplify the top and bottom of the complex fraction separately and then divide.

Answer
5 3

6 Simplify: $\dfrac{4 + \frac{2}{3}}{3 - \frac{1}{4}}$

EXAMPLE 6 Simplify: $\dfrac{3 + \frac{1}{2}}{2 - \frac{3}{4}}$

Solution The simplest approach here is to multiply both the top and bottom by the LCD for all fractions, which is 4:

$$\frac{3 + \frac{1}{2}}{2 - \frac{3}{4}} = \frac{\mathbf{4}\left(3 + \frac{1}{2}\right)}{\mathbf{4}\left(2 - \frac{3}{4}\right)} \qquad \text{Multiply the top and bottom by 4}$$

$$= \frac{\mathbf{4} \cdot 3 + \mathbf{4} \cdot \frac{1}{2}}{\mathbf{4} \cdot 2 - \mathbf{4} \cdot \frac{3}{4}} \qquad \text{Distributive property}$$

$$= \frac{12 + 2}{8 - 3} \qquad \text{Multiply each number by 4}$$

$$= \frac{14}{5} \qquad \text{Add on top and subtract on bottom}$$

$$= 2\tfrac{4}{5}$$

7 Simplify: $\dfrac{12\frac{1}{3}}{6\frac{2}{3}}$

EXAMPLE 7 Simplify: $\dfrac{10\frac{1}{3}}{8\frac{2}{3}}$

Solution The simplest way to simplify this complex fraction is to think of it as a division problem.

$$\frac{10\frac{1}{3}}{8\frac{2}{3}} = 10\tfrac{1}{3} \div 8\tfrac{2}{3} \qquad \text{Write with a} \div \text{symbol}$$

$$= \frac{31}{3} \div \frac{26}{3} \qquad \text{Change to improper fractions}$$

$$= \frac{31}{3} \cdot \frac{3}{26} \qquad \text{Write in terms of multiplication}$$

$$= \frac{31 \cdot \cancel{3}}{\cancel{3} \cdot 26} \qquad \text{Divide out the common factor 3}$$

$$= \frac{31}{26} \qquad \text{Answer as an improper fraction}$$

$$= 1\tfrac{5}{26} \qquad \text{Answer as a mixed number}$$

Answers
6 $1\frac{23}{33}$ **7** $1\frac{17}{20}$

CHAPTER 3 SUMMARY

Definition of Fractions [3.1]

A fraction is any number that can be written in the form $\frac{a}{b}$, where a and b are integers and b is not 0. The number a is called the *numerator* and the number b is called the *denominator.*

Properties of Fractions [3.1]

Multiplying the numerator and the denominator of a fraction by the same nonzero number will produce an equivalent fraction. The same is true for dividing the numerator and the denominator by the same nonzero number. In symbols the properties look like this: If a, b, and c are integers and b and c are not 0, then

Property 1 $\frac{a}{b} = \frac{a \cdot c}{b \cdot c}$ **Property 2** $\frac{a}{b} = \frac{a \div c}{b \div c}$

Fractions and Numbers [3.1]

If a represents any number, then

$$\frac{a}{1} = a \quad \text{and} \quad \frac{a}{a} = 1 \quad \text{(where } a \text{ is not 0)}$$

Reducing Fractions to Lowest Terms [3.2]

To reduce a fraction to lowest terms, factor the numerator and the denominator and then divide both the numerator and denominator by any factors they have in common.

Multiplying Fractions [3.3]

To multiply fractions, multiply numerators and multiply denominators.

The Area of a Triangle [3.3]

The formula for the area of a triangle with base b and height h is

$$A = \frac{1}{2}bh$$

h

b

Reciprocals [3.4]

Any two numbers whose product is 1 are called *reciprocals.* The numbers $\frac{2}{3}$ and $\frac{3}{2}$ are reciprocals, since their product is 1.

Division with Fractions [3.4]

To divide by a fraction, multiply by its reciprocal. That is, the quotient of two fractions is defined to be the product of the first fraction with the reciprocal of the second fraction (the divisor).

Examples

1 Each of the following is a fraction:

$$\frac{1}{2}, \quad \frac{3}{4}, \quad \frac{8}{1}, \quad \frac{7}{3}$$

2 Change $\frac{3}{4}$ to an equivalent fraction with denominator 12.

$$\frac{3}{4} = \frac{3 \cdot 3}{4 \cdot 3} = \frac{9}{12}$$

3 $\frac{5}{1} = 5, \quad \frac{5}{5} = 1$

4
$$\frac{90}{588} = \frac{\cancel{2} \cdot \cancel{3} \cdot 3 \cdot 5}{\cancel{2} \cdot 2 \cdot \cancel{3} \cdot 7 \cdot 7} = \frac{3 \cdot 5}{2 \cdot 7 \cdot 7} = \frac{15}{98}$$

5 $\frac{3}{5} \cdot \frac{4}{7} = \frac{3 \cdot 4}{5 \cdot 7} = \frac{12}{35}$

6 If the base of a triangle is 10 inches and the height is 7 inches, then the area is

$$A = \frac{1}{2}bh = \frac{1}{2} \cdot 10 \cdot 7 = 5 \cdot 7 = 35 \text{ square inches}$$

7 $\frac{3}{8} \div \frac{1}{3} = \frac{3}{8} \cdot \frac{3}{1} = \frac{9}{8}$

Least Common Denominator (LCD) [3.5]

The *least common denominator* (LCD) for a set of denominators is the smallest number that is exactly divisible by each denominator.

Addition and Subtraction of Fractions [3.5]

To add (or subtract) two fractions with a common denominator, add (or subtract) numerators and use the common denominator. In symbols: If a, b, and c are integers with c not equal to 0, then

$$\frac{a}{c} + \frac{b}{c} = \frac{a+b}{c} \quad \text{and} \quad \frac{a}{c} - \frac{b}{c} = \frac{a-b}{c}$$

8 $\frac{1}{8} + \frac{3}{8} = \frac{1+3}{8} = \frac{4}{8} = \frac{1}{2}$

Additional Facts about Fractions

1. In some books fractions are called *rational numbers*.
2. Every whole number can be written as a fraction with a denominator of 1.
3. The commutative, associative, and distributive properties are true for fractions.
4. The word *of* as used in the expression "$\frac{2}{3}$ *of* 12" indicates that we are to multiply $\frac{2}{3}$ and 12.
5. Two fractions with the same value are called *equivalent* fractions.

Mixed-Number Notation [3.6]

A mixed number is the sum of a whole number and a fraction. The + sign is not shown when we write mixed numbers; it is implied. The mixed number $4\frac{2}{3}$ is actually the sum $4 + \frac{2}{3}$.

Changing Mixed Numbers to Improper Fractions [3.6]

To change a mixed number to an improper fraction, we write the mixed number showing the + sign and add as usual. The result is the same if we multiply the denominator of the fraction by the whole number and add what we get to the numerator of the fraction, putting this result over the denominator of the fraction.

9 $4\frac{2}{3} = \frac{3 \cdot 4 + 2}{3} = \frac{14}{3}$

Mixed number ($4\frac{2}{3}$) — Improper fraction ($\frac{14}{3}$)

Changing an Improper Fraction to a Mixed Number [3.6]

To change an improper fraction to a mixed number, divide the denominator into the numerator. The quotient is the whole-number part of the mixed number. The fraction part is the remainder over the divisor.

10 Change $\frac{14}{3}$ to a mixed number.

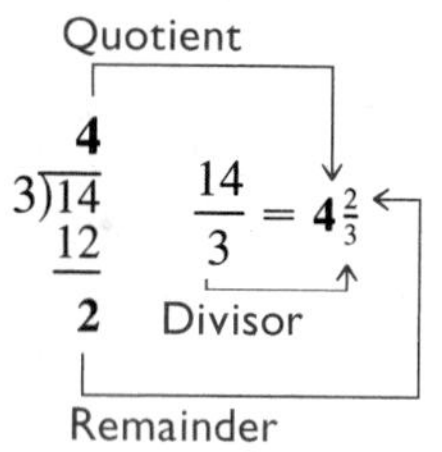

Multiplication and Division with Mixed Numbers [3.7]

To multiply or divide two mixed numbers, change each to an improper fraction and multiply or divide as usual.

11 $2\frac{1}{3} \cdot 1\frac{3}{4} = \frac{7}{3} \cdot \frac{7}{4} = \frac{49}{12} = 4\frac{1}{12}$

Addition and Subtraction with Mixed Numbers [3.8]

To add or subtract two mixed numbers, add or subtract the whole-number parts and the fraction parts separately. This is best done with the numbers written in columns.

12

$$\begin{array}{rcrcr} 3\frac{4}{9} & = & 3\frac{4}{9} & = & 3\frac{4}{9} \\ +\,2\frac{2}{3} & = & 2\frac{2\cdot 3}{3\cdot 3} & = & 2\frac{6}{9} \\ \hline & & & & 5\frac{10}{9} = 6\frac{1}{9} \end{array}$$

Common denominator; Add whole numbers; Add fractions

Borrowing in Subtraction with Mixed Numbers [3.8]

It is sometimes necessary to borrow when doing subtraction with mixed numbers. We always change to a common denominator before we actually borrow.

13

$$\begin{array}{rcrcr} 4\frac{1}{3} & = & 4\frac{2}{6} & = & 3\frac{8}{6} \\ -\,1\frac{5}{6} & = & -\,1\frac{5}{6} & = & -\,1\frac{5}{6} \\ \hline & & & & 2\frac{3}{6} = 2\frac{1}{2} \end{array}$$

COMMON MISTAKES

1 The most common mistake when working with fractions occurs when we try to add two fractions without using a common denominator. For example,

$$\frac{2}{3} + \frac{4}{5} \neq \frac{2+4}{3+5}$$

If the two fractions we are trying to add don't have the same denominators, then we *must* rewrite each one as an equivalent fraction with a common denominator. *We never add denominators when adding fractions.*

Note We do *not* need a common denominator when multiplying fractions.

2 A common mistake made with division of fractions occurs when we multiply by the reciprocal of the first fraction instead of the reciprocal of the divisor. For example,

$$\frac{2}{3} \div \frac{5}{6} \neq \frac{3}{2} \cdot \frac{5}{6}$$

Remember, we perform division by multiplying by the reciprocal of the divisor (the fraction to the right of the division symbol).

3 If the answer to a problem turns out to be a fraction, that fraction should always be written in lowest terms. It is a mistake not to reduce to lowest terms.

4 A common mistake when working with mixed numbers is to confuse mixed-number notation for multiplication of fractions. The notation $3\frac{2}{5}$ does *not* mean 3 *times* $\frac{2}{5}$. It means 3 *plus* $\frac{2}{5}$.

5 Another mistake occurs when multiplying mixed numbers. The mistake occurs when we don't change the mixed number to an improper fraction before multiplying, and instead try to multiply the whole numbers and fractions separately. Like this:

$$\begin{aligned} 2\tfrac{1}{2} \cdot 3\tfrac{1}{3} &= (2 \cdot 3) + \left(\frac{1}{2} \cdot \frac{1}{3}\right) \qquad \text{Mistake} \\ &= 6 + \frac{1}{6} \\ &= 6\tfrac{1}{6} \end{aligned}$$

Remember, the correct way to multiply mixed numbers is to first change to improper fractions and then multiply numerators and multiply denominators. This is correct:

$$2\tfrac{1}{2} \cdot 3\tfrac{1}{3} = \frac{5}{2} \cdot \frac{10}{3} = \frac{50}{6} = 8\tfrac{2}{6} = 8\tfrac{1}{3} \qquad \text{Correct}$$

CHAPTER 3 REVIEW

Reduce each of the following fractions to lowest terms. **[3.2]**

1 $\frac{6}{8}$ **2** $\frac{12}{36}$ **3** $\frac{110a^3}{70a}$ **4** $\frac{45xy}{75y}$

Multiply the following fractions. (That is, find the product in each case, and reduce to lowest terms.) **[3.3]**

5 $\frac{1}{5}(5x)$ **6** $\frac{80}{27}\left(-\frac{3}{20}\right)$ **7** $\frac{96}{25}\cdot\frac{15}{98}\cdot\frac{35}{54}$ **8** $\frac{3}{5}\cdot 75\cdot\frac{2}{3}$

Find the following quotients. (That is, divide and reduce to lowest terms.) **[3.4]**

9 $\frac{8}{9}\div\frac{4}{3}$ **10** $\frac{9}{10}\div(-3)$ **11** $\frac{a^3}{b}\div\frac{a}{b^2}$ **12** $-\frac{18}{49}\div\left(-\frac{36}{28}\right)$

Perform the indicated operations. Reduce all answers to lowest terms. **[3.5]**

13 $\frac{6}{8}-\frac{2}{8}$ **14** $\frac{9}{x}+\frac{11}{x}$ **15** $-3-\frac{1}{2}$ **16** $\frac{3}{x}-\frac{5}{6}$

17 $\frac{11}{126}-\frac{5}{84}$ **18** $\frac{3}{10}+\frac{7}{25}+\frac{3}{4}$

19 Change $3\frac{5}{8}$ to an improper fraction. **[3.6]**

20 Add: $3+\frac{2}{x}$ **[3.6]**

21 Subtract: $x-\frac{3}{4}$ **[3.6]**

22 Change $\frac{110}{8}$ to a mixed number. **[3.6]**

Perform the indicated operations. **[3.7, 3.8]**

23 $2\div 3\frac{1}{4}$ **24** $4\frac{7}{8}\div 2\frac{3}{5}$ **25** $6\cdot 2\frac{1}{2}\cdot\frac{4}{5}$ **26** $3\frac{1}{5}+4\frac{2}{5}$ **27** $8\frac{2}{3}+9\frac{1}{4}$ **28** $5\frac{1}{3}-2\frac{8}{9}$

Simplify each of the following as much as possible. **[3.9]**

29 $3 + 2(4\frac{1}{3})$

30 $\left(2\frac{1}{2} + \frac{3}{4}\right)\left(2\frac{1}{2} - \frac{3}{4}\right)$

Simplify each complex fraction as much as possible. **[3.9]**

31 $\dfrac{1 + \dfrac{2}{3}}{1 - \dfrac{2}{3}}$

32 $\dfrac{3 - \dfrac{3}{4}}{3 + \dfrac{3}{4}}$

33 $\dfrac{\dfrac{7}{8} - \dfrac{1}{2}}{\dfrac{1}{4} + \dfrac{1}{2}}$

34 $\dfrac{2\frac{1}{8} + 3\frac{1}{3}}{1\frac{1}{6} + 5\frac{1}{4}}$

35 If $\frac{1}{10}$ of the items in a shipment of 200 items are defective, how many are defective?

36 If 80 students took a math test and $\frac{3}{4}$ of them passed, then how many students passed the test?

37 What is 3 times the sum of $2\frac{1}{4}$ and $\frac{3}{4}$?

38 Subtract $\frac{5}{6}$ from the product of $1\frac{1}{2}$ with $\frac{2}{3}$.

39 If a recipe that calls for $2\frac{1}{2}$ cups of flour will make 48 cookies, how much flour is needed to make 36 cookies?

40 A piece of wood $10\frac{3}{4}$ inches long is divided into 6 equal pieces. How long is each piece?

41 A recipe that calls for $3\frac{1}{2}$ tablespoons of oil is tripled. How much oil must be used in the tripled recipe?

42 A farmer fed his sheep $10\frac{1}{2}$ pounds of feed on Monday, $9\frac{3}{4}$ pounds on Tuesday, and $12\frac{1}{4}$ pounds on Wednesday. What is the total number of pounds of feed he used on these 3 days?

43 Find the area and the perimeter of the triangle below.

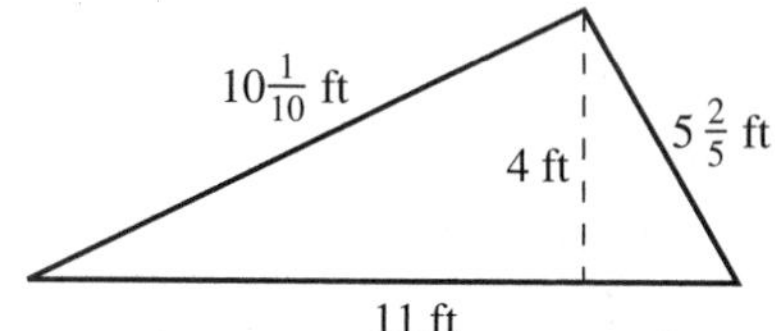

CHAPTER 3 TEST

Reduce each fraction to lowest terms.

1 $\frac{10}{15}$

2 $\frac{130xy}{50x}$

Find each product and reduce your answer to lowest terms.

3 $\frac{3}{5}(30x)$

4 $\frac{48}{49} \cdot \frac{35}{50} \cdot \frac{6}{18}$

Find each quotient and reduce your answer to lowest terms.

5 $\frac{5}{18} \div \frac{15}{16}$

6 $\frac{4}{5} \div -8$

Perform the indicated operations. Reduce all answers to lowest terms.

7 $\frac{3}{10} + \frac{1}{10}$

8 $\frac{5}{x} - \frac{2}{x}$

9 $-4 - \frac{3}{5}$

10 $\frac{3}{x} + \frac{2}{5}$

11 $\frac{5}{6} + \frac{2}{9} + \frac{1}{4}$

12 Change $5\frac{2}{7}$ to an improper fraction.

13 Change $\frac{43}{5}$ to a mixed number.

14 Add: $5 + \frac{4}{x}$

Perform the indicated operations.

15 $6 \div 1\frac{1}{3}$

16 $7\frac{1}{3} + 2\frac{3}{8}$

17 $5\frac{1}{6} - 1\frac{1}{2}$

Simplify each of the following as much as possible.

18 $4 + 3\left(4\frac{1}{4}\right)$

19 $\left(2\frac{1}{3} + \frac{1}{2}\right)\left(3\frac{2}{3} - \frac{1}{6}\right)$

20 $\dfrac{\frac{11}{12} - \frac{2}{3}}{\frac{1}{6} + \frac{1}{3}}$

21 If $\frac{1}{3}$ of a shipment of 120 grapefruit is spoiled, how many grapefruit are spoiled?

22 A dress that is $31\frac{1}{6}$ inches long is shortened by $3\frac{2}{3}$ inches. What is the length of the new dress?

23 A recipe that calls for $4\frac{2}{3}$ cups of sugar is doubled. How much sugar must be used in the doubled recipe?

24 A piece of rope $15\frac{2}{3}$ feet long is divided into 5 equal pieces. How long is each piece?

25 Find the area and the perimeter of the triangle below.

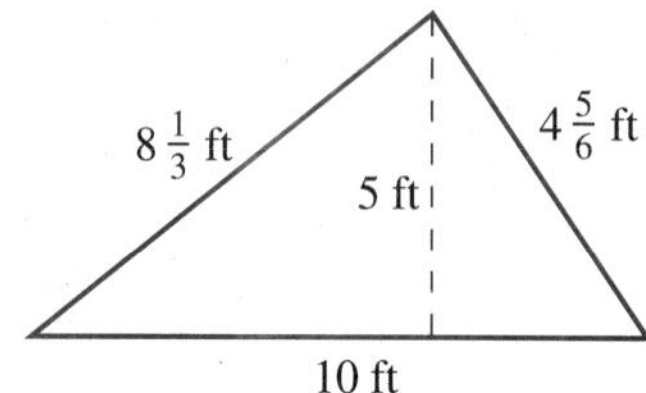

SOLVING EQUATIONS

INTRODUCTION

As we mentioned previously, in the U.S. system, temperature is measured on the Fahrenheit scale. In the metric system, temperature is measured on the Celsius scale. On this scale, water boils at 100 degrees and freezes at 0 degrees. To denote a temperature of 100 degrees on the Celsius scale, we write 100°C and read "100 degrees Celsius." The table below gives some corresponding temperatures on both temperature scales. The figure is a line graph of the information in the table.

Comparing temperatures on two scales (numeric)

Temperature in Degrees Celsius	Temperature in Degrees Fahrenheit
0°C	32°F
25°C	77°F
50°C	122°F
75°C	167°F
100°C	212°F

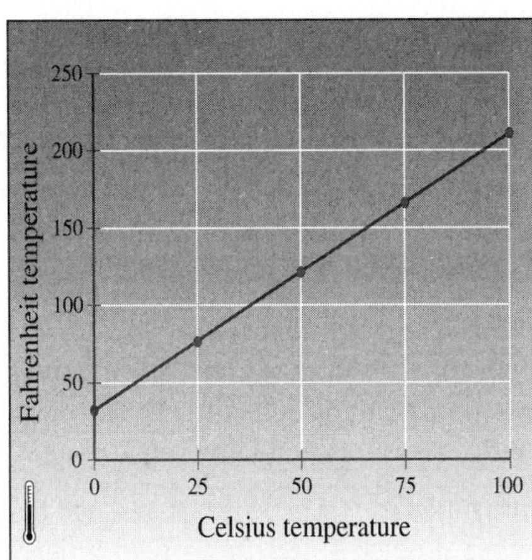

A line graph of the data in the table (geometric)

The information in the table is *numeric* in nature, whereas the information in the figure is *geometric*. In this chapter we will add a third category for presenting information and writing relationships. Information in this new category is *alge-*

braic in nature because it is presented using equations and formulas. The following equation is a formula that tells us how to convert from a temperature on the Celsius scale to the corresponding temperature on the Fahrenheit scale.

$$F = \frac{9}{5}C + 32$$

All three items above—the table, the line graph, and the equation—show the basic relationship between the two temperature scales. The differences among them are merely differences in form: The table is numeric, the line graph is geometric, and the equation is algebraic.

We begin this chapter with equations that contain one variable. The first three sections of the chapter cover the major steps necessary to solve these equations. Section 4.4 combines the ideas developed in the first three sections into a general method of solving these equations. Section 4.5 covers applications. These applications, or word problems, show how equations can be used as models for everyday situations. The last three sections of the chapter offer an introduction to graphing in two dimensions. To be successful in this chapter, you must be familiar with the material we covered in Chapter 2—that is, you must know how to add, subtract, multiply, and divide with negative numbers.

STUDY SKILLS

The study skills for this chapter are about attitude. They are points of view that point toward success.

1 Be focused, not distracted. I have students who begin their assignments by asking themselves, "Why am I taking this class?" or, "When am I ever going to use this stuff?" If you are asking yourself similar questions, you may be distracting yourself from doing the things that will produce the results you want in this course. Don't dwell on questions and evaluations of the class that can be used as excuses for not doing well. If you want to succeed in this course, focus your energy and efforts toward success.

2 Be resilient. Don't let setbacks keep you from your goals. You want to put yourself on the road to becoming someone who can succeed in this class or any class in college. Failing a test or quiz or having a difficult time on some topics is normal. No one goes through college without some setbacks. A low grade on a test or quiz is simply a signal that some reevaluation of your study habits needs to take place.

3 Intend to succeed. I always have a few students who simply go through the motions of studying without intending to master the material. It is more important to them to look like they are studying than to actually study. You need to study with the intention of being successful in the course no matter what it takes.

4 Begin to develop confidence with word problems. The main difference between people who are good at working word problems and those who are not seems to be confidence. People with confidence know that no matter how long it takes them, they will eventually be able to solve the problem they are working on. Those without confidence begin by saying to themselves, "I'll never be able to work this problem." If you are in this second group, then instead of telling yourself that you can't do word problems, that you don't like them, or that they're not good for anything anyway, decide to do whatever it takes to master them.

4.1 The Distributive Property and Algebraic Expressions

We begin this section by extending the work we have done previously with the distributive property. Our first example is a review of one type of problem we covered at the end of Chapter 2.

EXAMPLE 1 Apply the distributive property to the expression

$5(x + 3)$

Solution Distributing the 5 over x and 3, we have

$$\begin{aligned} 5(x + 3) &= 5(x) + 5(3) && \text{Distributive property} \\ &= 5x + 15 && \text{Multiplication} \end{aligned}$$

Remember, $5x$ means "5 times x."

The distributive property can be applied to more complicated expressions involving negative numbers.

EXAMPLE 2 Multiply: $-4(3x + 5)$

Solution Multiplying both the $3x$ and the 5 by -4, we have

$$\begin{aligned} -4(3x + 5) &= -4(3x) + (-4)5 && \text{Distributive property} \\ &= -12x + (-20) && \text{Multiplication} \\ &= -12x - 20 && \text{Definition of subtraction} \end{aligned}$$

Notice, first of all, that when we apply the distributive property here, we multiply through by -4. It is important to include the sign with the number when we use the distributive property. Second, when we multiply -4 and $3x$, the result is $-12x$ because

$$\begin{aligned} -4(3x) &= (-4 \cdot 3)x && \text{Associative property} \\ &= -12x && \text{Multiplication} \end{aligned}$$

Recall from Chapter 2 that we can also use the distributive property to simplify expressions like $4x + 3x$.

$$\begin{aligned} 4x + 3x &= (4 + 3)x && \text{Distributive property} \\ &= 7x && \text{Addition} \end{aligned}$$

EXAMPLE 3 Simplify: $5x - 2 + 3x + 7$

Solution We begin by changing subtraction to addition of the opposite and applying the commutative property to rearrange the order of the terms. We want similar terms to be written next to each other.

$$\begin{aligned} 5x - 2 + 3x + 7 &= 5x + 3x + (-2) + 7 && \text{Commutative property} \\ &= (5 + 3)x + (-2) + 7 && \text{Distributive property} \\ &= 8x + 5 && \text{Addition} \end{aligned}$$

Notice that we take the negative sign in front of the 2 with the 2 when we rearrange terms. How do we justify doing this?

Practice Problems

1 Apply the distributive property to the expression

$6(x + 4)$

2 Multiply: $-3(2x + 4)$

3 Simplify: $6x - 2 + 3x + 8$

Answers

1 $6x + 24$ 2 $-6x - 12$

3 $9x + 6$

4 Simplify: $2(4x + 3) + 7$

5 Simplify: $3(2x + 1) + 5(4x - 3)$

EXAMPLE 4 Simplify: $3(4x + 5) + 6$

Solution We begin by distributing the 3 across the sum of $4x$ and 5. Then we combine similar terms.

$$\begin{aligned} 3(4x + 5) + 6 &= 12x + 15 + 6 && \text{Distributive property} \\ &= 12x + 21 && \text{Add 15 and 6} \end{aligned}$$

EXAMPLE 5 Simplify: $2(3x + 1) + 4(2x - 5)$

Solution Again, we apply the distributive property first; then we combine similar terms. Here is the solution showing only the essential steps:

$$\begin{aligned} 2(3x + 1) + 4(2x - 5) &= 6x + 2 + 8x - 20 && \text{Distributive property} \\ &= 14x - 18 && \text{Combine similar terms} \end{aligned}$$

The Value of an Algebraic Expression

An expression such as $3x + 5$ will take on different values depending on what x is. If we were to let x equal 2, the expression $3x + 5$ would become 11. Here's how:

$$\begin{aligned} \text{When} \quad & x = 2 \\ \text{the expression} \quad & 3x + 5 \\ \text{becomes} \quad & 3(2) + 5 && \text{Substitute 2 for } x \\ & = 6 + 5 && \text{Multiply 3 and 2 to get 6} \\ & = 11 && \text{Add 6 and 5 to get 11} \end{aligned}$$

On the other hand, if x is 10, the same expression has a value of 35:

$$\begin{aligned} \text{When} \quad & x = 10 \\ \text{the expression} \quad & 3x + 5 \\ \text{becomes} \quad & 3(10) + 5 && \text{Substitute 10 for } x \\ & = 30 + 5 && \text{The product of 3 and 10 is 30} \\ & = 35 && \text{The sum of 30 and 5 is 35} \end{aligned}$$

Table 1 lists some other algebraic expressions, along with specific values for the variables and the corresponding value of the expression after the variable has been replaced with the given number.

Table 1

Original Expression	Value of the Variable	Value of the Expression
$5x + 2$	$x = 4$	$5(4) + 2 = 20 + 2$ $= 22$
$3x - 9$	$x = 2$	$3(2) - 9 = 6 - 9$ $= -3$
$-2a + 7$	$a = 3$	$-2(3) + 7 = -6 + 7$ $= 1$
$6x + 3$	$x = -2$	$6(-2) + 3 = -12 + 3$ $= -9$
$-4y + 9$	$y = -1$	$-4(-1) + 9 = 4 + 9$ $= 13$

Answers
4 $8x + 13$ 5 $26x - 12$

4.2 The Addition Property of Equality

In this section we will solve some simple equations. Solving an equation involves finding all replacements for the variable that make the equation a true statement.

DEFINITION The **solution** for an equation is the number that when used in place of the variable makes the equation a true statement.

For example, the equation $x + 3 = 7$ has as its solution the number 4, since replacing x with 4 in the equation gives a true statement:

When $x = 4$

the equation $x + 3 = 7$

becomes $4 + 3 = 7$

or $7 = 7$ A true statement

EXAMPLE 1 Is $x = 5$ the solution to the equation $3x + 2 = 17$?

Solution To see if it is we replace x with 5 in the equation and see if the result is a true statement:

When $x = 5$

the equation $3x + 2 = 17$

becomes $3(5) + 2 = 17$

$15 + 2 = 17$

$17 = 17$ A true statement

Since the result is a true statement, we can conclude that $x = 5$ is the solution to $3x + 2 = 17$.

EXAMPLE 2 Is $a = -2$ the solution to the equation $7a + 4 = 3a - 2$?

Solution When $a = -2$

the equation $7a + 4 = 3a - 2$

becomes $7(-2) + 4 = 3(-2) - 2$

$-14 + 4 = -6 - 2$

$-10 = -8$ A false statement

Since the result is a false statement, we must conclude that $a = -2$ is *not* a solution to the equation $7a + 4 = 3a - 2$.

Practice Problems

1 Show that $x = 3$ is a solution to the equation $5x - 4 = 11$.

2 Is $a = -3$ the solution to the equation $6a - 3 = 2a + 4$?

Answers

1 See solutions section. 2 No

We want to develop a process for solving equations with one variable. The most important property needed for solving the equations in this section is called the *addition property of equality*. The formal definition looks like this:

Addition Property of Equality

Let A, B, and C represent algebraic expressions.

If $\quad A = B$

then $\quad A + C = B + C$

In words: Adding the same quantity to both sides of an equation never changes the solution to the equation.

This property is extremely useful in solving equations. Our goal in solving equations is to isolate the variable on one side of the equation. We want to end up with an expression of the form

$x =$ A number

To do so we use the addition property of equality.

3 Solve for x: $\quad x + 5 = -2$

EXAMPLE 3 Solve for x: $\quad x + 4 = -2$

Solution We want to isolate x on one side of the equation. If we add -4 to both sides, the left side will be $x + 4 + (-4)$, which is $x + 0$ or just x.

$x + 4 = -2$	
$x + 4 + (\mathbf{-4}) = -2 + (\mathbf{-4})$	Add **−4** to both sides
$x + 0 = -6$	Addition
$x = -6$	$x + 0 = x$

The solution is -6. We can check it if we want to by replacing x with -6 in the original equation:

When	$x = -6$	
the equation	$x + 4 = -2$	
becomes	$-6 + 4 = -2$	
	$-2 = -2$	A true statement

Note
With some of the equations in this section, you will be able to see the solution just by looking at the equation. But it is important that you show all the steps used to solve the equations anyway. The equations you come across in the future will not be as easy to solve, so you should learn the steps involved very well.

4 Solve for a: $\quad a - 2 = 7$

EXAMPLE 4 Solve for a: $\quad a - 3 = 5$

Solution

$a - 3 = 5$	
$a - 3 + \mathbf{3} = 5 + \mathbf{3}$	Add **3** to both sides
$a + 0 = 8$	Addition
$a = 8$	$a + 0 = a$

The solution to $a - 3 = 5$ is $a = 8$.

Answers
3 -7 **4** 9

EXAMPLE 5 Solve for y: $y + 4 - 6 = 7 - 1$

Solution Before we apply the addition property of equality, we must simplify each side of the equation as much as possible:

$$\begin{aligned} y + 4 - 6 &= 7 - 1 \\ y - 2 &= 6 && \text{Simplify each side} \\ y - 2 + \mathbf{2} &= 6 + \mathbf{2} && \text{Add } \mathbf{2} \text{ to both sides} \\ y + 0 &= 8 && \text{Addition} \\ y &= 8 && y + 0 = y \end{aligned}$$

EXAMPLE 6 Solve for x: $3x - 2 - 2x = 4 - 9$

Solution Simplifying each side as much as possible, we have

$$\begin{aligned} 3x - 2 - 2x &= 4 - 9 \\ x - 2 &= -5 && 3x - 2x = x \\ x - 2 + \mathbf{2} &= -5 + \mathbf{2} && \text{Add } \mathbf{2} \text{ to both sides} \\ x + 0 &= -3 && \text{Addition} \\ x &= -3 && x + 0 = x \end{aligned}$$

EXAMPLE 7 Solve for x: $-3 - 6 = x + 4$

Solution The variable appears on the right side of the equation in this problem. This makes no difference; we can isolate x on either side of the equation. We can leave it on the right side if we like:

$$\begin{aligned} -3 - 6 &= x + 4 \\ -9 &= x + 4 && \text{Simplify the left side} \\ -9 + (\mathbf{-4}) &= x + 4 + (\mathbf{-4}) && \text{Add } \mathbf{-4} \text{ to both sides} \\ -13 &= x + 0 && \text{Addition} \\ -13 &= x && x + 0 = x \end{aligned}$$

The statement $-13 = x$ is equivalent to the statement $x = -13$. In either case the solution to our equation is -13.

EXAMPLE 8 Solve: $a - \frac{3}{4} = \frac{5}{8}$

Solution To isolate a we add $\frac{3}{4}$ to each side:

$$\begin{aligned} a - \frac{3}{4} &= \frac{5}{8} \\ a - \frac{3}{4} + \mathbf{\frac{3}{4}} &= \frac{5}{8} + \mathbf{\frac{3}{4}} \\ a &= \frac{11}{8} \end{aligned}$$

When solving equations we will leave answers like $\frac{11}{8}$ as improper fractions, rather than change them to mixed numbers.

5 Solve for y: $y + 6 - 2 = 8 - 9$

6 Solve for x:

$5x - 3 - 4x = 4 - 7$

7 Solve for x: $-5 - 7 = x + 2$

8 Solve: $a - \frac{2}{3} = \frac{5}{6}$

Answers

5 -5 **6** 0 **7** -14 **8** $\frac{3}{2}$

4.3 The Multiplication Property of Equality

In this section we will continue to solve equations in one variable. We will again use the addition property of equality, but we will also use another property—the *multiplication property of equality*—to solve the equations in this section. We state the multiplication property of equality and then see how it is used by looking at some examples.

> **Multiplication Property of Equality**
>
> Let A, B, and C represent algebraic expressions, with C not equal to 0.
>
> $$\begin{aligned} \text{If} \quad A &= B \\ \text{then} \quad AC &= BC \end{aligned}$$
>
> *In words:* Multiplying both sides of an equation by the same nonzero quantity never changes the solution to the equation.

Now, since division is defined as multiplication by the reciprocal, we are also free to divide both sides of an equation by the same nonzero quantity and always be sure we have not changed the solution to the equation.

EXAMPLE 1 Solve for x: $\frac{1}{2}x = 3$

Solution Our goal here is the same as it was in Section 4.2. We want to isolate x (that is, $1x$) on one side of the equation. We have $\frac{1}{2}x$ on the left side. If we multiply both sides by 2, we will have $1x$ on the left side. Here is how it looks:

$$\frac{1}{2}x = 3$$

$$\mathbf{2}\left(\frac{1}{2}x\right) = \mathbf{2}(3) \quad \text{Multiply both sides by } \mathbf{2}$$

$$x = 6 \quad \text{Multiplication}$$

To see how $2\left(\frac{1}{2}x\right)$ is equivalent to x, we use the associative property:

$$\begin{aligned} 2\left(\frac{1}{2}x\right) &= \left(2 \cdot \frac{1}{2}\right)x && \text{Associative property} \\ &= 1 \cdot x && 2 \cdot \frac{1}{2} = 1 \\ &= x && 1 \cdot x = x \end{aligned}$$

Although we will not show this step when solving problems, it is implied.

Practice Problems

1 Solve for x: $\frac{1}{3}x = 5$

Answer

1 15

2 Solve for a: $\frac{1}{5}a + 3 = 7$

EXAMPLE 2 Solve for a: $\frac{1}{3}a + 2 = 7$

Solution We begin by adding -2 to both sides to get $\frac{1}{3}a$ by itself. We then multiply by 3 to solve for a.

$$\frac{1}{3}a + 2 = 7$$

$$\frac{1}{3}a + 2 + (\mathbf{-2}) = 7 + (\mathbf{-2}) \quad \text{Add } \mathbf{-2} \text{ to both sides}$$

$$\frac{1}{3}a = 5 \quad \text{Addition}$$

$$\mathbf{3} \cdot \frac{1}{3}a = \mathbf{3} \cdot 5 \quad \text{Multiply both sides by } \mathbf{3}$$

$$a = 15 \quad \text{Multiplication}$$

We can check our solution to see that it is correct:

When $a = 15$

the equation $\frac{1}{3}a + 2 = 7$

becomes $\frac{1}{3}(15) + 2 = 7$

$$5 + 2 = 7$$

$$7 = 7 \quad \text{A true statement}$$

3 Solve: $\frac{3}{5}y = 6$

EXAMPLE 3 Solve for y: $\frac{2}{3}y = 12$

Solution In this case we multiply each side of the equation by the reciprocal of $\frac{2}{3}$, which is $\frac{3}{2}$.

$$\frac{2}{3}y = 12$$

$$\frac{\mathbf{3}}{\mathbf{2}}\left(\frac{2}{3}y\right) = \frac{\mathbf{3}}{\mathbf{2}}(12)$$

$$y = 18$$

The solution checks because $\frac{2}{3}$ of 18 is 12.

4 Solve: $-\frac{3}{4}x = \frac{6}{5}$

EXAMPLE 4 Solve: $-\frac{4}{5}x = \frac{8}{15}$

Solution The reciprocal of $-\frac{4}{5}$ is $-\frac{5}{4}$.

$$-\frac{4}{5}x = \frac{8}{15}$$

$$-\frac{\mathbf{5}}{\mathbf{4}}\left(-\frac{4}{5}x\right) = -\frac{\mathbf{5}}{\mathbf{4}}\left(\frac{8}{15}\right)$$

$$x = -\frac{2}{3}$$

Answers

2 20 **3** 10 **4** $-\frac{8}{5}$

Many times it is convenient to divide both sides by a nonzero number to solve an equation, as the next example shows.

EXAMPLE 5 Solve for x: $4x = -20$

Solution If we divide both sides by 4, the left side will be just x, which is what we want. It is okay to divide both sides by 4 since division by 4 is equivalent to multiplication by $\frac{1}{4}$, and the multiplication property of equality states that we can multiply both sides by any number so long as it isn't 0.

$4x = -20$

$\frac{4x}{\mathbf{4}} = \frac{-20}{\mathbf{4}}$ Divide both sides by **4**

$x = -5$ Division

Since $4x$ means "4 times x," the factors in the numerator of $\frac{4x}{4}$ are 4 and x. Since the factor 4 is common to the numerator and the denominator, we divide it out to get just x.

5 Solve for x: $6x = -42$

Note
If we multiply each side by $\frac{1}{4}$, the solution looks like this:

$$\frac{\mathbf{1}}{\mathbf{4}}(4x) = \frac{\mathbf{1}}{\mathbf{4}}(-20)$$
$$\left(\frac{1}{4} \cdot 4\right)x = -5$$
$$1x = -5$$
$$x = -5$$

EXAMPLE 6 Solve for x: $-3x + 7 = -5$

Solution We begin by adding -7 to both sides to reduce the left side to just $-3x$.

$-3x + 7 = -5$

$-3x + 7 + (\mathbf{-7}) = -5 + (\mathbf{-7})$ Add **−7** to both sides

$-3x = -12$ Addition

$\frac{-3x}{\mathbf{-3}} = \frac{-12}{\mathbf{-3}}$ Divide both sides by **−3**

$x = 4$ Division

6 Solve for x: $-5x + 6 = -14$

With more complicated equations we simplify both sides first before we do any addition or multiplication on both sides. The examples below illustrate.

EXAMPLE 7 Solve for x: $5x - 8x + 3 = 4 - 10$

Solution We combine similar terms to simplify each side and then solve as usual.

$5x - 8x + 3 = 4 - 10$

$-3x + 3 = -6$ Simplify each side

$-3x + 3 + (\mathbf{-3}) = -6 + (\mathbf{-3})$ Add **−3** to both sides

$-3x = -9$ Addition

$\frac{-3x}{\mathbf{-3}} = \frac{-9}{\mathbf{-3}}$ Divide both sides by **−3**

$x = 3$ Division

7 Solve for x:

$3x - 7x + 5 = 3 - 18$

Answers
5 −7 **6** 4 **7** 5

8 Solve for x:

$$-5 + 4 = 2x - 11 + 3x$$

Answer

8 2

EXAMPLE 8 Solve for x: $-8 + 11 = 4x - 11 + 3x$

Solution We begin by simplifying each side separately.

$$-8 + 11 = 4x - 11 + 3x$$

$$3 = 7x - 11 \qquad \text{Simplify both sides}$$

$$3 + \mathbf{11} = 7x - 11 + \mathbf{11} \qquad \text{Add } \mathbf{11} \text{ to both sides}$$

$$14 = 7x \qquad \text{Addition}$$

$$\frac{14}{\mathbf{7}} = \frac{7x}{\mathbf{7}} \qquad \text{Divide both sides by } \mathbf{7}$$

$$2 = x$$

$$x = 2$$

Again, it makes no difference which side of the equation x ends up on, so long as it is just one x. ■

COMMON MISTAKE

Before we end this section we should mention a very common mistake made by students when they first begin to solve equations. It involves trying to subtract away the number in front of the variable—like this:

$$7x = 21$$

$$7x - 7 = 21 - 7 \qquad \text{Add } -7 \text{ to both sides}$$

$$x = 14 \longleftarrow \text{Mistake}$$

The mistake is not in trying to subtract 7 from both sides of the equation. The mistake occurs when we say $7x - 7 = x$. It just isn't true. We can add and subtract only similar terms. The numbers $7x$ and 7 are not similar, since one contains x and the other doesn't. The correct way to do the problem is like this:

$$7x = 21$$

$$\frac{7x}{\mathbf{7}} = \frac{21}{\mathbf{7}} \qquad \text{Divide both sides by } \mathbf{7}$$

$$x = 3 \qquad \text{Division}$$

4.4 Linear Equations in One Variable

In this chapter we have been solving what are called *linear equations in one variable*. They are equations that contain only one variable, and that variable is always raised to the first power and never appears in a denominator. Here are some examples of linear equations in one variable:

$$3x + 2 = 17, \qquad 7a + 4 = 3a - 2, \qquad 2(3y - 5) = 6$$

Because of the work we have done in the first three sections of this chapter, we are now able to solve any linear equation in one variable. The steps outlined below can be used as a guide to solving these equations.

Steps to Solve a Linear Equation in One Variable

Step 1 Simplify each side of the equation as much as possible. This step is done using the commutative, associative, and distributive properties.

Step 2 Use the addition property of equality to get all *variable terms* on one side of the equation and all *constant terms* on the other. A *variable term* is any term that contains the variable. A *constant term* is any term that contains only a number.

Step 3 Use the multiplication property of equality to get x by itself on one side of the equation.

Step 4 Check your solution in the original equation if you think it is necessary.

Note
Once you have some practice at solving equations, these steps will seem almost automatic. Until that time, it is a good idea to pay close attention to these steps.

EXAMPLE 1 Solve: $3(x + 2) = -9$

Solution We begin by applying the distributive property to the left side:

Step 1
$$3(x + 2) = -9$$
$$3x + 6 = -9 \qquad \text{Distributive property}$$

Step 2
$$3x + 6 + (\mathbf{-6}) = -9 + (\mathbf{-6}) \qquad \text{Add } \mathbf{-6} \text{ to both sides}$$
$$3x = -15 \qquad \text{Addition}$$

Step 3
$$\frac{3x}{\mathbf{3}} = \frac{-15}{\mathbf{3}} \qquad \text{Divide both sides by } \mathbf{3}$$
$$x = -5 \qquad \text{Division}$$

Practice Problems
1 Solve: $4(x + 3) = -8$

This general method of solving linear equations involves using the two properties developed in Sections 4.2 and 4.3. We can add any number to both sides of an equation or multiply (or divide) both sides by the same nonzero number and always be sure we have not changed the solution to the equation. The equations may change in form, but the solution to the equation stays the same. Looking back to Example 1, we can see that each equation looks a little different from the preceding one. What is interesting, and useful, is that each of the equations says the same thing about x. They all say that x is -5. The last equation, of course, is the easiest to read. That is why our goal is to end up with x isolated on one side of the equation.

Answer
1 -5

2 Solve: $6a + 7 = 4a - 3$

EXAMPLE 2 Solve: $4a + 5 = 2a - 7$

Solution Neither side can be simplified any further. What we have to do is get the variable terms ($4a$ and $2a$) on the same side of the equation. We can eliminate the variable term from the right side by adding $-2a$ to both sides:

Step 2

$$4a + 5 = 2a - 7$$

$$4a + (\mathbf{-2a}) + 5 = 2a + (\mathbf{-2a}) - 7 \quad \text{Add } \mathbf{-2a} \text{ to both sides}$$

$$2a + 5 = -7 \quad \text{Addition}$$

$$2a + 5 + (\mathbf{-5}) = -7 + (\mathbf{-5}) \quad \text{Add } \mathbf{-5} \text{ to both sides}$$

$$2a = -12 \quad \text{Addition}$$

Step 3

$$\frac{2a}{\mathbf{2}} = \frac{-12}{\mathbf{2}} \quad \text{Divide by } \mathbf{2}$$

$$a = -6 \quad \text{Division}$$

3 Solve: $5(x - 2) + 3 = -12$

EXAMPLE 3 Solve: $2(x - 4) + 5 = -11$

Solution We begin by using the distributive property to multiply 2 and $x - 4$:

Step 1

$$2(x - 4) + 5 = -11$$

$$2x - 8 + 5 = -11 \quad \text{Distributive property}$$

$$2x - 3 = -11 \quad \text{Addition}$$

Step 2

$$2x - 3 + \mathbf{3} = -11 + \mathbf{3} \quad \text{Add } \mathbf{3} \text{ to both sides}$$

$$2x = -8 \quad \text{Addition}$$

Step 3

$$\frac{2x}{\mathbf{2}} = \frac{-8}{\mathbf{2}} \quad \text{Divide by } \mathbf{2}$$

$$x = -4 \quad \text{Division}$$

4 Solve: $3(4x - 5) + 6 = 3x + 9$

EXAMPLE 4 Solve: $5(2x - 4) + 3 = 4x - 5$

Solution We apply the distributive property to multiply 5 and $2x - 4$. We then combine similar terms and solve as usual:

Step 1

$$5(2x - 4) + 3 = 4x - 5$$

$$10x - 20 + 3 = 4x - 5 \quad \text{Distributive property}$$

$$10x - 17 = 4x - 5 \quad \text{Simplify the left side}$$

Step 2

$$10x + (\mathbf{-4x}) - 17 = 4x + (\mathbf{-4x}) - 5 \quad \text{Add } \mathbf{-4x} \text{ to both sides}$$

$$6x - 17 = -5 \quad \text{Addition}$$

$$6x - 17 + \mathbf{17} = -5 + \mathbf{17} \quad \text{Add } \mathbf{17} \text{ to both sides}$$

$$6x = 12 \quad \text{Addition}$$

Step 3

$$\frac{6x}{\mathbf{6}} = \frac{12}{\mathbf{6}} \quad \text{Divide by } \mathbf{6}$$

$$x = 2 \quad \text{Division}$$

Answers

2 -5 **3** -1 **4** 2

Equations Involving Fractions

In the rest of this section we will solve some equations that involve fractions. Since integers are usually easier to work with than fractions, we will begin each problem by clearing the equation we are trying to solve of all fractions. To do this we will use the multiplication property of equality to multiply each side of the equation by the LCD for all fractions appearing in the equation. Here is an example.

EXAMPLE 5 Solve the equation $\frac{x}{2} + \frac{x}{6} = 8$.

5 Solve: $\frac{x}{3} + \frac{x}{6} = 9$

Solution The LCD for the fractions $x/2$ and $x/6$ is 6. It has the property that both 2 and 6 divide it evenly. Therefore, if we multiply both sides of the equation by 6, we will be left with an equation that does not involve fractions.

$$6\left(\frac{x}{2} + \frac{x}{6}\right) = 6(8) \quad \text{Multiply each side by 6}$$

$$6\left(\frac{x}{2}\right) + 6\left(\frac{x}{6}\right) = 6(8) \quad \text{Apply the distributive property}$$

$$3x + x = 48 \quad \text{Multiplication}$$

$$4x = 48 \quad \text{Combine similar terms}$$

$$x = 12 \quad \text{Divide each side by 4}$$

We could check our solution by substituting 12 for x in the original equation. If we do so the result is a true statement. The solution is 12.

As you can see from Example 5, the most important step in solving an equation that involves fractions is the first step. In that first step we multiply both sides of the equation by the LCD for all the fractions in the equation. After we have done so, the equation is clear of fractions because the LCD has the property that all the denominators divide it evenly.

EXAMPLE 6 Solve the equation $2x + \frac{1}{2} = \frac{3}{4}$.

6 Solve: $3x + \frac{1}{4} = \frac{5}{8}$

Solution This time the LCD is 4. We begin by multiplying both sides of the equation by 4 to clear the equation of fractions.

$$4\left(2x + \frac{1}{2}\right) = 4\left(\frac{3}{4}\right) \quad \text{Multiply each side by the LCD, 4}$$

$$4(2x) + 4\left(\frac{1}{2}\right) = 4\left(\frac{3}{4}\right) \quad \text{Apply the distributive property}$$

$$8x + 2 = 3 \quad \text{Multiplication}$$

$$8x = 1 \quad \text{Add } -2 \text{ to each side}$$

$$x = \frac{1}{8} \quad \text{Divide each side by 8}$$

Answers

5 18 **6** $\frac{1}{8}$

7 Solve: $\frac{4}{x} + 3 = \frac{11}{5}$

EXAMPLE 7 Solve for x: $\frac{3}{x} + 2 = \frac{1}{2}$ (Assume x is not 0.)

Solution This time the LCD is $2x$. Following the steps we used in Examples 5 and 6, we have

$2x\left(\frac{3}{x} + 2\right) = 2x\left(\frac{1}{2}\right)$	Multiply through by the LCD, $2x$
$2x\left(\frac{3}{x}\right) + 2x(2) = 2x\left(\frac{1}{2}\right)$	Distributive property
$6 + 4x = x$	Multiplication
$6 = -3x$	Add $-4x$ to each side
$-2 = x$	Divide each side by -3

Answer
7 -5

4.5 Applications

Introduction . . . As you begin reading through the examples in this section, you may find yourself asking why some of these problems seem so contrived. The title of the section is "Applications," but many of the problems here don't seem to have much to do with real life. You are right about that. Example 4 is what we refer to as an "age problem." Realistically, it is not the kind of problem you would expect to find if you choose a career in which you use algebra. However, solving age problems is good practice for someone with little experience with application problems, since the solution process has a form that can be applied to all similar age problems.

To begin this section we list the steps used in solving application problems. We call this strategy *Blueprint for Problem Solving*. It is an outline that will overlay the solution process we use on all application problems.

Blueprint for Problem Solving

Step 1 **Read** the problem and then mentally **list** the items that are known and the items that are unknown.

Step 2 **Assign a variable** to one of the unknown items. (In most cases this will amount to letting $x =$ the item that is asked for in the problem.) Then **translate** the other **information** in the problem to expressions involving the variable.

Step 3 **Reread** the problem and then **write an equation**, using the items and variables listed in Steps 1 and 2, that describes the situation.

Step 4 **Solve the equation** found in Step 3.

Step 5 **Write** your **answer** using a complete sentence.

Step 6 **Reread** the problem and **check** your solution with the original words in the problem.

There are a number of substeps within each of the steps in our blueprint. For instance, with Steps 1 and 2 it is always a good idea to draw a diagram or picture if it helps you to visualize the relationship between the items in the problem.

Practice Problems

1 The sum of a number and 3 is 10. Find the number.

EXAMPLE 1 The sum of a number and 2 is 8. Find the number.

Solution Using our blueprint for problem solving as an outline, we solve the problem as follows:

Step 1 *Read* the problem and then mentally *list* the items that are known and the items that are unknown.

Known items: The numbers 2 and 8
Unknown item: The number in question

Step 2 *Assign a variable* to one of the unknown items. Then *translate* the other *information* in the problem to expressions involving the variable.

Let $x =$ the number asked for in the problem.
Then "The sum of a number and 2" translates to $x + 2$.

Step 3 *Reread* the problem and then *write an equation*, using the items and variables listed in Steps 1 and 2, that describes the situation.

With all word problems, the word "is" translates to $=$.

The sum of x and 2 is 8

$$x + 2 = 8$$

Answer

1 7

Step 4 *Solve the equation* found in Step 3.

$$x + 2 = 8$$
$$x + 2 + (\mathbf{-2}) = 8 + (\mathbf{-2}) \quad \text{Add } \mathbf{-2} \text{ to each side}$$
$$x = 6$$

Step 5 *Write* your *answer* using a complete sentence.

The number is 6.

Step 6 *Reread* the problem and *check* your solution with the original words in the problem.

The sum of **6** and 2 is 8. A true statement

To help with other problems of the type shown in Example 1, here are some common English words and phrases and their mathematical translations.

English	Algebra
The sum of a and b	$a + b$
The difference of a and b	$a - b$
The product of a and b	$a \cdot b$
The quotient of a and b	$\frac{a}{b}$
Of	$\cdot$ (multiply)
Is	$=$ (equals)
A number	x
4 more than x	$x + 4$
4 times x	$4x$
4 less than x	$x - 4$

You may find some examples and problems in this section, and the problem set that follows, that you can solve without using algebra or our blueprint. It is very important that you solve those problems using the methods we are showing here. The purpose behind these problems is to give you experience using the blueprint as a guide to solving problems written in words. Your answers are much less important than the work that you show in obtaining your answer.

2 If 4 is added to the sum of twice a number and three times the number, the result is 34. Find the number.

EXAMPLE 2 If 5 is added to the sum of twice a number and three times the number, the result is 25. Find the number.

Solution

Step 1 *Read and list.*

Known items: The numbers 5 and 25, twice a number, and three times a number

Unknown item: The number in question

Step 2 *Assign a variable and translate the information.*

Let $x =$ the number asked for in the problem.

Then "The sum of twice a number and three times the number" translates to $2x + 3x$.

Answer
2 6

Step 3 *Reread and write an equation.*

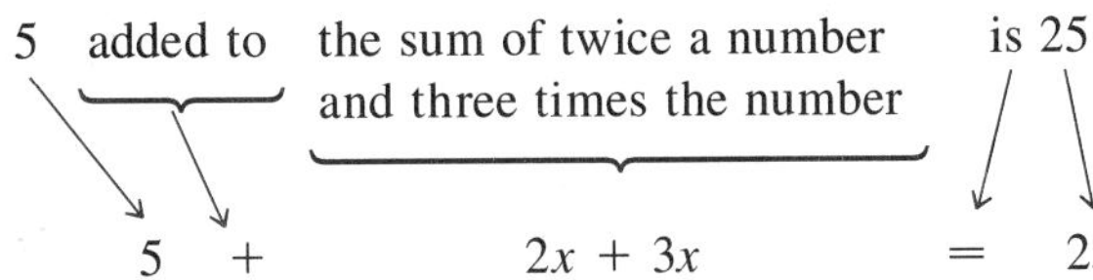

Step 4 *Solve the equation.*

$$5 + 2x + 3x = 25$$

$$5x + 5 = 25 \quad \text{Simplify the left side}$$

$$5x + 5 + (-5) = 25 + (-5) \quad \text{Add } -5 \text{ to both sides}$$

$$5x = 20 \quad \text{Addition}$$

$$\frac{5x}{5} = \frac{20}{5} \quad \text{Divide by } 5$$

$$x = 4 \quad \text{Division}$$

Step 5 *Write your answer.*

The number is 4.

Step 6 *Reread and check.*

Twice **4** is 8 and three times **4** is 12. Their sum is 8 + 12 = 20. Five added to this is 25. Therefore, 5 added to the sum of twice **4** and three times **4** is 25.

EXAMPLE 3 The length of a rectangle is three times the width. The perimeter is 72 centimeters. Find the width and the length.

3 The length of a rectangle is twice the width. The perimeter is 42 centimeters. Find the length and the width.

Solution

Step 1 *Read and list.*

Known items: The length is three times the width. The perimeter is 72 centimeters.

Unknown items: The length and the width

Step 2 *Assign a variable and translate the information.* We let x = the width. Since the length is three times the width, the length must be $3x$. A picture will help.

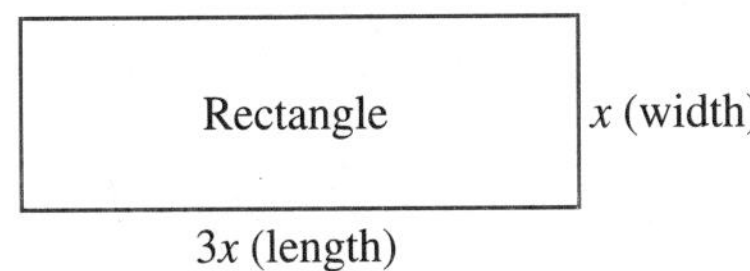

Figure 1

Step 3 *Reread and write an equation.* Since the perimeter is the sum of the sides, it must be $x + x + 3x + 3x$ (the sum of the four sides). But the perimeter is also given as 72 centimeters. Hence,

$$x + x + 3x + 3x = 72$$

Step 4 *Solve the equation.*

$$x + x + 3x + 3x = 72$$

$$8x = 72$$

$$x = 9$$

Answer

3 Width 7 cm, length 14 cm

Step 5 *Write your answer.* The width, x, is 9 centimeters. The length, $3x$, must be 27 centimeters.

Step 6 *Reread and check.* From the diagram below, we see that these solutions check:

Perimeter is 72	Length = 3 × Width
$9 + 9 + 27 + 27 = 72$	$27 = 3 \cdot 9$

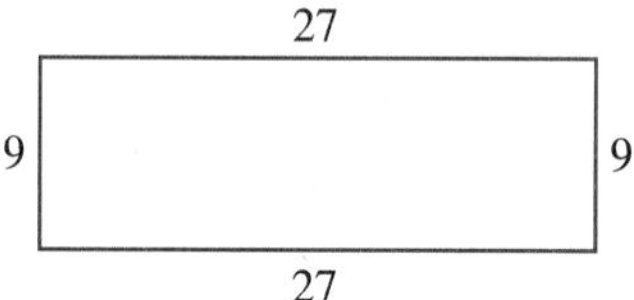

Figure 2

4 Joyce is 21 years older than her son Travis. In six years the sum of their ages will be 49. How old are they now?

EXAMPLE 4 Jo Ann is 22 years older than her daughter Stacey. In six years the sum of their ages will be 42. How old are they now?

Solution

Step 1 *Read and list.*

Known items: Jo Ann is 22 years older than Stacey. Six years from now their ages will add to 42.

Unknown items: Their ages now

Step 2 *Assign a variable and translate the information.* Let x = Stacey's age now. Since Jo Ann is 22 years older than Stacey, her age is $x + 22$.

Step 3 *Reread and write an equation.* As an aid in writing the equation we use the following table:

	Now	In six years
Stacey	x	$x + 6$
Jo Ann	$x + 22$	$x + 28$

Their ages in six years will be their ages now plus 6.

Since the sum of their ages six years from now is 42, we write the equation as

$$(x + 6) + (x + 28) = 42$$

↑ Stacey's age in 6 years ↑ Jo Ann's age in 6 years

Step 4 *Solve the equation.*

$$\begin{aligned} x + 6 + x + 28 &= 42 \\ 2x + 34 &= 42 \\ 2x &= 8 \\ x &= 4 \end{aligned}$$

Step 5 *Write your answer.* Stacey is now 4 years old, and Jo Ann is $4 + 22 = 26$ years old.

Step 6 *Reread and check.* To check, we see that in six years, Stacey will be 10 and Jo Ann will be 32. The sum of 10 and 32 is 42, which checks.

Answer

4 Travis is 8; Joyce is 29.

4.6 Evaluating Formulas

Introduction . . . In mathematics a formula is an equation that contains more than one variable. The equation $P = 2w + 2l$ is an example of a formula. As you know from Chapter 1, this formula tells us the relationship between the perimeter P of a rectangle, its length l, and its width w.

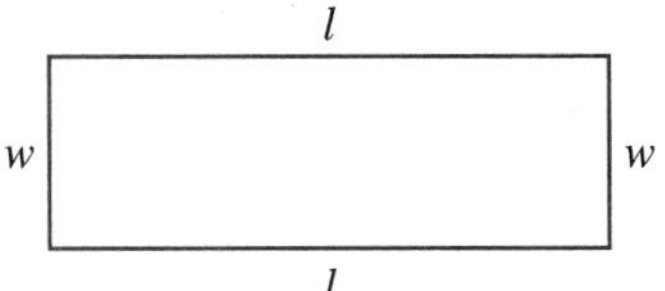

Figure 1

There are many formulas with which you may be familiar already. Perhaps you have used the formula $d = r \cdot t$ to find out how far you would go if you traveled at 50 miles an hour for 3 hours. If you take a chemistry class while you are in college, you will certainly use the formula that gives the relationship between the two temperature scales, Fahrenheit and Celsius, that we mentioned at the beginning of this chapter.

$$F = \frac{9}{5}C + 32$$

Although there are many kinds of problems we can work using formulas, we will limit ourselves to those that require only substitutions. The examples that follow illustrate this type of problem.

EXAMPLE 1 Use the formula $P = 2w + 2l$ to find the perimeter P of a rectangle if the width is $w = 8$ feet and the length is $l = 13$ feet.

Solution The perimeter P is the distance around the outside of the rectangle. To find P when w is 8 and l is 13, we substitute:

$$\begin{aligned} \text{When} \quad & w = 8 \text{ and } l = 13 \\ \text{the formula} \quad & P = 2w + 2l \\ \text{becomes} \quad & P = 2 \cdot 8 + 2 \cdot 13 \\ & = 16 + 26 \\ & = 42 \end{aligned}$$

The rectangle has a perimeter of 42 feet.

Practice Problems

1 If the length and the width of the rectangle in Example 1 are both increased by 2 feet, find the new perimeter of the rectangle.

Answer

1 50 feet

2 Use the formula in Example 2 to find C when F is 77 degrees.

Note

The formula we are using here,

$$C = \frac{5}{9}(F - 32)$$

is an alternative form of the formula we mentioned in the introduction,

$$F = \frac{9}{5}C + 32$$

Both formulas describe the same relationship between the two temperature scales. If you go on to take an algebra class, you will learn how to convert one formula into the other.

3 Use the formula in Example 3 to find y when x is 0.

4 Use the formula in Example 4 to find y when x is -3.

Answers

2 25 degrees Celsius **3** 6

4 $\frac{10}{3}$

EXAMPLE 2 Use the formula $C = \frac{5}{9}(F - 32)$ to find C when F is 95 degrees.

Solution Substituting 95 for F in the formula gives us the following.

$$\begin{aligned} \text{When} \quad F &= 95 \\ \text{the formula} \quad C &= \frac{5}{9}(F - 32) \\ \text{becomes} \quad C &= \frac{5}{9}(95 - 32) \\ &= \frac{5}{9}(63) \\ &= \frac{5}{9} \cdot \frac{63}{1} \\ &= \frac{315}{9} \\ &= 35 \end{aligned}$$

A temperature of 95 degrees Fahrenheit is the same as a temperature of 35 degrees Celsius.

EXAMPLE 3 Use the formula $y = 2x + 6$ to find y when x is -2.

Solution Proceeding as we have in the previous examples gives us the following.

$$\begin{aligned} \text{When} \quad x &= -2 \\ \text{the formula} \quad y &= 2x + 6 \\ \text{becomes} \quad y &= 2(-2) + 6 \\ &= -4 + 6 \\ &= 2 \end{aligned}$$

In some cases evaluating a formula also involves solving an equation, as the next example illustrates.

EXAMPLE 4 Find y when x is 3 in the formula $2x + 3y = 4$.

Solution First we substitute 3 for x; then we solve the resulting equation for y.

$$\begin{aligned} \text{When} \quad x &= 3 \\ \text{the equation} \quad 2x + 3y &= 4 \\ \text{becomes} \quad 2(3) + 3y &= 4 \\ 6 + 3y &= 4 \\ 3y &= -2 && \text{Add } -6 \text{ to each side} \\ y &= -\frac{2}{3} && \text{Divide each side by 3} \end{aligned}$$

4.7 Paired Data and Equations in Two Variables

Introduction . . . The introduction to this chapter contained a table and a line graph that showed the relationship between the Fahrenheit and Celsius temperature scales. We repeat them below.

Table 1 Comparing temperatures on two scales (numeric)

Temperature in Degrees Celsius	Temperature in Degrees Fahrenheit
0°C	32°F
25°C	77°F
50°C	122°F
75°C	167°F
100°C	212°F

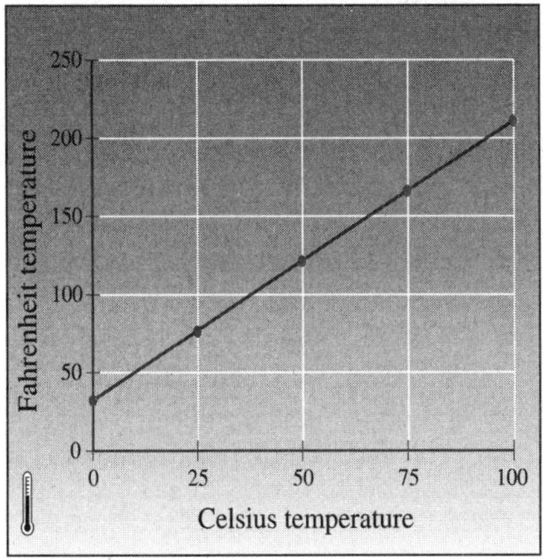

Figure 1 A line graph of the data in the table (geometric)

The data in Table 1 is called *paired data* because it is organized so that each number in the first column is paired with a specific number in the second column: A Celsius temperature of 100°C (first column) corresponds to a Fahrenheit temperature of 212°F (second column). The information in Figure 1 is also paired data because each dot on the line graph comes from one of the pairs of numbers in the table: The upper rightmost dot on the line graph corresponds to 100°C and 212°F.

We considered paired data again in Section 4.6, when we used formulas to produce pairs of numbers: When we substitute $C = 100$ into the formula below, we obtain $F = 212$, the same pair of numbers shown in Table 1 and again in the line graph in Figure 1.

$$F = \frac{9}{5}C + 32$$

The concept of paired data is very important in mathematics, so we want to continue our study of it. We start with a more detailed look at equations that contain two variables.

Recall that a solution to an equation in one variable is a single number that when replaced for the variable in the equation, turns the equation into a true statement. For example, $x = 5$ is a solution to the equation $2x - 4 = 6$, because replacing x with 5 turns the equation into $6 = 6$, a true statement.

Next, consider the equation $x + y = 5$. It has two variables instead of one. Therefore, a solution to $x + y = 5$ will have to consist of two numbers, one for x and one for y, that together make the equation a true statement. One pair of numbers is $x = 2$ and $y = 3$, because when we substitute 2 for x and 3 for y into the equation $x + y = 5$, the result is a true statement. That is,

When	$x = 2$ and $y = 3$
the equation	$x + y = 5$
becomes	$2 + 3 = 5$
	$5 = 5$ A true statement

To simplify our work, we write the pair of numbers $x = 2$, $y = 3$ in the shorthand form (2, 3). The expression (2, 3) is called an *ordered pair* of numbers. Here is the formal definition:

DEFINITION A pair of numbers enclosed in parentheses and separated by a comma, such as (2, 3), is called an **ordered pair** of numbers. The first number in the pair is called the ***x*-coordinate** of the ordered pair, while the second number is called the ***y*-coordinate**. For the ordered pair (2, 3), the x-coordinate is 2 and the y-coordinate is 3.

In the equation $x + y = 5$, we find that (2, 3) is not the only solution. Another solution is (0, 5), because when $x = 0$ and $y = 5$, then

$0 + 5 = 5$ A true statement

Still another solution is the ordered pair (−7, 12), because

When	$x = -7$ and $y = 12$
the equation	$x + y = 5$
becomes	$-7 + 12 = 5$
	$5 = 5$ A true statement

As you can imagine, there are many more ordered pairs that are solutions to the equation $x + y = 5$. As a matter of fact, for any number we choose for x, there is another number we can use for y that will make the equation a true statement. There is an infinite number of ordered pairs that are solutions to the equation $x + y = 5$.

EXAMPLE 1 Fill in the ordered pairs (0,   ), (, −2), and (3,   ) so they are solutions to the equation $2x + 3y = 6$.

Solution To complete the ordered pair (0,   ), we substitute 0 for x in the equation and then solve for y:

When	$x = 0$
the equation	$2x + 3y = 6$
becomes	$2 \cdot 0 + 3y = 6$
	$3y = 6$
	$y = 2$

Therefore, the ordered pair (0, 2) is a solution to $2x + 3y = 6$.

Practice Problems

1 Fill in the ordered pairs (0,   ), (, 0), and (−5,   ) so they are solutions to the equation $3x + 5y = 15$.

Answer

1 (0, 3), (5, 0), and (−5, 6)

To complete the ordered pair (, −2), we substitute −2 for y and then solve for x.

$$\begin{aligned} \text{When} \quad & y = -2 \\ \text{the equation} \quad & 2x + 3y = 6 \\ \text{becomes} \quad & 2x + 3(-2) = 6 \\ & 2x - 6 = 6 \\ & 2x = 12 \\ & x = 6 \end{aligned}$$

Therefore, the ordered pair (6, −2) is another solution to our equation.

Finally, to complete the ordered pair (3,), we substitute 3 for x and then solve for y. The result is $y = 0$. The ordered pair (3, 0) is a third solution to our equation.

EXAMPLE 2 Use the equation $5x - 2y = 20$ to complete the table below.

x	y
2	
0	
	5
	0

Solution Filling in the table is equivalent to completing the following ordered pairs: (2,), (0,), (, 5), and (, 0). We proceed as in Example 1.

When $x = 2$, we have

$$\begin{aligned} 5 \cdot 2 - 2y &= 20 \\ 10 - 2y &= 20 \\ -2y &= 10 \\ y &= -5 \end{aligned}$$

When $x = 0$, we have

$$\begin{aligned} 5 \cdot 0 - 2y &= 20 \\ 0 - 2y &= 20 \\ -2y &= 20 \\ y &= -10 \end{aligned}$$

When $y = 5$, we have

$$\begin{aligned} 5x - 2 \cdot 5 &= 20 \\ 5x - 10 &= 20 \\ 5x &= 30 \\ x &= 6 \end{aligned}$$

When $y = 0$, we have

$$\begin{aligned} 5x - 2 \cdot 0 &= 20 \\ 5x - 0 &= 20 \\ 5x &= 20 \\ x &= 4 \end{aligned}$$

Using these results, we complete our table.

x	y
2	−5
0	−10
6	5
4	0

2 Use the equation $5x +$ [illegible] 20 to complete the table b[illegible]

x	y
2	
0	

Answer

2

x	y
2	5
0	10
2	5
4	0

3 Complete the table below for the equation $y = \frac{1}{2}x + 1$.

x	y
0	
4	
	7
	−3

EXAMPLE 3 Complete the table below for the equation $y = 2x + 1$.

x	y
0	
5	
	7
	3

Solution When $x = 0$, we have

$$y = 2 \cdot 0 + 1$$
$$y = 0 + 1$$
$$y = 1$$

When $x = 5$, we have

$$y = 2 \cdot 5 + 1$$
$$y = 10 + 1$$
$$y = 11$$

When $y = 7$, we have

$$7 = 2x + 1$$
$$6 = 2x$$
$$3 = x$$

When $y = 3$, we have

$$3 = 2x + 1$$
$$2 = 2x$$
$$1 = x$$

The completed table looks like this:

x	y
0	1
5	11
3	7
1	3

4 Which of the ordered pairs (1, 5) and (2, 4) are solutions to the equation $y = 5x - 6$?

EXAMPLE 4 Which of the ordered pairs (1, 5) and (2, 4) are solutions to the equation $y = 3x + 2$?

Solution If an ordered pair is a solution to an equation, then it must yield a true statement when the coordinates of the ordered pair are substituted for x and y in the equation.

First, we try (1, 5) in the equation $y = 3x + 2$:

$$5 = 3 \cdot 1 + 2$$
$$5 = 3 + 2$$
$$5 = 5 \quad \text{A true statement}$$

Next, we try (2, 4) in the equation:

$$4 = 3 \cdot 2 + 2$$
$$4 = 6 + 2$$
$$4 = 8 \quad \text{A false statement}$$

The ordered pair (1, 5) is a solution to the equation $y = 3x + 2$, but (2, 4) is not a solution to the equation.

Answers

3

x	y
0	1
4	3
12	7
−8	−3

4 (2, 4)

4.8 The Rectangular Coordinate System

Introduction . . . The table and line graph below are similar to those that we have used previously in this chapter to discuss temperature. Note that we have extended both the table and the line graph to show some temperatures below zero on both scales.

Table 1 Comparing temperatures on two scales

Temperature in Degrees Celsius	Temperature in Degrees Fahrenheit
−100°C	−148°F
−75°C	−103°F
−50°C	−58°F
−25°C	−13°F
0°C	32°F
25°C	77°F
50°C	122°F
75°C	167°F
100°C	212°F

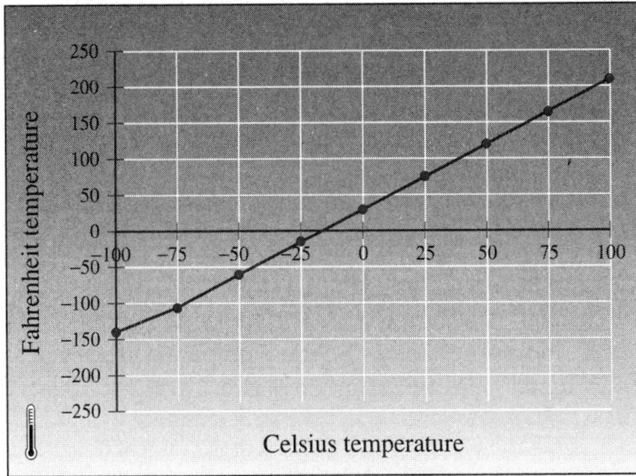

Figure 1 A line graph of the data in Table 1

We know from the previous section that the information in both the table and the line graph is paired data. We also know that solutions to equations in two variables consist of pairs of numbers that together satisfy the equations. What we want to do next is look at solutions to equations in two variables from a visual perspective. In order to do so, we need to standardize the way in which we present paired data visually. This is accomplished with the *rectangular coordinate system.*

The Rectangular Coordinate System

The rectangular coordinate system can be used to plot (or graph) pairs of numbers (see Figure 2). It consists of two number lines, called *axes*, which intersect at right angles. (A right angle is a 90° angle.) The point at which the axes intersect is called the *origin*.

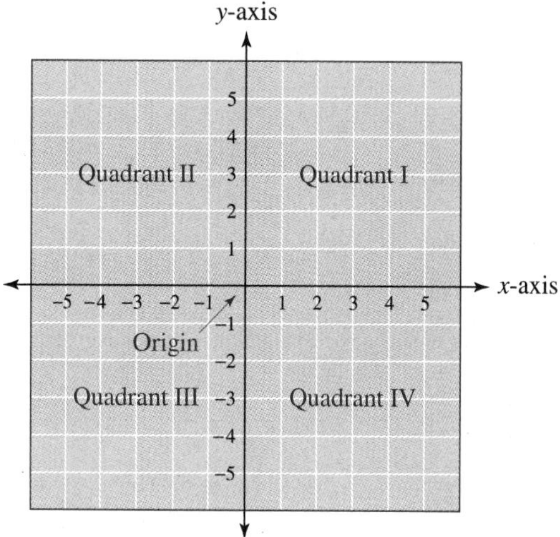

Figure 2

The horizontal number line is exactly the same as the real number line and is called the *x-axis*. The vertical number line is also the same as the real number line with the positive direction up and the negative direction down. It is called the *y-axis*. As you can see, the axes divide the plane into four regions, called *quadrants*, which are numbered I through IV in a counterclockwise direction.

Since the rectangular coordinate system consists of two number lines, one called the x-axis and the other called the y-axis, we can plot pairs of numbers such as $x = 2$ and $y = 3$. As a matter of fact, each point in the rectangular coordinate system is named by exactly one pair of numbers. We call the pair of numbers that name a point the *coordinates* of that point. To find the point that is associated with the pair of numbers $x = 2$ and $y = 3$, we start at the origin and move 2 units horizontally to the right and then 3 units vertically up (see Figure 3). The place where we end up is the point named by the pair of numbers $x = 2$, $y = 3$, which we write in shorthand form as the ordered pair (2, 3).

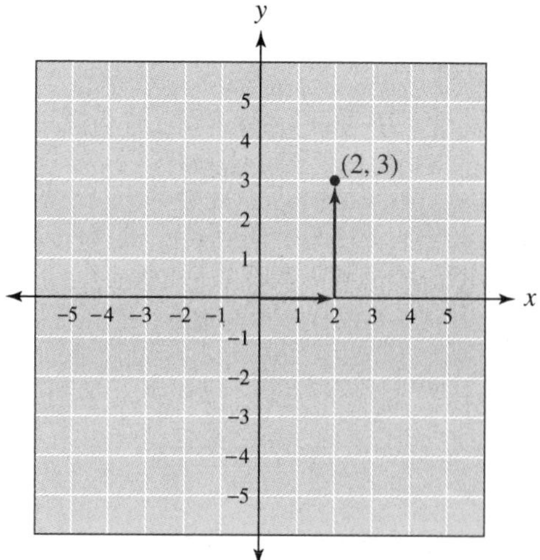

Figure 3

In general, to graph an ordered pair (a, b) on the rectangular coordinate system, we start at the origin and move a units right or left (right if a is positive, left if a is negative). From there we move b units up or down (up if b is positive, down if b is negative). The point where we end up is the graph of the ordered pair (a, b).

EXAMPLE 1 Plot (graph) the ordered pairs (2, 3), (−2, 3), (−2, −3), and (2, −3).

Solution To graph the ordered pair (2, 3), we start at the origin and move 2 units to the right, then 3 units up. We are now at the point whose coordinates are (2, 3). We plot the other three ordered pairs in the same manner (Figure 4).

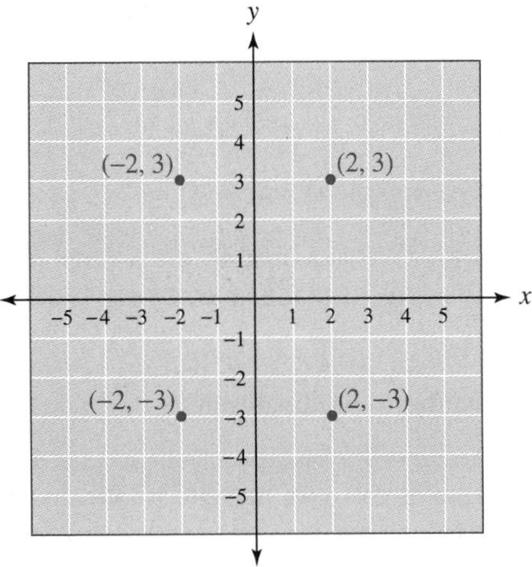

Figure 4

Note Looking at Example 1, we see that any point in quadrant I must have positive x- and y-coordinates (+, +). In quadrant II, x-coordinates are negative and y-coordinates are positive, (−, +). In quadrant III, both coordinates are negative (−, −). Finally, in quadrant IV, all ordered pairs must have the form (+, −).

EXAMPLE 2 Plot the ordered pairs (1, −4), $\left(\frac{1}{2}, 3\right)$, (2, 0), (0, −2), and (−3, 0).

Solution

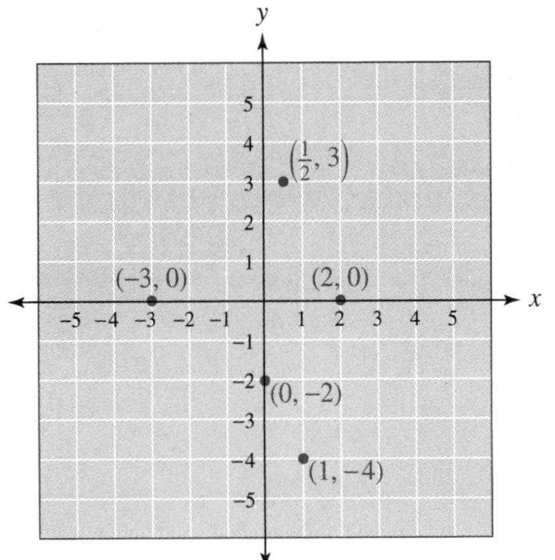

Practice Problems

1 Plot the ordered pairs (3, 2), (3, −2), (−3, −2), and (−3, 2) on the coordinate system used in Example 1.

2 Plot the ordered pairs (−1, −4), $\left(-\frac{1}{2}, 3\right)$, (0, 2), (5, 0), (0, −5), and (−1, 0) on the coordinate system used in Example 2.

Answers

1 See solutions section.
2 See solutions section.

3 Give the coordinates of each point in the figure below.

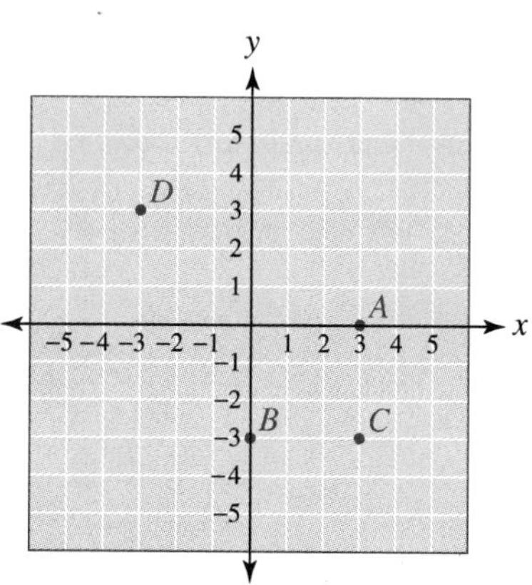

EXAMPLE 3 Give the coordinates of each point in Figure 5.

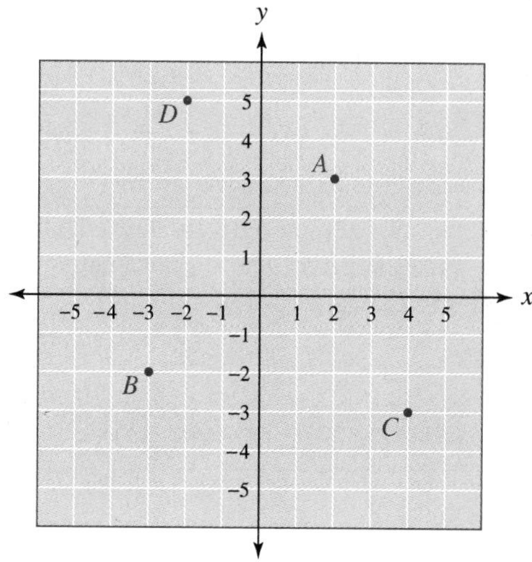

Figure 5

Solution A is named by the ordered pair (2, 3). B is named by the ordered pair (−3, −2). C is named by the ordered pair (4, −3). D is named by the ordered pair (−2, 5).

4 The points (2, 0), (−3, 0), $\left(\frac{5}{2}, 0\right)$, and (−8, 0) all lie on which axis?

EXAMPLE 4 Where are all the points that have coordinates of the form $(x, 0)$?

Solution Since the y-coordinate is 0, these points must lie on the x-axis. Remember, the y-coordinate tells us how far up or down we move to find the point in question. If the y-coordinate is 0, then we don't move up or down at all. Therefore, we must stay on the x-axis.

5 a Does the point (−3, −2) lie on the line shown in Example 5?

b Does the point (3, 2) lie on the line shown in Example 5?

EXAMPLE 5 Graph the points (1, 2) and (3, 4) and draw a line through them. Then use your result to answer these questions.

a Does the graph of (2, 3) lie on this line?

b Does the graph of (−3, −5) lie on this line?

Solution Figure 6 shows the graphs of (1, 2) and (3, 4) and the line that connects them. The line does not pass through the point (−3, −5) but does pass through (2, 3).

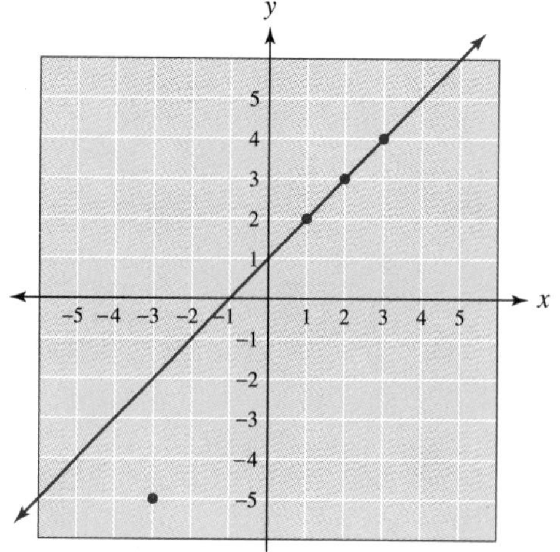

Figure 6

Answers

3 A is (3, 0); B is (0, −3); C is (3, −3); D is (−3, 3).

4 The x-axis

5 a Yes **b** No

Graphing Straight Lines

In this section we use what we have learned in the previous two sections to graph straight lines.

EXAMPLE 1 Graph the solution set for $x + y = 4$.

Solution We know from our work in the previous section that there is an infinite number of solutions to this equation, and that each solution is an ordered pair of numbers. Some of the ordered pairs that satisfy the equation $x + y = 4$ are (0, 4), (2, 2), (4, 0), and (5, −1). If we plot these ordered pairs, we have the points shown in Figure 1.

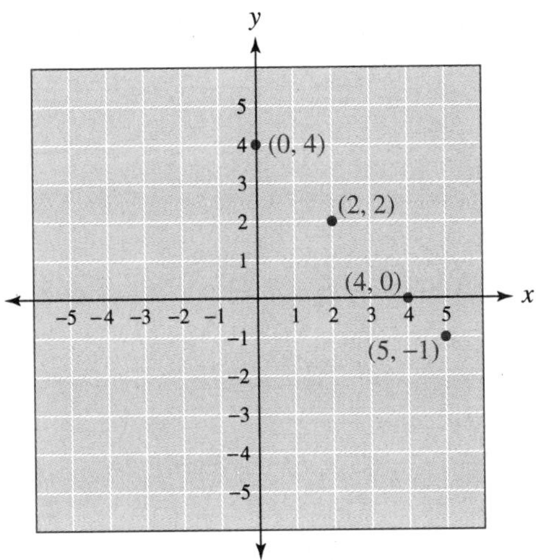

Figure 1 The graph of four solutions to $x + y = 4$

Notice that all four points lie in a straight line. If we were to find other solutions to the equation $x + y = 4$, we would find that they too would line up with the points shown in Figure 1. In fact, every solution to $x + y = 4$ is a point that lies in line with the points shown in Figure 1. Therefore, to graph the solution set to $x + y = 4$, we simply draw a line through the points in Figure 1, to obtain the graph shown in Figure 2.

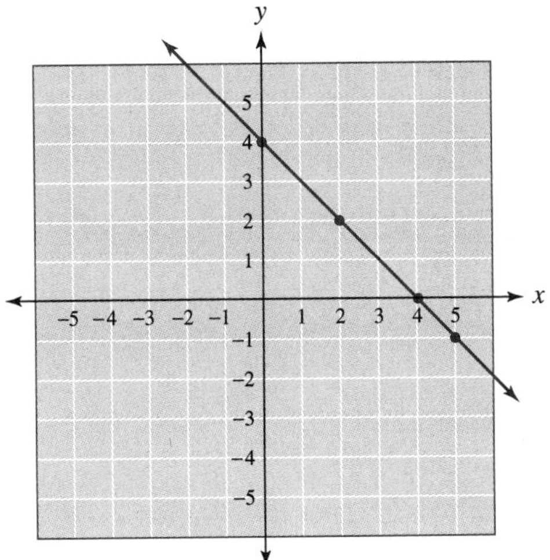

Figure 2 The graph of the line $x + y = 4$

Practice Problems

1 Graph the solution set for $x + y = 3$.

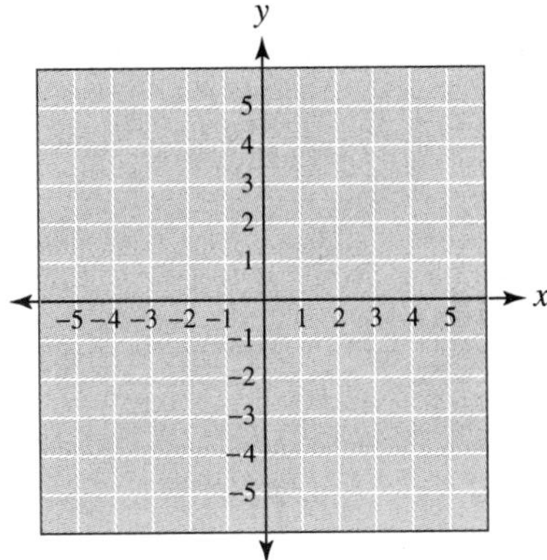

Note

In Figure 2, every ordered pair that satisfies $x + y = 4$ has its graph on the line and any point on the line has coordinates that satisfy the equation $x + y = 4$.

Answers

For answers to the Practice Problems in this section, see the solutions section.

To Graph a Straight Line

Step 1 Find any three ordered pairs that satisfy the equation. This is usually accomplished by substituting a number for one of the variables in the equation, and then solving for the other variable.

Step 2 Plot the three ordered pairs found in Step 1. (Actually, we need only two points to graph a straight line. The third point serves as a check. If all three points don't line up, then we have made a mistake in our work.)

Step 3 Draw a line through the three points you plotted in Step 2.

2 Graph the equation $y = -2x + 1$ by completing the ordered pairs (1,), (0,), and (−1,).

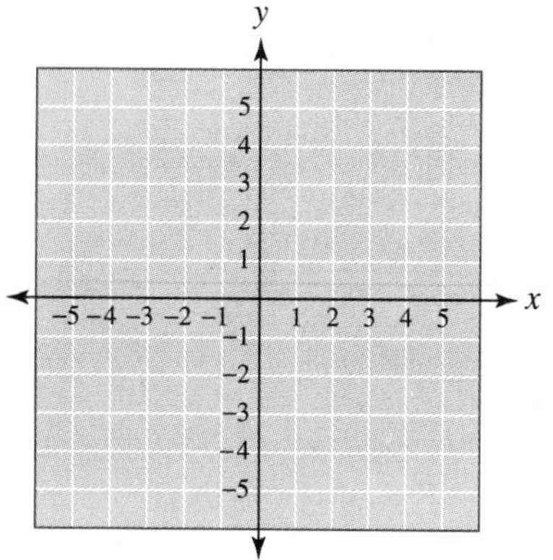

EXAMPLE 2 Graph the equation $y = 2x + 1$ by completing the ordered pairs (1,), (0,), and (−1,).

Solution To complete the ordered pair (1,), we let $x = 1$ in the equation $y = 2x + 1$:

$$y = 2 \cdot 1 + 1$$
$$y = 2 + 1$$
$$y = 3$$

To complete the ordered pair (0,), we let $x = 0$ in the equation:

$$y = 2 \cdot 0 + 1$$
$$y = 0 + 1$$
$$y = 1$$

To complete the ordered pair (−1,), we let $x = -1$ in the equation:

$$y = 2(-1) + 1$$
$$y = -2 + 1$$
$$y = -1$$

The ordered pairs (1, 3), (0, 1), and (−1, −1) each satisfy the equation $y = 2x + 1$. Graphing each ordered pair and then drawing a line through them, we have the graph shown in Figure 3.

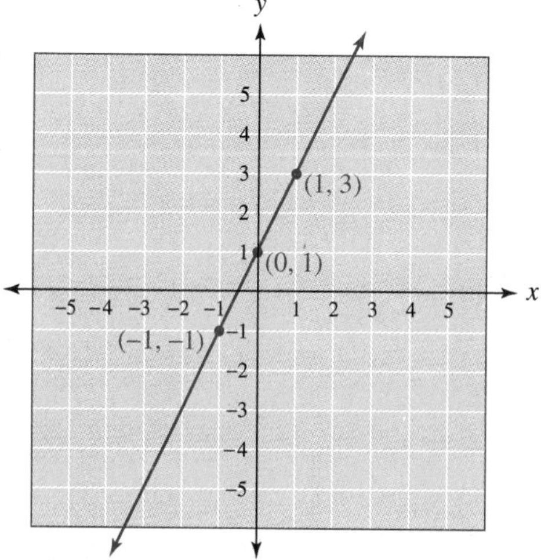

Figure 3 The graph of $y = 2x + 1$

EXAMPLE 3 Graph the equation $y = 3x - 2$.

Solution We need to find three solutions to the equation. We can do so by choosing some numbers to use for x and then seeing what values of y they yield. Since the choice of the numbers we will use for x is up to us, let's make it easy on ourselves and use numbers like $x = -1$, 0, and 2, that are easy to work with.

Let $x = -1$: $y = 3(-1) - 2$
$y = -3 - 2$
$y = -5$ $(-1, -5)$ is one solution.

Let $x = 0$: $y = 3 \cdot 0 - 2$
$y = 0 - 2$
$y = -2$ $(0, -2)$ is another solution.

Let $x = 2$: $y = 3 \cdot 2 - 2$
$y = 6 - 2$
$y = 4$ $(2, 4)$ is a third solution.

Now we plot the solutions we found above and then draw a line through them.

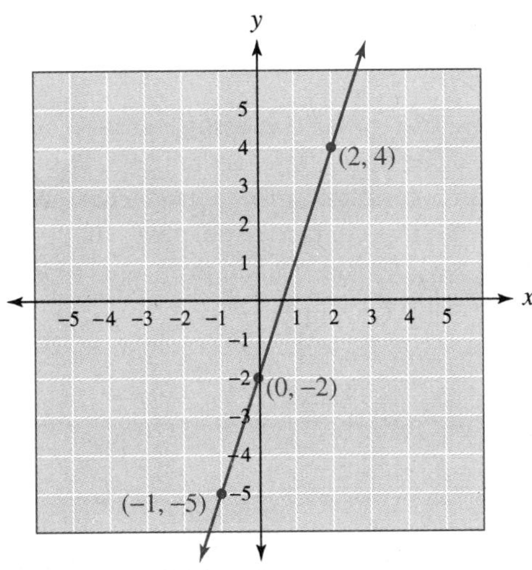

Figure 4 The graph of $y = 3x - 2$

3 Graph the equation $y = 2x - 3$.

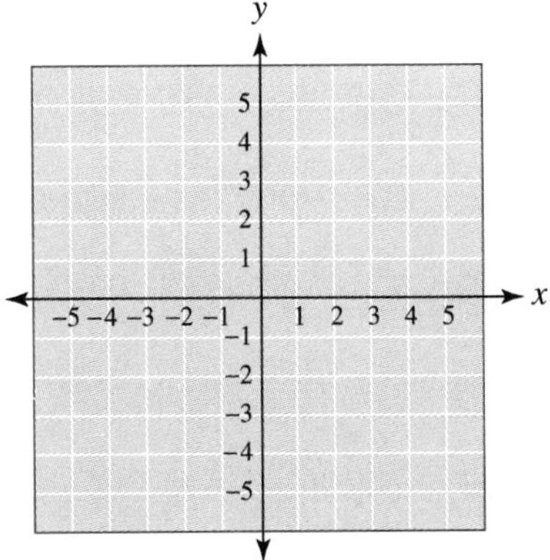

Note
Every point on the line shown in Figure 4 has coordinates that satisfy the equation $y = 3x - 2$. Likewise, every ordered pair that satisfies the equation $y = 3x - 2$ has its graph on the line shown in Figure 4. We say that there is a one-to-one correspondence between points on the graph of $y = 3x - 2$ and ordered pairs that satisfy the equation $y = 3x - 2$.

EXAMPLE 4 Graph the line $3x + 2y = 6$.

Solution This time, let's substitute values for both x and y to find the three solutions we need to draw the graph.

Let $x = 0$: $3 \cdot 0 + 2y = 6$
$0 + 2y = 6$
$2y = 6$
$y = 3$ $(0, 3)$ is one solution.

Let $y = 0$: $3x + 2 \cdot 0 = 6$
$3x + 0 = 6$
$3x = 6$
$x = 2$ $(2, 0)$ is a second solution.

Let $y = -3$: $3x + 2(-3) = 6$
$3x - 6 = 6$
$3x = 12$
$x = 4$ $(4, -3)$ is a third solution.

4 Graph the line $3x - 2y = 6$.

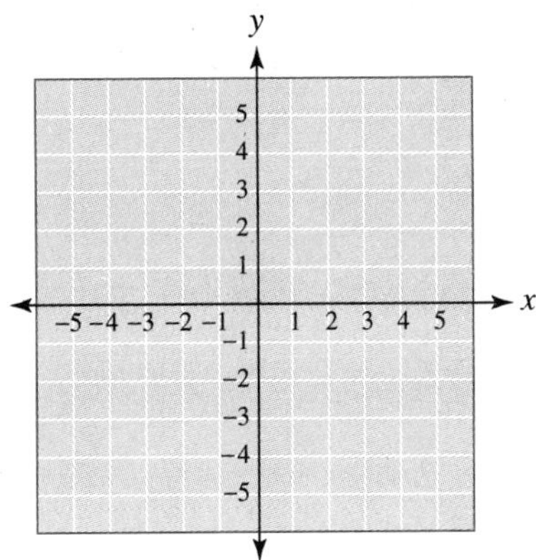

Plotting these three solutions and drawing a line through them, we have the graph shown in Figure 5.

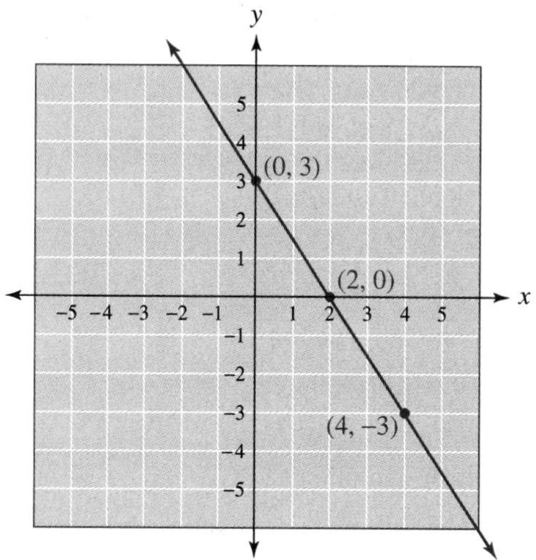

Figure 5 The graph of $3x + 2y = 6$

5 Graph the line $x = -3$.

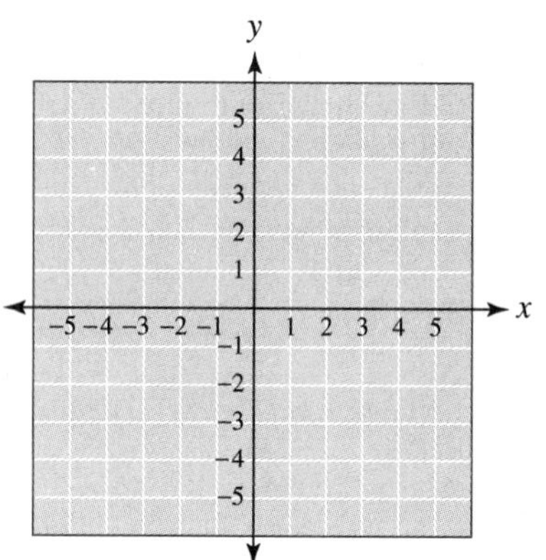

EXAMPLE 5 Graph the line $x = 3$.

Solution When we encounter an equation like $x = 3$, we assume that it looks like $x + 0y = 3$, so that no matter what we use for y, x will always be 3. A solution to $x = 3$ is any ordered pair that has an x-coordinate of 3; y can be any number, so long as x is 3. Some ordered pairs with this characteristic are (3, 0), (3, 1), (3, −1), (3, 4), and (3, −4), to name just a few. Plotting each of these, we see that they line up vertically in a straight line (Figure 6).

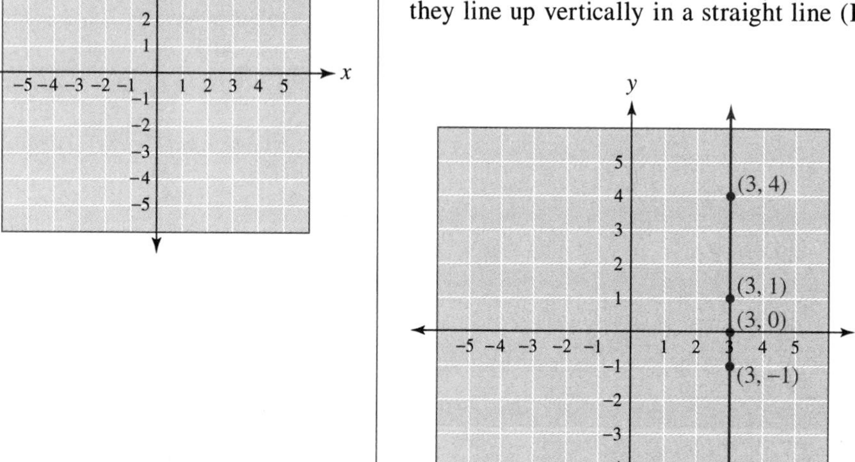

Figure 6 The graph of $x = 3$

As you can see from Example 5, the graph of $x = 3$ is a vertical line. If we take a similar approach to graphing the equation $y = -4$, we find that the graph is a *horizontal* line, since any solution to the equation $y = -4$ will be an ordered pair with a y-coordinate of -4. For example, the points $(0, -4)$, $(3, -4)$, $(-2, -4)$, and $(4, -4)$ are all solutions to the equation $y = -4$. If we were to graph these points, they would line up in a horizontal line 4 units below the x-axis. Plot these points on the coordinate system in Figure 6 and see for yourself.

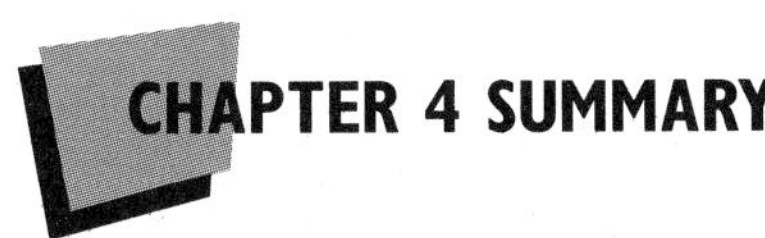

CHAPTER 4 SUMMARY

Combining Similar Terms [4.1]

Two terms are similar terms if they have the same variable part. The expressions $7x$ and $2x$ are similar because the variable part in each is the same. Similar terms are combined by using the distributive property.

Examples

1 $7x + 2x = (7 + 2)x$
$= 9x$

Finding the Value of an Algebraic Expression [4.1]

An algebraic expression is a mathematical expression that contains numbers and variables. Expressions that contain a variable will take on different values depending on the value of the variable.

2 When $x = 5$, the expression $2x + 7$ becomes

$$2(5) + 7 = 10 + 7 = 17$$

The Solution to an Equation [4.2]

The solution to an equation is the number that, when used in place of the variable, makes the equation a true statement.

The Addition Property of Equality [4.2]

Let A, B, and C represent algebraic expressions.

If $A = B$
then $A + C = B + C$

In words: Adding the same amount to both sides of an equation will not change the solution.

3 We solve $x - 4 = 9$ by adding 4 to each side.

$$x - 4 = 9$$
$$x - 4 + \mathbf{4} = 9 + \mathbf{4}$$
$$x + 0 = 13$$
$$x = 13$$

The Multiplication Property of Equality [4.3]

Let A, B, and C represent algebraic expressions with C not equal to 0.

If $A = B$
then $AC = BC$

In words: Multiplying both sides of an equation by the same nonzero number will not change the solution to the equation. This property holds for division as well.

4 Solve $\frac{1}{3}x = 5$.

$$\frac{1}{3}x = 5$$
$$\mathbf{3} \cdot \frac{1}{3}x = \mathbf{3} \cdot 5$$
$$x = 15$$

Steps Used to Solve a Linear Equation in One Variable [4.4]

Step 1 Simplify each side of the equation.

Step 2 Use the addition property of equality to get all variable terms on one side and all constant terms on the other side.

Step 3 Use the multiplication property of equality to get just one x isolated on either side of the equation.

Step 4 Check the solution in the original equation if necessary.

If the original equation contains fractions, you can begin by multiplying each side by the LCD for all fractions in the equation.

5 $2(x - 4) + 5 = -11$

$$2x - 8 + 5 = -11$$
$$2x - 3 = -11$$
$$2x - 3 + \mathbf{3} = -11 + \mathbf{3}$$
$$2x = -8$$
$$\frac{2x}{\mathbf{2}} = \frac{-8}{\mathbf{2}}$$
$$x = -4$$

6 When $w = 8$ and $l = 13$

the formula $P = 2w + 2l$

becomes $P = 2 \cdot 8 + 2 \cdot 13$

$= 16 + 26$

$= 42$

Evaluating Formulas [4.6]

In mathematics, a formula is an equation that contains more than one variable. For example, the formula for the perimeter of a rectangle is $P = 2l + 2w$. We evaluate a formula by substituting values for all but one of the variables, and then solving the resulting equation for that variable.

The Rectangular Coordinate System [4.8]

The rectangular coordinate system consists of two number lines, called axes, which intersect at right angles. The point at which the axes intersect is called the origin.

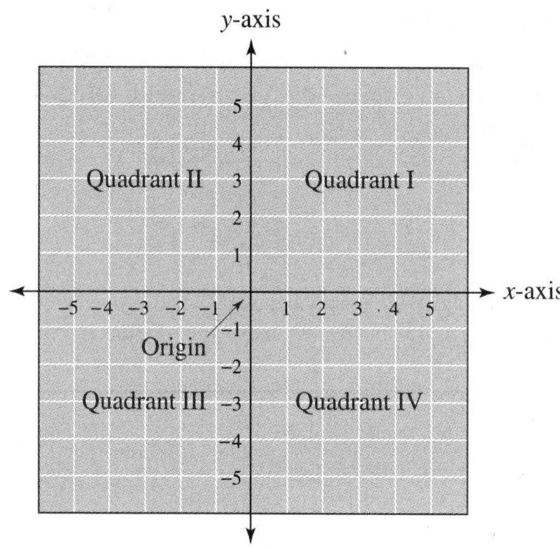

Graphing Ordered Pairs [4.8]

To graph an ordered pair (a, b) on the rectangular coordinate system, we start at the origin and move a units right or left (right if a is positive, left if a is negative). From there we move b units up or down (up if b is positive, down if b is negative). The point where we end up is the graph of the ordered pair (a, b).

7

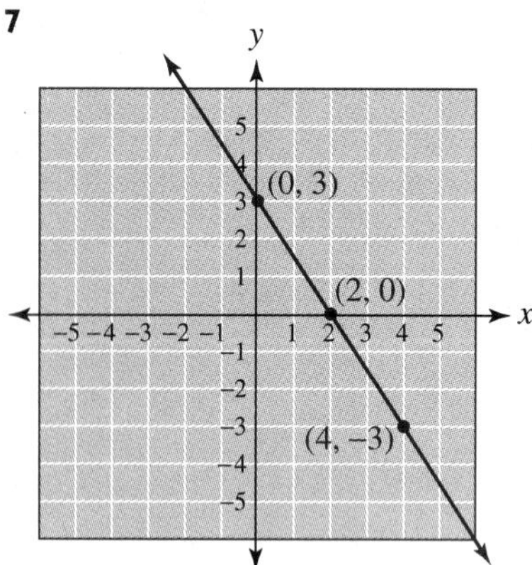

Figure 1 The graph of $3x + 2y = 6$

To Graph a Straight Line [4.9]

Step 1 Find any three ordered pairs that satisfy the equation. This is usually accomplished by substituting a number for one of the variables in the equation, and then solving for the other variable.

Step 2 Plot the three ordered pairs found in Step 1. (Actually, we need only two points to graph a straight line. The third point serves as a check. If all three points don't line up, then we have made a mistake in our work.)

Step 3 Draw a line through the three points you plotted in Step 2.

CHAPTER 4 REVIEW

Simplify the expressions by combining similar terms. **[4.1]**

1 $10x + 7x$

2 $8x - 12x$

3 $2a + 9a + 3 - 6$

4 $4y - 7y + 8 - 10$

5 $6x - x + 4$

6 $-5a + a + 4 - 3$

7 $2a - 6 + 8a + 2$

8 $12y - 4 + 3y - 9$

Find the value of each expression when x is 4. **[4.1]**

9 $10x + 2$

10 $5x - 12$

11 $-2x + 9$

12 $-x + 8$

13 Is $x = -3$ a solution to $5x - 2 = -17$? **[4.2]**

14 Is $x = 4$ a solution to $3x - 2 = 2x + 1$? **[4.2]**

Solve the equations. **[4.2, 4.3, 4.4]**

15 $x - 5 = 4$

16 $-x + 3 + 2x = 6 - 7$

17 $2x + 1 = 7$

18 $3x - 5 = 1$

19 $2x + 4 = 3x - 5$

20 $4x + 8 = 2x - 10$

21 $3(x - 2) = 9$

22 $4(x - 3) = -20$

23 $3(2x + 1) - 4 = -7$

24 $4(3x + 1) = -2(5x - 2)$

25 $5x + \frac{3}{8} = -\frac{1}{4}$

26 $\frac{7}{x} - \frac{2}{5} = 1$

27 The sum of a number and -3 is -5. Find the number. **[4.5]**

28 If twice a number is added to 3, the result is 7. Find the number. **[4.5]**

29 Three times the sum of a number and 2 is -6. Find the number. **[4.5]**

30 If 7 is subtracted from twice a number, the result is 5. Find the number. **[4.5]**

31 The length of a rectangle is twice its width. If the perimeter is 42 meters, find the length and the width. **[4.5]**

32 Patrick is 3 years older than Amy. In 5 years the sum of their ages will be 31. How old are they now? **[4.5]**

In Problems 33–36, use the equation $3x + 2y = 6$ to find y. **[4.6, 4.7]**

33 $x = -2$

34 $x = 6$

35 $x = 0$

36 $x = \frac{1}{3}$

In Problems 37–40, use the equation $3x + 2y = 6$ to find x. **[4.6, 4.7]**

37 $y = 3$

38 $y = -3$

39 $y = 0$

40 $y = -\frac{1}{2}$

In Problems 41–48, plot the points. (Put **41**, **43**, **45**, and **47** on one graph; put **42**, **44**, **46**, and **48** on the other.) **[4.8]**

41 (4, 2)

42 (4, −2)

43 (−4, 2)

44 (−4, −2)

45 (4, 0)

46 (0, 4)

47 (0, −4)

48 (−4, 0)

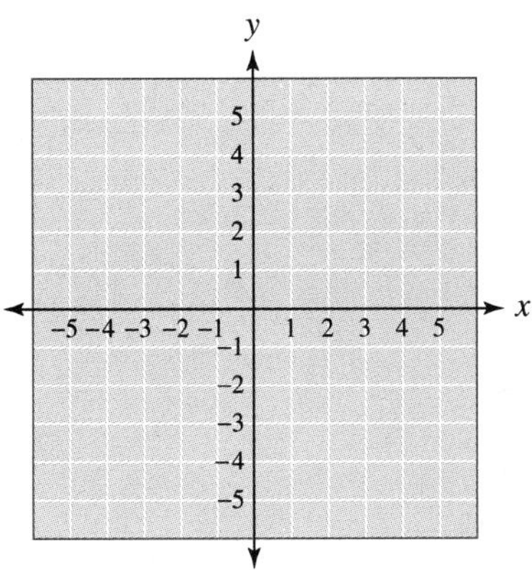

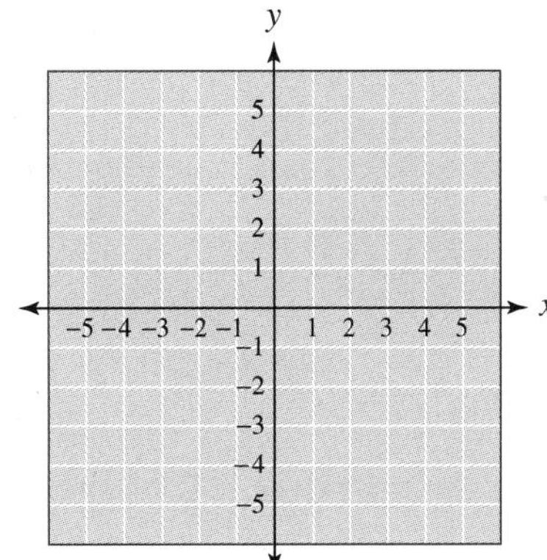

In Problems 49–58, graph each line. **[4.9]**

49 $x + y = 2$

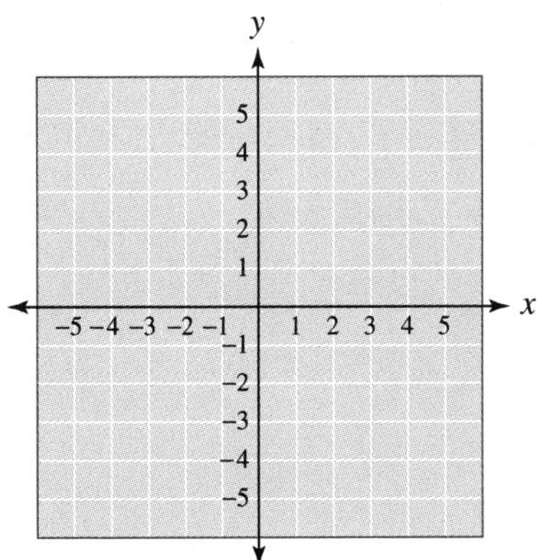

50 $x - y = 2$

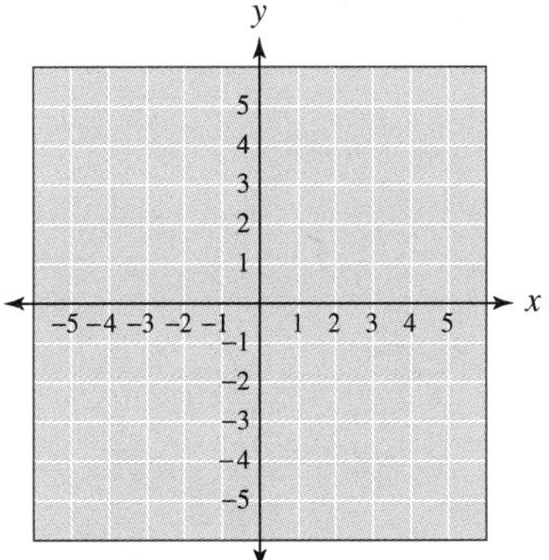

51 $3x + 2y = 6$

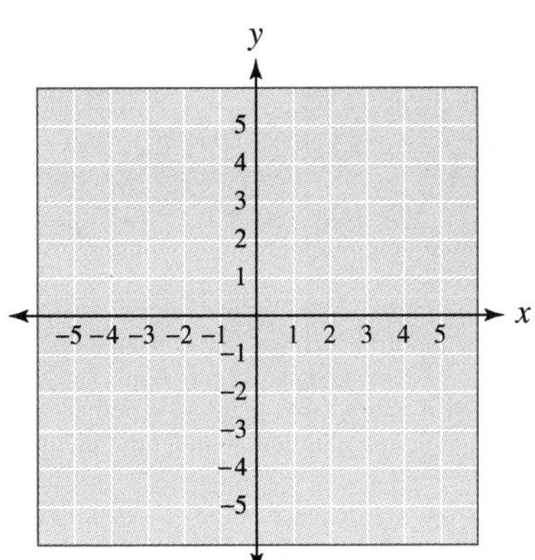

52 $2x + 3y = 6$

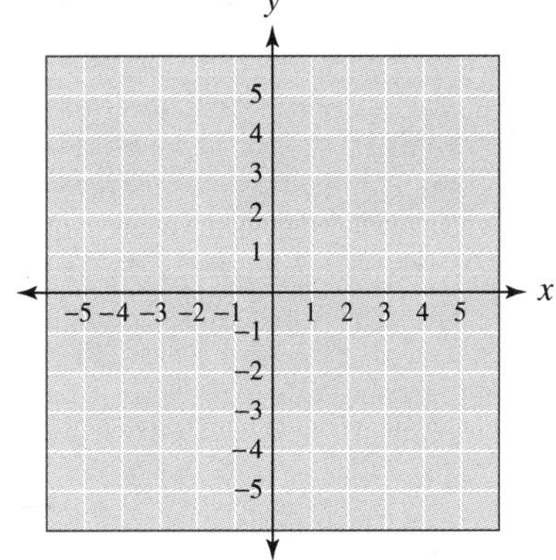

53 $y = \frac{1}{2}x$

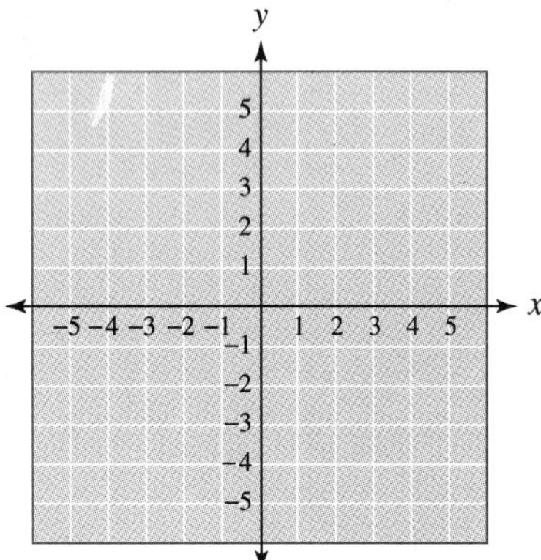

54 $y = 2x$

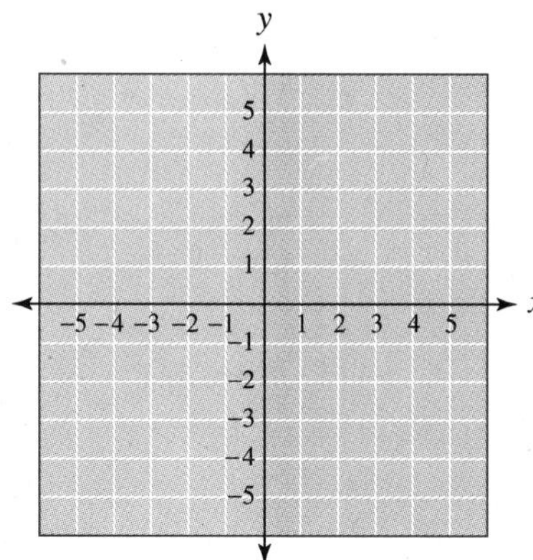

55 $y = 2x - 3$

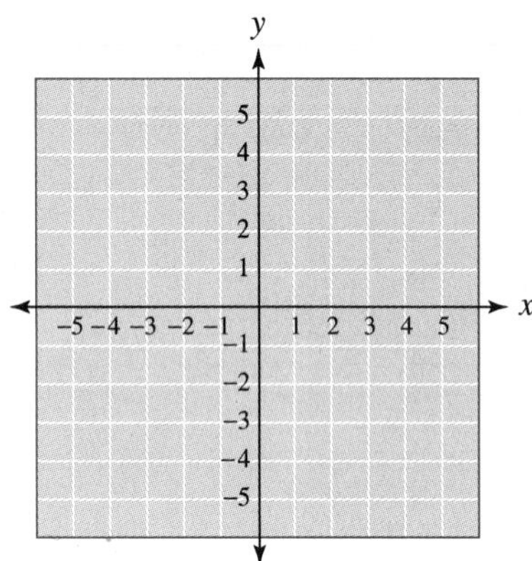

56 $y = \frac{1}{3}x + 1$

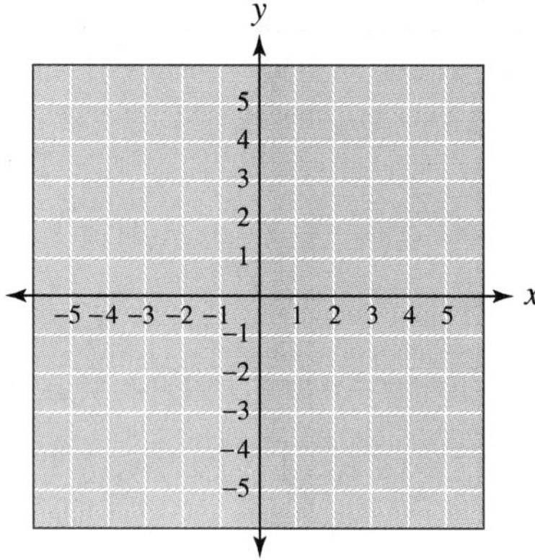

57 $x = -2$

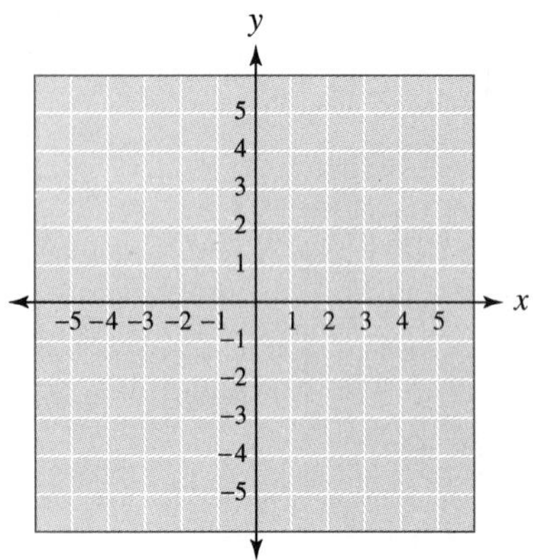

58 $y = 3$

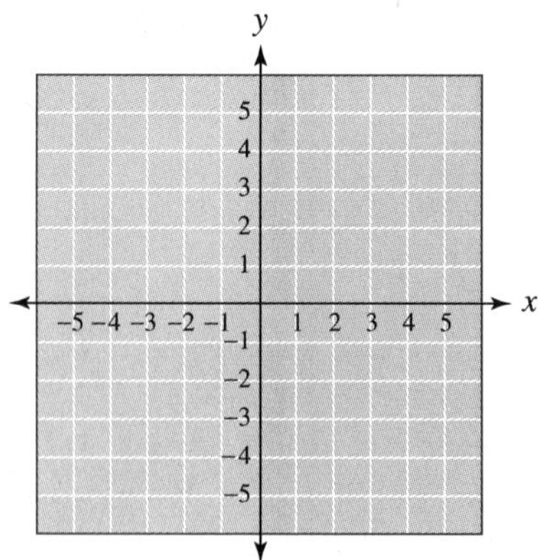

CHAPTER 4 TEST

Simplify each expression by combining similar terms.

1 $9x - 3x + 7 - 12$

2 $4b - 1 - b - 3$

Find the value of each expression when $x = 3$.

3 $3x - 12$

4 $-x + 12$

5 Is $x = -1$ a solution to $4x - 3 = -7$?

Solve each equation.

6 $x - 7 = -3$

7 $\frac{2}{3}y = 18$

8 $3x - 7 = 5x + 1$

9 $2(x - 5) = -8$

10 $3(2x + 3) = -3(x - 5)$

11 $6(3x - 2) - 8 = 4x - 6$

12 Twice the sum of a number and 3 is -10. Find the number.

13 If 8 is subtracted from three times a number, the result is 1. Find the number.

14 The length of a rectangle is 4 centimeters longer than its width. If the perimeter is 28 centimeters, find the length and the width.

15 Karen is 5 years younger than Susan. Three years ago, the sum of their ages was 11. How old are they now?

16 Use the equation $4x + 3y = 12$ to find y when $x = -3$.

17 Plot the following points: $(3, 2)$, $(-3, 2)$, $(3, 0)$, $(0, -2)$.

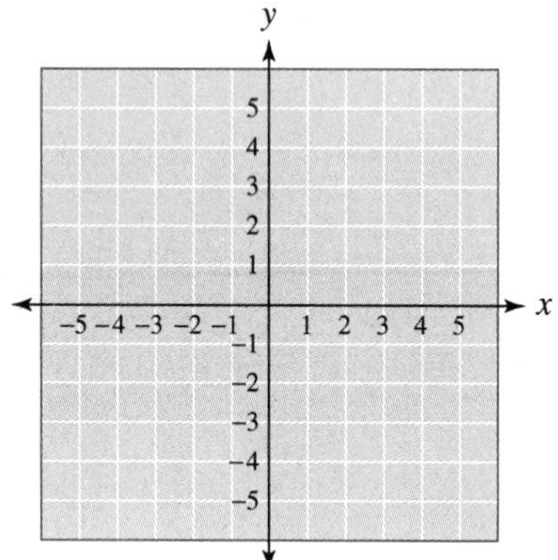

18 Graph $2x + y = 4$.

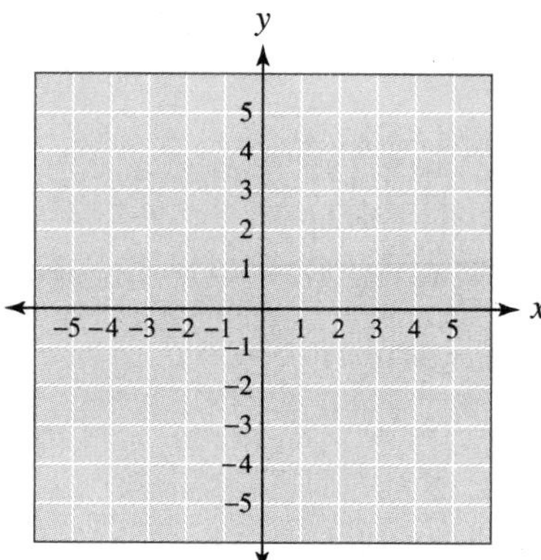

19 Graph $y = 2x - 1$.

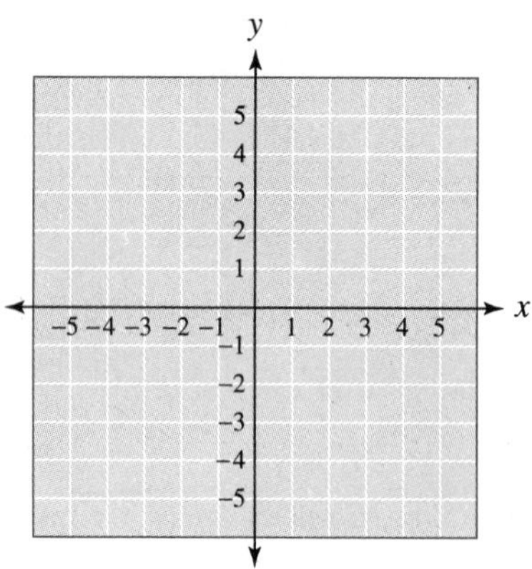

20 Graph $x = 2$.

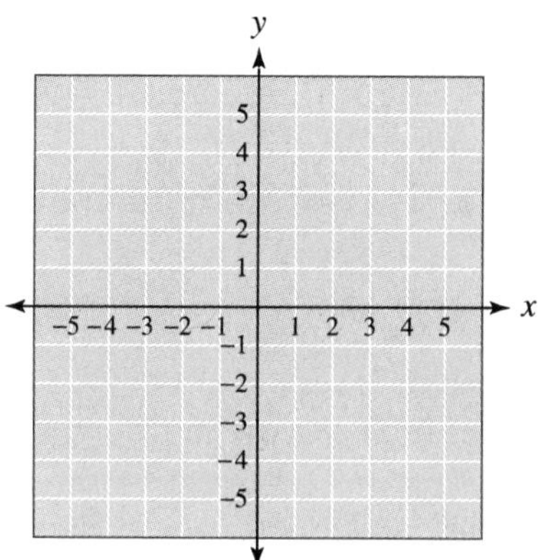

5

DECIMALS

INTRODUCTION

I have always been surprised by the number of my students who have difficulty calculating their grade point average (GPA). During her first semester in college, my daughter, Amy, earned the following grades:

Class	Units	Grade
Algebra	5	B
Chemistry	4	C
English	3	A
History	3	B

When her grades arrived in the mail, she told me she had a 3.0 grade point average, because the A and C grades averaged to a B. I told her that her GPA was a little less than a 3.0. What do you think? Can you calculate her GPA? If not, you will be able to after you finish this chapter.

In this chapter we will focus our attention on *decimals.* Anyone who has used money in the United States has worked with decimals already. For example, if you have been paid an hourly wage, such as

$6.25 per hour

↑ Decimal point

you have had experience with decimals. What is interesting and useful about decimals is their relationship to fractions and to powers of ten. The work we have done up to now—especially our work with fractions—can be used to develop the properties of decimal numbers.

STUDY SKILLS

This is the last chapter in which we will mention study skills. You know by now what works best for you, and what you have to do to achieve your goals for this course. From now on it is simply a matter of continuing to use the skills that work for you and avoiding those that do not. No one can do this for you: It is up to you to maintain the skills that will get you where you want to be in the course.

If you intend to take more classes in mathematics, and you want to ensure your success in those classes, then you can work toward this goal: *Become the type of student who can learn mathematics on their own.* Most people who have degrees in mathematics were students who could learn mathematics on their own. This doesn't mean that you have to learn it all on your own; it simply means that if you have to, you can. Attaining this goal gives you independence and puts you in control of your success in any mathematics class you take.

5.1 Decimal Notation and Place Value

In Chapter 1 we developed the idea of place value for the digits in a whole number. At that time we gave the name and the place value of each of the first seven columns in our number system, as follows:

Millions Column	Hundred Thousands Column	Ten Thousands Column	Thousands Column	Hundreds Column	Tens Column	Ones Column
1,000,000	100,000	10,000	1,000	100	10	1

As we move from right to left, we multiply by 10 each time. The value of each column is 10 times the value of the column on its right, with the rightmost column being 1. Up until now we have always looked at place value as increasing by a factor of 10 each time we move one column to the left:

Ten Thousands		Thousands		Hundreds		Tens		Ones
10,000	←	1,000	←	100	←	10	←	1
	Multiply by 10		Multiply by 10		Multiply by 10		Multiply by 10	

To understand the idea behind decimal numbers, we notice that moving in the opposite direction, from left to right, we *divide* by 10 each time:

Ten Thousands		Thousands		Hundreds		Tens		Ones
10,000	→	1,000	→	100	→	10	→	1
	Divide by 10		Divide by 10		Divide by 10		Divide by 10	

If we keep going to the right, the next column will have to be

$$1 \div 10 = \frac{1}{10} \qquad \text{Tenths}$$

The next one after that will be

$$\frac{1}{10} \div 10 = \frac{1}{10} \cdot \frac{1}{10} = \frac{1}{100} \qquad \text{Hundredths}$$

After that, we have

$$\frac{1}{100} \div 10 = \frac{1}{100} \cdot \frac{1}{10} = \frac{1}{1{,}000} \qquad \text{Thousandths}$$

We could continue this pattern as long as we wanted. We simply divide by 10 to move one column to the right. (And remember, dividing by 10 gives the same result as multiplying by $\frac{1}{10}$.)

To show where the ones column is, we use a *decimal point* between the ones column and the tenths column.

Thousands	Hundreds	Tens	Ones		Tenths	Hundredths	Thousandths	Ten Thousandths	Hundred Thousandths
1,000	100	10	1	.	$\frac{1}{10}$	$\frac{1}{100}$	$\frac{1}{1,000}$	$\frac{1}{10,000}$	$\frac{1}{100,000}$
				↑ Decimal point					

Note
Because the digits to the right of the decimal point have fractional place values, numbers with digits to the right of the decimal point are called *decimal fractions*. In this book we will also call them *decimal numbers*, or simply *decimals* for short.

The ones column can be thought of as the middle column, with columns larger than 1 to the left and columns smaller than 1 to the right. The first column to the right of the ones column is the tenths column, the next column to the right is the hundredths column, the next is the thousandths column, and so on. The decimal point is always written between the ones column and the tenths column.

We can use the place value of decimal fractions to write them in expanded form.

Practice Problems

1 Write 785.462 in expanded form.

EXAMPLE 1 Write 423.576 in expanded form.

Solution $423.576 = 400 + 20 + 3 + \frac{5}{10} + \frac{7}{100} + \frac{6}{1,000}$

2 Write in words.

a 0.06

b 0.7

c 0.008

EXAMPLE 2 Write each number in words.

a 0.4
b 0.04
c 0.004

Solution

a 0.4 is "four tenths."
b 0.04 is "four hundredths."
c 0.004 is "four thousandths."

When a decimal fraction contains digits to the left of the decimal point, we use the word "and" to indicate where the decimal point is when writing the number in words.

3 Write in words.

a 5.06

b 4.7

c 3.008

EXAMPLE 3 Write each number in words.

a 5.4
b 5.04
c 5.004

Solution

a 5.4 is "five and four tenths."
b 5.04 is "five and four hundredths."
c 5.004 is "five and four thousandths."

Answers
1–3 See solutions section.

Note Sometimes we name decimal fractions by simply reading the digits from left to right and using the word "point" to indicate where the decimal point is. For example, using this method the number 5.04 is read "five point zero four."

EXAMPLE 4 Write 3.64 in words.

Solution The number 3.64 is read "three and sixty-four hundredths." The place values of the digits are as follows:

$$\underset{\text{3 ones}}{\underset{\uparrow}{3}} \;.\; \underset{\text{6 tenths}}{\underset{\uparrow}{6}} \;\; \underset{\text{4 hundredths}}{\underset{\nwarrow}{4}}$$

We read the decimal part as "sixty-four hundredths" because

$$6 \text{ tenths} + 4 \text{ hundredths} = \frac{6}{10} + \frac{4}{100} = \frac{60}{100} + \frac{4}{100} = \frac{64}{100}$$

EXAMPLE 5 Write 25.4936 in words.

Solution Using the idea given in Example 4, we write 25.4936 in words as "twenty-five and four thousand, nine hundred thirty-six ten thousandths."

In order to understand addition and subtraction of decimals in the next section, we need to be able to convert decimal numbers to fractions or mixed numbers.

EXAMPLE 6 Write each number as a fraction or a mixed number. Do not reduce to lowest terms.

a 0.004

b 3.64

c 25.4936

Solution

a Since 0.004 is 4 thousandths, we write

$$0.004 = \frac{4}{1{,}000}$$

Three digits after the decimal point (↑ under 0.004); Three zeros (↖ at 1,000)

b Looking over the work in Example 4, we can write

$$3.64 = 3\tfrac{64}{100}$$

Two digits after the decimal point (↑ under 3.64); Two zeros (↖ at 100)

c From the way in which we wrote 25.4936 in words in Example 5, we have

$$25.4936 = 25\tfrac{4936}{10{,}000}$$

Four digits after the decimal point (↑ under 25.4936); Four zeros (↖ at 10,000)

4 Write 5.98 in words.

5 Write 305.406 in words.

6 Write each number as a fraction or a mixed number. Do not reduce to lowest terms.

a 0.06

b 5.98

c 305.406

Answers

4–5 See solutions section.

6 a $\frac{6}{100}$ **b** $5\frac{98}{100}$ **c** $305\frac{406}{1{,}000}$

Rounding Decimal Numbers

The rule for rounding decimal numbers is similar to the rule for rounding whole numbers. If the digit in the column to the right of the one we are rounding to is 5 or more, we add 1 to the digit in the column we are rounding to; otherwise, we leave it alone. We then replace all digits to the right of the column we are rounding to with zeros if they are to the left of the decimal point; otherwise, we simply delete them. Table 1 illustrates the procedure.

Table 1

	ROUNDED TO THE NEAREST		
Number	Whole Number	Tenth	Hundredth
24.785	25	24.8	24.79
2.3914	2	2.4	2.39
0.98243	1	1.0	0.98
14.0942	14	14.1	14.09
0.545	1	0.5	0.55

7 Round 8,935.042 to the nearest hundred.

EXAMPLE 7 Round 9,235.492 to the nearest hundred.

Solution The number next to the hundreds column is 3, which is less than 5. We change all digits to the right to 0, and we can drop all digits to the right of the decimal point, so we write

9,200

8 Round 0.05067 to the nearest ten thousandth.

EXAMPLE 8 Round 0.00346 to the nearest ten thousandth.

Solution Since the number to the right of the ten thousandths column is more than 5, we add 1 to the 4 and get

0.0035

Answers
7 8,900 **8** 0.0507

PROBLEM SET 5.1

Write out the name of each number in words.

1 0.3

2 0.03

3 0.015

4 0.0015

5 3.4

6 2.04

7 52.7

8 46.8

Write each number as a fraction or a mixed number. Do not reduce your answers.

9 405.36

10 362.78

11 9.009

12 60.06

13 1.234

14 12.045

15 0.00305

16 2.00106

Give the place value of the 5 in each of the following numbers.

17 458.327

18 327.458

19 29.52

20 25.92

21 0.00375

22 0.00532

23 275.01

24 0.356

25 539.76

26 0.123456

Write each of the following as a decimal number.

27 Fifty-five hundredths

28 Two hundred thirty-five ten thousandths

29 Six and nine tenths

30 Forty-five thousand and six hundred twenty-one thousandths

31 Eleven and eleven hundredths

32 Twenty-six thousand, two hundred forty-five and sixteen hundredths

33 One hundred and two hundredths

34 Seventy-five and seventy-five hundred thousandths

35 Three thousand and three thousandths

36 One thousand, one hundred eleven and one hundred eleven thousandths

Complete the following table.

	ROUNDED TO THE NEAREST			
Number	Whole Number	Tenth	Hundredth	Thousandth
37 47.5479	________	________	________	________
38 100.9256	________	________	________	________
39 0.8175	________	________	________	________
40 29.9876	________	________	________	________
41 0.1562	________	________	________	________
42 128.9115	________	________	________	________
43 2,789.3241	________	________	________	________
44 0.8743	________	________	________	________
45 99.9999	________	________	________	________
46 71.7634	________	________	________	________

47 If you have a penny dated anytime from 1959 through 1982, its original weight was 3.11 grams. If the penny has a date of 1983 or later, the original weight was 2.5 grams. Write the two weights in words.

48 A local savings and loan company advertises savings notes that give a 7.78 percent interest rate. Round 7.78 to the nearest tenth.

49 The speed of light is 186,282.3976 miles per second. Round this number to the nearest hundredth.

50 Halley's comet was seen from the earth during 1986. It will be another 76.1 years before it returns. Write 76.1 in words.

51 A 50-gram egg contains 0.15 milligram of riboflavin. Write 0.15 in words.

52 One medium banana contains 0.64 milligram of vitamin B-6. Write 0.64 in words.

53 Write the following numbers in order from smallest to largest.
0.02 0.05 0.025 0.052 0.005 0.002

54 Write the following numbers in order from smallest to largest.
0.2 0.02 0.4 0.04 0.42 0.24

55 Which of the following numbers will round to 7.5?
7.451 7.449 7.54 7.56

56 Which of the following numbers will round to 3.2?
3.14999 3.24999 3.279 3.16111

Change each decimal to a fraction, and then reduce to lowest terms.

57 0.25

58 0.75

59 0.5

60 0.50

61 0.35

62 0.65

63 0.125

64 0.375

65 0.625

66 0.0625

67 0.875

68 0.1875

Estimating

For each pair of numbers, choose the number that is closest to 10.

69 9.9 and 9.99

70 8.5 and 8.05

71 10.5 and 10.05

72 10.9 and 10.99

For each pair of numbers, choose the number that is closest to 0.

73 0.5 and 0.05

74 0.10 and 0.05

75 0.01 and 0.02

76 0.1 and 0.01

Review Problems

In the next section we will do addition and subtraction with decimals. To understand the process of addition and subtraction, we need to understand the process of addition and subtraction with mixed numbers from Section 3.8.

Find each of the following sums and differences. (Add and subtract.)

77 $4\frac{3}{10} + 2\frac{1}{100}$

78 $5\frac{35}{100} + 2\frac{3}{10}$

79 $8\frac{5}{10} - 2\frac{4}{100}$

80 $6\frac{3}{100} - 2\frac{125}{1,000}$

81 $5\frac{1}{10} + 6\frac{2}{100} + 7\frac{3}{1,000}$

82 $4\frac{27}{100} + 6\frac{3}{10} + 7\frac{123}{1,000}$

83 $2 + \dfrac{3}{x}$

84 $3 + \dfrac{2}{x}$

85 $x - \dfrac{3}{4}$

86 $x - \dfrac{5}{6}$

One Step Further: Extending the Concepts

87 List all the three-digit numbers that round to 0.29.

88 List all the three-digit numbers that round to 2.5.

89 List all the two-digit numbers that round to 3.

90 List all the two-digit numbers that round to 5.

91 In your own words, give a step-by-step explanation of how you would round the number 3.456 to the nearest hundredth.

92 In your own words, give a step-by-step explanation of how you would round the number 3.456 to the nearest whole number.

5.2 Addition and Subtraction with Decimals

Introduction . . . If you are earning \$8.50 per hour and you get a raise of \$1.25 per hour, then your new hourly rate of pay is

$$\begin{array}{r} \$8.50 \\ +\ \$1.25 \\ \hline \$9.75 \end{array}$$

To add the two rates of pay, we align the decimal points and then add in columns.

To see why this is true in general, we can use mixed-number notation:

$$\begin{array}{rl} 8.50 = & 8\frac{50}{100} \\ +\ 1.25 = & 1\frac{25}{100} \\ \hline & 9\frac{75}{100} = 9.75 \end{array}$$

EXAMPLE 1 Add by first changing to fractions: 25.43 + 2.897 + 379.6

Solution We first change each decimal to a mixed number. We then write each fraction using the least common denominator and add as usual:

$$\begin{array}{rcrcr} 25.43 & = & 25\frac{43}{100} & = & 25\frac{430}{1,000} \\ 2.897 & = & 2\frac{897}{1,000} & = & 2\frac{897}{1,000} \\ +\ 379.6 & = & 379\frac{6}{10} & = & 379\frac{600}{1,000} \\ \hline & & & & 406\frac{1,927}{1,000} = 407\frac{927}{1,000} = 407.927 \end{array}$$

Again, the result is the same if we just line up the decimal points and add as if we were adding whole numbers:

$$\begin{array}{r} 25.430 \\ 2.897 \\ +\ 379.600 \\ \hline 407.927 \end{array}$$

Notice that we can fill in zeros on the right to help keep the numbers in the correct columns. Doing this does not change the value of any of the numbers.

Note: The decimal point in the answer is directly below the decimal points in the problem.

The same thing would happen if we were to subtract two decimal numbers. We can use these facts to write a rule for addition and subtraction of decimal numbers.

RULE To add (or subtract) decimal numbers, we line up the decimal points and add (or subtract) as usual. The decimal point in the result is written directly below the decimal points in the problem.

We will use this rule for the rest of the examples in this section.

Practice Problems

1 Change each decimal to a fraction, and then add. Write your answer as a decimal.

38.45 + 456.073

Answer

1 494.523

2 Subtract: $78.674 - 23.431$

EXAMPLE 2 Subtract: $39.812 - 14.236$

Solution We write the numbers vertically, with the decimal points lined up, and subtract as usual.

$$\begin{array}{r} 39.812 \\ -\ 14.236 \\ \hline 25.576 \end{array}$$

3 Add: $16 + 0.033 + 4.6 + 0.08$

EXAMPLE 3 Add: $8 + 0.002 + 3.1 + 0.04$

Solution To make sure we keep the digits in the correct columns, we can write zeros to the right of the rightmost digits:

$$8 = 8.000$$
$$3.1 = 3.100$$
$$0.04 = 0.040$$

Writing the extra zeros here is really equivalent to finding a common denominator for the fractional parts of the original four numbers—now we have a thousandths column in all the numbers

This doesn't change the value of any of the numbers, and it makes our task easier. Now we have

$$\begin{array}{r} 8.000 \\ 0.002 \\ 3.100 \\ +\ \ 0.040 \\ \hline 11.142 \end{array}$$

4 Subtract: $6.7 - 2.0563$

EXAMPLE 4 Subtract: $5.9 - 3.0814$

Solution In this case it is very helpful to write 5.9 as 5.9000, since we will have to borrow in order to subtract.

$$\begin{array}{r} 5.9000 \\ -\ 3.0814 \\ \hline 2.8186 \end{array}$$

5 Subtract 5.89 from the sum of 7 and 3.567.

EXAMPLE 5 Subtract 3.09 from the sum of 9 and 5.472.

Solution Writing the problem in symbols, we have

$$\begin{aligned} (9 + 5.472) - 3.09 &= 14.472 - 3.09 \\ &= 11.382 \end{aligned}$$

Answers
2 55.243 3 20.713 4 4.6437
5 4.677

EXAMPLE 6 Add: $2.89 + (-5.93)$

Solution Recall from Chapter 2 that to add two numbers with different signs, we subtract the smaller absolute value from the larger. The sign of the answer is the same as the sign of the number with the larger absolute value.

$2.89 + (-5.93) = -3.04$

The answer has the sign of the number with the larger absolute value

6 Add: $4.93 + (-7.85)$

EXAMPLE 7 Subtract: $-8 - 1.37$

Solution Since subtraction can be thought of as addition of the opposite, instead of subtracting 1.37 we can add -1.37.

$-8 - 1.37 = -8 + (-1.37)$

Now, to add two numbers with the same sign, we add their absolute values. The answer has the same sign as the original two numbers.

$-8 + (-1.37) = -9.37$

The answer has the same sign as the original two numbers

$$\begin{array}{r} 8.00 \\ +\ 1.37 \\ \hline 9.37 \end{array}$$

Add the absolute values of the two numbers

7 Subtract: $-4.09 - 3$

EXAMPLE 8 While I was writing this section of the book, I stopped to have lunch with a friend at a coffee shop near my office. The bill for lunch was $15.64. I gave the person at the cash register a $20 bill. For change, I received four $1 bills, a quarter, a nickel, and a penny. Was my change correct?

Solution To find the total amount of money I received in change, we add:

Four $1 bills	= $4.00
One quarter	= 0.25
One nickel	= 0.05
One penny	= 0.01
Total	= $4.31

To find out if this is the correct amount, we subtract the amount of the bill from $20.00.

$$\begin{array}{r} \$20.00 \\ -\ 15.64 \\ \hline \$\ 4.36 \end{array}$$

The change was not correct. It is off by 5 cents. Instead of the nickel, I should have been given a dime.

8 If you pay for a purchase of $9.56 with a $10 bill, how much money should you receive in change? What will you do if the change that is given to you is one quarter, two dimes, and four pennies?

Answers
6 -2.92 7 -7.09
8 $0.44; Tell the clerk that you have been given too much change. Instead of two dimes, you should have received one dime and one nickel.

9 Below are two more cars and the time it takes each to accelerate from 0 miles per hour to 60 miles per hour. Extend the histogram in Figure 1 to include these two cars.

Car	Time (in seconds)
Chevrolet Corvette LT1	5.7
Honda Prelude	7.1

EXAMPLE 9 Table 1 lists the number of seconds it takes various 1994 model cars to accelerate from 0 miles per hour to 60 miles per hour. Construct a histogram from the information in the table. Then find the difference in elapsed time between the slowest car and the fastest car.

Table 1 Elapsed time from 0 mph to 60 mph

Car	Time (in seconds)
BMW 325i	7.4
Ford Mustang GT	6.7
Mazda Miata	8.8
Porsche 911 Carrera	5.2
Toyota Supra Turbo	5.3
Volkswagen GTI	7.5

Solution We construct a histogram (Figure 1) with the car names on the horizontal axis. The numbers on the vertical axis represent the time, in seconds.

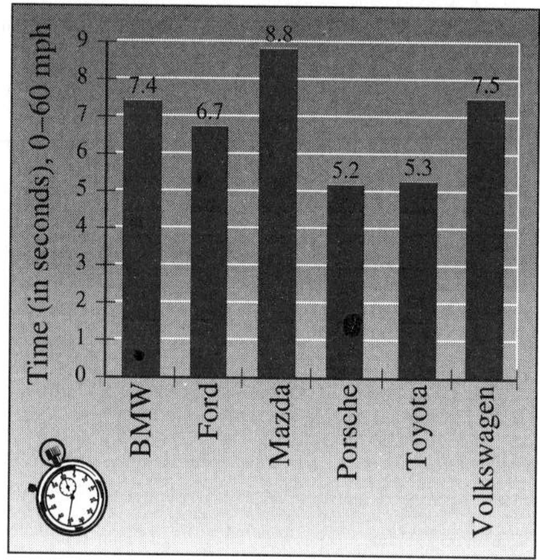

Figure 1 A histogram of the data in Table 1

The slowest car is the Mazda at 8.8 seconds, whereas the fastest car is the Porsche at 5.2 seconds. The difference between these two times is

$$8.8 - 5.2 = 3.6 \text{ seconds}$$

If the two cars started at the same time, the Porsche would reach a speed of 60 miles per hour 3.6 seconds before the Mazda.

Answer
9 See solutions section.

PROBLEM SET 5.2

Find each of the following sums. (Add.)

1 2.91 + 3.28

2 8.97 + 2.04

3 0.04 + 0.31 + 0.78

4 0.06 + 0.92 + 0.65

5 3.89 + (−2.4)

6 7.65 + (−3.8)

7 4.532 + 1.81 + 2.7

8 9.679 + 3.49 + 6.5

9 0.081 + (−5) + 2.94

10 0.396 + (−7) + 3.96

11 5.0003 + 6.78 + 0.004

12 27.0179 + 7.89 + 0.009

13 $\begin{array}{l} 7.123 \\ 8.12 \\ \underline{9.1} \end{array}$

14 $\begin{array}{l} 5.432 \\ 4.32 \\ \underline{3.2} \end{array}$

15 $\begin{array}{l} 9.001 \\ 8.01 \\ \underline{7.1} \end{array}$

16 $\begin{array}{l} 6.003 \\ 5.02 \\ \underline{4.1} \end{array}$

17 $\begin{array}{r} 89.7854 \\ 3.4 \\ 65.35 \\ \underline{100.006} \end{array}$

18 $\begin{array}{r} 57.4698 \\ 9.89 \\ 32.032 \\ \underline{572.0079} \end{array}$

19 $\begin{array}{r} 543.21 \\ \underline{123.45} \end{array}$

20 $\begin{array}{r} 987.654 \\ \underline{456.789} \end{array}$

Find each of the following differences. (Subtract.)

21 99.34 − 88.23

22 47.69 − 36.58

23 2.4 − 5.97

24 1.04 − 9.87

25 6.3 − 2.08

26 7.5 − 3.04

27 28.96 − 149.37

28 32.68 − 796.45

29 45 − 0.067

30 48 − 0.075

31 −8 − 0.327

32 −12 − 0.962

33 765.432 − 234.567

34 654.321 − 123.456

Subtract.

35 $\begin{array}{r} 34.07 \\ -\ 6.18 \\ \hline \end{array}$

36 $\begin{array}{r} 25.008 \\ -\ 3.119 \\ \hline \end{array}$

37 $\begin{array}{r} 40.04 \\ -\ 4.4 \\ \hline \end{array}$

38 $\begin{array}{r} 50.05 \\ -\ 5.5 \\ \hline \end{array}$

39 $\begin{array}{r} 768.436 \\ -\ 356.998 \\ \hline \end{array}$

40 $\begin{array}{r} 495.237 \\ -\ 247.668 \\ \hline \end{array}$

Add and subtract as indicated.

41 (7.8 − 4.3) + 2.5

42 (8.3 − 1.2) + 3.4

43 7.8 − (4.3 + 2.5)

44 8.3 − (1.2 + 3.4)

45 (9.7 − 5.2) − 1.4

46 (7.8 − 3.2) − 1.5

47 9.7 − (5.2 − 1.4)

48 7.8 − (3.2 − 1.5)

49 Subtract 5 from the sum of 8.2 and 0.072.

50 Subtract 8 from the sum of 9.37 and 2.5.

51 What number is added to 0.035 to obtain 4.036?

52 What number is added to 0.043 to obtain 6.054?

53 A family buying school clothes for their two children spends $25.37 at one store, $39.41 at another, and $52.04 at a third store. What is the total amount spent at the three stores?

54 A 4-H Club member is raising a lamb to take to the county fair. If she spent $75 for the lamb, $25.60 for feed, and $35.89 for shearing tools, what was the total cost of the project?

55 A waiter making $5.75 per hour is given a raise to $6.04 per hour. How much is the raise?

56 A person making $5.19 per hour is given a raise to $5.72 per hour. How much is the raise?

57 A college professor making $2,105.96 per month has deducted from her check $311.93 for federal income tax, $158.21 for retirement, and $64.72 for state income tax. How much does the professor take home after the deductions have been taken from her monthly income?

58 A cook making $1,504.75 a month has deductions of $157.32 for federal income tax, $58.52 for social security, and $45.12 for state income tax. How much does the cook take home after the deductions have been taken from his check?

59 A person buys \$4.57 worth of candy. If he pays for the candy with a \$10 bill, how much change should he receive?

60 If a person buys \$15.37 worth of groceries and pays for them with a \$20 bill, how much change should she receive?

61 A three-piece suit that usually sells for \$179.95 is marked down by \$25.83. What is the new price of the suit?

62 A dress that usually sells for \$49.95 is marked down by \$9.99. What is the new price of the dress?

63 A checking account contains \$42.35. If checks are written for \$9.25, \$3.77, and \$24.50, how much money is left in the account?

64 A checking account contains \$342.38. If checks are written for \$25.04, \$36.71, and \$210, how much money is left in the account?

65 The table below gives the number of seconds it takes various 1995 model cars to accelerate from 0 miles per hour to 60 miles per hour. Construct a histogram from the information in the table. Then find the difference in elapsed time between the slowest car and the fastest car.

Car	Time (in seconds)
BMW M3	5.4
Chrysler Cirrus LXi	9.4
Ford Contour SE	9.5
Honda Civic EX	8.8
Nissan Sentra GLE	11.0
Toyota Corolla DX	10.2

66 The table below gives the number of seconds it takes various sport-utility vehicles to accelerate from 0 miles per hour to 60 miles per hour. Construct a histogram from the information in the table. Then find the difference in elapsed time between the slowest vehicle and the fastest vehicle.

Car	Time (in seconds)
Chevrolet Blazer	9.1
Ford Explorer	10.7
Isuzu Trooper	10.9
Jeep Cherokee	9.7
Nissan Pathfinder	12.3
Toyota 4Runner	15.7

Estimating

67 Which number below is closest to 0.005 + 0.09?

a 1.0 **b** 0.1 **c** 0.01 **d** 0.001

68 Which number below is closest to 0.05 + 0.009?

a 1.0 **b** 0.1 **c** 0.01 **d** 0.001

Calculator Problems

Work the following problems on your calculator.

69 39.0715 + 6.5498 + 173.4629

70 5.00073 − 2.98995

71 (28.40962 + 77.35674) − 36.42099

72 (37.929394 − 16.50732) − 18.79765

73 Add:

$$\begin{array}{r} 279.8024 \\ 429.07 \\ 22.9009 \\ +\ 304.021 \\ \hline \end{array}$$

74 Add:

$$\begin{array}{r} 46.35772 \\ 4.0404 \\ 32.6262 \\ +\ 0.0457 \\ \hline \end{array}$$

Review Problems

To understand how to multiply decimals, we need to understand multiplication with whole numbers and fractions. The following problems review these two concepts from Sections 1.7, 2.4, and 3.3.

Multiply.

75 $\frac{1}{10} \cdot \frac{3}{10}$

76 $-\frac{5}{10} \cdot \frac{6}{10}$

77 $\frac{3}{100} \cdot \frac{17}{100}$

78 $\frac{7}{100} \cdot \frac{31}{100}$

79 $-5\left(-\frac{3}{10}\right)$

80 $7 \cdot \frac{7}{10}$

81 $56 \cdot 25$

82 $39(-48)$

One Step Further: Extending the Concepts

83 A rectangle has a perimeter of 9.5 inches. If the length is 2.75 inches, find the width.

84 A rectangle has a perimeter of 11 inches. If the width is 2.5 inches, find the length.

85 Suppose you eat dinner in a restaurant and the bill comes to $16.76. If you give the cashier a $20 bill and a penny, how much change should you receive? List the bills and coins you should receive for change.

86 Suppose you buy some tools at the hardware store and the bill comes to $37.87. If you give the cashier two $20 bills and 2 pennies, how much change should you receive? List the bills and coins you should receive for change.

Find the next number in each sequence.

87 2.5, 2.75, 3, . . .

88 3.125, 3.375, 3.625, . . .

Multiplication with Decimals; Circumference and Area of a Circle

Introduction . . . During a half-price sale a calendar that usually sells for \$6.42 is priced at \$3.21. Therefore it must be true that

$$\frac{1}{2} \text{ of } 6.42 \text{ is } 3.21$$

But, since $\frac{1}{2}$ can be written as 0.5 and *of* translates to *multiply*, we can write this problem again as

$$0.5 \times 6.42 = 3.21$$

If we were to ignore the decimal points in this problem and simply multiply 5 and 642, the result would be 3,210. So, multiplication with decimal numbers is similar to multiplication with whole numbers. The difference lies in deciding where to place the decimal point in the answer. To find out how this is done, we can use fraction notation.

EXAMPLE 1 Change each decimal to a fraction and multiply:

0.5×0.3 — To indicate multiplication we are using a $\times$ sign here instead of a dot so we won't confuse the decimal points with the multiplication symbol.

Solution Changing each decimal to a fraction and multiplying, we have

$$\begin{aligned} 0.5 \times 0.3 &= \frac{5}{10} \times \frac{3}{10} && \text{Change to fractions} \\ &= \frac{15}{100} && \text{Multiply numerators and multiply denominators} \\ &= 0.15 && \text{Write the answer in decimal form} \end{aligned}$$

The result is 0.15, which has two digits to the right of the decimal point.

What we want to do now is find a shortcut that will allow us to multiply decimals without first having to change each decimal number to a fraction. Let's look at another example.

EXAMPLE 2 Change each decimal to a fraction and multiply: 0.05×0.003

Solution

$$\begin{aligned} 0.05 \times 0.003 &= \frac{5}{100} \times \frac{3}{1,000} && \text{Change to fractions} \\ &= \frac{15}{100,000} && \text{Multiply numerators and multiply denominators} \\ &= 0.00015 && \text{Write the answer in decimal form} \end{aligned}$$

The result is 0.00015, which has a total of five digits to the right of the decimal point.

Looking over these first two examples, we can see that the digits in the result are just what we would get if we simply forgot about the decimal points and

Practice Problems

1 Change each decimal to a fraction and multiply:

0.4×0.6

Write your answer as a decimal.

2 Change each decimal to a fraction and multiply:

0.5×0.007

Write your answer as a decimal.

Answers

1 0.24 **2** 0.0035

multiplied; that is, $3 \times 5 = 15$. The decimal point in the result is placed so that the total number of digits to its right is the same as the total number of digits to the right of both decimal points in the original two numbers. The reason this is true becomes clear when we look at the denominators after we have changed from decimals to fractions.

3 Change to fractions and multiply:

3.5×0.04

EXAMPLE 3 Multiply: 2.1×0.07

Solution

$$\begin{aligned} 2.1 \times 0.07 &= 2\tfrac{1}{10} \times \frac{7}{100} && \text{Change to fractions} \\ &= \frac{21}{10} \times \frac{7}{100} \\ &= \frac{147}{1{,}000} && \text{Multiply numerators and multiply denominators} \\ &= 0.147 && \text{Write the answer as a decimal} \end{aligned}$$

Again, the digits in the answer come from multiplying $21 \times 7 = 147$. The decimal point is placed so that there are three digits to its right, since that is the total number of digits to the right of the decimal points in 2.1 and 0.07.

We summarize this discussion with a rule.

RULE To multiply two decimal numbers:

1. Multiply as you would if the decimal points were not there.
2. Place the decimal point in the answer so that the number of digits to its right is equal to the total number of digits to the right of the decimal points in the original two numbers in the problem.

4 How many digits will be to the right of the decimal point in the following product?

3.705×55.88

EXAMPLE 4 How many digits will be to the right of the decimal point in the following product?

2.987×24.82

Solution There are three digits to the right of the decimal point in 2.987, and two digits to the right in 24.82. Therefore, there will be $3 + 2 = 5$ digits to the right of the decimal point in their product.

5 Multiply: 4.03×5.22

EXAMPLE 5 Multiply: 3.05×4.36

Solution We can set this up as if it were a multiplication problem with whole numbers. We multiply and then place the decimal point in the correct position in the answer.

```
   3.05 <— 2 digits to right of decimal point
 × 4.36 <— 2 digits to right of decimal point
 ------
   1830
   915
 12 20
 ------
13.2980
 ↑
 └ The decimal point is placed
   so that there are 2 + 2 = 4
   digits to its right
```

Answers
3 0.14 **4** 5 **5** 21.0366

As you can see, multiplying decimal numbers is just like multiplying whole numbers, except that we must place the decimal point in the result in the correct position.

EXAMPLE 6 Multiply: $-1.3(-5.6)$

Solution In this case we are multiplying two negative numbers. To do so we simply multiply their absolute values. The answer is positive, since the original two numbers had the same sign.

$$-1.3(-5.6) = 7.28$$

EXAMPLE 7 Multiply: $4.56(-100)$

Solution The product of two numbers with different signs is negative.

$$4.56(-100) = -456$$

Estimating

Look back to Example 5. We could have placed the decimal point in the answer by rounding the two numbers to the nearest whole number and then multiplying them. Since 3.05 rounds to 3 and 4.36 rounds to 4, and the product of 3 and 4 is 12, we estimate that the answer to 3.05×4.36 will be close to 12. We then place the decimal point in the product 132980 between the 3 and the 2 in order to make it into a number close to 12.

EXAMPLE 8 Estimate the answer to each of the following products.

a 29.4×8.2 **b** 68.5×172 **c** $(6.32)^2$

Solution

a Since 29.4 is approximately 30 and 8.2 is approximately 8, we estimate this product to be about $30 \times 8 = 240$. (If we were to multiply 29.4 and 8.2, we would find the product to be exactly 241.08.)

b Rounding 68.5 to 70 and 172 to 170, we estimate this product to be $70 \times 170 = 11{,}900$. (The exact answer is 11,782.) Note here that we do not always round the numbers to the nearest whole number when making estimates. The idea is to round to numbers that will be easy to multiply.

c Since 6.32 is approximately 6 and $6^2 = 36$, we estimate our answer to be close to 36. (The actual answer is 39.9424.)

Combined Operations

We can use the rule for order of operations to simplify expressions involving decimal numbers and addition, subtraction, and multiplication.

EXAMPLE 9 Perform the indicated operations: $0.05(4.2 + 0.03)$

Solution We begin by adding inside the parentheses:

$$\begin{aligned} 0.05(4.2 + 0.03) &= 0.05(4.23) && \text{Add} \\ &= 0.2115 && \text{Multiply} \end{aligned}$$

Notice that we could also have used the distributive property first, and the result would be unchanged:

$$\begin{aligned} 0.05(4.2 + 0.03) &= 0.05(4.2) + 0.05(0.03) && \text{Distributive property} \\ &= 0.210 + 0.0015 && \text{Multiply} \\ &= 0.2115 && \text{Add} \end{aligned}$$

6 Multiply: $1.3(-5.6)$

7 Multiply: $-4.56(-100)$

8 Estimate the answer to each product.

a 82.3×5.8

b 37.5×178

c $(8.21)^2$

9 Perform the indicated operations: $0.03(5.5 + 0.02)$

Answers
6 -7.28 **7** 456
8 a 480 **b** 7,200 **c** 64
9 0.1656

10 Simplify: $5.7 + 14(2.4)^2$

EXAMPLE 10 Simplify: $4.8 + 12(3.2)^2$

Solution According to the rule for order of operations, we must first evaluate the number with an exponent, then multiply, and finally add.

$$\begin{aligned} 4.8 + 12(3.2)^2 &= 4.8 + 12(10.24) && (3.2)^2 = 10.24 \\ &= 4.8 + 122.88 && \text{Multiply} \\ &= 127.68 && \text{Add} \end{aligned}$$

Applications

11 How much will the phone company in Example 11 charge for a 20-minute call?

EXAMPLE 11 Suppose a phone company charges \$0.43 for the first minute and then \$0.32 for each additional minute for a long-distance phone call. How much will it cost for a 10-minute call?

Solution For this 10-minute call, the first minute will cost \$0.43 and the next 9 minutes will cost \$0.32 each. The total cost of the call will be

$$\begin{aligned} \underbrace{0.43}_{\text{Cost of the first minute}} + \underbrace{9(0.32)}_{\text{Cost of the next 9 minutes}} &= 0.43 + 2.88 \\ &= 3.31 \end{aligned}$$

The call will cost \$3.31.

12 How much will Sally make if she works 50 hours in one week?

EXAMPLE 12 Sally earns \$6.32 for each of the first 36 hours she works in one week and \$9.48 in overtime pay for each additional hour she works in the same week. How much money will she make if she works 42 hours in one week?

Solution The difference between 42 and 36 is 6 hours of overtime pay. The total amount of money she will make is

$$\begin{aligned} \underbrace{6.32(36)}_{\text{Pay for the first 36 hours}} + \underbrace{9.48(6)}_{\text{Pay for the next 6 hours}} &= 227.52 + 56.88 \\ &= 284.40 \end{aligned}$$

She will make \$284.40 for working 42 hours in one week.

Note
To estimate the answer to Example 12 before doing the actual calculations, we would do the following:

$$6(40) + 9(6) = 240 + 54 = 294$$

The Circumference of a Circle

The *circumference* of a circle is the distance around the outside, just as the perimeter of a polygon is the distance around the outside. The circumference of a circle can be found by measuring its radius or diameter and then using the appropriate formula. The *radius* of a circle is the distance from the center of the circle to the circle itself. The radius is denoted by the letter r. The *diameter* of a circle is the distance from one side to the other, through the center. The diameter is denoted by the letter d. In Figure 1 we can see that the diameter is twice the radius, or

$$d = 2r$$

Answers
10 86.34 **11** \$6.51
12 \$360.24

The relationship between the circumference and the diameter or radius is not as obvious. As a matter of fact, it takes some fairly complicated mathematics to show just what the relationship between the circumference and the diameter is.

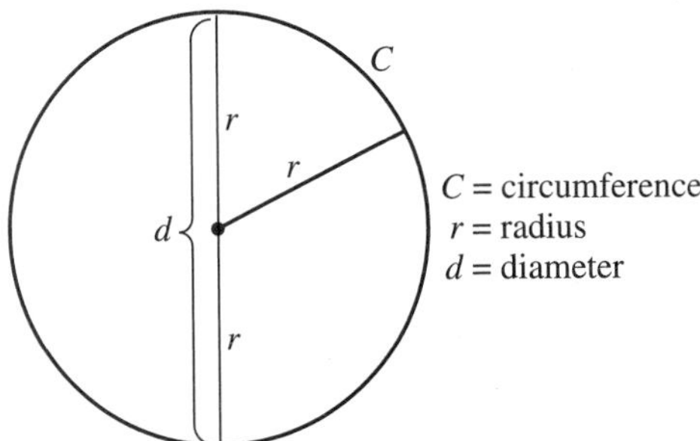

Figure 1

If you took a string and actually measured the circumference of a circle by wrapping the string around the circle, and then measured the diameter of the same circle, you would find that the ratio of the circumference to the diameter, C/d, would be approximately equal to 3.14. The actual ratio of C to d in any circle is an irrational number. It can't be written in decimal form. We use the symbol π (Greek pi) to represent this ratio. In symbols the relationship between the circumference and the diameter in any circle is

$$\frac{C}{d} = \pi$$

Knowing what we do about the relationship between division and multiplication, we can rewrite this formula as

$$C = \pi d$$

This is the formula for the circumference of a circle. When we do the actual calculations, we will use the approximation 3.14 for π.

Since $d = 2r$, the same formula written in terms of the radius is

$$C = 2\pi r$$

Here are some examples that show how we use the formulas given above to find the circumference of a circle.

EXAMPLE 13 Find the circumference of a circle with a diameter of 5 feet.

Solution Substituting 5 for d in the formula $C = \pi d$, and using 3.14 for π, we have

$$\begin{aligned} C &= 3.14(5) \\ &= 15.7 \text{ feet} \end{aligned}$$

EXAMPLE 14 Find the circumference of a circle with a radius of 8 centimeters.

Solution We use the formula $C = 2\pi r$ this time and substitute 8 for r to get

$$C = 2(3.14)8 = 50.24 \text{ centimeters}$$

13 Find the circumference of a circle with a diameter of 3 centimeters.

14 Find the circumference of a circle with a radius of 5 feet.

Answers
13 9.42 cm **14** 31.4 ft

Other Formulas Involving π

Two figures are presented here, along with some important formulas that are associated with each figure. As you can see, each of the formulas contains the number π.

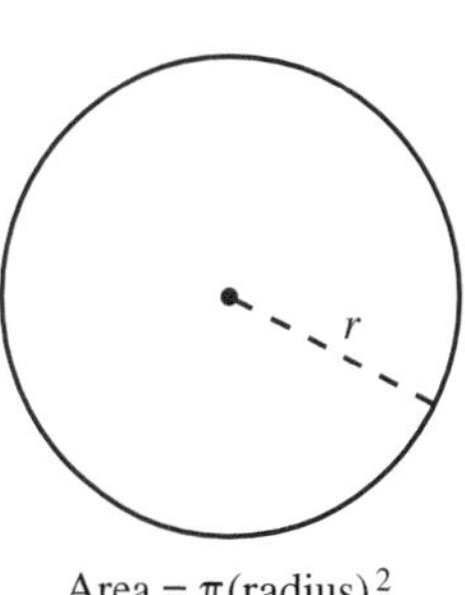

Area = π(radius)^2
$A = \pi r^2$

Figure 2 Circle

Volume = π(radius)^2(height)
$V = \pi r^2 h$

Figure 3 Right circular cylinder

15 Find the area of a circle with a diameter of 20 feet.

EXAMPLE 15 Find the area of a circle with a diameter of 10 feet.

Solution The formula for the area of a circle is $A = \pi r^2$. Since the radius r is half the diameter and the diameter is 10 feet, the radius is 5 feet. Therefore,

$$A = \pi r^2 = (3.14)(5)^2 = (3.14)(25) = 78.5 \text{ ft}^2$$

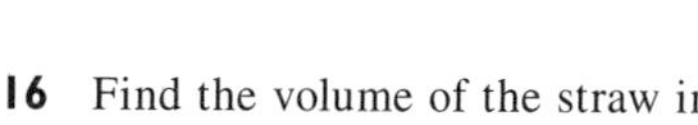

16 Find the volume of the straw in Example 16, if the radius is doubled. Round your answer to the nearest thousandth.

EXAMPLE 16 The drinking straw shown in Figure 4 has a radius of 0.125 inch and a length of 6 inches. To the nearest thousandth, find the volume of liquid that it will hold.

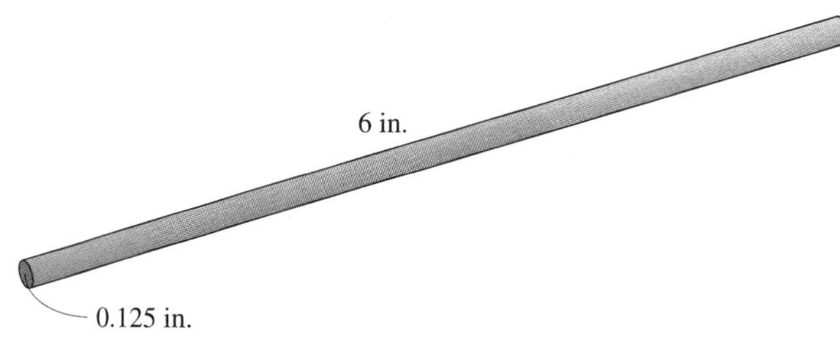

Figure 4

Solution The total volume is found from the formula for the volume of a right circular cylinder. In this case the radius is $r = 0.125$ and the height is $h = 6$. We approximate π with 3.14.

$$\begin{aligned} V &= \pi r^2 h \\ &= (3.14)(0.125)^2(6) \\ &= (3.14)(0.015625)(6) \\ &= 0.294 \text{ in}^3 \text{ to the nearest thousandth} \end{aligned}$$

Answers
15 314 ft^2 **16** 1.178 in^3

PROBLEM SET 5.3

Find each of the following products. (Multiply.)

1 0.7×0.4

2 0.8×0.3

3 0.07×0.4

4 0.8×0.03

5 0.03×0.09

6 0.07×0.002

7 $-2.6(0.3)$

8 $-8.9(0.2)$

9 0.9×0.88

10 0.8×0.99

11 3.12×0.005

12 4.69×0.006

13 4.003×6.07

14 7.0001×3.04

15 $-5(0.006)$

16 $-7(0.005)$

17 75.14×2.5

18 963.8×0.24

19 0.1×0.02

20 0.3×0.02

21 $2.796(-10)$

22 $97.531(-100)$

23 0.0043×100

24 $12.345 \times 1{,}000$

25 $49.94 \times 1{,}000$

26 $157.02 \times 10{,}000$

27 $987.654 \times 10{,}000$

28 $1.23 \times 100{,}000$

Perform the following operations according to the rule for order of operations.

29 $2.1(3.5 - 2.6)$

30 $5.4(9.9 - 6.6)$

31 $-0.05(0.02 + 0.03)$

32 $-0.04(0.07 + 0.09)$

33 $2.02(0.03 - 2.5)$

34 $4.04(0.05 - 6.6)$

35 $(2.1 + 0.03)(3.4 + 0.05)$

36 $(9.2 + 0.01)(3.5 + 0.03)$

37 $(2.1 - 0.1)(2.1 + 0.1)$

38 $(9.6 - 0.5)(9.6 + 0.5)$

39 $3.08 - 0.2(5 + 0.03)$

40 $4.09 + 0.5(6 + 0.02)$

41 $4.23 - 5(0.04 + 0.09)$

42 $7.89 - 2(0.31 + 0.76)$

43 $2.5 + 10(4.3)^2$

44 $3.6 + 15(2.1)^2$

45 $100(1 + 0.08)^2$

46 $500(1 + 0.12)^2$

47 $(1.5)^2 + (2.5)^2 + (3.5)^2$

48 $(1.1)^2 + (2.1)^2 + (3.1)^2$

Solve each of the following word problems. Note that not all of the problems are solved by simply multiplying the numbers in the problems. Many of the problems involve addition and subtraction as well as multiplication.

49 What is the product of 6 and the sum of 0.001 and 0.02?

50 Find the product of 8 and the sum of 0.03 and 0.002.

51 What does multiplying a decimal number by 100 do to the decimal point?

52 What does multiplying a decimal number by 1,000 do to the decimal point?

53 Ground beef is priced at $1.98 per pound. How much will it cost for 3.5 pounds?

54 If 1 cup of regular coffee contains 105 milligrams of caffeine, how much caffeine is contained in 3.5 cups of coffee?

55 If gasoline costs 119.9¢ per gallon, how much does it cost to fill an 11-gallon tank?

56 If gasoline costs $1.37 per gallon when you pay with a credit card, but $0.06 per gallon less if you pay with cash, how much do you save by filling up a 12-gallon tank and paying for it with cash?

57 On a certain home mortgage, there is a monthly payment of $9.66 for every $1,000 that is borrowed. What is the monthly payment on this type of loan if $43,000 is borrowed?

58 If the loan mentioned in Problem 57 is for $53,000, how much more will the payments be each month?

59 If a phone company charges $0.45 for the first minute and $0.35 for each additional minute for a long-distance call, how much will a 20-minute long-distance call cost?

60 Suppose a phone company charges $0.53 for the first minute and $0.27 for each additional minute for a long-distance call. How much will a 15-minute long-distance call cost?

61 Suppose it costs $15 per day and $0.12 per mile to rent a car. What is the total bill if a car is rented for 2 days and is driven 120 miles?

62 Suppose it costs $20 per day and $0.08 per mile to rent a car. What is the total bill if the car is rented for 2 days and is driven 120 miles?

63 A man earns $5.92 for each of the first 36 hours he works in one week and $8.88 in overtime pay for each additional hour he works in the same week. How much money will he make if he works 45 hours in one week?

64 A student earns $4.56 for each of the first 40 hours she works in one week and $6.84 in overtime pay for each additional hour she works in the same week. How much money will she make if she works 44 hours in one week?

Find the circumference and the area of each circle. Use 3.14 for π.

65

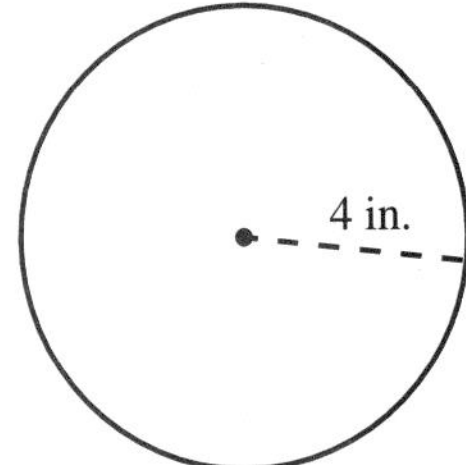

66

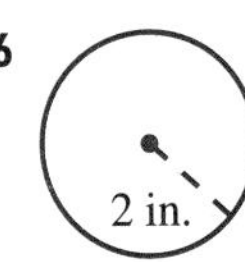

67 A dinner plate has a radius of 6 inches. Find the circumference and the area.

68 A salad plate has a radius of 3 inches. Find the circumference and the area.

69 The radius of the earth is approximately 3,900 miles. Find the circumference of the earth at the equator. (The equator is a circle around the earth that divides the earth into two equal halves.)

70 The radius of the moon is approximately 1,100 miles. Find the circumference of the moon around its equator.

Find the volume of each cylinder.

71

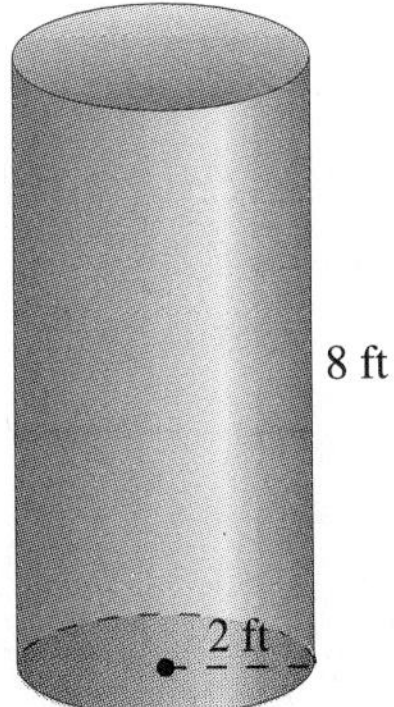

72

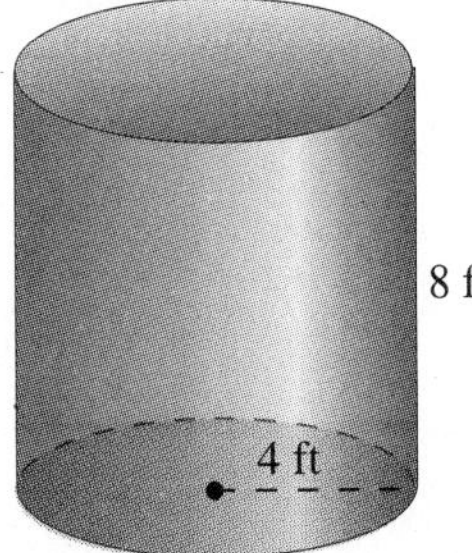

73

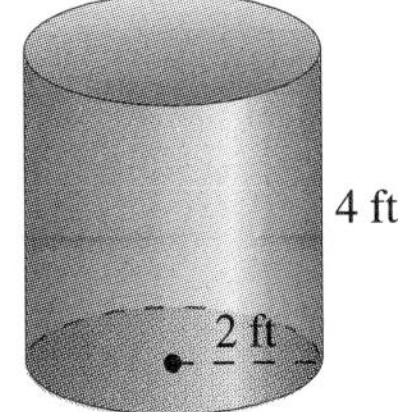

74

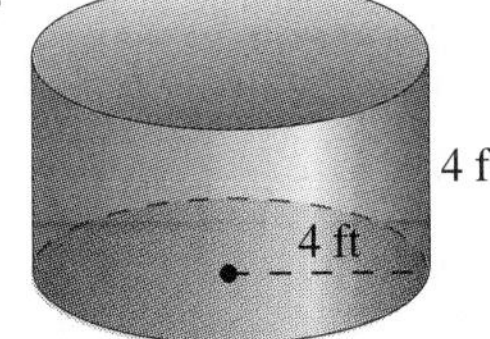

Calculator Problems

Find each of the following products by using a calculator.

75 (429.5)(246.7)

76 (0.0799)(8.437)

77 (6.59)(7.32)(4.2)

78 (0.04)(0.35)(0.172)

79 $(0.75)^3$

80 $(7.5)^3$

81 Find the area and the circumference of the circle, to the nearest hundredth.

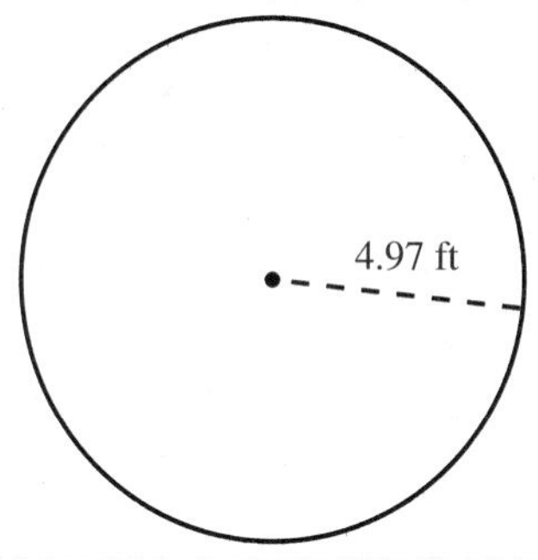

82 Find the volume of the cylinder, to the nearest hundredth.

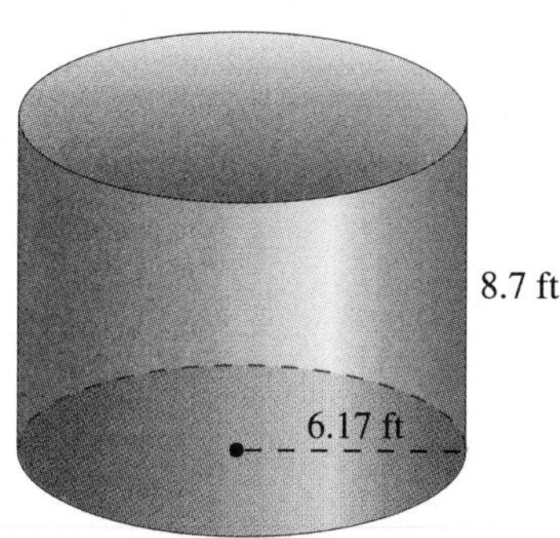

Review Problems

To get ready for the next section, which covers division with decimals, we will review division with whole numbers and fractions from Sections 1.8 and 3.4.

Perform each of the following divisions. (Find the quotients.)

83 $3{,}758 \div 2$

84 $9{,}900 \div 22$

85 $50{,}032 \div 33$

86 $90{,}902 \div 5$

87 $\frac{3}{10} \div \frac{4}{10}$

88 $\frac{6}{10} \div \frac{3}{100}$

89 $-\frac{6}{100} \div \left(-\frac{3}{10}\right)$

90 $\frac{27}{100} \div \left(-\frac{9}{10}\right)$

91 $\frac{45}{100} \div \frac{5}{1{,}000}$

92 $\frac{225}{1{,}000} \div \frac{25}{1{,}000}$

93 $\frac{x^3}{y^5} \div \frac{x}{y^8}$

94 $\frac{x^4}{y^3} \div \frac{x}{y^9}$

95 $\frac{a^2b^3}{c^4} \div \frac{a^2b}{c^7}$

96 $\frac{a^3b^2}{c} \div \frac{a^3b}{c^4}$

97 $\frac{10x^5}{21y^3} \div \frac{5x^3}{42y^5}$

98 $\frac{15x^7}{14y^3} \div \frac{3x^5}{28y^8}$

One Step Further: Number Sequences

Each sequence below is either arithmetic or geometric. Find the next number in each sequence, and then give a written description of the sequence. For example, the sequence 0.5, 0.6, 0.7, . . . can be described as follows: "An arithmetic sequence, starting with 0.5, in which each term is found by adding 0.1 to the term before it."

99 0.1, 0.2, 0.3, . . .

100 5.2, 6.3, 7.4, . . .

101 0.1, 0.01, 0.001, . . .

102 50, 5, 0.5, . . .

A sequence of numbers is referred to as *increasing* if each term is larger than the term that came before it. Likewise, a sequence is *decreasing* if each term is smaller than the term that came before it. Simplify each term in the following sequences, and then identify the sequence as increasing or decreasing.

103 $(0.2)^2, (0.2)^3, (0.2)^4$; . . .

104 $(1.2)^2, (1.2)^3, (1.2)^4$, . . .

105 $1.1, (1.1)^2, (1.1)^3$, . . .

106 $0.1, (0.1)^2, (0.1)^3$, . . .

5.4 Division with Decimals

Introduction . . . Suppose three friends go out for lunch and their total bill comes to \$18.75. If they decide to split the bill equally, how much does each person owe? To find out, they will have to divide 18.75 by 3. If you think about the dollars and cents separately, you will see that each person owes \$6.25. Therefore,

$$\$18.75 \div 3 = \$6.25$$

In this section we will find out how to do division with any combination of decimal numbers.

EXAMPLE 1 Divide: 5,974 ÷ 20

Solution

$$\begin{array}{r} 298 \\ 20\overline{)5,974} \\ \underline{4\,0} \\ 1\,97 \\ \underline{1\,80} \\ 174 \\ \underline{160} \\ 14 \end{array}$$

In the past we have written this answer as $298\frac{14}{20}$ or, after reducing the fraction, $298\frac{7}{10}$. Since $\frac{7}{10}$ can be written as 0.7, we could also write our answer as 298.7. This last form of our answer is exactly the same result we obtain if we write 5,974 as 5,974.0 and continue the division until we have no remainder. Here is how it looks:

$$\begin{array}{r} 298.7 \\ 20\overline{)5,974.0} \\ \underline{4\,0} \\ 1\,97 \\ \underline{1\,80} \\ 174 \\ \underline{160} \\ 14\,0 \\ \underline{14\,0} \\ 0 \end{array}$$

Notice that we place the decimal point in the answer directly above the decimal point in the problem

Let's try another division problem. This time one of the numbers in the problem will be a decimal.

EXAMPLE 2 Divide: 34.8 ÷ 4

Solution We can use the ideas from Example 1 and divide as usual. The decimal point in the answer will be placed directly above the decimal point in the problem.

$$\begin{array}{r} 8.7 \\ 4\overline{)34.8} \\ \underline{32} \\ 2\,8 \\ \underline{2\,8} \\ 0 \end{array} \qquad \textit{Check:} \quad \begin{array}{r} 8.7 \\ \underline{\times\quad 4} \\ 34.8 \end{array}$$

The answer is 8.7.

Practice Problems

1 Divide: 4,626 ÷ 30

Note

We can estimate the answer to Example 1 by rounding 5,974 to 6,000 and dividing by 20:

$$\frac{6,000}{20} = 300$$

2 Divide: 33.5 ÷ 5

Note

We never need to make a mistake with division, since we can always check our results with multiplication.

Answers

1 154.2 2 6.7

We can use these facts to write a rule for dividing decimal numbers.

RULE To divide a decimal by a whole number, we do the usual long division as if there were no decimal point involved. The decimal point in the answer is placed directly above the decimal point in the problem.

Here are some more examples to illustrate the procedure.

3 Divide: 47.448 ÷ 18

EXAMPLE 3 Divide: 49.896 ÷ 27

Solution

```
       1.848
  27)49.896
     27
     22 8
     21 6
      1 29
      1 08
        216
        216
          0
```

Check this result by multiplication:

```
   1.848
 ×    27
  12 936
  36 96
  49.896
```

We can write as many zeros as we choose after the rightmost digit in a decimal number without changing the value of the number. For example,

6.91 = 6.910 = 6.9100 = 6.91000

There are times when this can be very useful, as Example 4 shows.

4 Divide: 1,138.5 ÷ 25

EXAMPLE 4 Divide: 1,138.9 ÷ 35

Solution

```
         32.54
  35)1,138.90
     1 05
        88
        70
        18 9
        17 5
         1 40
         1 40
            0
```

Write 0 after the 9. It doesn't change the original number, but it gives us another digit to bring down.

Check:

```
     32.54
 ×      35
    162 70
    976 2
  1,138.90
```

Until now we have considered only division by whole numbers. Extending division to include division by decimal numbers is a matter of knowing what to do about the decimal point in the divisor.

5 Divide: 13.23 ÷ 4.2

EXAMPLE 5 Divide: 31.35 ÷ 3.8

Solution In fraction form, this problem is equivalent to

$$\frac{31.35}{3.8}$$

Answers
3 2.636 **4** 45.54 **5** 3.15

If we want to write the divisor as a whole number, we can multiply the numerator and the denominator of this fraction by 10:

$$\frac{31.35 \times \mathbf{10}}{3.8 \times \mathbf{10}} = \frac{313.5}{38}$$

So, since this fraction is equivalent to the original fraction, our original division problem is equivalent to

```
      8.25
  38)313.50     Put 0 after the last digit
     304
       9 5
       7 6
       1 90
       1 90
          0
```

We can summarize division with decimal numbers by listing the following points, as illustrated by the first five examples.

Summary of Division with Decimals

1 We divide decimal numbers by the same process used in Chapter 1 to divide whole numbers. The decimal point in the answer is placed directly above the decimal point in the dividend.
2 We are free to write as many zeros after the last digit in a decimal number as we need.
3 If the divisor is a decimal, we can change it to a whole number by moving the decimal point to the right as many places as necessary so long as we move the decimal point in the dividend the same number of places.

Note
We do not always use the rules for rounding numbers to make estimates. For example, to estimate the answer to Example 5, 31.35 ÷ 3.8, we can get a rough estimate of the answer by reasoning that 3.8 is close to 4 and 31.35 is close to 32. Therefore, our answer will be approximately 32 ÷ 4 = 8.

EXAMPLE 6 Divide, and round the answer to the nearest hundredth: 0.3778 ÷ 0.25

Solution

```
          1.5112
  0.25.).37.7800
          25
          12 7
          12 5
             28
             25
              30
              25
               50
               50
                0
```

Rounding to the nearest hundredth, we have 1.51. We actually did not need to have this many digits to round to the hundredths column. We could have stopped at the thousandths column and rounded off.

6 Divide, and round your answer to the nearest hundredth:

0.4553 ÷ 0.32

Note
Moving the decimal point two places in both the divisor and the dividend is justified like this:

$$\frac{0.3778 \times \mathbf{100}}{0.25 \times \mathbf{100}} = \frac{37.78}{25}$$

Answer
6 1.42

7 Divide, and round to the nearest tenth:

$19 \div 0.06$

EXAMPLE 7 Divide, and round to the nearest tenth: $17 \div 0.03$

Solution Since we are rounding to the nearest tenth, we will continue dividing until we have a digit in the hundredths column. We don't have to go any further to round to the tenths column.

```
          5 66.66
0.03.)17.00.00
        15
         2 0
         1 8
           20
           18
            2 0
            1 8
              20
              18
               2
```

Rounding to the nearest tenth, we have 566.7.

8 A woman earning \$6.54 an hour receives a paycheck for \$186.39. How many hours did the woman work?

EXAMPLE 8 If a man earning \$5.26 an hour receives a paycheck for \$170.95, how many hours did he work?

Solution To find the number of hours the man worked, we divide \$170.95 by \$5.26.

```
             32.5
5.26.)170.95.0
       157 8
        13 15
        10 52
         2 63 0
         2 63 0
              0
```

The man worked 32.5 hours.

9 If the phone company in Example 9 charged \$4.39 for a call, how long was the call?

EXAMPLE 9 A telephone company charges \$0.43 for the first minute and then \$0.33 for each additional minute for a long-distance call. If a long-distance call costs \$3.07, how many minutes was the call?

Solution To solve this problem we need to find the number of additional minutes for the call. To do so, we first subtract the cost of the first minute from the total cost, and then we divide the result by the cost of each additional minute. Without showing the actual arithmetic involved, the solution looks like this:

Total cost of the call ↓ (3.07); Cost of the first minute ↙ (0.43)

$$\text{The number of additional minutes} = \frac{3.07 - 0.43}{0.33} = \frac{2.64}{0.33} = 8$$

↑ Cost of each additional minute (0.33)

The call was 9 minutes long. (The number 8 is the number of additional minutes past the first minute.)

Answers
7 316.7 **8** 28.5 hours
9 13 minutes

Grade Point Average

When you calculate your grade point average (GPA), you are calculating what is called a *weighted average*. To calculate your grade point average, you must first calculate the number of grade points you have earned in each class that you have completed. The number of grade points for a class is the product of the number of units the class is worth times the value of the grade received. The table below shows the value that is assigned to each grade.

Grade	Value
A	4
B	3
C	2
D	1
F	0

If you earn a B in a 4-unit class, you earn $4 \times 3 = 12$ grade points. A grade of C in the same class gives you $4 \times 2 = 8$ grade points. To find your grade point average for one term (a semester or quarter), you must add your grade points and divide that total by the number of units. Round your answer to the nearest hundredth.

EXAMPLE 10 At the end of her first semester at college, Amy had the following grades:

Class	Units	Grade
Algebra	5	B
Chemistry	4	C
English	3	A
History	3	B

Calculate her grade point average.

Solution We begin by writing in two more columns, one for the value of each grade (4 for an A, 3 for a B, 2 for a C, 1 for a D, and 0 for an F), and another for the grade points earned for each class. To fill in the grade points column, we multiply the number of units by the value of the grade:

Class	Units	Grade	Value	Grade Points
Algebra	5	B	3	$5 \times 3 = 15$
Chemistry	4	C	2	$4 \times 2 = 8$
English	3	A	4	$3 \times 4 = 12$
History	3	B	3	$3 \times 3 = 9$
Total Units:	15		Total Grade Points:	44

To find her grade point average, we divide 44 by 15 and round (if necessary) to the nearest hundredth:

$$\text{Grade point average} = \frac{44}{15} = 2.93$$

10 If Amy had earned a B in chemistry, instead of a C, what grade point average would she have?

Answer
10 3.20

PROBLEM SET 5.4

Perform each of the following divisions.

1 $394 \div 20$

2 $486 \div 30$

3 $248 \div 40$

4 $372 \div 80$

5 $5\overline{)26}$

6 $8\overline{)36}$

7 $25\overline{)276}$

8 $50\overline{)276}$

9 $28.8 \div (-6)$

10 $15.5 \div (-5)$

11 $-77.6 \div (-8)$

12 $-31.48 \div (-4)$

13 $35\overline{)92.05}$

14 $26\overline{)146.38}$

15 $45\overline{)190.8}$

16 $55\overline{)342.1}$

17 $86.7 \div 34$

18 $411.4 \div 44$

19 $-29.7 \div 22$

20 $-488.4 \div 88$

21 $4.5\overline{)29.25}$

22 $3.3\overline{)21.978}$

23 $0.11\overline{)1.089}$

24 $0.75\overline{)2.40}$

25 $2.3\overline{)0.115}$

26 $6.6\overline{)0.198}$

27 $0.012\overline{)1.068}$

28 $0.052\overline{)0.23712}$

29 $1.1\overline{)2.42}$

30 $2.2\overline{)7.26}$

Carry out each of the following divisions only so far as needed to round the results to the nearest hundredth.

31 $26\overline{)35}$ **32** $18\overline{)47}$ **33** $3.3\overline{)56}$ **34** $4.4\overline{)75}$

35 $0.1234 \div 0.5$ **36** $0.543 \div 2.1$ **37** $19 \div 7$ **38** $16 \div 6$

39 $0.059\overline{)0.69}$ **40** $0.048\overline{)0.49}$

In Problems 41–44, indicate whether the statements are *True* or *False*.

41 $6.99 \div 0.3 = 6.99 \div 3$ **42** $3.5 \div 7 = 35 \div 70$ **43** $42 \div 0.06 = 420 \div 0.6$ **44** $75 \div 0.25 = 7.5 \div 25$

45 If a six-pack of soft drinks costs $1.98, how many six-packs were purchased for $9.90?

46 Patrick owns 15 sheep. If his monthly feed bill is $103.50, how much does it cost him to feed one sheep for a month? If he buys two more sheep, how much will his monthly feed bill increase?

47 Suppose a man earns $33.20 for working an 8-hour shift. How much does the man earn per hour?

48 If a woman earns $33.90 for working 6 hours, how much does she earn per hour?

49 How many hours does a person making $4.76 per hour have to work in order to earn $121.38?

50 How many hours does a person making $6.78 per hour have to work in order to earn $257.64?

51 If a car travels 336 miles on 15 gallons of gas, how far will the car travel on 1 gallon of gas?

52 If a car travels 392 miles on 16 gallons of gas, how far will the car travel on 1 gallon of gas?

53 Suppose a woman earns $6.78 an hour for the first 36 hours she works in a week and then $10.17 an hour in overtime pay for each additional hour she works in the same week. If she makes $294.93 in one week, how many hours did she work overtime?

54 Suppose a woman makes $286.08 in one week. If she is paid $5.96 an hour for the first 36 hours she works and then $8.94 an hour in overtime pay for each additional hour she works in the same week, how many hours did she work overtime that week?

55 Suppose a telephone company charges $0.41 for the first minute and then $0.32 for each additional minute for a long-distance call. If a long-distance call costs $2.33, how many minutes was the call?

56 Suppose a telephone company charges $0.45 for the first three minutes and then $0.29 for each additional minute for a long-distance call. If a long-distance call costs $2.77, how many minutes was the call?

The following grades were earned by Steve during his first term in college. Use these data to answer Problems 57–60.

Class	Units	Grade
Basic mathematics	3	A
Health	2	B
History	3	B
English	3	C
Chemistry	4	C

57 Calculate Steve's GPA.

58 If his grade in chemistry had been a B instead of a C, by how much would his GPA have increased?

59 If his grade in health had been a C instead of a B, by how much would his grade point average have dropped?

60 If his grades in both English and chemistry had been B's, what would his GPA have been?

Calculator Problems

Work each of the following problems on your calculator. If rounding is necessary, round to the nearest hundred thousandth.

61 $7 \div 9$

62 $11 \div 13$

63 $243 \div 0.791$

64 $67.8 \div 37.92$

65 $0.0503 \div 0.0709$

66 $429.87 \div 16.925$

Review Problems

In the next section we will consider the relationship between fractions and decimals in more detail. The problems below review some of the material from Sections 3.1 and 3.2 that is necessary to make a successful start in the next section.

Reduce to lowest terms.

67 $\frac{75}{100}$

68 $\frac{220}{1,000}$

69 $\frac{12x}{18xy}$

70 $\frac{15xy}{30x}$

71 $\frac{75x^2y^3}{100xy^2}$

72 $\frac{220x^3y^2}{1,000x^2y}$

Write each fraction as an equivalent fraction with denominator 10.

73 $\frac{3}{5}$

74 $\frac{1}{2}$

Write each fraction as an equivalent fraction with denominator 100.

75 $\frac{3}{5}$

76 $\frac{17}{20}$

Write each fraction as an equivalent fraction with denominator $15x$.

77 $\frac{4}{5}$

78 $\frac{2}{3}$

79 $\frac{4}{x}$

80 $\frac{2}{x}$

81 $\frac{6}{5x}$

82 $\frac{7}{3x}$

One Step Further: Extending the Concepts

83 A rectangle has an area of 2.4675 square feet. If the width is 1.05 feet, find the length.

84 A rectangle has an area of 35.0343 square feet. If the length is 6.03 feet, find the width.

For each of the following problems, use 3.14 for π.

85 The circumference of a circle is 7.85 meters. Find the diameter and the radius.

86 The circumference of a circle is 23.236 meters. Find the diameter and the radius.

87 The area of a circle is 50.24 square meters. Find the radius and the diameter.

88 The area of a circle is 78.5 square meters. Find the radius and the diameter.

5.5 Fractions and Decimals, and the Volume of a Sphere

Introduction . . . If you are shopping for clothes and a store has a sale advertising $\frac{1}{3}$ off the regular price, how much can you expect to pay for a pair of pants that normally sells for \$31.95? If the sale price of the pants is \$22.30, have they really been marked down by $\frac{1}{3}$? To answer questions like these, we need to know how to solve problems that involve fractions and decimals together.

We begin this section by showing how to convert back and forth between fractions and decimals.

Converting Fractions to Decimals

You may recall that the notation we use for fractions can be interpreted as implying division. That is, the fraction $\frac{3}{4}$ can be thought of as meaning "3 divided by 4." We can use this idea to convert fractions to decimals.

EXAMPLE 1 Write $\frac{3}{4}$ as a decimal.

Solution Dividing 3 by 4, we have

$$\begin{array}{r} .75 \\ 4\overline{)3.00} \\ \underline{2\,8}\downarrow \\ 20 \\ \underline{20} \\ 0 \end{array}$$

The fraction $\frac{3}{4}$ is equal to the decimal 0.75.

EXAMPLE 2 Write $\frac{7}{12}$ as a decimal correct to the thousandths column.

Solution Since we want the decimal to be rounded to the thousandths column, we divide to the ten thousandths column and round off to the thousandths column:

$$\begin{array}{r} .5833 \\ 12\overline{)7.0000} \\ \underline{6\,0} \\ 1\,00 \\ \underline{96} \\ 40 \\ \underline{36} \\ 40 \\ \underline{36} \\ 4 \end{array}$$

Rounding off to the thousandths column, we have 0.583. Since $\frac{7}{12}$ is not exactly the same as 0.583, we write

$$\frac{7}{12} \approx 0.583$$

where the symbol $\approx$ is read "is approximately."

Practice Problems

1 Write $\frac{3}{5}$ as a decimal.

2 Write $\frac{11}{12}$ as a decimal correct to the thousandths column.

Answers

1 0.6 2 0.917

If we wrote more zeros after 7.0000 in Example 2, the pattern of 3's would continue for as many places as we could want. When we get a sequence of digits that repeat like this, 0.58333 . . . , we can indicate it by writing

$0.58\overline{3}$ The bar over the 3 indicates that the 3 repeats from there on.

3 Write $\frac{5}{11}$ as a decimal.

EXAMPLE 3 Write $\frac{3}{11}$ as a decimal.

Solution Dividing 3 by 11, we have

$$
\begin{array}{r}
.272727 \\
11\overline{)3.000000} \\
\underline{2\,2} \\
80 \\
\underline{77} \\
30 \\
\underline{22} \\
80 \\
\underline{77} \\
30 \\
\underline{22} \\
80 \\
\underline{77} \\
3
\end{array}
$$

Note
The bar over the 2 and the 7 in $0.\overline{27}$ is used to indicate that the pattern repeats itself forever.

No matter how long we continue the division, the remainder will never be 0, and the pattern will continue. We write the decimal form of $\frac{3}{11}$ as $0.\overline{27}$, where

$0.\overline{27} = 0.272727\ldots$ The dots mean "and so on"

Converting Decimals to Fractions

To convert decimals to fractions, we take advantage of the place values we assigned to the digits to the right of the decimal point.

4 Write 0.48 as a fraction in lowest terms.

EXAMPLE 4 Write 0.38 as a fraction in lowest terms.

Solution 0.38 is 38 hundredths, or

$$0.38 = \frac{38}{100}$$

$$= \frac{19}{50}$$ Divide the numerator and the denominator by 2 to reduce to lowest terms

The decimal 0.38 is equal to the fraction $\frac{19}{50}$.

We could check our work here by converting $\frac{19}{50}$ back to a decimal. We do this by dividing 19 by 50. That is,

$$
\begin{array}{r}
.38 \\
50\overline{)19.00} \\
\underline{15\,0} \\
4\,00 \\
\underline{4\,00} \\
0
\end{array}
$$

Answers

3 $0.\overline{45}$ 4 $\frac{12}{25}$

EXAMPLE 5 Convert 0.075 to a fraction.

Solution We have 75 thousandths, or

$$0.075 = \frac{75}{1{,}000}$$

$$= \frac{3}{40} \quad \text{Divide the numerator and the denominator by 25 to reduce to lowest terms}$$

EXAMPLE 6 Write 15.6 as a mixed number.

Solution Converting 0.6 to a fraction, we have

$$0.6 = \frac{6}{10} = \frac{3}{5} \quad \text{Reduce to lowest terms}$$

Since $0.6 = \frac{3}{5}$, we have $15.6 = 15\frac{3}{5}$.

Problems Containing Both Fractions and Decimals

We will end this section by working some problems that involve both fractions and decimals.

EXAMPLE 7 Simplify: $\frac{19}{50}(1.32 + 0.48)$

Solution In Example 4, we found that $0.38 = \frac{19}{50}$. Therefore we can rewrite the problem as

$$\frac{19}{50}(1.32 + 0.48) = 0.38(1.32 + 0.48) \quad \text{Convert all numbers to decimals}$$

$$= 0.38(1.80) \quad \text{Add: } 1.32 + 0.48$$

$$= 0.684 \quad \text{Multiply: } 0.38 \times 1.80$$

EXAMPLE 8 Simplify: $\frac{1}{2} + (0.75)\left(\frac{2}{5}\right)$

Solution We could do this problem one of two different ways. First, we could convert all fractions to decimals and then simplify:

$$\frac{1}{2} + (0.75)\left(\frac{2}{5}\right) = 0.5 + 0.75(0.4) \quad \text{Convert to decimals}$$

$$= 0.5 + 0.300 \quad \text{Multiply: } 0.75 \times 0.4$$

$$= 0.8$$

Or, we could convert 0.75 to $\frac{3}{4}$ and then simplify:

$$\frac{1}{2} + 0.75\left(\frac{2}{5}\right) = \frac{1}{2} + \frac{3}{4}\left(\frac{2}{5}\right)$$

$$= \frac{1}{2} + \frac{3}{10} \quad \text{Multiply: } \frac{3}{4} \times \frac{2}{5}$$

$$= \frac{5}{10} + \frac{3}{10} \quad \text{The common denominator is 10}$$

$$= \frac{8}{10} \quad \text{Add numerators}$$

$$= \frac{4}{5}$$

The answers are equivalent. That is, $0.8 = \frac{8}{10} = \frac{4}{5}$. Either method can be used with problems of this type.

5 Convert 0.025 to a fraction.

6 Write 12.8 as a mixed number.

7 Simplify: $\frac{14}{25}(2.43 + 0.27)$

8 Simplify: $\frac{1}{4} + 0.25\left(\frac{3}{5}\right)$

Answers

5 $\frac{1}{40}$ 6 $12\frac{4}{5}$ 7 1.512

8 $\frac{2}{5}$, or 0.4

9 Simplify:

$$\left(\frac{1}{3}\right)^3(5.4) + \left(\frac{1}{5}\right)^2(2.5)$$

EXAMPLE 9 Simplify: $\left(\frac{1}{2}\right)^3(2.4) + \left(\frac{1}{4}\right)^2(3.2)$

Solution This expression can be simplified without any conversions between fractions and decimals. To begin, we evaluate all numbers that contain exponents. Then we multiply. After that, we add.

$$\begin{aligned}\left(\frac{1}{2}\right)^3(2.4) + \left(\frac{1}{4}\right)^2(3.2) &= \frac{1}{8}(2.4) + \frac{1}{16}(3.2) && \text{Evaluate exponents}\\ &= 0.3 + 0.2 && \text{Multiply by } \tfrac{1}{8} \text{ and } \tfrac{1}{16}\\ &= 0.5 && \text{Add}\end{aligned}$$

10 A shirt that normally sells for \$35.50 is on sale for $\frac{1}{4}$ off. What is the sale price of the shirt?

EXAMPLE 10 If a shirt that normally sells for \$27.99 is on sale for $\frac{1}{3}$ off, how much is the sale price of the shirt?

Solution To find out how much the shirt is marked down, we must find $\frac{1}{3}$ of 27.99. That is, we multiply $\frac{1}{3}$ and 27.99, which is the same as dividing 27.99 by 3.

$$\frac{1}{3}(27.99) = \frac{27.99}{3} = 9.33$$

The shirt is marked down \$9.33. The sale price is the original price minus the amount it is marked down:

$$\text{Sale price} = 27.99 - 9.33 = 18.66$$

The sale price is \$18.66. We also could have solved this problem by simply multiplying the original price by $\frac{2}{3}$, since, if the shirt is marked $\frac{1}{3}$ off, then the sale price must be $\frac{2}{3}$ of the original price. Multiplying by $\frac{2}{3}$ is the same as dividing by 3 and then multiplying by 2. The answer would be the same.

The Volume of a Sphere

Figure 1 shows a sphere and the formula for its volume. Since the formula contains both the fraction $\frac{4}{3}$ and the number π, and we have been using 3.14 for π, we can think of the formula as containing both a fraction and a decimal.

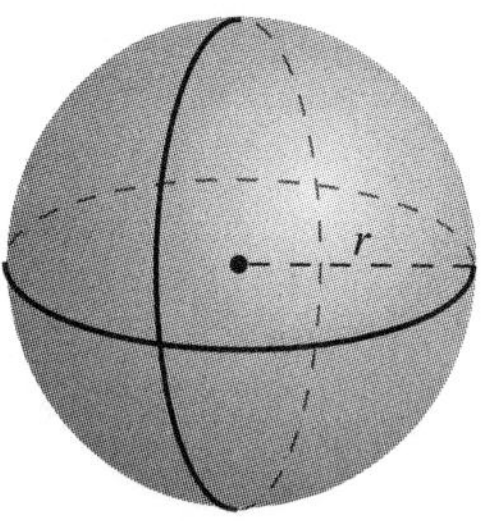

$$\text{Volume} = \tfrac{4}{3}\pi(\text{radius})^3 = \tfrac{4}{3}\pi r^3$$

Figure 1 Sphere

Answers
9 0.3 **10** \$26.63

EXAMPLE 11 Figure 2 is composed of a right circular cylinder with half a sphere on top. (A half-sphere is called a *hemisphere*.) To the nearest tenth, find the total volume enclosed by the figure.

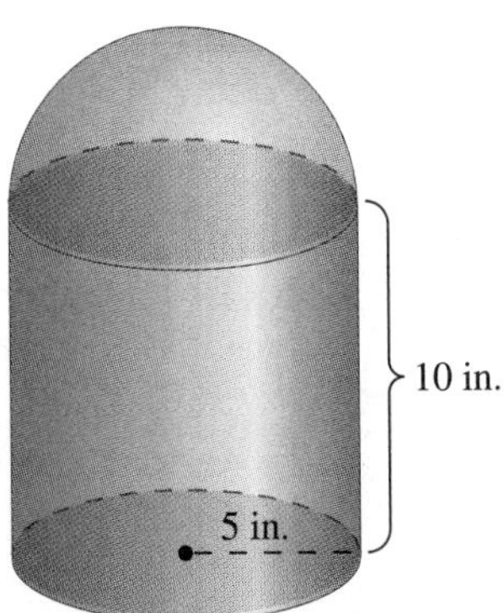

Figure 2

Solution The total volume is found by adding the volume of the cylinder to the volume of the hemisphere.

$$V = \text{volume of cylinder} + \text{volume of hemisphere}$$

$$= \pi r^2 h + \frac{1}{2} \cdot \frac{4}{3} \pi r^3$$

$$= (3.14)(5)^2(10) + \frac{1}{2} \cdot \frac{4}{3}(3.14)(5)^3$$

$$= (3.14)(25)(10) + \frac{1}{2} \cdot \frac{4}{3}(3.14)(125)$$

$$= 785 + \frac{2}{3}(392.5) \qquad \text{Multiply: } \frac{1}{2} \cdot \frac{4}{3} = \frac{4}{6} = \frac{2}{3}$$

$$= 785 + \frac{785}{3} \qquad \text{Multiply: } 2(392.5) = 785$$

$$= 785 + 261.7 \qquad \text{Divide 785 by 3 and round to the nearest tenth}$$

$$= 1{,}046.7 \text{ in}^3$$

11 If the radius in Figure 2 is doubled so that it becomes 10 inches instead of 5 inches, what is the new volume of the figure? Round your answer to the nearest tenth.

Answer

11 $5{,}233.3 \text{ in}^3$

PROBLEM SET 5.5

Convert each of the following fractions to a decimal.

1 $\frac{1}{2}$ **2** $\frac{3}{4}$ **3** $\frac{3}{8}$ **4** $\frac{1}{8}$ **5** $\frac{7}{8}$ **6** $\frac{1}{4}$

7 $\frac{12}{25}$ **8** $\frac{14}{25}$ **9** $\frac{14}{32}$ **10** $\frac{18}{32}$ **11** $\frac{3}{5}$ **12** $\frac{4}{5}$

Write each fraction as a decimal correct to the hundredths column.

13 $\frac{12}{13}$ **14** $\frac{17}{19}$ **15** $\frac{3}{11}$ **16** $\frac{5}{11}$

17 $\frac{2}{23}$ **18** $\frac{3}{28}$ **19** $\frac{12}{43}$ **20** $\frac{15}{51}$

Write each decimal as a fraction in lowest terms.

21 0.64 **22** 0.98 **23** 0.15 **24** 0.45

25 0.08 **26** 0.06 **27** 0.375 **28** 0.475

Write each decimal as a mixed number.

29 5.6 **30** 8.4 **31** 5.06 **32** 8.04

33 17.26 **34** 39.35 **35** 1.22 **36** 2.11

Simplify each of the following as much as possible and write all answers as decimals.

37 $\frac{1}{2}(2.3 + 2.5)$ **38** $\frac{3}{4}(1.8 + 7.6)$ **39** $\frac{3}{8}(4.7)$ **40** $\frac{5}{8}(1.2)$

41 $3.4 - \frac{1}{2}(0.76)$

42 $6.7 - \frac{1}{5}(0.45)$

43 $\frac{2}{5}(0.3) + \frac{3}{5}(0.3)$

44 $\frac{1}{8}(0.7) + \frac{3}{8}(0.7)$

45 $6\left(\frac{3}{5}\right)(0.02)$

46 $8\left(\frac{4}{5}\right)(0.03)$

47 $\frac{5}{8} + 0.35\left(\frac{1}{2}\right)$

48 $\frac{7}{8} + 0.45\left(\frac{3}{4}\right)$

49 $\left(\frac{1}{3}\right)^2(5.4) + \left(\frac{1}{2}\right)^3(3.2)$

50 $\left(\frac{1}{5}\right)^2(7.5) + \left(\frac{1}{4}\right)^2(6.4)$

51 $\left(\frac{1}{2}\right)^2\left(\frac{2}{5}\right)^3(12.5)$

52 $\left(\frac{1}{4}\right)^2\left(\frac{2}{3}\right)^3(10.8)$

53 $(0.25)^2 + \left(\frac{1}{4}\right)^2(3)$

54 $(0.75)^2 + \left(\frac{1}{4}\right)^2(7)$

55 If each pound of beef costs \$2.59, how much does $3\frac{1}{4}$ pounds cost?

56 What does it cost to fill a $15\frac{1}{2}$-gallon gas tank if the gasoline is priced at 129.9¢ per gallon?

57 How many $6\frac{1}{2}$-ounce glasses can be filled from a 32-ounce container of orange juice?

58 How many $8\frac{1}{2}$-ounce glasses can be filled from a 32-ounce container of orange juice?

59 A dress that costs \$57.99 is on sale for $\frac{1}{3}$ off. What is the sale price of the dress?

60 A suit that normally sells for \$121 is on sale for $\frac{1}{4}$ off. What is the sale price of the suit?

61 A computer programmer has a yearly salary of \$26,958. If $\frac{1}{4}$ of his salary is used to make house payments, what is his monthly house payment?

62 A self-employed woman earns \$32,789 in a year. If she spends $\frac{1}{8}$ of this amount for health insurance, how much does she pay every month for health insurance?

63 If 1 ounce of ground beef contains 50.75 calories and 1 ounce of halibut contains 27.5 calories, what is the difference in calories between a $4\frac{1}{2}$-ounce serving of ground beef and a $4\frac{1}{2}$-ounce serving of halibut?

64 If a 1-ounce serving of baked potato contains 48.3 calories and a 1-ounce serving of chicken contains 24.6 calories, how many calories are in a meal of $5\frac{1}{4}$ ounces of chicken and a $3\frac{1}{3}$-ounce baked potato?

Find the volume of each sphere. Round to the nearest hundredth.

65

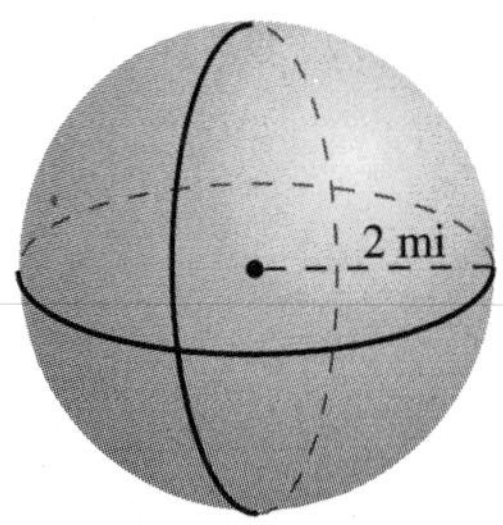

66

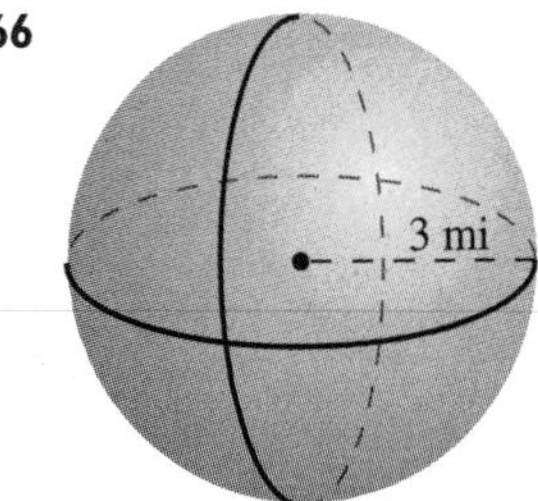

67 The radius of a sphere is 3.9 inches. Find the volume to the nearest hundredth.

68 The radius of a sphere is 1.1 inches. Find the volume to the nearest hundredth.

Find the total area enclosed by each figure below.

69

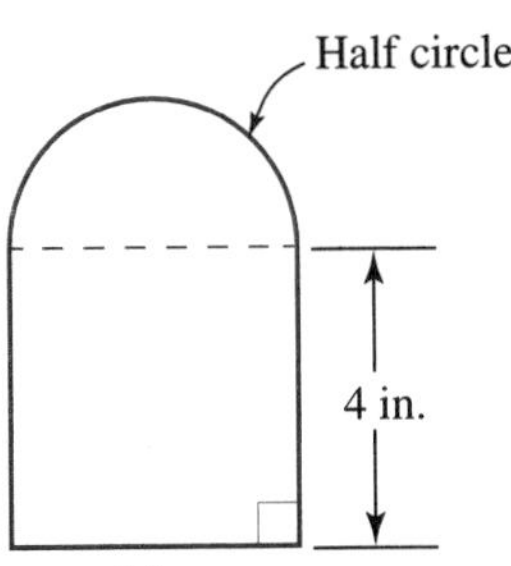

70

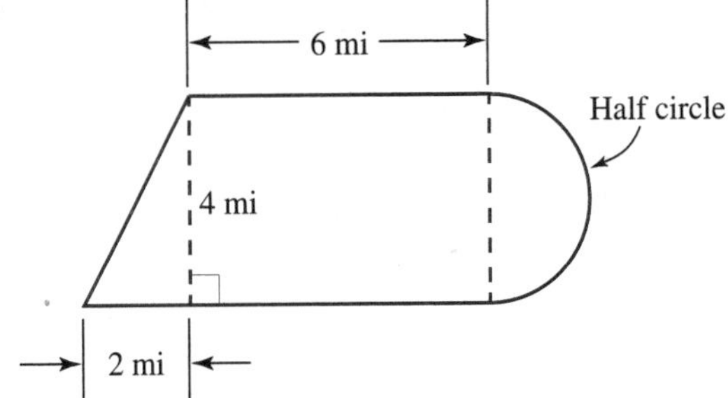

Find the volume of each figure. Round your answers to the nearest tenth.

71

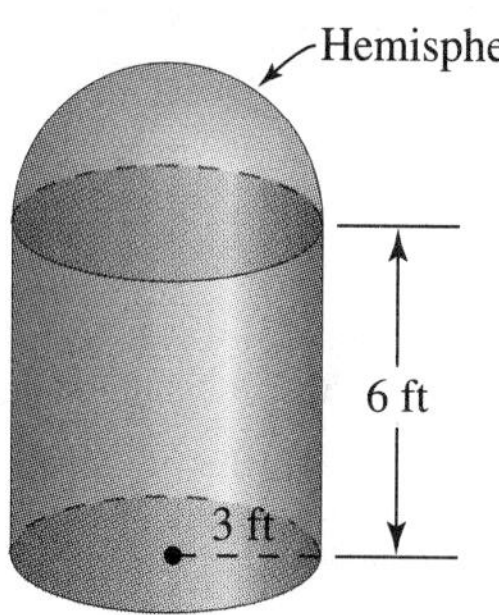

72

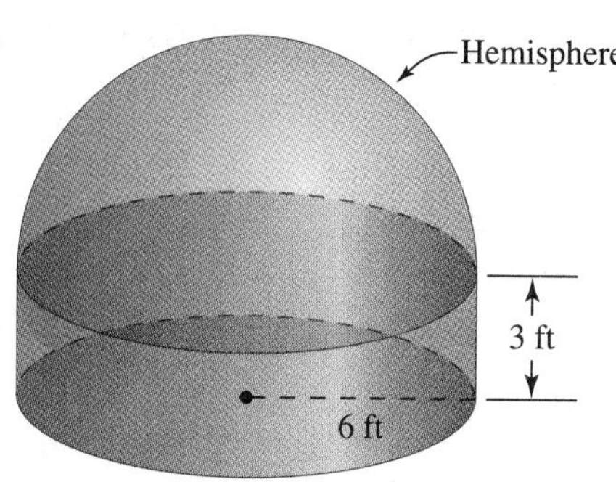

Calculator Problems

Work each of the following problems using a calculator.

73 Write $\frac{7}{9}$ as a decimal.

74 Write $\frac{12}{99}$ as a decimal.

75 Write $\frac{123}{999}$ as a decimal.

76 Write $\frac{73}{99}$ as a decimal.

77 Simplify: $(0.469)(24.2 + 89.06)$

78 Simplify: $(36.92)(0.043) + (75.87)(0.52)$

Review Problems

The problems below review some of the material on solving equations that we covered in Chapter 4. Reviewing these problems will help you with the next section.

79 $x + 3 = -6$

80 $x - 2 = -9$

81 $-4y = 28$

82 $-8y = -64$

83 $\frac{1}{7}a = -7$

84 $\frac{1}{5}a = -3$

85 $5n + 4 = -26$

86 $6n - 2 = 40$

87 $5x + 8 = 3x + 2$

88 $7x - 3 = 5x + 9$

89 $2(x + 3) = 10$

90 $3(x - 2) = 6$

91 $3(y - 4) + 5 = -4$

92 $5(y - 1) + 6 = -9$

One Step Further: Number Sequences

Give the next two numbers in each sequence.

93 $5.125,\ 6\frac{3}{8},\ 7.625,\ 8\frac{7}{8}, \ldots$

94 $3.2,\ 5\frac{3}{5},\ 8,\ 10\frac{2}{5}, \ldots$

95 $0.1,\ \frac{1}{100},\ 0.001, \ldots$

96 $5.5,\ \frac{11}{20},\ 0.055, \ldots$

5.6 Equations Containing Decimals

In this section we will continue our work with equations by considering some equations that involve decimals. We will also look at some application problems whose solutions come from equations with decimals.

EXAMPLE 1 Solve the equation $x + 8.2 = 5.7$.

Solution We use the addition property of equality to add -8.2 to each side of the equation.

$$x + 8.2 = 5.7$$

$$x + 8.2 + (\mathbf{-8.2}) = 5.7 + (\mathbf{-8.2}) \quad \text{Add } \mathbf{-8.2} \text{ to each side}$$

$$x + 0 = -2.5 \quad \text{Simplify each side}$$

$$x = -2.5$$

EXAMPLE 2 Solve: $3y = 2.73$

Solution To get just one y on the left side, we divide each side by 3.

$$3y = 2.73$$

$$\frac{3y}{\mathbf{3}} = \frac{2.73}{\mathbf{3}} \quad \text{Divide each side by } \mathbf{3}$$

$$y = 0.91 \quad \text{Division}$$

EXAMPLE 3 Solve: $\frac{1}{2}x - 3.78 = 2.52$

Solution We begin by adding 3.78 to each side of the equation. Then we multiply each side by 2.

$$\frac{1}{2}x - 3.78 = 2.52$$

$$\frac{1}{2}x - 3.78 + \mathbf{3.78} = 2.52 + \mathbf{3.78} \quad \text{Add } \mathbf{3.78} \text{ to each side}$$

$$\frac{1}{2}x = 6.30$$

$$\mathbf{2}\left(\frac{1}{2}x\right) = \mathbf{2}(6.30) \quad \text{Multiply each side by } \mathbf{2}$$

$$x = 12.6$$

EXAMPLE 4 Solve: $5a - 0.42 = -3a + 0.98$

Solution We can isolate a on the left side of the equation by adding $3a$ to each side.

$$5a + \mathbf{3a} - 0.42 = -3a + \mathbf{3a} + 0.98 \quad \text{Add } \mathbf{3a} \text{ to each side}$$

$$8a - 0.42 = 0.98$$

$$8a - 0.42 + \mathbf{0.42} = 0.98 + \mathbf{0.42} \quad \text{Add } \mathbf{0.42} \text{ to each side}$$

$$8a = 1.40$$

$$\frac{8a}{\mathbf{8}} = \frac{1.40}{\mathbf{8}} \quad \text{Divide each side by } \mathbf{8}$$

$$a = 0.175$$

Practice Problems

1 Solve: $x - 3.4 = 6.7$

2 Solve: $4y = 3.48$

3 Solve: $\frac{1}{5}x - 2.4 = 8.3$

4 Solve: $7a - 0.18 = 2a + 0.77$

Answers

1 10.1 2 0.87 3 53.5
4 0.19

5 If a car was rented from the company in Example 5 for 2 days and the total charge was \$41, how many miles was the car driven?

EXAMPLE 5 A car rental company charges \$11 per day and 16 cents per mile for their cars. If a car was rented for 1 day and the charge was \$25.40, how many miles was it driven?

Solution If we let $x =$ the number of miles driven, then the charge for the miles driven will be $0.16x$, the cost per mile times the number of miles. To find the total cost to rent the car, we add 11 to $0.16x$. Here is the equation that describes the situation:

$$\underbrace{\text{\$11 per day}}_{11} + \underbrace{\text{16 cents per mile}}_{0.16x} = \underbrace{\text{Total cost}}_{25.40}$$

To solve the equation, we add -11 to each side and then divide each side by 0.16.

$$11 + (\mathbf{-11}) + 0.16x = 25.40 + (\mathbf{-11}) \qquad \text{Add } \mathbf{-11} \text{ to each side}$$

$$0.16x = 14.40$$

$$\frac{0.16x}{\mathbf{0.16}} = \frac{14.40}{\mathbf{0.16}} \qquad \text{Divide each side by } \mathbf{0.16}$$

$$x = 90 \qquad 14.40 \div 0.16 = 90$$

The car was driven 90 miles.

6 Amy has \$1.75 in dimes and quarters. If she has 7 more dimes than quarters, how many of each coin does she have?

EXAMPLE 6 Diane has \$1.60 in dimes and nickels. If she has 7 more dimes than nickels, how many of each coin does she have?

Solution If we let $x =$ the number of nickels, then the number of dimes must be $x + 7$, because she has 7 more dimes than nickels. Since each nickel is worth 5 cents, the amount of money she has in nickels is $0.05x$. Similarly, since each dime is worth 10 cents, the amount of money she has in dimes is $0.10(x + 7)$. Here is a table that summarizes what we have so far:

	Nickels	Dimes
Number of	x	$x + 7$
Value of	$0.05x$	$0.10(x + 7)$

Since the total value of all the coins is \$1.60, the equation that describes this situation is

$$\underbrace{\text{Amount of money in nickels}}_{0.05x} + \underbrace{\text{Amount of money in dimes}}_{0.10(x+7)} = \underbrace{\text{Total amount of money}}_{1.60}$$

This time, let's show only the essential steps in the solution.

$$0.05x + 0.10x + 0.70 = 1.60 \qquad \text{Distributive property}$$

$$0.15x + 0.70 = 1.60 \qquad \text{Add } 0.05x \text{ and } 0.10x \text{ to get } 0.15x$$

$$0.15x = 0.90 \qquad \text{Add } -0.70 \text{ to each side}$$

$$x = 6 \qquad \text{Divide each side by } 0.15$$

Since $x = 6$, she has 6 nickels. To find the number of dimes, we add 7 to the number of nickels (she has 7 more dimes than nickels). The number of dimes is $6 + 7 = 13$. Here is a check of our results.

6 nickels are worth 6(\$0.05) = \$0.30
13 dimes are worth 13(\$0.10) = \$1.30
The total value is \$1.60

Answers
5 118.75 miles
6 3 quarters, 10 dimes

PROBLEM SET 5.6

Solve each equation.

1 $x + 3.7 = 2.2$

2 $x + 4.8 = 9.1$

3 $x - 0.45 = 0.32$

4 $x - 23.3 = -4.5$

5 $8a = 1.2$

6 $6a = 18.6$

7 $-4y = 1.4$

8 $-7y = -0.63$

9 $0.5n = -0.4$

10 $0.6n = -0.12$

11 $4x - 4.7 = 3.5$

12 $2x + 3.8 = -7.7$

13 $0.02 + 5y = -0.3$

14 $0.8 + 10y = -0.7$

15 $\frac{1}{3}x - 2.99 = 1.02$

16 $\frac{1}{7}x + 2.87 = -3.01$

17 $7n - 0.32 = 5n + 0.56$

18 $6n + 0.88 = 2n - 0.77$

19 $3a + 4.6 = 7a + 5.3$

20 $2a - 3.3 = 7a - 5.2$

21 $0.5x + 0.1(x + 20) = 3.2$

22 $0.1x + 0.5(x + 8) = 7$

23 $0.08x + 0.09(x + 2000) = 690$

24 $0.11x + 0.12(x + 4000) = 940$

25 A car rental company charges $10 a day and 16 cents per mile to rent their cars. If a car was rented for 1 day for a total charge of $23.92, how many miles was it driven?

26 A car rental company charges $12 a day and 18 cents per mile to rent their cars. If the total charge for a 1-day rental was $33.78, how many miles was the car driven?

27 A car rental company charges $9 per day and 15 cents a mile to rent their cars. If a car was rented for 2 days for a total charge of $40.05, how many miles was it driven?

28 A car rental company charges $11 a day and 18 cents per mile to rent their cars. If the total charge for a 2-day rental was $61.60, how many miles was it driven?

29 Mary has \$2.20 in dimes and nickels. If she has 10 more dimes than nickels, how many of each coin does she have?

30 Bob has \$1.65 in dimes and nickels. If he has 9 more nickels than dimes, how many of each coin does he have?

31 Suppose you have \$9.60 in dimes and quarters. How many of each coin do you have if you have twice as many quarters as dimes?

32 A collection of dimes and quarters has a total value of \$2.75. If there are three times as many dimes as quarters, how many of each coin is in the collection?

Review Problems

The problems below review the material on exponents we have covered previously.

Expand and simplify.

33 5^3 **34** 2^5 **35** $(-3)^2$ **36** $(-2)^3$

37 $\left(\frac{1}{3}\right)^4$ **38** $\left(\frac{3}{4}\right)^3$ **39** $\left(-\frac{5}{6}\right)^2$ **40** $\left(-\frac{3}{5}\right)^3$

41 $(0.5)^2$ **42** $(0.1)^3$ **43** $(-1.2)^2$ **44** $(-2.1)^2$

Use the properties of exponents from Section 2.7 to simplify each expression.

45 $x^5 \cdot x^8$ **46** $x^3 \cdot x^5 \cdot x^7$ **47** $5a^3 \cdot 9a^4$ **48** $8a^5 \cdot 6a^4$

49 $(15x^5y^3)^2$ **50** $(6x^3y^2)^3$

5.7 Square Roots and the Pythagorean Theorem

Introduction . . . Figure 1 shows the front view of the roof of a tool shed. How do we find the length d of the diagonal part of the roof? (Imagine that you are drawing the plans for the shed. Since the shed hasn't been built yet, you can't just measure the diagonal. But you need to know how long it will be so you can buy the correct amount of material to build the shed.)

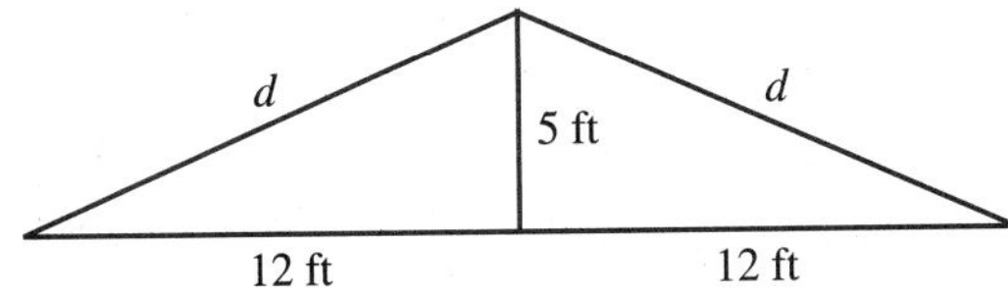

Figure 1

There is a formula from geometry that gives the length d:

$$d = \sqrt{12^2 + 5^2}$$

where $\sqrt{}$ is called the *square root symbol.* If we simplify what is under the square root symbol, we have this:

$$\begin{aligned} d &= \sqrt{144 + 25} \\ &= \sqrt{169} \end{aligned}$$

The expression $\sqrt{169}$ stands for the number we *square* to get 169. Since $13 \cdot 13 = 169$, that number is 13. Therefore the length d in our original diagram is 13 feet.

Here is a more detailed discussion of square roots. In Chapter 1, we did some work with exponents. In particular, we spent some time finding squares of numbers. For example, we considered expressions like this:

$$\begin{aligned} 5^2 &= 5 \cdot 5 = 25 \\ 7^2 &= 7 \cdot 7 = 49 \\ x^2 &= x \cdot x \end{aligned}$$

We say that "the square of 5 is 25" and "the square of 7 is 49." To square a number, we multiply it by itself. When we ask for the *square root* of a given number, we want to know what number we *square* in order to obtain the given number. We say that the square root of 49 is 7, because 7 is the number we square to get 49. Likewise, the square root of 25 is 5, because $5^2 = 25$. The symbol we use to denote square root is $\sqrt{}$, which is also called a *radical sign*. Here is the precise definition of square root.

Note

The square root we are describing here is actually the principal square root. There is another square root that is a negative number. We won't see it in this book, but if you go on to take an algebra course you will see it there.

DEFINITION The **square root** of a positive number a, written $\sqrt{a}$, is the number we square to get a. *In symbols:*

If $\sqrt{a} = b$, then $b^2 = a$.

We list some common square roots in Table 1.

Table 1

Statement	In Words	Reason
$\sqrt{0} = 0$	The square root of 0 is 0	Because $0^2 = 0$
$\sqrt{1} = 1$	The square root of 1 is 1	Because $1^2 = 1$
$\sqrt{4} = 2$	The square root of 4 is 2	Because $2^2 = 4$
$\sqrt{9} = 3$	The square root of 9 is 3	Because $3^2 = 9$
$\sqrt{16} = 4$	The square root of 16 is 4	Because $4^2 = 16$
$\sqrt{25} = 5$	The square root of 25 is 5	Because $5^2 = 25$

Numbers like 1, 9, and 25, whose square roots are whole numbers, are called *perfect squares*. To find the square root of a perfect square, we look for the whole number that is squared to get the perfect square. The following examples involve square roots of perfect squares.

EXAMPLE 1 Simplify: $7\sqrt{64}$

Solution The expression $7\sqrt{64}$ means 7 times $\sqrt{64}$. To simplify this expression, we write $\sqrt{64}$ as 8 and multiply:

$$7\sqrt{64} = 7 \cdot 8 = 56$$

We know $\sqrt{64} = 8$, because $8^2 = 64$.

EXAMPLE 2 Simplify: $\sqrt{9} + \sqrt{16}$

Solution We write $\sqrt{9}$ as 3 and $\sqrt{16}$ as 4. Then we add:

$$\sqrt{9} + \sqrt{16} = 3 + 4 = 7$$

EXAMPLE 3 Simplify: $\sqrt{\frac{25}{81}}$

Solution We are looking for the number we square (multiply times itself) to get $\frac{25}{81}$. We know that when we multiply two fractions, we multiply the numerators and multiply the denominators. Since $5 \cdot 5 = 25$ and $9 \cdot 9 = 81$, the square root of $\frac{25}{81}$ must be $\frac{5}{9}$:

$$\sqrt{\frac{25}{81}} = \frac{5}{9} \quad \text{because} \quad \left(\frac{5}{9}\right)^2 = \frac{5}{9} \cdot \frac{5}{9} = \frac{25}{81}$$

In Examples 4–6, we simplify each expression as much as possible.

EXAMPLE 4 Simplify: $12\sqrt{25} = 12 \cdot 5 = 60$

EXAMPLE 5 Simplify: $\sqrt{100} - \sqrt{36} = 10 - 6 = 4$

EXAMPLE 6 Simplify: $\sqrt{\frac{49}{121}} = \frac{7}{11}$ because $\left(\frac{7}{11}\right)^2 = \frac{7}{11} \cdot \frac{7}{11} = \frac{49}{121}$

Practice Problems

1 Simplify: $4\sqrt{25}$

2 Simplify: $\sqrt{36} + \sqrt{4}$

3 Simplify: $\sqrt{\frac{36}{100}}$

Simplify each expression as much as possible.

4 $14\sqrt{36}$

5 $\sqrt{81} - \sqrt{25}$

6 $\sqrt{\frac{64}{121}}$

Answers

1 20 **2** 8 **3** $\frac{3}{5}$ **4** 84

5 4 **6** $\frac{8}{11}$

So far in this section we have been concerned only with square roots of perfect squares. The next question is, "What about square roots of numbers that are not perfect squares, like $\sqrt{7}$, for example?" We know that

$$\sqrt{4} = 2 \quad \text{and} \quad \sqrt{9} = 3$$

And since 7 is between 4 and 9, $\sqrt{7}$ should be between $\sqrt{4}$ and $\sqrt{9}$. That is, $\sqrt{7}$ should be between 2 and 3. But what is it exactly? The answer is, we cannot write it exactly in decimal or fraction form. Because of this, it is called an *irrational number.* We can approximate it with a decimal but we can never write it exactly with a decimal. Table 2 gives some decimal approximations for $\sqrt{7}$. The decimal approximations were obtained by using a calculator. We could continue the list to any accuracy we desired. However, we would never reach a number in decimal form whose square was exactly 7.

Table 2 Approximations for the square root of 7

Accurate to the Nearest	The Square Root of 7 Is	Check by Squaring
Tenth	$\sqrt{7} = 2.6$	$(2.6)^2 = 6.76$
Hundredth	$\sqrt{7} = 2.65$	$(2.65)^2 = 7.0225$
Thousandth	$\sqrt{7} = 2.646$	$(2.646)^2 = 7.001316$
Ten thousandth	$\sqrt{7} = 2.6458$	$(2.6458)^2 = 7.00025764$

EXAMPLE 7 Give a decimal approximation for the expression $5\sqrt{12}$ that is accurate to the nearest ten thousandth.

Solution Let's agree not to round to the nearest ten thousandth until we have first done all the calculations. Using a calculator, we find $\sqrt{12} \approx 3.4641016$. Therefore,

$$\begin{aligned} 5\sqrt{12} &\approx 5(3.4641016) && \sqrt{12} \text{ on calculator} \\ &= 17.320508 && \text{Multiplication} \\ &= 17.3205 && \text{To the nearest ten thousandth} \end{aligned}$$

EXAMPLE 8 Approximate $\sqrt{301} + \sqrt{137}$ to the nearest hundredth.

Solution Using a calculator to approximate the square roots, we have

$$\sqrt{301} + \sqrt{137} \approx 17.349352 + 11.704700 = 29.054052$$

To the nearest hundredth, the answer is 29.05.

EXAMPLE 9 Approximate $\sqrt{\frac{7}{11}}$ to the nearest thousandth.

Solution Since we are using calculators, we first change $\frac{7}{11}$ to a decimal and then find the square root:

$$\sqrt{\frac{7}{11}} \approx \sqrt{0.6363636} \approx 0.7977240$$

To the nearest thousandth, the answer is 0.798.

7 Give a decimal approximation for the expression $5\sqrt{14}$ that is accurate to the nearest ten thousandth.

8 Approximate $\sqrt{405} + \sqrt{147}$ to the nearest hundredth.

9 Approximate $\sqrt{\frac{7}{12}}$ to the nearest thousandth.

Answers
7 18.7083 **8** 32.25 **9** 0.764

The Pythagorean Theorem

At the beginning of this section we mentioned a formula from geometry that we used to find the length of the diagonal of the roof of a tool shed. That formula is called the Pythagorean theorem and it applies to right triangles. A *right triangle* is a triangle that contains a 90° (or right) angle. The longest side in a right triangle is called the *hypotenuse*, and we use the letter c to denote it. The two shorter sides are denoted by the letters a and b. The Pythagorean theorem states that the hypotenuse is the square root of the sum of the squares of the two shorter sides. In symbols:

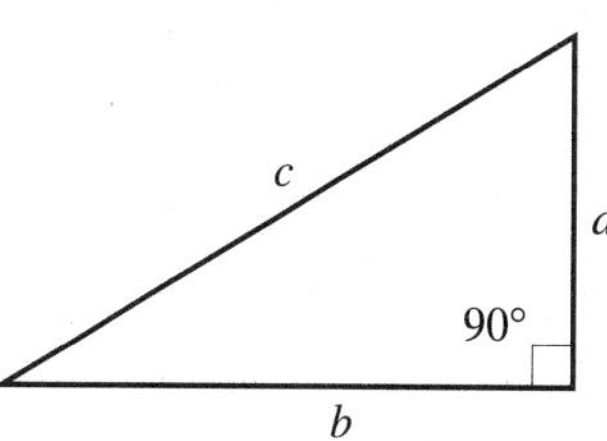

$$c = \sqrt{a^2 + b^2}$$

10 Find the length of the hypotenuse in each right triangle.

a

c
12 cm
16 cm

b

5 ft
c
5 ft

EXAMPLE 10 Find the length of the hypotenuse in each right triangle.

a

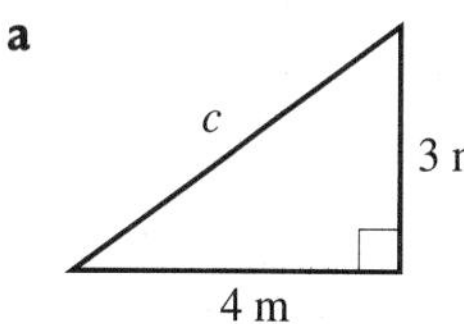

b

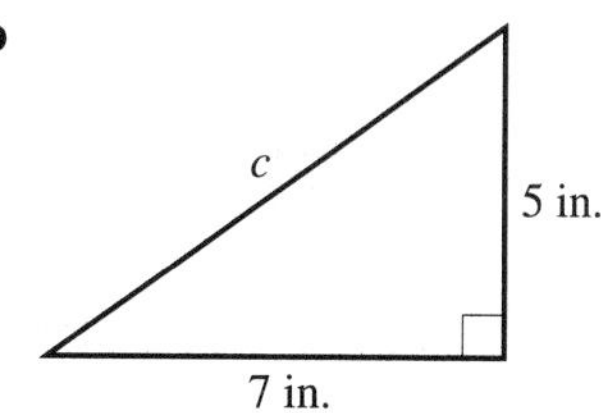

Solution We apply the formula given above.

a When $a = 3$ and $b = 4$:

$$\begin{aligned} c &= \sqrt{3^2 + 4^2} \\ &= \sqrt{9 + 16} \\ &= \sqrt{25} \\ c &= 5 \text{ meters} \end{aligned}$$

b When $a = 5$ and $b = 7$:

$$\begin{aligned} c &= \sqrt{5^2 + 7^2} \\ &= \sqrt{25 + 49} \\ &= \sqrt{74} \\ c &\approx 8.60 \text{ inches} \end{aligned}$$

In part (a), the solution is a whole number, whereas in part (b), we must use a calculator to get 8.60 as an approximation to $\sqrt{74}$.

11 A wire from the top of a 12-foot pole is fastened to the ground by a stake that is 5 feet from the bottom of the pole. What is the length of the wire?

EXAMPLE 11 A ladder is leaning against the top of a 6-foot wall. If the bottom of the ladder is 8 feet from the wall, how long is the ladder?

Solution A picture of the situation is shown in Figure 2. We let c denote the length of the ladder. Applying the Pythagorean theorem, we have

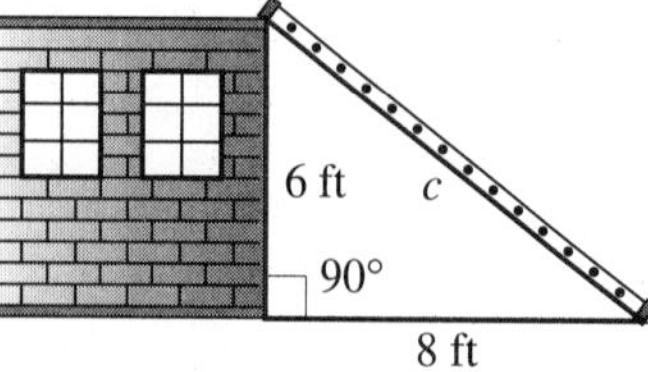

Figure 2

$$\begin{aligned} c &= \sqrt{6^2 + 8^2} \\ &= \sqrt{36 + 64} \\ &= \sqrt{100} \\ &= 10 \text{ feet} \end{aligned}$$

The ladder is 10 feet long.

Answers
10 a 20 cm **b** 7.07 ft
11 13 ft

PROBLEM SET 5.7

Find each of the following square roots without using a calculator.

1 $\sqrt{64}$ **2** $\sqrt{100}$ **3** $\sqrt{81}$ **4** $\sqrt{49}$

5 $\sqrt{36}$ **6** $\sqrt{144}$ **7** $\sqrt{25}$ **8** $\sqrt{169}$

Simplify each of the following expressions without using a calculator.

9 $3\sqrt{25}$ **10** $9\sqrt{49}$ **11** $6\sqrt{64}$ **12** $11\sqrt{100}$

13 $15\sqrt{9}$ **14** $8\sqrt{36}$ **15** $16\sqrt{9}$ **16** $9\sqrt{16}$

17 $\sqrt{49} + \sqrt{64}$ **18** $\sqrt{1} + \sqrt{0}$ **19** $\sqrt{16} - \sqrt{9}$ **20** $\sqrt{25} - \sqrt{4}$

21 $3\sqrt{25} + 9\sqrt{49}$ **22** $6\sqrt{64} + 11\sqrt{100}$ **23** $15\sqrt{9} - 9\sqrt{16}$ **24** $7\sqrt{49} - 2\sqrt{4}$

25 $\sqrt{\frac{16}{49}}$ **26** $\sqrt{\frac{100}{121}}$ **27** $\sqrt{\frac{36}{64}}$ **28** $\sqrt{\frac{81}{144}}$

Indicate whether each of the expressions in Problems 29–32 is *True* or *False*.

29 $\sqrt{4} + \sqrt{9} = \sqrt{4 + 9}$ **30** $\sqrt{\frac{16}{25}} = \frac{\sqrt{16}}{\sqrt{25}}$

31 $\sqrt{25 \cdot 9} = \sqrt{25} \cdot \sqrt{9}$ **32** $\sqrt{100} - \sqrt{36} = \sqrt{100 - 36}$

Find the length of the hypotenuse in each right triangle. Round to the nearest hundredth, if rounding is necessary.

33

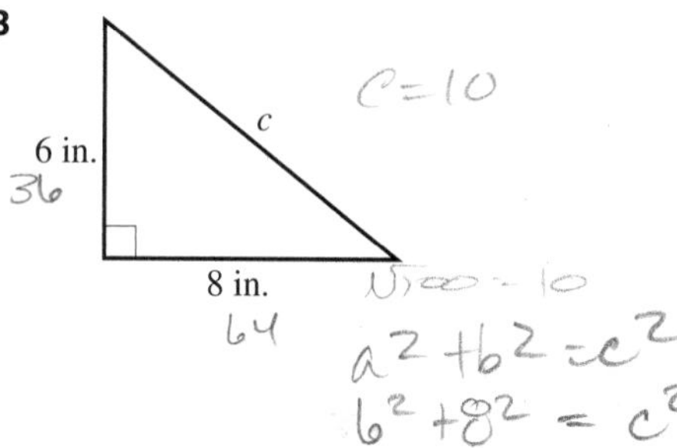

34

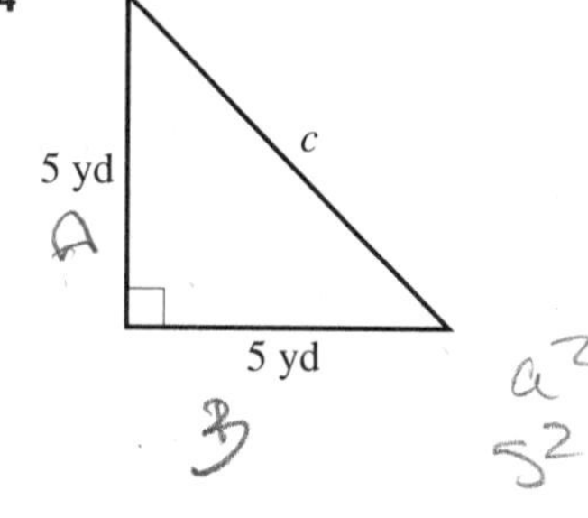

35

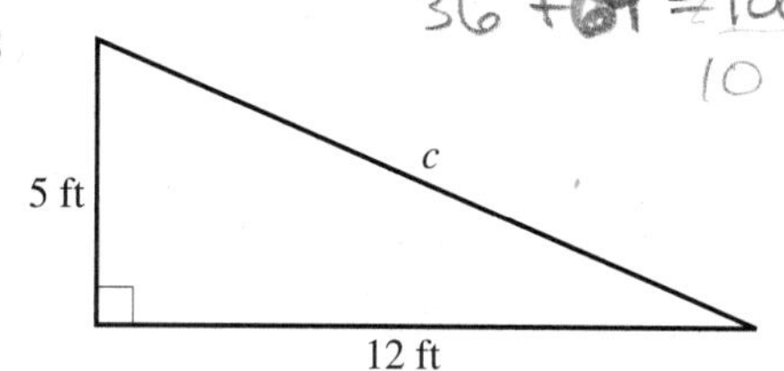

36

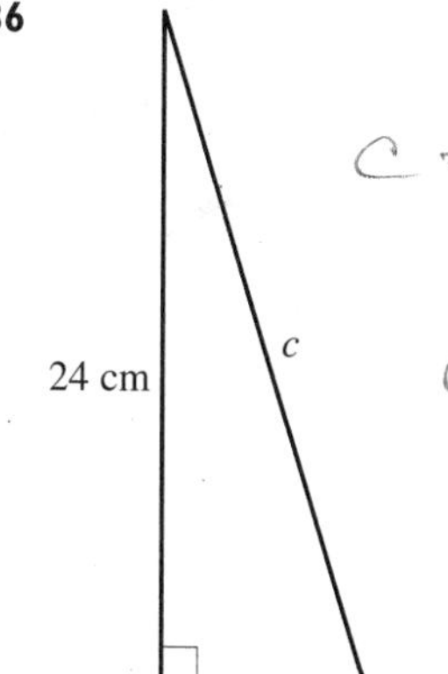

37

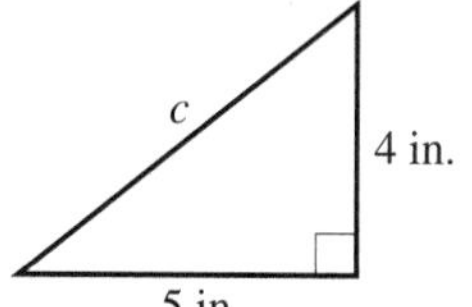

38

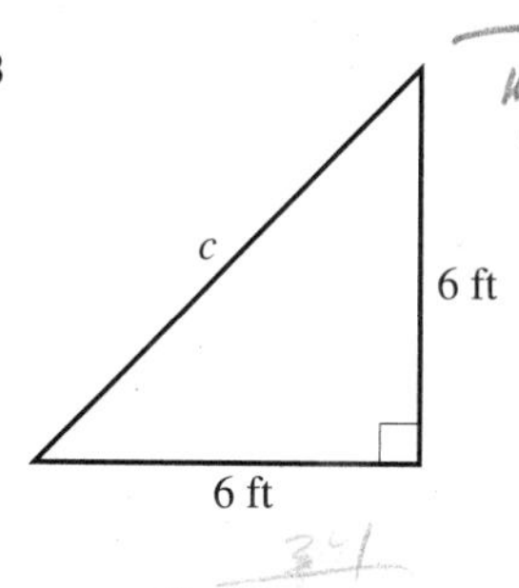

39

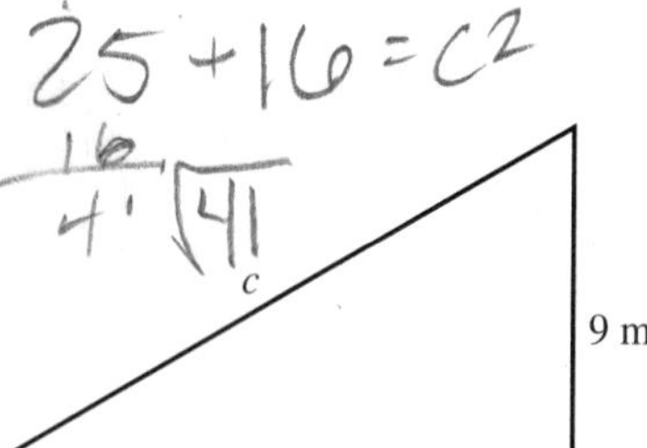

40

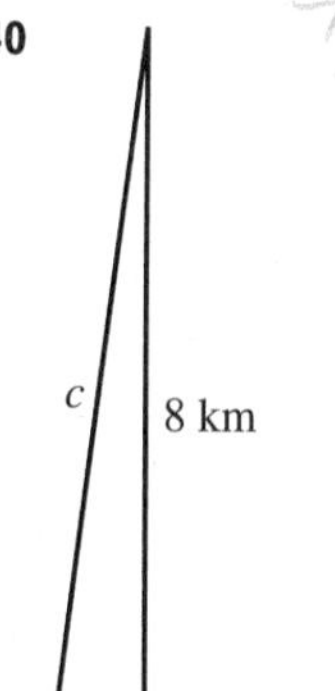

41 A ladder is leaning against the top of a 15-foot wall. If the bottom of the ladder is 20 feet from the wall, how long is the ladder?

42 A wire from the top of a 24-foot pole is fastened to the ground by a stake that is 10 feet from the bottom of the pole. How long is the wire?

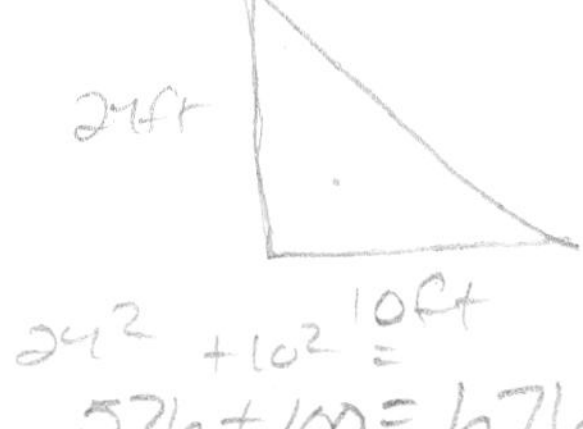

Calculator Problems

Use a calculator to work each of the following problems.

Approximate each of the following square roots to the nearest ten thousandth.

43 $\sqrt{2}$ **44** $\sqrt{5}$ **45** $\sqrt{12}$ **46** $\sqrt{18}$

47 $\sqrt{125}$ **48** $\sqrt{75}$ **49** $\sqrt{324}$ **50** $\sqrt{1296}$

Approximate each of the following expressions to the nearest hundredth.

51 $2\sqrt{3}$ **52** $3\sqrt{2}$ **53** $5\sqrt{5}$ **54** $5\sqrt{3}$

55 $\frac{\sqrt{3}}{3}$ **56** $\frac{\sqrt{2}}{2}$ **57** $\sqrt{\frac{1}{3}}$ **58** $\sqrt{\frac{1}{2}}$

Approximate each of the following expressions to the nearest thousandth.

59 $\sqrt{12} + \sqrt{75}$ **60** $\sqrt{18} + \sqrt{50}$ **61** $\sqrt{87}$ **62** $\sqrt{68}$

63 $2\sqrt{3} + 5\sqrt{3}$ **64** $3\sqrt{2} + 5\sqrt{2}$ **65** $7\sqrt{3}$ **66** $8\sqrt{2}$

67 Find the length of the hypotenuse of a right triangle (to the nearest hundredth) if the other two sides are 4.33 inches and 6.17 inches long.

68 Find the length of the hypotenuse of a right triangle (to the nearest hundredth) if the lengths of the other two sides are 4.5 feet and 16.3 feet.

Review Problems

The problems below review some of the material we have covered in Chapter 3, involving fractions and mixed numbers.

Perform the indicated operations. Write your answers as whole numbers, proper fractions, or mixed numbers.

69 $\frac{5}{7} \cdot \frac{14}{25}$

70 $1\frac{1}{4} \div 2\frac{1}{8}$

71 $4\frac{3}{10} + 5\frac{2}{100}$

72 $8\frac{1}{5} + 1\frac{1}{10}$

73 $3\frac{2}{10} \cdot 2\frac{5}{10}$

74 $6\frac{9}{10} \div 2\frac{3}{10}$

75 $7\frac{1}{10} - 4\frac{3}{10}$

76 $3\frac{7}{10} - 1\frac{97}{100}$

One Step Further: Extending the Concepts

77 The hypotenuse of a right triangle is 10 inches long. If one of the sides is 8 inches long, find the length of the other side and the area of the triangle.

78 The hypotenuse of a right triangle is 13 inches long. If one of the sides is 5 inches long, find the length of the other side and the area of the triangle.

79 The area of a right triangle is 6 square feet. If one of the shorter sides is 3 feet long, how long is the hypotenuse?

80 The area of a right triangle is 54 square feet. If one of the shorter sides is 12 feet long, how long is the hypotenuse?

5.8 Simplifying Square Roots

Do you know that $\sqrt{50}$ and $5 \cdot \sqrt{2}$ are the same number? One way to convince yourself that this is true is with a calculator. To find a decimal approximation to the first expression, we enter 50 and then press the $\sqrt{\ }$ key:

50 [$\sqrt{\ }$] The calculator shows 7.0710678

To find a decimal approximation to the second expression, we multiply 5 and $\sqrt{2}$:

5 [×] 2 [$\sqrt{\ }$] [=] The calculator shows 7.0710678

Although a calculator will give the same result for both $\sqrt{50}$ and $5 \cdot \sqrt{2}$, it does not tell us *why* the answers are the same. The discussion below shows why the results are the same.

First, notice that the expressions $\sqrt{4 \cdot 9}$ and $\sqrt{4} \cdot \sqrt{9}$ have the same value:

$$\sqrt{4 \cdot 9} = \sqrt{36} = 6 \quad \text{and} \quad \sqrt{4} \cdot \sqrt{9} = 2 \cdot 3 = 6$$

Both are equal to 6. When we are multiplying and taking square roots, we can either multiply first and then take the square root of what we get, or we can take square roots first and then multiply. In symbols, we write it this way:

Multiplication Property for Square Roots

If a and b are positive numbers, then

$$\sqrt{a \cdot b} = \sqrt{a} \cdot \sqrt{b}$$

In words: The square root of a product is the product of the square roots.

Second, when a number occurs twice as a factor under a square root, then it can be taken out from under the square root. For example, $\sqrt{5 \cdot 5}$ is really $\sqrt{25}$, which is the same as just 5. Therefore, $\sqrt{5 \cdot 5} = 5$.

Repeated Factor Property for Square Roots

If a is a positive number, then

$$\sqrt{a \cdot a} = a$$

But how do these two properties help us simplify expressions such as $\sqrt{50}$? To see the answer to this question, we must factor 50 into the product of its prime factors:

$$\sqrt{50} = \sqrt{5 \cdot 5 \cdot 2}$$

The factor 5 occurs twice, meaning that we have a perfect square $(5 \cdot 5 = 25)$ under the radical. Writing this as two separate square roots, we have

$$\begin{aligned} \sqrt{5 \cdot 5 \cdot 2} &= \sqrt{5 \cdot 5} \cdot \sqrt{2} \\ &= 5 \cdot \sqrt{2} \end{aligned}$$

RULE When the number under a square root is factored completely, any factor that occurs twice can be taken out from under the square root symbol.

Practice Problems

1 Simplify: $\sqrt{63}$

EXAMPLE 1 Simplify: $\sqrt{45}$

Solution To begin we factor 45 into the product of prime factors:

$$\begin{aligned} \sqrt{45} &= \sqrt{3 \cdot 3 \cdot 5} && \text{Factor} \\ &= \sqrt{3 \cdot 3} \cdot \sqrt{5} && \text{Multiplication property} \\ &= 3 \cdot \sqrt{5} && \text{Repeated factor property} \end{aligned}$$

The expressions $\sqrt{45}$ and $3 \cdot \sqrt{5}$ are equivalent. The expression $3 \cdot \sqrt{5}$ is said to be in *simplified form* because the number under the radical is as small as possible.

Note

When we work with square roots, the expressions $3\sqrt{5}$ and $3 \cdot \sqrt{5}$ are the same. Both of them represent the product of 3 and $\sqrt{5}$. For simplicity, we usually omit the multiplication dot.

In our next example, a variable appears under the square root symbol. We will assume that all variables that appear under a radical represent positive numbers.

2 Simplify: $\sqrt{45x^2}$

EXAMPLE 2 Simplify: $\sqrt{18x^2}$

Solution We factor $18x^2$ into $3 \cdot 3 \cdot 2 \cdot x \cdot x$. Since the factor 3 occurs twice, it can be taken out from under the radical. Likewise, since the factor x appears twice, it also can be taken out from under the radical. Therefore,

$$\begin{aligned} \sqrt{18x^2} &= \sqrt{3 \cdot 3 \cdot 2 \cdot x \cdot x} \\ &= 3 \cdot x \cdot \sqrt{2} \\ &= 3x\sqrt{2} \end{aligned}$$

You may be wondering if we can check this answer on a calculator. The answer is yes, but we need to substitute a value for x first. Suppose x is 5. Then

$$\sqrt{18x^2} = \sqrt{18 \cdot 25} = \sqrt{450} \approx 21.213203$$

and

$$3x\sqrt{2} = 3 \cdot 5\sqrt{2} = 15\sqrt{2} \approx 15(1.4142136) = 21.213203$$

3 Simplify: $\sqrt{300}$

EXAMPLE 3 Simplify: $\sqrt{180}$

Solution We factor and then look for factors occurring twice.

$$\begin{aligned} \sqrt{180} &= \sqrt{2 \cdot 2 \cdot 3 \cdot 3 \cdot 5} \\ &= 2 \cdot 3 \cdot \sqrt{5} \\ &= 6\sqrt{5} \end{aligned}$$

4 Simplify: $\sqrt{50x^3}$

EXAMPLE 4 Simplify: $\sqrt{48x^3}$

Solution
$$\begin{aligned} \sqrt{48x^3} &= \sqrt{2 \cdot 2 \cdot 2 \cdot 2 \cdot 3 \cdot x \cdot x \cdot x} \\ &= 2 \cdot 2 \cdot x \cdot \sqrt{3 \cdot x} \\ &= 4x\sqrt{3x} \end{aligned}$$

Answers

1 $3\sqrt{7}$ **2** $3x\sqrt{5}$ **3** $10\sqrt{3}$
4 $5x\sqrt{2x}$

PROBLEM SET 5.8

Simplify each expression by taking as much out from under the radical as possible. You may assume that all variables represent positive numbers.

1 $\sqrt{12}$ **2** $\sqrt{18}$ **3** $\sqrt{20}$ **4** $\sqrt{27}$ **5** $\sqrt{72}$ **6** $\sqrt{48}$

7 $\sqrt{98}$ **8** $\sqrt{75}$ **9** $\sqrt{28}$ **10** $\sqrt{44}$ **11** $\sqrt{200}$ **12** $\sqrt{300}$

13 $\sqrt{12x^2}$ **14** $\sqrt{18x^2}$ **15** $\sqrt{50x^2}$ **16** $\sqrt{45x^2}$ **17** $\sqrt{75x^3}$ **18** $\sqrt{8x^3}$

19 $\sqrt{50x^3}$ **20** $\sqrt{45x^3}$ **21** $\sqrt{32x^2y^3}$ **22** $\sqrt{90x^2y^3}$ **23** $\sqrt{243x^4}$ **24** $\sqrt{288x^4}$

25 $\sqrt{72x^2y^4}$ **26** $\sqrt{72x^4y^2}$ **27** $\sqrt{12x^3y^3}$ **28** $\sqrt{20x^3y^3}$

The triangles below are called *isosceles* right triangles because the two shorter sides are the same length. In each case, use the Pythagorean theorem to find the length of the hypotenuse. Simplify your answers, but do not use a calculator to approximate them.

29

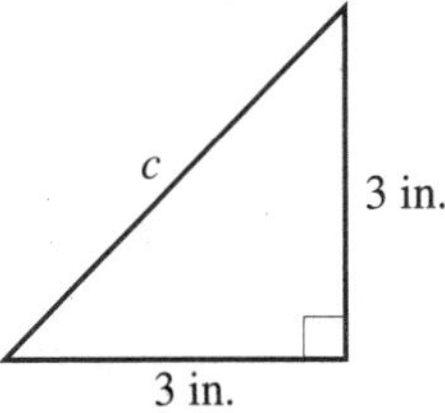

30

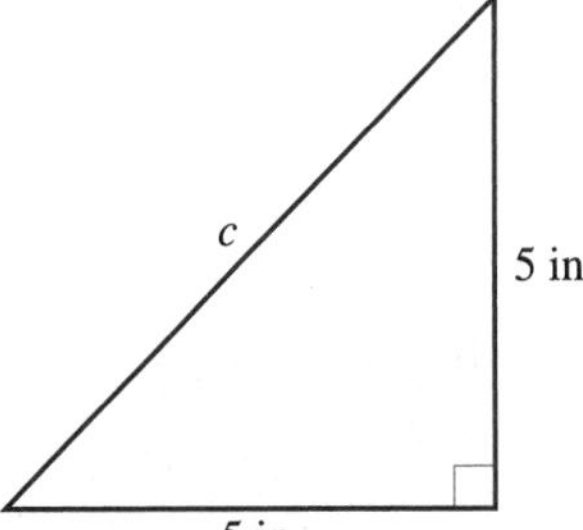

31

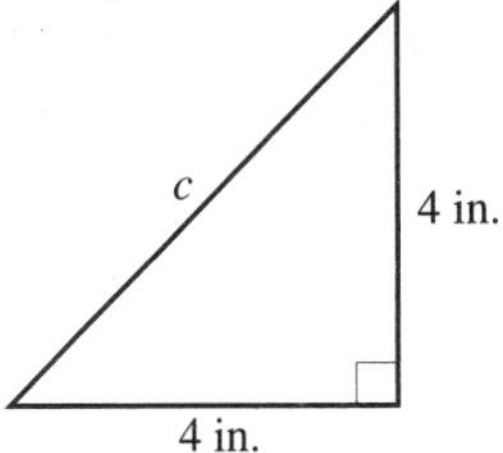

32

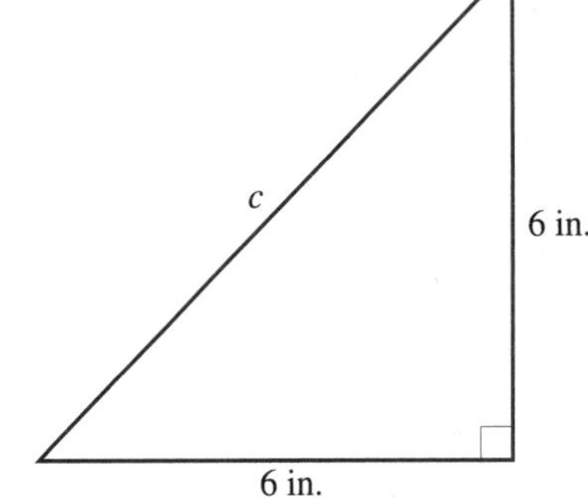

33

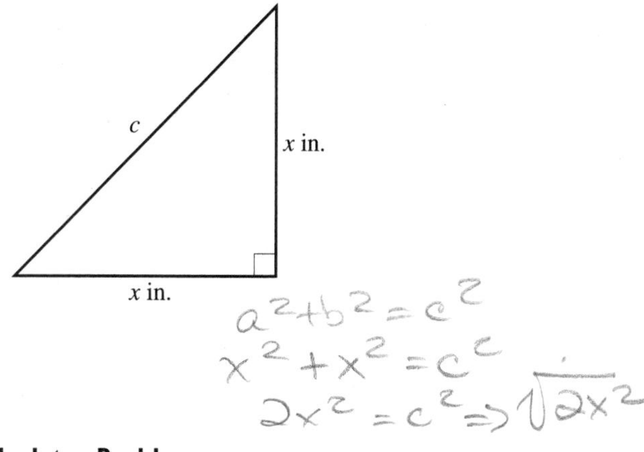

34

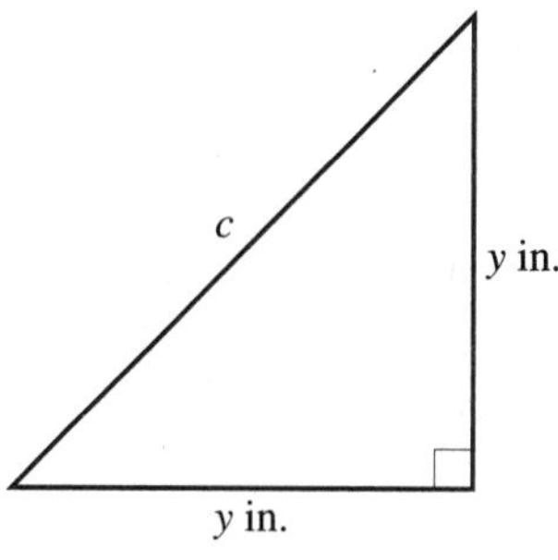

Calculator Problems

Use a calculator to find decimal approximations for each of the following numbers.

35 $\sqrt{72}$ and $6\sqrt{2}$ **36** $\sqrt{75}$ and $5\sqrt{3}$ **37** $\sqrt{24}$ and $2\sqrt{6}$ **38** $\sqrt{45}$ and $3\sqrt{5}$

Substitute $x = 5$ into each of the following expressions, and then use a calculator to obtain a decimal approximation to each.

39 $x\sqrt{x}$ **40** $\sqrt{x^3}$ **41** $x^2\sqrt{x}$ **42** $\sqrt{x^5}$

Review Problems

The problems below review material we covered in Section 4.6.

In Problems 43–48, use the formula $y = \frac{1}{2}x - 3$ to find y if:

43 $x = 0$ **44** $x = 1$ **45** $x = -4$ **46** $x = 4$

47 $x = 2$ **48** $x = -2$

In Problems 49–52, use the formula $2x + 5y = 10$ to find y if:

49 $x = 0$ **50** $x = -5$ **51** $x = 5$ **52** $x = \frac{5}{2}$

In Problems 53–56, use the formula $2x + 5y = 10$ to find x if:

53 $y = 0$ **54** $y = 2$ **55** $y = -2$ **56** $y = \frac{2}{5}$

5.9 Adding and Subtracting Roots

In the past we have combined similar terms by applying the distributive property. Here are some examples that will remind you of that procedure. The middle step in each example shows the distributive property.

$$7x + 3x = (7 + 3)x = 10x$$
$$5a^2 - 2a^2 = (5 - 2)a^2 = 3a^2$$
$$6xy + 3xy = (6 + 3)xy = 9xy$$
$$4x + 3x - 2x = (4 + 3 - 2)x = 5x$$

The distributive property is the only property that allows us to combine similar terms as we have done above. In order for the distributive property to be applied, the variable parts in each expression must be the same.

We add radical expressions in the same way we add similar terms, that is, by applying the distributive property. Here is how we use the distributive property to add $7\sqrt{2}$ and $3\sqrt{2}$:

$$7\sqrt{2} + 3\sqrt{2} = (7 + 3)\sqrt{2} \quad \text{Distributive property}$$
$$= 10\sqrt{2} \quad \text{Add 7 and 3}$$

To understand the steps shown here, you must remember that $7\sqrt{2}$ means 7 times $\sqrt{2}$; the 7 and the $\sqrt{2}$ are not "stuck together." Here are some additional examples. Compare them with the problems we did at the beginning of this section. Combine using the distributive property. (Assume all variables represent positive numbers.)

EXAMPLE 1 $5\sqrt{6} - 2\sqrt{6} = (5 - 2)\sqrt{6} = 3\sqrt{6}$

EXAMPLE 2 $6\sqrt{x} + 3\sqrt{x} = (6 + 3)\sqrt{x} = 9\sqrt{x}$

EXAMPLE 3 $4\sqrt{3} + 3\sqrt{3} - 2\sqrt{3} = (4 + 3 - 2)\sqrt{3} = 5\sqrt{3}$

Practice Problems

Use the distributive property to combine each of the following.

1 $5\sqrt{3} - 2\sqrt{3}$

2 $7\sqrt{y} + 3\sqrt{y}$

3 $8\sqrt{5} - 2\sqrt{5} + 9\sqrt{5}$

As you can see, it is easy to combine radical expressions when each term contains the same square root.

Next, suppose we try to add $\sqrt{12}$ and $\sqrt{75}$. How should we go about it? You may think we should add 12 and 75 to get $\sqrt{87}$, but notice that we have not done that in any of the examples above. In fact, in Examples 1–3, the square root in the answer is the same square root we started with; we never added the number under the square roots!

A calculator can help us decide if $\sqrt{12} + \sqrt{75}$ is the same as $\sqrt{87}$. Here are the decimal approximations a calculator will give us:

$$\sqrt{12} + \sqrt{75} \approx 3.4641016 + 8.6602540 = 12.1243556$$
$$\sqrt{12 + 75} = \sqrt{87} \approx 9.3273791$$

As you can see, the two results are quite different, so we can assume that it would be a mistake to add the numbers under the square roots. That is:

$$\sqrt{12} + \sqrt{75} \neq \sqrt{12 + 75}$$

The correct way to add $\sqrt{12}$ and $\sqrt{75}$ is to simplify each expression by taking as much out from under each square root as possible. Then, if the square roots in the

Answers

1 $3\sqrt{3}$ 2 $10\sqrt{y}$ 3 $15\sqrt{5}$

resulting expressions are the same, we can add using the distributive property. Here is the way the problem is done correctly.

$$\begin{aligned} \sqrt{12} + \sqrt{75} &= \sqrt{2 \cdot 2 \cdot 3} + \sqrt{5 \cdot 5 \cdot 3} && \text{Simplify each square root} \\ &= 2\sqrt{3} + 5\sqrt{3} \\ &= (2 + 5)\sqrt{3} && \text{Distributive property} \\ &= 7\sqrt{3} && \text{Add 2 and 5} \end{aligned}$$

On a calculator, $7\sqrt{3} \approx 7(1.7320508) = 12.1243556$, which matches the approximation a calculator gives for $\sqrt{12} + \sqrt{75}$.

Note You may be thinking "Why did he show us the wrong way to do the problem first?" The reason is simple: Many people will try to add $\sqrt{12}$ and $\sqrt{75}$ by adding 12 and 75—it is a natural thing to want to do. The reason we don't is that it gives us the wrong answer every time! One of the things you need to know about learning algebra is that your intuition may lead you to a mistake.

4 Combine, if possible: $\sqrt{27} + \sqrt{75} - \sqrt{12}$

EXAMPLE 4 Combine, if possible: $\sqrt{18} + \sqrt{50} - \sqrt{8}$

Solution First we simplify each term by taking as much out from under the square root as possible. Then we use the distributive property to combine terms if they contain the same square root.

$$\begin{aligned} \sqrt{18} + \sqrt{50} - \sqrt{8} &= \sqrt{3 \cdot 3 \cdot 2} + \sqrt{5 \cdot 5 \cdot 2} - \sqrt{2 \cdot 2 \cdot 2} \\ &= 3\sqrt{2} + 5\sqrt{2} - 2\sqrt{2} \\ &= (3 + 5 - 2)\sqrt{2} \\ &= 6\sqrt{2} \end{aligned}$$

5 Subtract: $5\sqrt{45} - 3\sqrt{20}$

EXAMPLE 5 Subtract: $5\sqrt{54} - 3\sqrt{24}$

Solution Proceeding as we did in the previous example, we simplify each expression first; then we subtract by applying the distributive property.

$$\begin{aligned} 5\sqrt{54} - 3\sqrt{24} &= 5\sqrt{3 \cdot 3 \cdot 6} - 3\sqrt{2 \cdot 2 \cdot 6} \\ &= 5 \cdot 3\sqrt{6} - 3 \cdot 2\sqrt{6} \\ &= 15\sqrt{6} - 6\sqrt{6} \\ &= (15 - 6)\sqrt{6} \\ &= 9\sqrt{6} \end{aligned}$$

6 Combine, if possible: $7\sqrt{18x^3} - 3\sqrt{50x^3}$

EXAMPLE 6 Assume x is a positive number and combine, if possible:

$$5\sqrt{12x^3} - 3\sqrt{75x^3}$$

Solution We simplify each square root, and then we subtract.

$$\begin{aligned} 5\sqrt{12x^3} - 3\sqrt{75x^3} &= 5\sqrt{2 \cdot 2 \cdot 3 \cdot x \cdot x \cdot x} - 3\sqrt{5 \cdot 5 \cdot 3 \cdot x \cdot x \cdot x} \\ &= 5 \cdot 2 \cdot x\sqrt{3x} - 3 \cdot 5 \cdot x\sqrt{3x} \\ &= 10x\sqrt{3x} - 15x\sqrt{3x} \\ &= -5x\sqrt{3x} \end{aligned}$$

Answers
4 $6\sqrt{3}$ **5** $9\sqrt{5}$ **6** $6x\sqrt{2x}$

PROBLEM SET 5.9

Combine by applying the distributive property. Assume all variables represent positive numbers.

1 $2\sqrt{3} + 8\sqrt{3}$

2 $2\sqrt{3} - 8\sqrt{3}$

3 $7\sqrt{5} - 3\sqrt{5}$

4 $7\sqrt{5} + 3\sqrt{5}$

5 $9\sqrt{x} + 3\sqrt{x} - 5\sqrt{x}$

6 $6\sqrt{x} + 10\sqrt{x} - 3\sqrt{x}$

7 $8\sqrt{7} + \sqrt{7}$

8 $9\sqrt{7} + \sqrt{7}$

9 $2\sqrt{y} + \sqrt{y} + 3\sqrt{y}$

10 $7\sqrt{y} + \sqrt{y} + 2\sqrt{y}$

Simplify each square root, and then combine if possible. Assume all variables represent positive numbers.

11 $\sqrt{18} + \sqrt{32}$

12 $\sqrt{12} + \sqrt{27}$

13 $\sqrt{75} + \sqrt{27}$

14 $\sqrt{50} + \sqrt{8}$

15 $2\sqrt{75} - 4\sqrt{27}$

16 $4\sqrt{50} - 5\sqrt{8}$

17 $2\sqrt{90} + 3\sqrt{40} - 4\sqrt{10}$

18 $5\sqrt{40} - 2\sqrt{90} + 3\sqrt{10}$

19 $\sqrt{72x^2} - \sqrt{50x^2}$

20 $\sqrt{98x^2} - \sqrt{72x^2}$

21 $4\sqrt{20x^3} + 3\sqrt{45x^3}$

22 $8\sqrt{48x^3} + 2\sqrt{12x^3}$

In each diagram below, find the distance from A to B. Simplify your answers, but do not use a calculator.

23

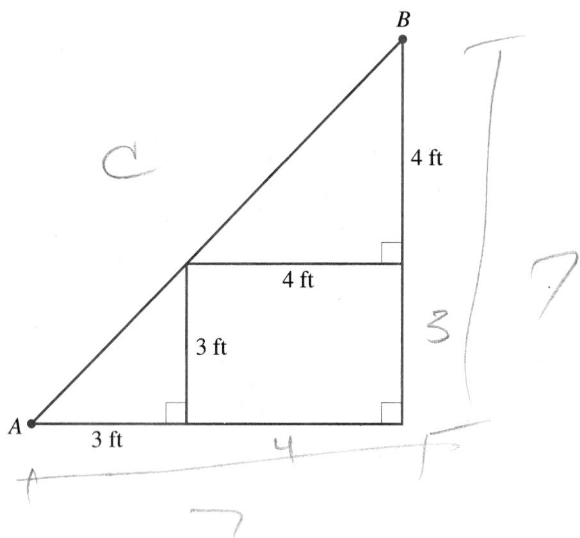

24

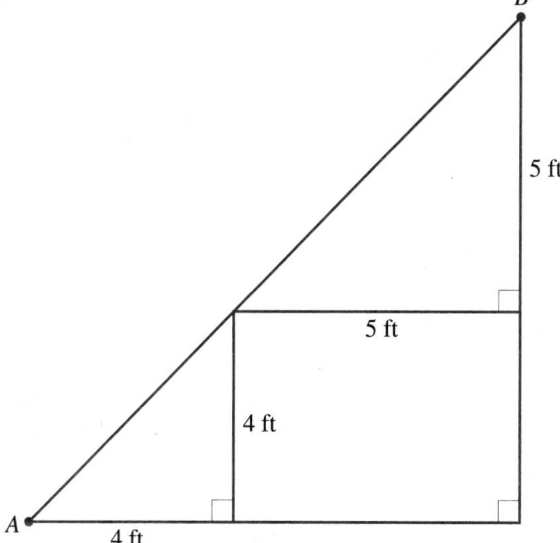

Calculator Problems

25 Use a calculator to show that $\sqrt{2} + \sqrt{3}$ is not the same as $\sqrt{5}$.

26 Use a calculator to show that $\sqrt{5} - \sqrt{2}$ is not the same as $\sqrt{3}$.

27 Use a calculator to show that $\sqrt{16} - \sqrt{10}$ is not the same as $\sqrt{6}$.

28 Use a calculator to show that $\sqrt{9} - \sqrt{5}$ is not the same as $\sqrt{4}$.

29 Use a calculator to show that the expression $\sqrt{8} + \sqrt{2}$ is equal to the expression $\sqrt{18}$. Then simplify each square root and show that the two expressions are equal without using a calculator.

30 Use a calculator to show that the expression $\sqrt{18} + \sqrt{2}$ is equal to the expression $\sqrt{32}$. Then simplify each square root and show that the two expressions are equal without using a calculator.

Review Problems

The problems below review material we covered in Section 4.9.

Graph each equation.

31 $x + y = 3$

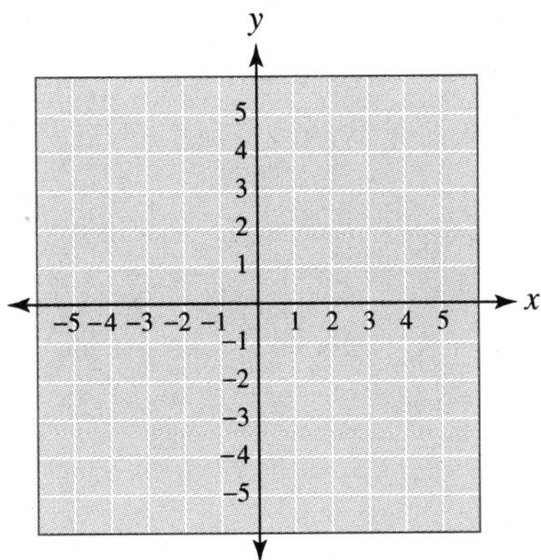

32 $x - y = 3$

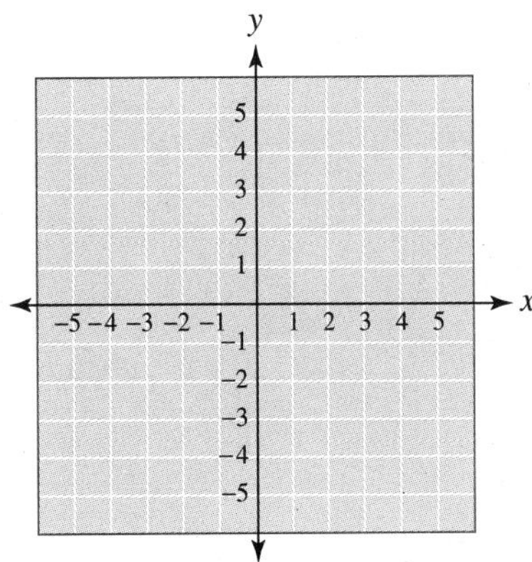

33 $y = \frac{1}{2}x - 3$

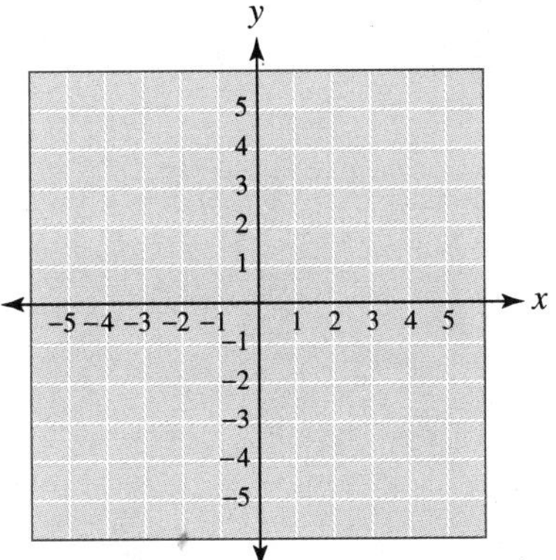

34 $y = \frac{1}{2}x + 3$

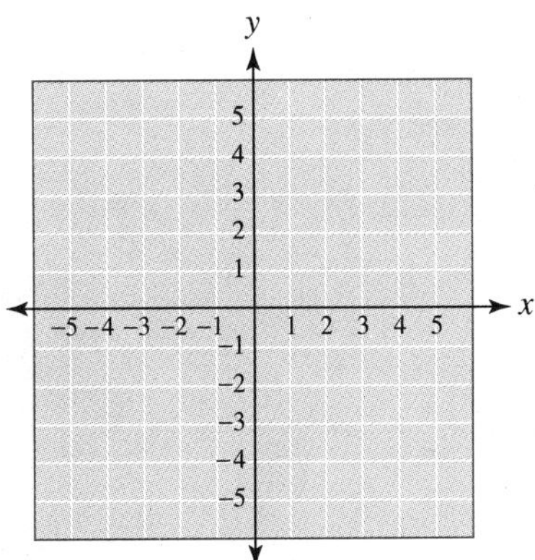

35 $2x + 5y = 10$

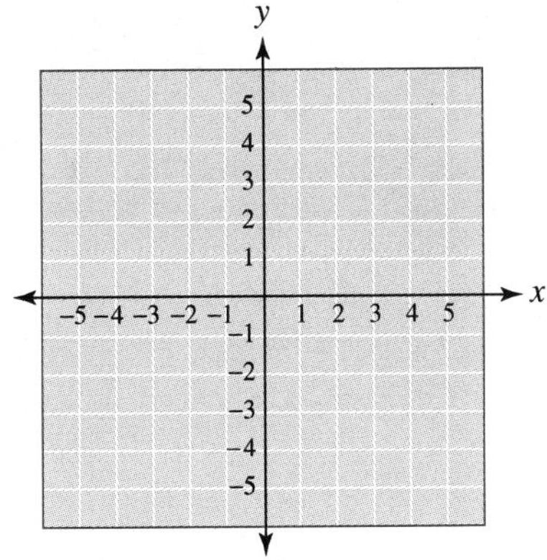

36 $5x + 2y = 10$

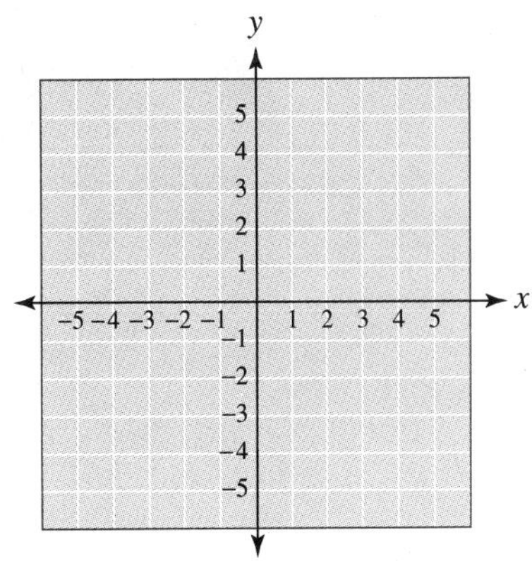

37 $2x - 5y = 10$

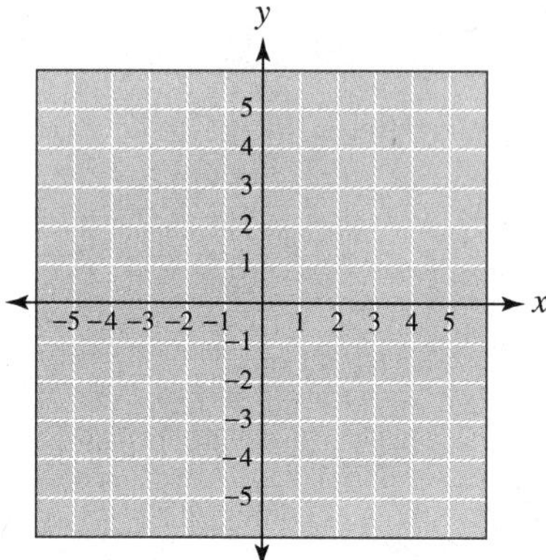

38 $5x - 2y = 10$

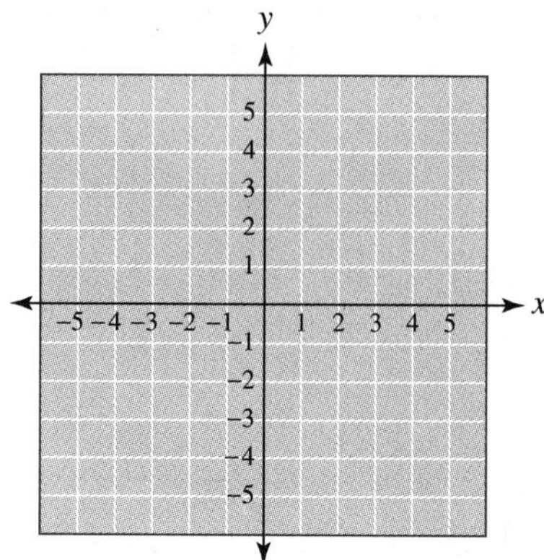

39 $y = -2x$

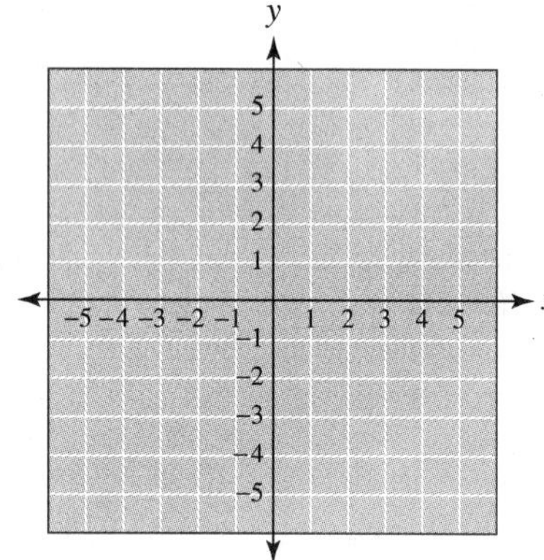

40 $y = 2x$

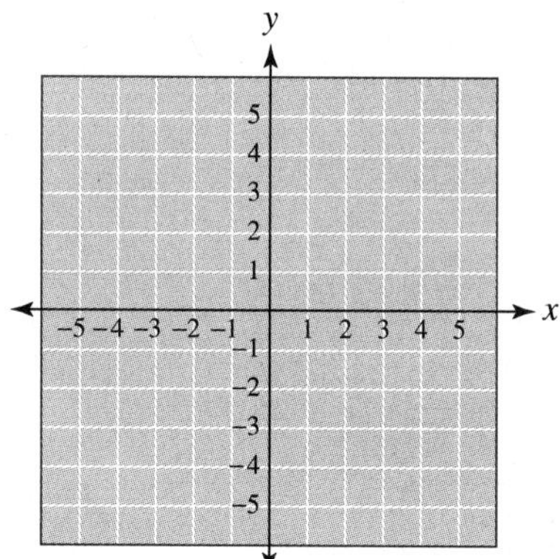

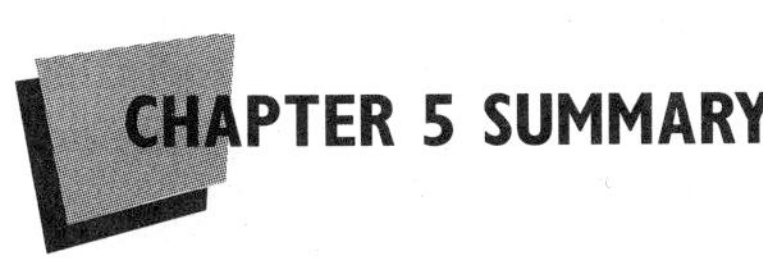

CHAPTER 5 SUMMARY

Place Value [5.1]

The place values for the first five places to the right of the decimal point are

Decimal Point	Tenths	Hundredths	Thousandths	Ten Thousandths	Hundred Thousandths
.	$\frac{1}{10}$	$\frac{1}{100}$	$\frac{1}{1{,}000}$	$\frac{1}{10{,}000}$	$\frac{1}{100{,}000}$

Examples

1 The number 4.123 in words is "four and one hundred twenty-three thousandths."

Rounding Decimals [5.1]

If the digit in the column to the right of the one we are rounding to is 5 or more, we add 1 to the digit in the column we are rounding to; otherwise, we leave it alone. We then replace all digits to the right of the column we are rounding to with zeros if they are to the left of the decimal point; otherwise, we simply delete them.

2 357.753 rounded to the nearest

Tenth: 357.8
Ten : 360

Addition and Subtraction with Decimals [5.2]

To add (or subtract) decimal numbers, we align the decimal points and add (or subtract) as if we were adding (or subtracting) whole numbers. The decimal point in the answer goes directly below the decimal points in the problem.

3
$$\begin{array}{r} 3.400 \\ 25.060 \\ +\ 0.347 \\ \hline 28.807 \end{array}$$

Multiplication with Decimals [5.3]

To multiply two decimal numbers, we multiply as if the decimal points were not there. The decimal point in the product has as many digits to the right as there are total digits to the right of the decimal points in the two original numbers.

4 If we multiply 3.49×5.863, there will be a total of $2 + 3 = 5$ digits to the right of the decimal point in the answer.

Division with Decimals [5.4]

To begin a division problem with decimals, we make sure that the divisor is a whole number. If it is not, we move the decimal point in the divisor to the right as many places as it takes to make it a whole number. We must then be sure to move the decimal point in the dividend the same number of places to the right. Once the divisor is a whole number, we divide as usual. The decimal point in the answer is placed directly above the decimal point in the dividend.

5
$$\begin{array}{r} 1.39 \\ 2.5.\overline{)3.4.75} \\ 2\,5 \\ \hline 9\,7 \\ 7\,5 \\ \hline 2\,25 \\ 2\,25 \\ \hline 0 \end{array}$$

Changing Fractions to Decimals [5.5]

To change a fraction to a decimal, we divide the numerator by the denominator.

6 $\frac{4}{15} = 0.2\overline{6}$ because

$$\begin{array}{r} 266 \\ 15\overline{)4.000} \\ 3\,0 \\ \hline 1\,00 \\ 90 \\ \hline 100 \\ 90 \\ \hline 10 \end{array}$$

Changing Decimals to Fractions [5.5]

To change a decimal to a fraction, we write the digits to the right of the decimal point over the appropriate power of 10.

7 $0.781 = \frac{781}{1{,}000}$

8 $\frac{1}{2}x - 3.78 = 2.52$

$\frac{1}{2}x - 3.78 + \mathbf{3.78} = 2.52 + \mathbf{3.78}$

$\frac{1}{2}x = 6.30$

$\mathbf{2}\left(\frac{1}{2}x\right) = \mathbf{2}(6.30)$

$x = 12.6$

9 $\sqrt{49} = 7$ because $7^2 = 7 \cdot 7 = 49$

10
$\sqrt{50} = \sqrt{5 \cdot 5 \cdot 2}$ Factor 50
$= \sqrt{5 \cdot 5} \cdot \sqrt{2}$ Multiplication property
$= 5 \cdot \sqrt{2}$ Repeated factor property

11 $\sqrt{12} + \sqrt{75}$
$= \sqrt{2 \cdot 2 \cdot 3} + \sqrt{5 \cdot 5 \cdot 3}$
$= 2\sqrt{3} + 5\sqrt{3}$
$= (2 + 5)\sqrt{3}$
$= 7\sqrt{3}$

Equations Containing Decimals [5.6]

We solve equations that contain decimals by applying the addition property of equality and the multiplication property of equality, as we did with the equations in Chapter 4.

Square Roots [5.7]

The square root of a positive number a, written $\sqrt{a}$, is the number we square to get a.

Pythagorean Theorem [5.7]

In any right triangle, the length of the longest side (the hypotenuse) is equal to the square root of the sum of the squares of the two shorter sides.

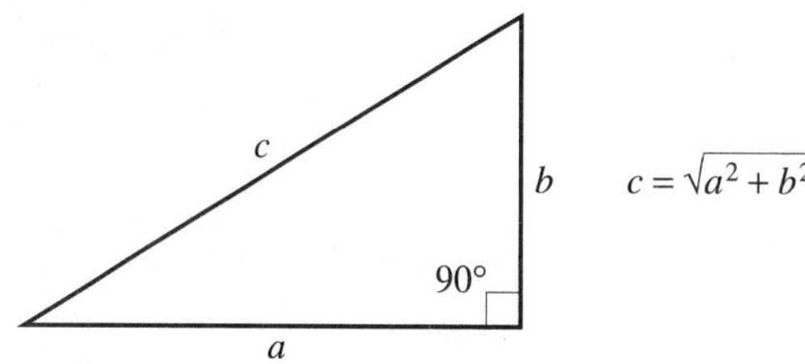

Multiplication Property for Square Roots [5.8]

If a and b are positive numbers, then

$$\sqrt{a \cdot b} = \sqrt{a} \cdot \sqrt{b}$$

In words: The square root of a product is the product of the square roots.

Repeated Factor Property for Square Roots [5.8]

If a is a positive number, then

$$\sqrt{a \cdot a} = a$$

Simplifying Square Roots [5.8]

To simplify a square root we factor the number under the square root symbol into the product of prime factors. Then we use the two properties of square roots shown above to take as much out from under the square root symbol as possible.

Adding and Subtracting Roots [5.9]

We add expressions containing square roots by first simplifying each square root and then applying the distributive property.

CHAPTER 5 REVIEW

Give the place value of the 7 in each of the following numbers. **[5.1]**

1 36.007

2 121.379

Write each of the following with a decimal number. **[5.1]**

3 Thirty-seven and forty-two ten thousandths

4 One hundred and two hundred two hundred thousandths

5 Round 98.7654 to the nearest hundredth. **[5.1]**

Perform the following operations. **[5.2, 5.3, 5.4]**

6 $3.78 + 2.036$

7 $11.076 - 3.297$

8 6.7×5.43

9 $0.89(-24.24)$

10 $-29.07 \div (-3.8)$

11 $0.7134 \div 0.58$

12 Write $\frac{7}{8}$ as a decimal. **[5.5]**

13 Write 0.705 as a fraction in lowest terms. **[5.5]**

14 Write 14.125 as a mixed number. **[5.5]**

Simplify each of the following expressions as much as possible. **[5.5]**

15 $3.3 - 4(0.22)$

16 $54.987 - 2(3.05 + 0.151)$

17 $125\left(\frac{3}{5}\right) + 4$

18 $\frac{3}{5}(0.9) + \frac{2}{5}(0.4)$

Solve each equation. **[5.6]**

19 $x + 9.8 = 3.9$

20 $5x = 23.4$

21 $0.5y - 0.2 = 3$

22 $5x - 7.2 = 3x + 3.8$

Simplify each expression as much as possible. **[5.7]**

23 $3\sqrt{25}$

24 $\sqrt{64} - \sqrt{36}$

25 $4\sqrt{25} + 3\sqrt{81}$

26 $\sqrt{\frac{16}{49}}$

27 A student has bills of \$19.48 for heating and electricity, \$6.72 for the telephone, and \$241.50 for rent each month. What is the total of these three bills?

28 A person purchases \$7.23 worth of goods at a drugstore. If a \$10 bill is used to pay for the purchases, how much change is received?

29 What is the product of $\frac{3}{4}$ and the sum of 2.8 and 3.7?

30 If a person earns $223.60 for working 40 hours, what is the person's hourly wage?

Jayma earns the following grades during the Spring term of 1995.

Class	Units	Grade
College algebra	4	C
Speech	3	A
Accounting	3	B
Marketing	3	B
Real estate	2	C

31 Calculate her GPA.

32 If her grade in college algebra had been a B instead of a C, by how much would her GPA have increased?

Find the length of the hypotenuse in each right triangle. Round your answers to the nearest tenth. **[5.7]**

33

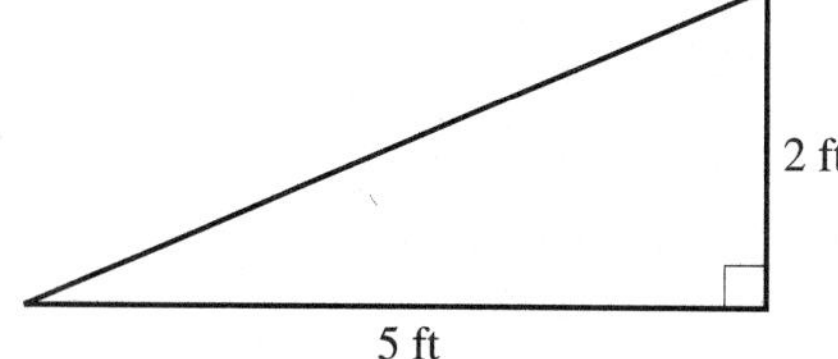

34

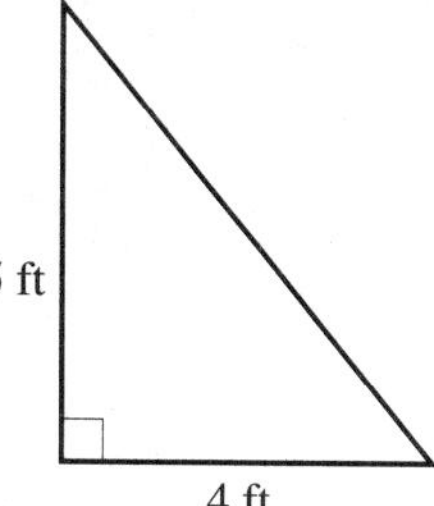

35 Find the perimeter (to the nearest tenth) and the area of the triangle in Problem 33.

36 Find the perimeter (to the nearest tenth) and the area of the triangle in Problem 34.

Simplify each expression by taking as much out from under the radical as possible. **[5.8]**

37 $\sqrt{12}$ **38** $\sqrt{50}$ **39** $\sqrt{32}$ **40** $\sqrt{24}$

41 $\sqrt{20x^2}$ **42** $\sqrt{20x^3}$ **43** $\sqrt{18x^2y^3}$ **44** $\sqrt{75x^3y^2}$

Combine. **[5.9]**

45 $8\sqrt{3} + 2\sqrt{3}$ **46** $7\sqrt{6} + 10\sqrt{6}$ **47** $4\sqrt{3} + \sqrt{3}$ **48** $9\sqrt{2} + \sqrt{2}$

Simplify each square root, and then combine if possible. **[5.9]**

49 $\sqrt{24} + \sqrt{54}$ **50** $\sqrt{18} + \sqrt{8}$ **51** $3\sqrt{75} - 8\sqrt{27}$ **52** $7\sqrt{20} - 2\sqrt{45}$

CHAPTER 5 TEST

1 Write the decimal number 5.053 in words.

2 Give the place value of the 4 in the number 53.0543.

3 Write seventeen and four hundred six ten thousandths as a decimal number.

4 Round 46.7549 to the nearest hundredth.

Perform the following operations.

5 $7 + 0.6 + 0.58$

6 $12.032 - 5.976$

7 $5.7(6.24)$

8 $-22.672 \div (-2.6)$

9 Write $\frac{23}{25}$ as a decimal.

10 Write 0.56 as a fraction in lowest terms.

Simplify each expression as much as possible.

11 $5.2(2.8 + 0.02)$

12 $5.2 - 3(0.17)$

13 $23.852 - 3(2.01 + 0.231)$

14 $\frac{3}{5}(0.6) - \frac{2}{3}(0.15)$

15 Solve: $6a - 0.18 = a + 0.77$

Simplify each expression as much as possible.

16 $2\sqrt{36} + 3\sqrt{64}$

17 $\sqrt{\dfrac{25}{81}}$

18 A person purchases $8.47 worth of goods at a drugstore. If a $20 bill is used to pay for the purchases, how much change is received?

19 If coffee sells for $5.44 per pound, how much will 3.5 pounds of coffee cost?

20 If a person earns $262 for working 40 hours, what is the person's hourly wage?

21 Find the length of the hypotenuse of the right triangle below.

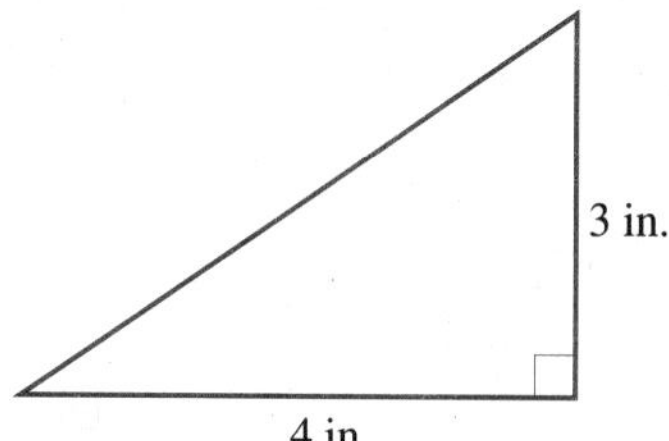

Simplify by taking as much out from under the radical as possible.

22 $\sqrt{72}$

23 $\sqrt{48x^2y}$

24 Combine $5\sqrt{7} + 3\sqrt{7}$.

25 Simplify each square root, and then combine, if possible:

$$6\sqrt{12} - 5\sqrt{48}$$

RATIO, PROPORTION, AND UNIT ANALYSIS

INTRODUCTION

In 1993, a ski resort in Vermont advertised their new high-speed chairlift as "the world's fastest chairlift with a speed of 1,100 feet per second." Most of us are not familiar with speeds given in feet per second; thus we don't have an intuitive feeling for a speed of 1,100 feet per second. However, knowing that there are 5,280 feet in 1 mile, 60 seconds in a minute, and 60 minutes in an hour, we can convert feet per second into miles per hour, a more intuitive measure of speed. The work necessary to do so is one of the things we will learn in this chapter. It looks like this:

$$\begin{aligned} 1{,}100 \text{ ft/sec} &= \frac{1{,}100 \text{ ft}}{1 \text{ sec}} \cdot \frac{1 \text{ mi}}{5{,}280 \text{ ft}} \cdot \frac{60 \text{ sec}}{1 \text{ min}} \cdot \frac{60 \text{ min}}{1 \text{ hr}} \\ &= \frac{1{,}100 \cdot 60 \cdot 60 \text{ mi}}{5{,}280 \text{ hr}} \\ &= 750 \text{ mi/hr} \end{aligned}$$

This result tells us that 1,100 feet per second is equivalent to a speed of 750 miles per hour. We must conclude that there is a mistake in the advertisement for this ski resort!

This chapter is about ratios, proportions, and unit analysis. *Ratios* are used to compare quantities. For example, when we talk about the gas mileage of a car, we use miles per gallon. The rate of pay for many jobs is given in dollars per hour. Both miles per gallon and dollars per hour are ratios. *Unit analysis* is a method of solving problems by using special ratios called *conversion factors.* We will use unit analysis many times in this chapter. The last section of this chapter is an introduction to *proportions*. Proportions are equal ratios. We will use proportions to solve a variety of word problems. In the process of learning about ratios, proportions, and unit analysis, we will learn more about length, area, volume, weight, and temperature, in both the metric system and the U.S. system of measurement.

6.1 Ratios

The *ratio* of two numbers is a way of comparing them. If we say that the ratio of two numbers is 2 to 1, then the first number is twice as large as the second number. For example, if there are 10 men and 5 women enrolled in a math class, then the ratio of men to women is 10 to 5. Since 10 is twice as large as 5, we can also say that the ratio of men to women is 2 to 1.

We can define the ratio of two numbers in terms of fractions.

DEFINITION The **ratio** of two numbers can be expressed as a fraction, where the first number in the ratio is the numerator and the second number in the ratio is the denominator. *In symbols*:

If a and b are any two numbers,

then the ratio of a to b is $\frac{a}{b}$ $\quad (b \neq 0)$

We handle ratios the same way we handle fractions. For example, when we said that the ratio of 10 men to 5 women was the same as the ratio 2 to 1, we were actually saying

$$\frac{10}{5} = \frac{2}{1} \quad \text{Reducing to lowest terms}$$

Since we have already studied fractions in detail, much of the introductory material on ratio will seem like review.

EXAMPLE 1 Express the ratio of 16 to 48 as a fraction in lowest terms.

Solution Since the ratio is 16 to 48, the numerator of the fraction is 16 and the denominator is 48:

$$\frac{16}{48} = \frac{1}{3} \quad \text{In lowest terms}$$

Notice that the first number in the ratio becomes the numerator of the fraction, and the second number in the ratio becomes the denominator.

EXAMPLE 2 Give the ratio of $\frac{2}{3}$ to $\frac{4}{9}$ as a fraction in lowest terms.

Solution We begin by writing the ratio of $\frac{2}{3}$ to $\frac{4}{9}$ as a complex fraction. The numerator is $\frac{2}{3}$ and the denominator is $\frac{4}{9}$. Then we simplify.

$$\frac{\frac{2}{3}}{\frac{4}{9}} = \frac{2}{3} \cdot \frac{9}{4} \quad \text{Division by } \tfrac{4}{9} \text{ is the same as multiplication by } \tfrac{9}{4}$$

$$= \frac{18}{12} \quad \text{Multiply}$$

$$= \frac{3}{2} \quad \text{Reduce to lowest terms}$$

Practice Problems

1 Express the ratio of 32 to 48 as a fraction in lowest terms.

2 Give the ratio of $\frac{3}{5}$ to $\frac{9}{10}$ as a fraction in lowest terms.

Answers

1 $\frac{2}{3}$ **2** $\frac{2}{3}$

EXAMPLE 3 Write the ratio of 0.08 to 0.12 as a fraction in lowest terms.

Solution When the ratio is in reduced form, it is customary to write it with whole numbers and not decimals. For this reason we multiply the numerator and the denominator of the ratio by 100 to clear it of decimals. Then we reduce to lowest terms.

$$\frac{0.08}{0.12} = \frac{0.08 \times 100}{0.12 \times 100}$$ Multiply the numerator and the denominator by 100 to clear the ratio of decimals

$$= \frac{8}{12}$$ Multiply

$$= \frac{2}{3}$$ Reduce to lowest terms

Table 1 shows several more ratios and their fractional equivalents. Notice that in each case the fraction has been reduced to lowest terms. Also, the ratio that contains decimals has been rewritten as a fraction that does not contain decimals.

Table 1

Ratio	Fraction	Fraction in Lowest Terms
25 to 35	$\frac{25}{35}$	$\frac{5}{7}$
35 to 25	$\frac{35}{25}$	$\frac{7}{5}$
8 to 2	$\frac{8}{2}$	$\frac{4}{1}$ We can also write this as just 4.
$\frac{1}{4}$ to $\frac{3}{4}$	$\frac{\frac{1}{4}}{\frac{3}{4}}$	$\frac{1}{3}$ because $\frac{\frac{1}{4}}{\frac{3}{4}} = \frac{1}{4} \cdot \frac{4}{3} = \frac{1}{3}$
0.6 to 1.7	$\frac{0.6}{1.7}$	$\frac{6}{17}$ because $\frac{0.6 \times 10}{1.7 \times 10} = \frac{6}{17}$

EXAMPLE 4 During a game, a basketball player makes 12 out of the 18 free throws he attempts. Write the ratio of the number of free throws he makes to the number of free throws he attempts as a fraction in lowest terms.

Solution Since he makes 12 out of 18, we want the ratio 12 to 18, or

$$\frac{12}{18} = \frac{2}{3}$$

Since the ratio is 2 to 3, we can say that, in this particular game, he made 2 out of every 3 free throws he attempted.

EXAMPLE 5 A solution of alcohol and water contains 15 milliliters of water and 5 milliliters of alcohol. Find the ratio of alcohol to water, water to alcohol, water to total solution, and alcohol to total solution. Write each ratio as a fraction and reduce to lowest terms.

Solution There is 5 milliliters of alcohol and 15 milliliters of water, so there is 20 milliliters of solution (alcohol + water). The ratios are as follows:

3 Write the ratio of 0.06 to 0.12 as a fraction in lowest terms.

Note

Another symbol used to denote ratio is the colon (:). The ratio of, say, 5 to 4 can be written as 5:4. Although we will not use it here, this notation is fairly common.

4 Suppose the basketball player in Example 4 makes 12 out of 16 free throws. Write the ratio again using these new numbers.

5 A solution of alcohol and water contains 12 milliliters of water and 4 milliliters of alcohol. Find the ratio of alcohol to water, water to alcohol, and water to total solution. Write each ratio as a fraction and reduce to lowest terms.

Answers

3 $\frac{1}{2}$ **4** $\frac{3}{4}$ **5** $\frac{1}{3}, \frac{3}{1}, \frac{3}{4}$

The ratio of alcohol to water is 5 to 15, or

$$\frac{5}{15} = \frac{1}{3}$$ In lowest terms

The ratio of water to alcohol is 15 to 5, or

$$\frac{15}{5} = \frac{3}{1}$$ In lowest terms

The ratio of water to total solution is 15 to 20, or

$$\frac{15}{20} = \frac{3}{4}$$ In lowest terms

The ratio of alcohol to total solution is 5 to 20, or

$$\frac{5}{20} = \frac{1}{4}$$ In lowest terms

Rates

When a ratio is used to compare two different kinds of quantities, the ratio is called a *rate*. For example, if we were to travel 120 miles in 3 hours, then our average rate of speed expressed as the ratio of miles to hours would be

$$\frac{120 \text{ miles}}{3 \text{ hours}} = \frac{40 \text{ miles}}{1 \text{ hour}}$$ Divide the numerator and the denominator by 3 to reduce to lowest terms

The ratio $\frac{40 \text{ miles}}{1 \text{ hour}}$ can be expressed as

$$40 \frac{\text{miles}}{\text{hour}} \quad \text{or} \quad 40 \text{ miles/hour} \quad \text{or} \quad 40 \text{ miles per hour}$$

Whenever a ratio compares two quantities that have different units (and neither unit can be converted to the other), then the ratio is called a rate. A rate is expressed in simplest form when the numerical part of the denominator is 1. To accomplish this we use division.

EXAMPLE 6 A train travels 125 miles in 2 hours. What is the train's rate in miles per hour?

Solution The ratio of miles to hours is

$$\frac{125 \text{ miles}}{2 \text{ hours}}$$

Dividing 125 by 2, we have

$$\frac{125 \text{ miles}}{2 \text{ hours}} = 62.5 \frac{\text{miles}}{\text{hour}}$$
$$= 62.5 \text{ miles per hour}$$

If the train travels 125 miles in 2 hours, then its average rate of speed is 62.5 miles per hour.

6 A car travels 107 miles in 2 hours. What is the car's rate in miles per hour?

Answer
6 53.5 miles/hour

7 A car travels 192 miles on 6 gallons of gas. Give the ratio of miles to gallons as a rate in miles per gallon.

EXAMPLE 7 A car travels 90 miles on 5 gallons of gas. Give the ratio of miles to gallons as a rate in miles per gallon.

Solution The ratio of miles to gallons is

$$\frac{90 \text{ miles}}{5 \text{ gallons}} = 18\,\frac{\text{miles}}{\text{gallon}} \qquad \text{Divide 90 by 5}$$

$$= 18 \text{ miles/gallon}$$

The gas mileage of the car is 18 miles per gallon.

Unit Pricing

One kind of rate that is very common is *unit pricing*. Unit pricing is the ratio of price to quantity. Suppose a 1-liter bottle of a certain soft drink costs \$0.53, whereas a 2-liter bottle of the same drink costs \$1.23. Which is the better buy? That is, which has the lower price per liter?

$$\frac{\$0.53}{1 \text{ liter}} = \frac{53¢}{1 \text{ liter}} = 53¢ \text{ per liter}$$

$$\frac{\$1.23}{2 \text{ liters}} = \frac{123¢}{2 \text{ liters}} = 61.5¢ \text{ per liter}$$

The unit price for the 1-liter bottle is 53¢ per liter, whereas the unit price for the 2-liter bottle is 61.5¢ per liter. The 1-liter bottle is a better buy.

8 A supermarket sells vegetable juice in three different containers at the following prices:

6 ounces	18¢
12 ounces	60¢
32 ounces	\$1.28

Give the unit price in cents per ounce for each one.

EXAMPLE 8 A supermarket sells low-fat milk in three different containers at the following prices:

1 gallon	\$1.73	
$\frac{1}{2}$ gallon	88¢	
1 quart	46¢	(1 quart = $\frac{1}{4}$ gallon)

Give the unit price in cents per gallon for each one.

Solution Since 1 quart = $\frac{1}{4}$ gallon, and \$1.73 = 173¢, we have

1-gallon container $\qquad \frac{\$1.73}{1 \text{ gallon}} = \frac{173¢}{1 \text{ gallon}} = 173¢ \text{ per gallon}$

$\frac{1}{2}$-gallon container $\qquad \frac{88¢}{\frac{1}{2} \text{ gallon}} = \frac{88¢}{0.5 \text{ gallon}} = 176¢ \text{ per gallon}$

1-quart container $\qquad \frac{46¢}{1 \text{ quart}} = \frac{46¢}{0.25 \text{ gallon}} = 184¢ \text{ per gallon}$

The 1-gallon container has the lowest unit price, whereas the 1-quart container has the highest unit price.

Answers
7 32 miles/gallon
8 3¢/ounce, 5¢/ounce, 4¢/ounce

PROBLEM SET 6.1

Write each of the following ratios as a fraction in lowest terms. None of the answers should contain decimals.

1 8 to 6

2 6 to 8

3 64 to 12

4 12 to 64

5 100 to 250

6 250 to 100

7 13 to 26

8 36 to 18

9 $\frac{3}{4}$ to $\frac{1}{4}$

10 $\frac{5}{8}$ to $\frac{3}{8}$

11 $\frac{7}{3}$ to $\frac{6}{3}$

12 $\frac{9}{5}$ to $\frac{11}{5}$

13 $\frac{6}{5}$ to $\frac{6}{7}$

14 $\frac{5}{3}$ to $\frac{1}{3}$

15 $2\frac{1}{2}$ to $3\frac{1}{2}$

16 $5\frac{1}{4}$ to $1\frac{3}{4}$

17 $2\frac{2}{3}$ to $\frac{5}{3}$

18 $\frac{1}{2}$ to $3\frac{1}{2}$

19 0.05 to 0.15

20 0.21 to 0.03

21 0.3 to 3

22 0.5 to 10

23 1.2 to 10

24 6.4 to 0.8

25 $\frac{1}{2}$ to 1.5

26 $\frac{1}{4}$ to 0.75

27 2.5 to $\frac{3}{2}$

28 1.25 to $\frac{3}{4}$

29 A family of four budgeted the following amounts for some of their monthly bills:

House payment	\$250
Food bill	400
Gas bill	100
Utilities bills	150

a What is the ratio of the house payment to the food bill?

b What is the ratio of the gas bill to the food bill?

c What is the ratio of the utilities bills to the food bill?

d What is the ratio of the house payment to the utilities bills?

30 One cup of breakfast cereal was found to contain the following nutrients:

Carbohydrates	21.0 grams
Minerals	4.4 grams
Vitamins	0.6 gram
Water	1.0 gram
Protein	2.0 grams

a Find the ratio of water to protein.

b Find the ratio of carbohydrates to protein.

c Find the ratio of vitamins to minerals.

d Find the ratio of protein to vitamins and minerals.

31 A car travels 220 miles in 4 hours. What is the rate of the car in miles per hour?

32 A train travels 360 miles in 5 hours. What is the rate of the train in miles per hour?

33 It takes a car 3 hours to travel 252 kilometers. What is the rate in kilometers per hour?

34 In 6 hours an airplane travels 4,200 kilometers. What is the rate of the airplane in kilometers per hour?

35 The flow of water from a water faucet can fill a 3-gallon container in 15 seconds. Give the ratio of gallons to seconds as a rate in gallons per second.

36 A 225-gallon drum is filled in 3 minutes. What is the rate in gallons per minute?

37 It takes 4 minutes to fill a 56-liter gas tank. What is the rate in liters per minute?

38 The gas tank on a car holds 60 liters of gas. At the beginning of a 6-hour trip, the tank is full. At the end of the trip, it contains only 12 liters. What is the rate at which the car uses gas in liters per hour?

39 A car travels 95 miles on 5 gallons of gas. Give the ratio of miles to gallons as a rate in miles per gallon.

40 On a 384-mile trip, an economy car uses 8 gallons of gas. Give this as a rate in miles per gallon.

41 The gas tank on a car has a capacity of 75 liters. On a full tank of gas, the car travels 325 miles. What is the gas mileage in miles per liter?

42 A car pulling a trailer can travel 105 miles on 70 liters of gas. What is the gas mileage in miles per liter?

43 A 6-ounce can of frozen orange juice costs 48¢. Give the unit price in cents per ounce.

44 A 2-liter bottle of root beer costs $1.25. Give the unit price in cents per liter.

45 A 20-ounce package of frozen peas is priced at 69¢. Give the unit price in cents per ounce.

46 A 4-pound bag of cat food costs $2.03. Give the unit price in cents per pound.

Calculator Problems

Write each of the following ratios as a fraction, and then use a calculator to change the fraction to a decimal. Round all answers to the nearest hundredth. Do not reduce fractions.

47 The total number of students attending a community college in the Midwest is 4,722. Of these students, 2,314 are male and 2,408 are female.

a Give the ratio of males to females as a fraction and as a decimal.

b Give the ratio of females to males as a fraction and as a decimal.

c Give the ratio of males to total number of students as a fraction and as a decimal.

d Give the ratio of total number of students to females as a fraction and as a decimal.

48 The following table gives the number of kilocalories of energy contained in an average serving of a number of foods:

Food	Kilocalories of Energy
Bacon	155
Wheat bread	55
Tuna	195
Ice cream	205
Apple pie	410

Write each of the following ratios as a fraction and as a decimal. Each ratio refers to the number of kilocalories of energy in one serving of the food. (Do not reduce the fractions. Round to the nearest hundredth.)

a Bacon to tuna

b Wheat bread to ice cream

c Apple pie to ice cream

d Wheat bread to apple pie and ice cream

49 A car travels 675.4 miles in $12\frac{1}{2}$ hours. Give the rate in miles per hour to the nearest hundredth.

50 At the beginning of a trip, the odometer on a car read 32,567.2 miles. At the end of the trip, it read 32,741.8 miles. If the trip took $4\frac{1}{4}$ hours, what was the rate of the car in miles per hour to the nearest tenth?

51 If a truck travels 128.4 miles on 13.8 gallons of gas, what is the gas mileage in miles per gallon? (Round to the nearest tenth.)

52 If a 15-day supply of vitamins costs \$1.62, what is the price in cents per day?

Review Problems

The problems that follow review multiplication with fractions from Section 3.3, along with some material covered in Section 5.5. Reviewing these problems will help you with the next section.

Multiply.

53 $8 \cdot \frac{1}{3}$

54 $9 \cdot \frac{1}{3}$

55 $25 \cdot \frac{1}{1{,}000}$

56 $25 \cdot \frac{1}{100}$

57 $36.5 \cdot \frac{1}{100} \cdot 10$

58 $36.5 \cdot \frac{1}{1{,}000} \cdot 100$

59 $248 \cdot \frac{1}{10} \cdot \frac{1}{10}$

60 $969 \cdot \frac{1}{10} \cdot \frac{1}{10}$

61 $48 \cdot \frac{1}{12} \cdot \frac{1}{3}$

62 $56 \cdot \frac{1}{12} \cdot \frac{1}{2}$

6.2 Unit Analysis I: Length

In this section we will become more familiar with the units used to measure length. We will look at the U.S. system of measurement and the metric system of measurement.

Measuring the length of an object is done by assigning a number to its length. To let other people know what that number represents, we include with it a unit of measure. The most common units used to represent length in the U.S. system are inches, feet, yards, and miles. The basic unit of length is the foot. The other units are defined in terms of feet, as Table 1 shows.

Table 1

12 inches (in.)	=	1 foot (ft)
1 yard (yd)	=	3 feet
1 mile (mi)	=	5,280 feet

As you can see from the table, the abbreviations for inches, feet, yards, and miles are in., ft, yd, and mi, respectively. What we haven't indicated, even though you may not have realized it, is what 1 foot represents. We have defined all our units associated with length in terms of feet, but we haven't said what a foot is.

There is a long history of the evolution of what is now called a foot. At different times in the past, a foot has represented different arbitrary lengths. Currently, a foot is defined to be exactly 0.3048 meter (the basic measure of length in the metric system), where a meter is 1,650,763.73 wavelengths of the orange–red line in the spectrum of krypton-86 in a vacuum (this doesn't mean much to me either). The reason a foot and a meter are defined this way is that we always want them to measure the same length. Since the wavelength of the orange–red line in the spectrum of krypton-86 will always remain the same, so will the length that a foot represents.

Now that we have said what we mean by 1 foot (even though we may not understand the technical definition), we can go on and look at some examples that involve converting from one kind of unit to another.

Practice Problems

1 Convert 8 feet to inches.

EXAMPLE 1 Convert 5 feet to inches.

Solution Since 1 foot = 12 inches, we can multiply 5 by 12 inches to get

$$\begin{aligned} 5 \text{ feet} &= 5 \times 12 \text{ inches} \\ &= 60 \text{ inches} \end{aligned}$$

This method of converting from feet to inches probably seems fairly simple. But as we go further in this chapter, the conversions from one kind of unit to another will become more complicated. For these more complicated problems, we need another way to show conversions so that we can be certain to end them with the correct unit of measure. For example, since 1 ft = 12 in., we can say that there are 12 in. per 1 ft or 1 ft per 12 in. That is:

$$\frac{12 \text{ in.}}{1 \text{ ft}} \longleftarrow \text{Per} \qquad \text{or} \qquad \frac{1 \text{ ft}}{12 \text{ in.}} \longleftarrow \text{Per}$$

We call the expressions $\frac{12 \text{ in.}}{1 \text{ ft}}$ and $\frac{1 \text{ ft}}{12 \text{ in.}}$ *conversion factors*. The fraction bar is

Answer

1 96 in.

read as "per." Both these conversion factors are really just the number 1. That is:

$$\frac{12 \text{ in.}}{1 \text{ ft}} = \frac{12 \text{ in.}}{12 \text{ in.}} = 1$$

We already know that multiplying a number by 1 leaves the number unchanged. So, to convert from one unit to the other, we can multiply by one of the conversion factors without changing value. Both the conversion factors above say the same thing about the units feet and inches. They both indicate that there are 12 inches in every foot. The one we choose to multiply by depends on what units we are starting with and what units we want to end up with. If we start with feet and we want to end up with inches, we multiply by the conversion factor

$$\frac{12 \text{ in.}}{1 \text{ ft}}$$

The feet will divide out and leave us with inches.

$$5 \text{ ft} = 5 \cancel{\text{ft}} \times \frac{12 \text{ in.}}{1 \cancel{\text{ft}}}$$ This is the number 1. Notice how $\cancel{\text{ft}}$ divide out.

$$= 5 \times 12 \text{ in.}$$

$$= 60 \text{ in.}$$

Note
We will use this method of converting from one kind of unit to another throughout the rest of this chapter. You should practice using it until you are comfortable with it and can use it correctly. However, it is not the only method of converting units. You may see shortcuts that will allow you to get results more quickly. Use shortcuts if you wish so long as you can consistently get correct answers and are not using your shortcuts because you don't understand our method of conversion. Use the method of conversion as given here until you are good at it; then use shortcuts if you want to.

The key to this method of conversion lies in setting the problem up so that the correct units divide out to simplify the expression. We are treating units such as feet in the same way we treated factors when reducing fractions. If a factor is common to the numerator and the denominator, we can divide it out and simplify the fraction. The same idea holds for units such as feet.

We can rewrite Table 1 so that it shows the conversion factors associated with units of length, as shown in Table 2.

Table 2 Units of length in the U.S. system

The Relationship between	Is	To Convert from One to the Other, Multiply by
feet and inches	12 in. = 1 ft	$\frac{12 \text{ in.}}{1 \text{ ft}}$ or $\frac{1 \text{ ft}}{12 \text{ in.}}$
feet and yards	1 yd = 3 ft	$\frac{3 \text{ ft}}{1 \text{ yd}}$ or $\frac{1 \text{ yd}}{3 \text{ ft}}$
feet and miles	1 mi = 5,280 ft	$\frac{5{,}280 \text{ ft}}{1 \text{ mi}}$ or $\frac{1 \text{ mi}}{5{,}280 \text{ ft}}$

2 The roof of a two-story house is 26 feet above the ground. How many yards is this?

EXAMPLE 2 The most common ceiling height in houses is 8 feet. How many yards is this?

Solution To convert 8 feet to yards, we multiply by the conversion factor $\frac{1 \text{ yd}}{3 \text{ ft}}$ so that feet will divide out and we will be left with yards.

$$8 \text{ ft} = 8 \cancel{\text{ft}} \times \frac{1 \text{ yd}}{3 \cancel{\text{ft}}}$$ Multiply by correct conversion factor

$$= \frac{8}{3} \text{ yd}$$ $8 \times \frac{1}{3} = \frac{8}{3}$

$$= 2\tfrac{2}{3} \text{ yd}$$ Or 2.67 yd to the nearest hundredth

Answer
2 $8\frac{2}{3}$ yd, or 8.67 yd

EXAMPLE 3 A football field is 100 yards long. How many inches long is a football field?

Solution In this example we must convert yards to feet and then feet to inches. (To make this example more interesting, we are pretending we don't know that there are 36 inches in a yard.) We choose the conversion factors that will allow all the units except inches to divide out.

$$\begin{aligned} 100 \text{ yd} &= 100 \text{ yd} \times \frac{3 \text{ ft}}{1 \text{ yd}} \times \frac{12 \text{ in.}}{1 \text{ ft}} \\ &= 100 \times 3 \times 12 \text{ in.} \\ &= 3{,}600 \text{ in.} \end{aligned}$$

3 How many inches are in 220 yards?

Metric Units of Length

In the metric system the standard unit of length is a meter. A meter is a little longer than a yard (about 3.4 inches longer). The other units of length in the metric system are written in terms of a meter. The metric system uses prefixes to indicate what part of the basic unit of measure is being used. For example, in *milli*meter the prefix *milli* means "one thousandth" of a meter. Table 3 gives the meanings of the most common metric prefixes.

Table 3 The meaning of metric prefixes

Prefix	Meaning
milli	0.001
centi	0.01
deci	0.1
deka	10
hecto	100
kilo	1,000

We can use these prefixes to write the other units of length and conversion factors for the metric system, as given in Table 4.

Table 4 Metric units of length

The Relationship between	Is	To Convert from One to the Other, Multiply by
millimeters (mm) and meters (m)	1,000 mm = 1 m	$\frac{1{,}000 \text{ mm}}{1 \text{ m}}$ or $\frac{1 \text{ m}}{1{,}000 \text{ mm}}$
centimeters (cm) and meters	100 cm = 1 m	$\frac{100 \text{ cm}}{1 \text{ m}}$ or $\frac{1 \text{ m}}{100 \text{ cm}}$
decimeters (dm) and meters	10 dm = 1 m	$\frac{10 \text{ dm}}{1 \text{ m}}$ or $\frac{1 \text{ m}}{10 \text{ dm}}$
dekameters (dam) and meters	1 dam = 10 m	$\frac{10 \text{ m}}{1 \text{ dam}}$ or $\frac{1 \text{ dam}}{10 \text{ m}}$
hectometers (hm) and meters	1 hm = 100 m	$\frac{100 \text{ m}}{1 \text{ hm}}$ or $\frac{1 \text{ hm}}{100 \text{ m}}$
kilometers (km) and meters	1 km = 1,000 m	$\frac{1{,}000 \text{ m}}{1 \text{ km}}$ or $\frac{1 \text{ km}}{1{,}000 \text{ m}}$

Answer
3 7,920 in.

We use the same method to convert between units in the metric system as we did with the U.S. system. We choose the conversion factor that will allow the units we start with to cancel, leaving the units we want to end up with.

4 Convert 67 centimeters to meters.

EXAMPLE 4 Convert 25 millimeters to meters.

Solution To convert from millimeters to meters, we multiply by the conversion factor $\frac{1\text{ m}}{1{,}000\text{ mm}}$:

$$\begin{aligned} 25\text{ mm} &= 25\text{ mm} \times \frac{1\text{ m}}{1{,}000\text{ mm}} \\ &= \frac{25\text{ m}}{1{,}000} \\ &= 0.025\text{ m} \end{aligned}$$

5 Convert 78.4 mm to decimeters.

EXAMPLE 5 Convert 36.5 centimeters to decimeters.

Solution We convert centimeters to meters and then meters to decimeters:

$$\begin{aligned} 36.5\text{ cm} &= 36.5\text{ cm} \times \frac{1\text{ m}}{100\text{ cm}} \times \frac{10\text{ dm}}{1\text{ m}} \\ &= \frac{36.5 \times 10}{100}\text{ dm} \\ &= 3.65\text{ dm} \end{aligned}$$

The most common units of length in the metric system are millimeters, centimeters, meters, and kilometers. The other units of length we have listed in our table of metric lengths are not as widely used. The method we have used to convert from one unit of length to another in Examples 2–5 is called *unit analysis*. If you take a chemistry class, you will see it used many times. The same is true of many other science classes as well.

We can summarize the procedure used in unit analysis with the following steps:

Steps Used in Unit Analysis

1 Identify the units you are starting with.

2 Identify the units you want to end with.

3 Find conversion factors that will bridge the starting units and the ending units.

4 Set up the multiplication problem so that all units except the units you want to end with will divide out.

Answers

4 0.67 m **5** 0.784 dm

EXAMPLE 6 A sheep farmer is making new lambing pens for the upcoming lambing season. Each pen is a rectangle 6 feet wide and 8 feet long. The fencing material he wants to use sells for \$1.36 per foot. If he is planning to build five separate lambing pens (they are separate because he wants a walkway between them), how much will he have to spend for fencing material?

Solution To find the amount of fencing material he needs for one pen, we find the perimeter of a pen.

Lambing pen

6 ft

8 ft

$$\text{Perimeter} = 6 + 6 + 8 + 8 = 28 \text{ feet}$$

We set up the solution to the problem using unit analysis. Our starting unit is *pens* and our ending unit is *dollars*. Here are the conversion factors that will form a bridge between pens and dollars:

$$1 \text{ pen} = 28 \text{ feet of fencing}$$

$$1 \text{ foot of fencing} = 1.36 \text{ dollars}$$

Next we write the multiplication problem, using the conversion factors, that will allow all the units except dollars to divide out:

$$\begin{aligned} 5 \text{ pens} &= 5\ \cancel{\text{pens}} \times \frac{28\ \cancel{\text{feet of fencing}}}{1\ \cancel{\text{pen}}} \times \frac{1.36 \text{ dollars}}{1\ \cancel{\text{foot of fencing}}} \\ &= 5 \times 28 \times 1.36 \text{ dollars} \\ &= \$190.40 \end{aligned}$$

6 The farmer in Example 6 decides to build six pens instead of five and upgrades his fencing material so that it costs \$1.72 per foot. How much does it cost him to build the six pens?

EXAMPLE 7 The speed of sound is 1,088 feet/second. Convert the speed of sound to miles/hour. Round your answer to the nearest whole number.

Solution In this case, we must convert feet to miles and seconds to hours. Here are the conversion factors we will use:

$$1 \text{ mile} = 5{,}280 \text{ feet}$$

$$1 \text{ hour} = 60 \text{ minutes}$$

$$1 \text{ minute} = 60 \text{ seconds}$$

$$\begin{aligned} 1{,}088 \text{ feet/second} &= \frac{1{,}088\ \cancel{\text{feet}}}{1\ \cancel{\text{second}}} \times \frac{1 \text{ mile}}{5{,}280\ \cancel{\text{feet}}} \times \frac{60\ \cancel{\text{seconds}}}{1\ \cancel{\text{minute}}} \times \frac{60\ \cancel{\text{minutes}}}{1 \text{ hour}} \\ &= \frac{1{,}088 \times 60 \times 60 \text{ miles}}{5{,}280 \text{ hours}} \\ &= 742 \text{ miles/hour} \qquad \text{To the nearest whole number} \end{aligned}$$

7 A car is traveling at 50 miles/hour. What is the speed of the car in feet/second? (Round your answer to the nearest tenth.)

Answers

6 \$288.96 **7** 73.3 feet/second

PROBLEM SET 6.2

Make the following conversions in the U.S. system by multiplying by the appropriate conversion factor. Write your answers as whole numbers or mixed numbers.

1 5 ft to inches

2 9 ft to inches

3 10 ft to inches

4 20 ft to inches

5 2 yd to feet

6 8 yd to feet

7 4.5 yd to inches

8 9.5 yd to inches

9 27 in. to feet

10 36 in. to feet

11 19 ft to yards

12 100 ft to yards

13 48 in. to yards

14 56 in. to yards

Make the following conversions in the metric system by multiplying by the appropriate conversion factor. Write your answers as whole numbers or decimals.

15 18 m to centimeters

16 18 m to millimeters

17 4.8 km to meters

18 8.9 km to meters

19 5 dm to centimeters

20 12 dm to millimeters

21 248 m to kilometers

22 969 m to kilometers

23 67 cm to millimeters

24 67 mm to centimeters

25 3,498 cm to meters

26 4,388 dm to meters

27 63.4 cm to decimeters

28 89.5 cm to decimeters

29 If the distance between first and second base in softball is 60 feet, how many yards is it from first to second base?

30 A transmitting tower is 100 feet tall. How many inches is that?

31 If a person high jumps 6 feet 8 inches, how many inches is the jump?

32 A desk is 48 inches wide. What is the width in yards?

33 Suppose the ceiling of a home is 2.44 meters above the floor. Express the height of the ceiling in centimeters.

34 Standard-sized notebook paper is 21.6 centimeters wide. Express this width in millimeters.

35 A dollar bill is about 6.5 centimeters wide. Express this width in millimeters.

36 Most new pencils are 19 centimeters long. Express this length in meters.

37 A unit of measure sometimes used in surveying is the *chain.* There are 80 chains in 1 mile. How many chains are in 37 miles?

38 Another unit of measure used in surveying is a *link*; 1 link is about 8 inches. About how many links are there in 5 feet?

39 A very small unit of measure in the metric system is the *micron* (abbreviated μm). There are 1,000 μm in 1 millimeter. How many microns are in 12 centimeters?

40 Another very small unit of measure in the metric system is the *angstrom* (abbreviated Å). There are 10,000,000 Å in 1 millimeter. How many angstroms are in 15 decimeters?

41 In horse racing, 1 *furlong* is 220 yards. How many feet are in 12 furlongs?

42 A *fathom* is 6 feet. How many yards are in 19 fathoms?

43 The maximum speed limit on Highway 101 in California is 55 miles/hour. Convert 55 miles/hour to feet/second. (Round to the nearest tenth.)

44 The maximum speed limit on Highway 5 in California is 65 miles/hour. Convert 65 miles/hour to feet/second. (Round to the nearest tenth.)

45 A person who runs the 100-yard dash in 10.5 seconds has an average speed of 9.52 yards/second. Convert 9.52 yards/second to miles/hour. (Round to the nearest tenth.)

46 A person who runs a mile in 8 minutes has an average speed of 0.125 miles/minute. Convert 0.125 miles/minute to miles/hour. (Round to the nearest tenth.)

47 The bullet from a rifle leaves the barrel traveling 1,500 feet/second. Convert 1,500 feet/second to miles/hour. (Round to the nearest whole number.)

48 A bullet fired from a machine gun on a B-17 Flying Fortress in World War II had a muzzle speed of 1,750 feet/second. Convert 1,750 feet/second to miles/hour. (Round to the nearest whole number.)

49 A farmer is fencing a pasture that is $\frac{1}{2}$ mile wide and 1 mile long. If the fencing material sells for \$1.15 per foot, how much will it cost him to fence all four sides of the pasture?

50 A family with a swimming pool puts up a chain-link fence around the pool. The fence forms a rectangle 12 yards wide and 24 yards long. If the chain-link fence sells for \$2.50 per foot, how much will it cost to fence all four sides of the pool?

51 A 4-H Club group is raising lambs to show at the County Fair. Each lamb eats $\frac{1}{8}$ of a bale of alfalfa a day. If the alfalfa costs \$5.25 per bale, how much will it cost to feed one lamb for 120 days?

52 A 4-H Club group is raising pigs to show at the County Fair. Each pig eats 2.4 pounds of grain a day. If the grain costs \$5.25 per pound, how much will it cost to feed one pig for 60 days?

Calculator Problems

Set up the following conversions as you have been doing. Then perform the calculations on a calculator.

53 Change 751 miles to feet.

54 Change 639.87 centimeters to meters.

55 Change 4,982 yards to inches.

56 Change 379 millimeters to kilometers.

57 Mount Whitney is the highest point in California. It is 14,494 feet above sea level. Give its height in miles to the nearest tenth.

58 The tallest mountain in the United States is Mount McKinley in Alaska. It is 20,320 feet tall. Give its height in miles to the nearest tenth.

59 California has 3,427 miles of shoreline. How many feet is this?

60 The tip of the television tower at the top of the Empire State Building in New York City is 1,472 feet above the ground. Express this height in miles to the nearest hundredth.

Review Problems

The review problems in this chapter cover the work we have done previously with fractions and mixed numbers. We will begin with a review of multiplication and division of fractions and mixed numbers. Write your answers as whole numbers, proper fractions, or mixed numbers.

Find each product. (Multiply.)

61 $\frac{2}{3} \cdot \frac{1}{2}$ **62** $\frac{7}{9} \cdot \frac{3}{14}$ **63** $-8\left(-\frac{3}{4}\right)$ **64** $-12 \cdot \frac{1}{3}$ **65** $1\frac{1}{2} \cdot 2\frac{1}{3}$ **66** $\frac{1}{6} \cdot 4\frac{2}{3}$

Find each quotient. (Divide.)

67 $\frac{3}{4} \div \frac{1}{8}$ **68** $\frac{3}{5} \div \frac{6}{25}$ **69** $-4 \div \left(-\frac{2}{3}\right)$ **70** $1 \div \left(-\frac{1}{3}\right)$ **71** $1\frac{3}{4} \div 2\frac{1}{2}$ **72** $\frac{9}{8} \div 1\frac{7}{8}$

One Step Further: Extending the Concepts

A Schwinn Airdyne exercise bicycle has a digital display that gives information to the rider, such as the number of calories per hour (Cal/hour) being used.

73 Find the total number of calories burned by riding the exercise bike at a rate of 300 Cal/hour for 25 minutes.

74 Find the total number of calories burned by riding the exercise bike at a rate of 240 Cal/hour for 50 minutes.

75 Mary Jo receives a Schwinn Airdyne exercise bicycle for her fifteenth wedding anniversary. She knows that she must burn 3,500 calories to lose 1 pound of body weight. If she exercises 25 minutes each day at a rate of 300 Cal/hour, how many days will it take her to lose 1 pound?

76 If Mary Jo exercises 50 minutes a day at the rate of 240 Cal/hour, how long will it take her to lose 2 pounds?

77 Suppose Mary Jo's husband, David, has his 20-year high school reunion in 12 weeks. He decides he must lose 6 pounds before the reunion. If he rides the exercise bike every day for 40 minutes, at what rate (in calories per hour) should he ride in order to lose the 6 pounds?

78 David decides he just can't peddle that fast, so he changes his goal to 4 pounds instead of 6. If he rides for 40 minutes each day, at what rate (in calories per hour) should he peddle in order to lose the 4 pounds?

6.3 Unit Analysis II: Area and Volume

Figure 1 below gives a summary of the geometric objects we have worked with in previous chapters, along with the formulas for finding the area of each object.

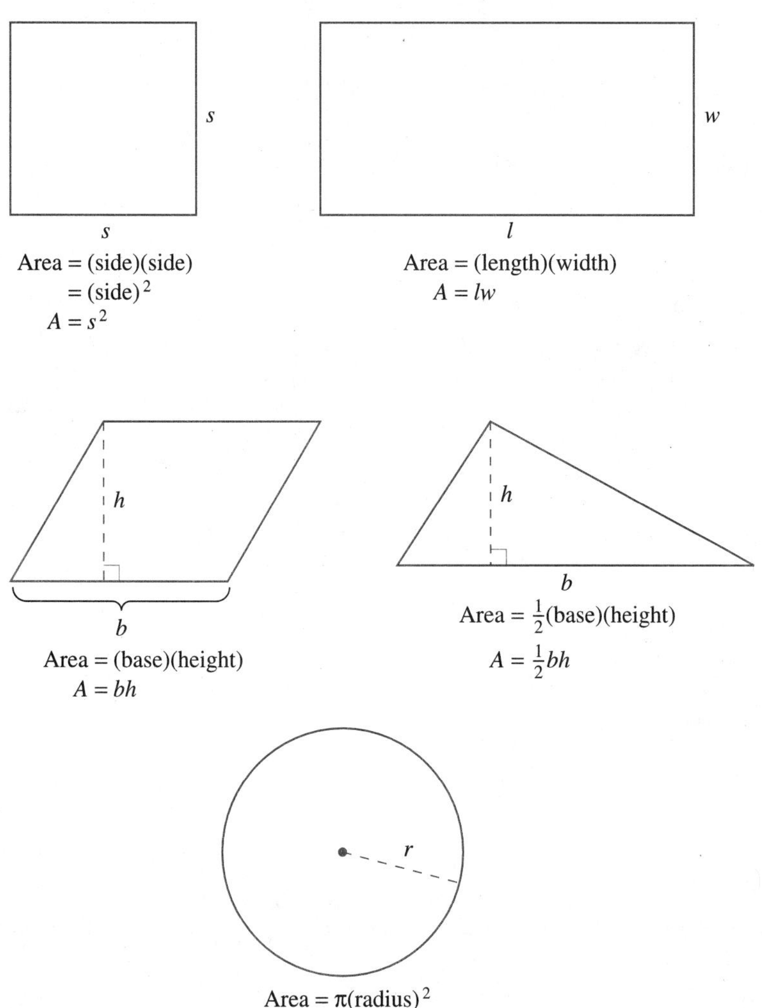

Figure 1 Areas of common geometric objects

In Example 2 of Section 1.10, we found the number of square inches in one square foot. Here is the same problem again, but this time we work it using unit analysis.

EXAMPLE 1 Find the number of square inches in 1 square foot.

Solution We can think of 1 square foot as $1 \text{ ft}^2 = 1 \text{ ft} \times \text{ft}$. To convert from feet to inches, we use the conversion factor 1 foot = 12 inches. Since the unit foot appears twice in 1 ft^2, we multiply by our conversion factor twice.

$$1 \text{ ft}^2 = 1 \text{ ft} \times \text{ft} \times \frac{12 \text{ in.}}{1 \text{ ft}} \times \frac{12 \text{ in.}}{1 \text{ ft}} = 12 \times 12 \text{ in.} \times \text{in.} = 144 \text{ in}^2$$

Practice Problems

1 Find the number of square feet in 1 square yard.

Answer

1 $1 \text{ yd}^2 = 9 \text{ ft}^2$

Now that we know that 1 ft^2 is the same as 144 in^2, we can use this fact as a conversion factor to convert between square feet and square inches. Depending on which units we are converting from, we would use either

$$\frac{144 \text{ in}^2}{1 \text{ ft}^2} \quad \text{or} \quad \frac{1 \text{ ft}^2}{144 \text{ in}^2}$$

2 If the poster in Example 2 is surrounded by a frame 6 inches wide, find the number of square feet of wall space covered by the framed poster.

EXAMPLE 2 A rectangular poster measures 36 inches by 24 inches. How many square feet of wall space will the poster cover?

Solution One way to work this problem is to find the number of square inches the poster covers, and then convert square inches to square feet.

$$\text{Area of poster} = \text{length} \times \text{width} = 36 \text{ in.} \times 24 \text{ in.} = 864 \text{ in}^2$$

To finish the problem, we convert square inches to square feet:

$$\begin{aligned} 864 \text{ in}^2 &= 864 \text{ in}^2 \times \frac{1 \text{ ft}^2}{144 \text{ in}^2} \\ &= \frac{864}{144} \text{ ft}^2 \\ &= 6 \text{ ft}^2 \end{aligned}$$

Table 1 gives the most common units of area in the U.S. system of measurement, along with the corresponding conversion factors.

Table 1 U.S. units of area

The Relationship between	Is	To Convert from One to the Other, Multiply by
square inches and square feet	$144 \text{ in}^2 = 1 \text{ ft}^2$	$\frac{144 \text{ in}^2}{1 \text{ ft}^2}$ or $\frac{1 \text{ ft}^2}{144 \text{ in}^2}$
square yards and square feet	$9 \text{ ft}^2 = 1 \text{ yd}^2$	$\frac{9 \text{ ft}^2}{1 \text{ yd}^2}$ or $\frac{1 \text{ yd}^2}{9 \text{ ft}^2}$
acres and square feet	$1 \text{ acre} = 43{,}560 \text{ ft}^2$	$\frac{43{,}560 \text{ ft}^2}{1 \text{ acre}}$ or $\frac{1 \text{ acre}}{43{,}560 \text{ ft}^2}$
acres and square miles	$640 \text{ acres} = 1 \text{ mi}^2$	$\frac{640 \text{ acres}}{1 \text{ mi}^2}$ or $\frac{1 \text{ mi}^2}{640 \text{ acres}}$

3 The same dressmaker orders a roll of material that is 1.5 yards wide and 45 yards long. How many square feet of material were ordered?

EXAMPLE 3 A dressmaker orders a roll of material that is 1.5 yards wide and 30 yards long. How many square feet of material were ordered?

Solution The area of the material in square yards is

$$\begin{aligned} A &= 1.5 \times 30 \\ &= 45 \text{ yd}^2 \end{aligned}$$

Converting this to square feet, we have

$$\begin{aligned} 45 \text{ yd}^2 &= 45 \text{ yd}^2 \times \frac{9 \text{ ft}^2}{1 \text{ yd}^2} \\ &= 405 \text{ ft}^2 \end{aligned}$$

Answers
2 12 ft^2 3 607.5 ft^2

EXAMPLE 4 A farmer has 75 acres of land. How many square feet of land does the farmer have?

Solution Changing acres to square feet, we have

$$\begin{aligned} 75 \text{ acres} &= 75 \text{ acres} \times \frac{43{,}560 \text{ ft}^2}{1 \text{ acre}} \\ &= 75 \times 43{,}560 \text{ ft}^2 \\ &= 3{,}267{,}000 \text{ ft}^2 \end{aligned}$$

4 A farmer has 55 acres of land. How many square feet of land does the farmer have?

EXAMPLE 5 A new shopping center is to be constructed on 256 acres of land. How many square miles is this?

Solution Multiplying by the conversion factor that will allow acres to divide out, we have

$$\begin{aligned} 256 \text{ acres} &= 256 \text{ acres} \times \frac{1 \text{ mi}^2}{640 \text{ acres}} \\ &= \frac{256}{640} \text{ mi}^2 \\ &= 0.4 \text{ mi}^2 \end{aligned}$$

5 A school is to be constructed on 960 acres of land. How many square miles is this?

Units of area in the metric system are considerably simpler than those in the U.S. system because metric units are given in terms of powers of 10. Table 2 lists the conversion factors that are most commonly used.

Table 2 Metric units of area

The Relationship between	Is	To Convert from One to the Other, Multiply by
square millimeters and square centimeters	$1 \text{ cm}^2 = 100 \text{ mm}^2$	$\frac{100 \text{ mm}^2}{1 \text{ cm}^2}$ or $\frac{1 \text{ cm}^2}{100 \text{ mm}^2}$
square centimeters and square decimeters	$1 \text{ dm}^2 = 100 \text{ cm}^2$	$\frac{100 \text{ cm}^2}{1 \text{ dm}^2}$ or $\frac{1 \text{ dm}^2}{100 \text{ cm}^2}$
square decimeters and square meters	$1 \text{ m}^2 = 100 \text{ dm}^2$	$\frac{100 \text{ dm}^2}{1 \text{ m}^2}$ or $\frac{1 \text{ m}^2}{100 \text{ dm}^2}$
square meters and ares (a)	$1 \text{ a} = 100 \text{ m}^2$	$\frac{100 \text{ m}^2}{1 \text{ a}}$ or $\frac{1 \text{ a}}{100 \text{ m}^2}$
ares and hectares (ha)	$1 \text{ ha} = 100 \text{ a}$	$\frac{100 \text{ a}}{1 \text{ ha}}$ or $\frac{1 \text{ ha}}{100 \text{ a}}$

EXAMPLE 6 How many square millimeters are in 1 square meter?

Solution We start with 1 m^2 and end up with square millimeters:

$$\begin{aligned} 1 \text{ m}^2 &= 1 \text{ m}^2 \times \frac{100 \text{ dm}^2}{1 \text{ m}^2} \times \frac{100 \text{ cm}^2}{1 \text{ dm}^2} \times \frac{100 \text{ mm}^2}{1 \text{ cm}^2} \\ &= 100 \times 100 \times 100 \text{ mm}^2 \\ &= 1{,}000{,}000 \text{ mm}^2 \end{aligned}$$

6 How many square centimeters are in 1 square meter?

Answers
4 2,395,800 ft^2 5 1.5 mi^2
6 10,000 cm^2

Units of Measure for Volume

Table 3 lists the units of volume in the U.S. system and their conversion factors.

Table 3 Units of volume in the U.S. system

The Relationship between	Is	To Convert from One to the Other, Multiply by		
cubic inches (in^3) and cubic feet (ft^3)	$1\ ft^3 = 1{,}728\ in^3$	$\frac{1{,}728\ in^3}{1\ ft^3}$	or	$\frac{1\ ft^3}{1{,}728\ in^3}$
cubic feet and cubic yards (yd^3)	$1\ yd^3 = 27\ ft^3$	$\frac{27\ ft^3}{1\ yd^3}$	or	$\frac{1\ yd^3}{27\ ft^3}$
fluid ounces (fl oz) and pints (pt)	1 pt = 16 fl oz	$\frac{16\ \text{fl oz}}{1\ \text{pt}}$	or	$\frac{1\ \text{pt}}{16\ \text{fl oz}}$
pints and quarts (qt)	1 qt = 2 pt	$\frac{2\ \text{pt}}{1\ \text{qt}}$	or	$\frac{1\ \text{qt}}{2\ \text{pt}}$
quarts and gallons (gal)	1 gal = 4 qt	$\frac{4\ \text{qt}}{1\ \text{gal}}$	or	$\frac{1\ \text{gal}}{4\ \text{qt}}$

7 How many pints are in a 5-gallon pail?

EXAMPLE 7 What is the capacity (volume) in pints of a 1-gallon container of milk?

Solution We change from gallons to quarts and then quarts to pints by multiplying by the appropriate conversion factors as given in Table 3.

$$\begin{aligned} 1\ \text{gal} &= 1\ \cancel{\text{gal}} \times \frac{4\ \cancel{\text{qt}}}{1\ \cancel{\text{gal}}} \times \frac{2\ \text{pt}}{1\ \cancel{\text{qt}}} \\ &= 1 \times 4 \times 2\ \text{pt} \\ &= 8\ \text{pt} \end{aligned}$$

A 1-gallon container has the same capacity as 8 1-pint containers.

8 A dairy herd produces 2,000 quarts of milk each day. How many 10-gallon containers will this milk fill?

EXAMPLE 8 A dairy herd produces 1,800 quarts of milk each day. How many gallons is this equivalent to?

Solution Converting 1,800 quarts to gallons, we have

$$\begin{aligned} 1{,}800\ \text{qt} &= 1{,}800\ \cancel{\text{qt}} \times \frac{1\ \text{gal}}{4\ \cancel{\text{qt}}} \\ &= \frac{1{,}800}{4}\ \text{gal} \\ &= 450\ \text{gal} \end{aligned}$$

We see that 1,800 quarts is equivalent to 450 gallons.

In the metric system the basic unit of measure for volume is the liter. A liter is the volume enclosed by a cube that is 10 cm on each edge, as shown in Figure 2. We can see that a liter is equivalent to 1,000 cm^3.

Answers
7 40 pt **8** 50 containers

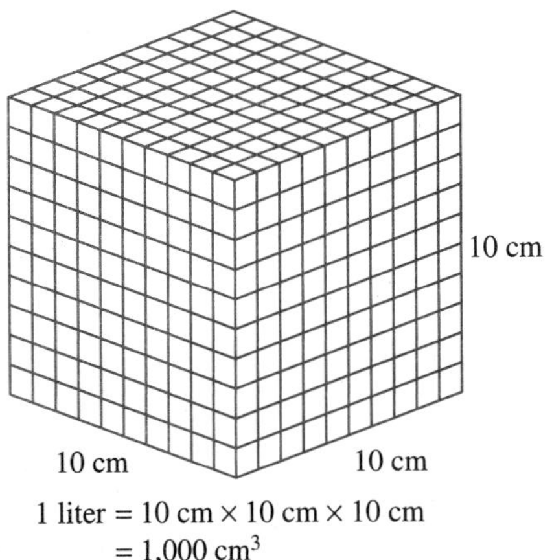

Figure 2

The other units of volume in the metric system use the same prefixes we encountered previously. The units with prefixes centi, deci, and deka are not as common as the others, so in Table 4 we include only liters, milliliters, hectoliters, and kiloliters.

Note
As you can see from the table and the discussion above, a cubic centimeter (cm^3) and a milliliter (mL) are equal. Both are one thousandth of a liter. It is also common in some fields (like medicine) to abbreviate the term cubic centimeter as cc. Although we will use the notation mL when discussing volume in the metric system, you should be aware that $1 \text{ mL} = 1 \text{ cm}^3 = 1 \text{ cc}$.

Table 4 Metric units of volume

The Relationship between	Is	To Convert from One to the Other, Multiply by
milliliters (mL) and liters	1 liter (L) = 1,000 mL	$\frac{1{,}000 \text{ mL}}{1 \text{ liter}}$ or $\frac{1 \text{ liter}}{1{,}000 \text{ mL}}$
hectoliters (hL) and liters	100 liters = 1 hL	$\frac{100 \text{ liters}}{1 \text{ hL}}$ or $\frac{1 \text{ hL}}{100 \text{ liters}}$
kiloliters (kL) and liters	1,000 liters (L) = 1 kL	$\frac{1{,}000 \text{ liters}}{1 \text{ kL}}$ or $\frac{1 \text{ kL}}{1{,}000 \text{ liters}}$

Here is an example of conversion from one unit of volume to another in the metric system.

EXAMPLE 9 A sports car has a 2.2-liter engine. What is the displacement (volume) of the engine in milliliters?

Solution Using the appropriate conversion factor from Table 4, we have

$$\begin{aligned} 2.2 \text{ liters} &= 2.2 \text{ liters} \times \frac{1{,}000 \text{ mL}}{1 \text{ liter}} \\ &= 2.2 \times 1{,}000 \text{ mL} \\ &= 2{,}200 \text{ mL} \end{aligned}$$

9 A 3.5-liter engine will have a volume of how many milliliters?

Answer
9 3,500 mL

PROBLEM SET 6.3

Use the tables given in this section to make the following conversions. Be sure to show the conversion factor used in each case.

1 3 ft^2 to square inches

2 5 ft^2 to square inches

3 288 in^2 to square feet

4 720 in^2 to square feet

5 30 acres to square feet

6 92 acres to square feet

7 2 mi^2 to acres

8 7 mi^2 to acres

9 1,920 acres to square miles

10 3,200 acres to square miles

11 12 yd^2 to square feet

12 20 yd^2 to square feet

13 17 cm^2 to square millimeters

14 150 mm^2 to square centimeters

15 2.8 m^2 to square centimeters

16 10 dm^2 to square millimeters

17 1,200 mm^2 to square meters

18 19.79 cm^2 to square meters

19 5 a to square meters

20 12 a to square centimeters

21 7 ha to ares

22 3.6 ha to ares

23 342 a to hectares

24 986 a to hectares

25 A rectangular athletic field has a width of 20 yards and a length of 100 yards. What is the area of the field in square feet?

26 A house sits on a lot that measures 108 feet by 50 feet. If the lot is in the shape of a rectangle, what is the area of the lot in square yards?

27 A public swimming pool measures 100 meters by 30 meters and is rectangular. What is the area of the pool in ares?

28 A family decides to put tiles in the entryway of their home. The entryway has an area of 6 square meters. If each tile is 5 centimeters by 5 centimeters, how many tiles will it take to cover the entryway?

29 A landscaper is putting in a brick patio. The area of the patio is 110 square meters. If the bricks measure 10 centimeters by 20 centimeters, how many bricks will it take to make the patio? Assume no space between bricks.

30 A dressmaker is using a pattern that requires 2 square yards of material. If the material is on a roll that is 54 inches wide, how long a piece of material must be cut from the roll to be sure there is enough material for the pattern?

Make the following conversions using the conversion factors given in Tables 3 and 4.

31 5 yd^3 to cubic feet

32 3.8 yd^3 to cubic feet

33 3 pt to fluid ounces

34 8 pt to fluid ounces

35 2 gal to quarts

36 12 gal to quarts

37 2.5 gal to pints

38 7 gal to pints

39 15 qt to fluid ounces

40 5.9 qt to fluid ounces

41 64 pt to gallons

42 256 pt to gallons

43 12 pt to quarts

44 18 pt to quarts

45 243 ft^3 to cubic yards

46 864 ft^3 to cubic yards

47 5 L to milliliters

48 9.6 L to milliliters

49 127 mL to liters

50 93.8 mL to liters

51 4 kL to milliliters

52 3 kL to milliliters

53 14.92 kL to liters

54 4.71 kL to liters

55 If a regular-size coffee cup holds about $\frac{1}{2}$ pint, about how many cups can be filled from a 1-gallon coffee maker?

56 If a regular-size drinking glass holds about 0.25 liter of liquid, how many glasses can be filled from a 750-milliliter container?

57 A refrigerator has a capacity of 20 cubic feet. What is the capacity of the refrigerator in cubic inches?

58 The gasoline tank on a car holds 18 gallons of gas. What is the volume of the tank in quarts?

59 How many 8-fluid ounce glasses of water will it take to fill a 3-gallon aquarium?

60 How many 5-milliliter test tubes filled with water will it take to fill a 1-liter container?

Calculator Problems

Set up the following problems as you have been doing. Then use a calculator to perform the actual calculations. Round off to two decimal places where appropriate.

61 Lake Superior is the largest of the Great Lakes. It covers 31,700 square miles of area. What is the area of Lake Superior in acres?

62 The state of California consists of 156,360 square miles of land and 2,330 square miles of water. Write the total area (both land and water) in acres.

63 Death Valley National Monument contains 2,067,795 acres of land. How many square miles is this?

64 The Badlands National Monument in South Dakota was established in 1929. It covers 243,302 acres of land. What is the area in square miles?

65 Convert 93.4 qt to gallons.

66 Convert 7,362 fl oz to gallons.

67 How many cubic feet are contained in 796 cubic yards?

68 The engine of a car has a displacement of 440 cubic inches. What is the displacement in cubic feet?

69 The Grand Coulee Dam holds 10,585,000 cubic yards of water. What is the volume of water in cubic feet?

70 Hoover Dam was built in 1936 on the Colorado River in Nevada. It holds a volume of 4,400,000 cubic yards of water. What is this volume in cubic feet?

Review Problems

The following problems review addition and subtraction with fractions and mixed numbers.

Find each sum. (Add.)

71 $\frac{3}{8} + \frac{1}{4}$

72 $\frac{1}{2} + \frac{1}{4}$

73 $3\frac{1}{2} + 5\frac{1}{2}$

74 $6\frac{7}{8} + 1\frac{5}{8}$

Find each difference. (Subtract.)

75 $\frac{7}{15} - \frac{2}{15}$

76 $\frac{5}{8} - \frac{1}{4}$

77 $\frac{5}{36} - \frac{1}{48}$

78 $\frac{7}{39} - \frac{2}{65}$

79 $5\frac{1}{6} - 2\frac{1}{9}$

80 $8\frac{1}{9} - 3\frac{1}{6}$

6.4 Weight

The most common units of weight in the U.S. system are ounces, pounds, and tons. The relationships among these units are given in Table 1.

Table 1 Units of weight in the U.S. system

The Relationship between	Is	To Convert from One to the Other, Multiply by
ounces (oz) and pounds (lb)	1 lb = 16 oz	$\frac{16\text{ oz}}{1\text{ lb}}$ or $\frac{1\text{ lb}}{16\text{ oz}}$
pounds and tons (T)	1 T = 2,000 lb	$\frac{2{,}000\text{ lb}}{1\text{ T}}$ or $\frac{1\text{ T}}{2{,}000\text{ lb}}$

Practice Problems

1 Convert 15 pounds to ounces.

2 A certain kind of cheese costs \$2.09 per pound. How much will 2 lb 4 oz cost?

EXAMPLE 1 Convert 12 pounds to ounces.

Solution Using the conversion factor from the table, and applying the method we have been using, we have

$$\begin{aligned} 12\text{ lb} &= 12\cancel{\text{lb}} \times \frac{16\text{ oz}}{1\cancel{\text{lb}}} \\ &= 12 \times 16\text{ oz} \\ &= 192\text{ oz} \end{aligned}$$

12 pounds is equivalent to 192 ounces.

EXAMPLE 2 If a certain cut of meat costs \$2.49 per pound, how much will 3 lb 12 oz of this meat cost?

Solution Since the cost is given in terms of pounds, we must convert 3 lb 12 oz to pounds. We can do this by first changing 12 oz to pounds:

$$\begin{aligned} 12\text{ oz} &= 12\cancel{\text{oz}} \times \frac{1\text{ lb}}{16\cancel{\text{oz}}} \\ &= \frac{12}{16}\text{ lb} \\ &= 0.75\text{ lb} \end{aligned}$$

Since 12 oz is equivalent to 0.75 lb, we have

3 lb 12 oz = 3.75 lb

Multiplying by the price per pound, we have

$$\underset{\substack{\uparrow \\ \text{Price per pound}}}{\$2.49} \times \underset{\substack{\uparrow \\ \text{Number of pounds}}}{3.75\text{ lb}} = \underset{\substack{\uparrow \\ \text{Total cost}}}{\$9.34} \qquad \text{Rounded to the nearest cent}$$

It costs \$9.34 for 3 lb 12 oz of meat that sells for \$2.49 per pound.

Answers
1 240 oz 2 \$4.70

In the metric system the basic unit of weight is a gram. We use the same prefixes we have already used to write the other units of weight in terms of grams. Table 2 lists the most common metric units of weight and their conversion factors.

Table 2 Metric units of weight

The Relationship between	Is	To Convert from One to the Other, Multiply by
milligrams (mg) and grams (g)	1 g = 1,000 mg	$\frac{1{,}000\text{ mg}}{1\text{ g}}$ or $\frac{1\text{ g}}{1{,}000\text{ mg}}$
centigrams (cg) and grams	1 g = 100 cg	$\frac{100\text{ cg}}{1\text{ g}}$ or $\frac{1\text{ g}}{100\text{ cg}}$
kilograms (kg) and grams	1,000 g = 1 kg	$\frac{1{,}000\text{ g}}{1\text{ kg}}$ or $\frac{1\text{ kg}}{1{,}000\text{ g}}$
metric tons (t) and kilograms	1,000 kg = 1 t	$\frac{1{,}000\text{ kg}}{1\text{ t}}$ or $\frac{1\text{ t}}{1{,}000\text{ kg}}$

3 Convert 5 kilograms to milligrams.

EXAMPLE 3 Convert 3 kilograms to centigrams.

Solution We convert kilograms to grams and then grams to centigrams:

$$\begin{aligned} 3\text{ kg} &= 3\cancel{\text{kg}} \times \frac{1{,}000\cancel{\text{g}}}{1\cancel{\text{kg}}} \times \frac{100\text{ cg}}{1\cancel{\text{g}}} \\ &= 3 \times 1{,}000 \times 100\text{ cg} \\ &= 300{,}000\text{ cg} \end{aligned}$$

4 A bottle of vitamin C contains 75 tablets. If each tablet contains 200 milligrams of vitamin C, what is the total number of grams of vitamin C in the bottle?

EXAMPLE 4 A bottle of vitamin C contains 50 tablets. Each tablet contains 250 milligrams of vitamin C. What is the total number of grams of vitamin C in the bottle?

Solution We begin by finding the total number of milligrams of vitamin C in the bottle. Since there are 50 tablets, and each contains 250 mg of vitamin C, we can multiply 50 by 250 to get the total number of milligrams of vitamin C:

$$\begin{aligned} \text{Milligrams of vitamin C} &= 50 \times 250\text{ mg} \\ &= 12{,}500\text{ mg} \end{aligned}$$

Next we convert 12,500 mg to grams:

$$\begin{aligned} 12{,}500\text{ mg} &= 12{,}500\cancel{\text{mg}} \times \frac{1\text{ g}}{1{,}000\cancel{\text{mg}}} \\ &= \frac{12{,}500}{1{,}000}\text{ g} \\ &= 12.5\text{ g} \end{aligned}$$

The bottle contains 12.5 g of vitamin C.

Answers
3 5,000,000 mg **4** 15 g

PROBLEM SET 6.4

Use the conversion factors given in Tables 1 and 2 to make the following conversions.

1 8 lb to ounces

2 5 lb to ounces

3 2 T to pounds

4 5 T to pounds

5 192 oz to pounds

6 176 oz to pounds

7 1,800 lb to tons

8 10,200 lb to tons

9 1 T to ounces

10 3 T to ounces

11 $3\frac{1}{2}$ lb to ounces

12 $5\frac{1}{4}$ lb to ounces

13 $6\frac{1}{2}$ T to pounds

14 $4\frac{1}{5}$ T to pounds

15 2 kg to grams

16 5 kg to grams

17 4 cg to milligrams

18 3 cg to milligrams

19 2 kg to centigrams

20 5 kg to centigrams

21 5.08 g to centigrams

22 7.14 g to centigrams

23 450 cg to grams

24 979 cg to grams

25 478.95 mg to centigrams

26 659.43 mg to centigrams

27 1,578 mg to grams

28 1,979 mg to grams

29 42,000 cg to kilograms

30 97,000 cg to kilograms

31 9 lb 4 oz to pounds

32 4 lb 8 oz to pounds

33 If meat costs \$2.03 per pound, how much will 4 lb 8 oz cost?

34 If meat costs \$2.09 per pound, how much will 9 lb 4 oz cost?

35 A bottle contains 30 vitamin tablets. If each tablet contains 500 milligrams of vitamins, how many total grams of vitamins are in the bottle?

36 A box contains six cans of soup. Each can weighs 305 grams. What is the total weight of the cans in kilograms?

Calculator Problems

Do any calculations required to work the following problems on a calculator.

37 Convert 3.98 lb to ounces.

38 Convert 59.46 oz to pounds. (Round to the nearest hundredth.)

39 Convert 12 lb 9 oz to pounds.

40 Convert 12 lb 7 oz to pounds.

41 A special cut of meat costs \$4.95 per pound. How much will 15 lb 7 oz of this meat cost?

42 If a man weighs $157\frac{1}{4}$ lb, what is his weight in ounces?

Review Problems

The following problems review material from Section 5.5.

Write each decimal as a fraction in lowest terms.

43 0.18 **44** 0.04 **45** 0.09 **46** 0.045 **47** 0.8

Write each fraction as a decimal.

48 $\frac{3}{4}$ **49** $\frac{9}{10}$ **50** $\frac{17}{20}$ **51** $\frac{1}{8}$ **52** $\frac{3}{5}$

Converting between the Two Systems

Since most of us have always used the U.S. system of measurement in our everyday lives, we are much more familiar with it on an intuitive level than we are with the metric system. We have an intuitive idea of how long feet and inches are, how much a pound weighs, and what a square yard of material looks like. The metric system is actually much easier to use than the U.S. system. The reason some of us have such a hard time with the metric system is that we don't have the feel for it that we do for the U.S. system. We have trouble visualizing how long a meter is or how much a gram weighs. The following list is intended to give you something to associate with each basic unit of measurement in the metric system:

1. A meter is just a little longer than a yard.
2. The length of the edge of a sugar cube is about 1 centimeter.
3. A liter is just a little larger than a quart.
4. A sugar cube has a volume of approximately 1 milliliter.
5. A paper clip weighs about 1 gram.
6. A 2-pound can of coffee weighs about 1 kilogram.

Table 1 Actual conversion factors between the metric and U.S. systems of measurement

The Relationship between	Is	To Convert from One to the Other, Multiply by
Length		
inches and centimeters	2.54 cm = 1 in.	$\frac{2.54\text{ cm}}{1\text{ in.}}$ or $\frac{1\text{ in.}}{2.54\text{ cm}}$
feet and meters	1 m = 3.28 ft	$\frac{3.28\text{ ft}}{1\text{ m}}$ or $\frac{1\text{ m}}{3.28\text{ ft}}$
miles and kilometers	1.61 km = 1 mi	$\frac{1.61\text{ km}}{1\text{ mi}}$ or $\frac{1\text{ mi}}{1.61\text{ km}}$
Area		
square inches and square centimeters	$6.45\text{ cm}^2 = 1\text{ in}^2$	$\frac{6.45\text{ cm}^2}{1\text{ in}^2}$ or $\frac{1\text{ in}^2}{6.45\text{ cm}^2}$
square meters and square yards	$1.196\text{ yd}^2 = 1\text{ m}^2$	$\frac{1.196\text{ yd}^2}{1\text{ m}^2}$ or $\frac{1\text{ m}^2}{1.196\text{ yd}^2}$
acres and hectares	1 ha = 2.47 acres	$\frac{2.47\text{ acres}}{1\text{ ha}}$ or $\frac{1\text{ ha}}{2.47\text{ acres}}$
Volume		
cubic inches and milliliters	$16.39\text{ mL} = 1\text{ in}^3$	$\frac{16.39\text{ mL}}{1\text{ in}^3}$ or $\frac{1\text{ in}^3}{16.39\text{ mL}}$
liters and quarts	1.06 qt = 1 liter	$\frac{1.06\text{ qt}}{1\text{ liter}}$ or $\frac{1\text{ liter}}{1.06\text{ qt}}$
gallons and liters	3.79 liters = 1 gal	$\frac{3.79\text{ liters}}{1\text{ gal}}$ or $\frac{1\text{ gal}}{3.79\text{ liters}}$
Weight		
ounces and grams	28.3 g = 1 oz	$\frac{28.3\text{ g}}{1\text{ oz}}$ or $\frac{1\text{ oz}}{28.3\text{ g}}$
kilograms and pounds	2.20 lb = 1 kg	$\frac{2.20\text{ lb}}{1\text{ kg}}$ or $\frac{1\text{ kg}}{2.20\text{ lb}}$

There are many other conversion factors that we could have included in Table 1. We have listed only the most common ones. Almost all of them are approximations. That is, most of the conversion factors are decimals that have been rounded to the nearest hundredth. If we want more accuracy, we obtain a table that has more digits in the conversion factors.

Practice Problems

1 Convert 10 inches to centimeters.

EXAMPLE 1 Convert 8 inches to centimeters.

Solution Choosing the appropriate conversion factor from Table 1, we have

$$\begin{aligned}8 \text{ in.} &= 8 \cancel{\text{in.}} \times \frac{2.54 \text{ cm}}{1 \cancel{\text{in.}}}\\ &= 8 \times 2.54 \text{ cm}\\ &= 20.32 \text{ cm}\end{aligned}$$

2 Convert 9 meters to feet.

EXAMPLE 2 Convert 80.5 kilometers to miles.

Solution Using the conversion factor that takes us from kilometers to miles, we have

$$\begin{aligned}80.5 \text{ km} &= 80.5 \cancel{\text{km}} \times \frac{1 \text{ mi}}{1.61 \cancel{\text{km}}}\\ &= \frac{80.5}{1.61} \text{ mi}\\ &= 50 \text{ mi}\end{aligned}$$

So 50 miles is equivalent to 80.5 kilometers. If we travel at 50 miles per hour in a car, we are moving at the rate of 80.5 kilometers per hour.

3 Convert 15 gallons to liters.

EXAMPLE 3 Convert 3 liters to pints.

Solution Since Table 1 doesn't list a conversion factor that will take us directly from liters to pints, we first convert liters to quarts, and then convert quarts to pints.

$$\begin{aligned}3 \text{ liters} &= 3 \cancel{\text{liters}} \times \frac{1.06 \cancel{\text{qt}}}{1 \cancel{\text{liter}}} \times \frac{2 \text{ pt}}{1 \cancel{\text{qt}}}\\ &= 3 \times 1.06 \times 2 \text{ pt}\\ &= 6.36 \text{ pt}\end{aligned}$$

4 The engine in a car has a 2.2-liter displacement. What is the displacement in cubic inches (to the nearest cubic inch)?

EXAMPLE 4 The engine in a car has a 2-liter displacement. What is the displacement in cubic inches?

Solution We convert liters to milliliters and then milliliters to cubic inches:

$$\begin{aligned}2 \text{ liters} &= 2 \cancel{\text{liters}} \times \frac{1{,}000 \cancel{\text{mL}}}{1 \cancel{\text{liter}}} \times \frac{1 \text{ in}^3}{16.39 \cancel{\text{mL}}}\\ &= \frac{2 \times 1{,}000}{16.39} \text{ in}^3 && \text{This calculation should be done on a calculator}\\ &= 122 \text{ in}^3 && \text{To the nearest cubic inch}\end{aligned}$$

Answers

1 25.4 cm **2** 29.52 ft
3 56.85 liters **4** 134 in^3

EXAMPLE 5 Suppose a person weighs 125 pounds. What is this person's weight in kilograms?

Solution Converting from pounds to kilograms, we have

$$125 \text{ lb} = 125 \cancel{\text{lb}} \times \frac{1 \text{ kg}}{2.20 \cancel{\text{lb}}}$$

$$= \frac{125}{2.20} \text{ kg}$$

$$= 56.8 \text{ kg} \quad \text{To the nearest tenth}$$

5 A person who weighs 165 pounds weighs how many kilograms?

Temperature

We end this section with a review of temperature in both systems of measurement.

In the U.S. system we measure temperature on the Fahrenheit scale. On this scale water boils at 212 degrees and freezes at 32 degrees. When we write 32 degrees measured on the Fahrenheit scale, we use the notation

32°F read "32 degrees Fahrenheit"

In the metric system the scale we use to measure temperature is the Celsius scale (formerly called the centigrade scale). On this scale water boils at 100 degrees and freezes at 0 degrees. When we write 100 degrees measured on the Celsius scale, we use the notation

100°C read "100 degrees Celsius"

Table 2 is intended to give you a feel for the relationship between the two temperature scales. Table 3 gives the formulas, in both symbols and words, that are used to convert between the two scales.

Table 2

Situation	Temperature Fahrenheit	Temperature Celsius
Water freezes	32°F	0°C
Room temperature	68°F	20°C
Normal body temperature	98.6°F	37°C
Water boils	212°F	100°C
Bake cookies	365°F	185°C
Broil meat	550°F	290°C

Table 3

To Convert from	Formula in Symbols	Formula in Words
Fahrenheit to Celsius	$C = \frac{5(F - 32)}{9}$	Subtract 32, multiply by 5, and then divide by 9.
Celsius to Fahrenheit	$F = \frac{9}{5}C + 32$	Multiply by $\frac{9}{5}$, and then add 32.

Answer
5 75 kg

6 Convert 40°C to degrees Fahrenheit.

7 A person is running a temperature of 101.6°F. What is the person's temperature, to the nearest tenth of a degree, on the Celsius scale?

Answers
6 104°F 7 38.7°C

The following examples show how we use the formulas given in Table 3.

EXAMPLE 6 Convert 120°C to degrees Fahrenheit.

Solution We use the formula

$$F = \frac{9}{5}C + 32$$

and replace C with 120:

$$\begin{aligned} \text{When} \quad C &= 120 \\ \text{the formula} \quad F &= \frac{9}{5}C + 32 \\ \text{becomes} \quad F &= \frac{9}{5}(120) + 32 \\ F &= 216 + 32 \\ F &= 248 \end{aligned}$$

We see that 120°C is equivalent to 248°F; they both mean the same temperature.

EXAMPLE 7 A person with the flu has a temperature of 102°F. What is the person's temperature on the Celsius scale?

Solution

$$\begin{aligned} \text{When} \quad F &= 102 \\ \text{the formula} \quad C &= \frac{5(F - 32)}{9} \\ \text{becomes} \quad C &= \frac{5(102 - 32)}{9} \\ C &= \frac{5(70)}{9} \\ C &= 38.9 \qquad \text{Rounded to the nearest tenth} \end{aligned}$$

The person's temperature, rounded to the nearest tenth, is 38.9°C on the Celsius scale.

PROBLEM SET 6.5

Use Tables 1 and 3 to make the following conversions.

1 6 in. to centimeters

2 1 ft to centimeters

3 4 m to feet

4 2 km to feet

5 6 m to yards

6 15 mi to kilometers

7 20 mi to meters (to the nearest meter)

8 600 m to yards

9 5 m^2 to square yards

10 2 in^2 to square centimeters

11 10 ha to acres

12 50 a to acres

13 500 in^3 to milliliters

14 400 in^3 to liters

15 2 L to quarts

16 15 L to quarts

17 20 gal to liters

18 15 gal to liters

19 12 oz to grams

20 1 lb to grams

21 15 kg to pounds

22 10 kg to ounces

23 185°C to degrees Fahrenheit

24 32°C to degrees Fahrenheit

25 86°F to degrees Celsius

26 122°F to degrees Celsius

Calculator Problems

Set up the following problems as we have set up the examples in this section. Then use a calculator for the calculations and round your answers to the nearest hundredth.

27 10 cm to inches

28 100 mi to kilometers

29 25 ft to meters

30 400 mL to cubic inches

31 49 qt to liters

32 65 L to gallons

33 500 g to ounces

34 100 lb to kilograms

35 Give your weight in kilograms.

36 Give your height in meters and centimeters.

37 The 100-yard dash is a popular race in track. How far is 100 yards in meters?

38 A 351-cubic-inch engine has a displacement of how many liters?

39 25 square yards of material is how many square meters?

40 How many grams does a 5 lb 4 oz roast weigh?

41 55 miles per hour is equivalent to how many kilometers per hour?

42 A 1-quart container holds how many liters?

43 A high jumper jumps 6 ft 8 in. How many meters is this?

44 A farmer owns 57 acres of land. How many ares is that?

45 A person has a temperature of 101°F. What is the person's temperature, to the nearest tenth, on the Celsius scale?

46 If the temperature outside is 30°C, is it a better day for water skiing or for snow skiing?

Review Problems

The problems below review the relationship between fractions and decimals from Section 5.5.

Write each fraction or mixed number as an equivalent decimal number.

47 $\frac{3}{4}$

48 $\frac{2}{5}$

49 $5\frac{1}{2}$

50 $8\frac{1}{4}$

51 $\frac{3}{100}$

52 $\frac{2}{50}$

Write each decimal as an equivalent proper fraction or mixed number.

53 0.34

54 0.08

55 2.4

56 5.05

57 1.75

58 3.125

6.6 Proportions

In this section we will extend our work with ratios to include *proportions*. When two ratios are equal, such as 3 to 2 and 6 to 4, we say that they are *proportional*. We define a proportion as follows:

DEFINITION A statement that two ratios are equal is called a **proportion**. If $\frac{a}{b}$ and $\frac{c}{d}$ are two equal ratios, then the statement

$$\frac{a}{b} = \frac{c}{d}$$

is called a proportion.

Each of the four numbers in a proportion is called a *term* of the proportion. We number the terms of a proportion as follows:

First term ⟶ a, Second term ⟶ b, c ⟵ Third term, d ⟵ Fourth term

$$\frac{a}{b} = \frac{c}{d}$$

The first and fourth terms of a proportion are called the *extremes*, and the second and third terms of a proportion are called the *means*.

Means ⟶ b, c; Extremes ⟵ a, d

$$\frac{a}{b} = \frac{c}{d}$$

EXAMPLE 1 In the proportion $\frac{3}{4} = \frac{6}{8}$, name the four terms, the means, and the extremes.

Solution The terms are numbered as follows:

First term = 3 Third term = 6

Second term = 4 Fourth term = 8

The means are 4 and 6; the extremes are 3 and 8.

The only additional thing we need to know about proportions is the following property.

Fundamental Property of Proportions

In any proportion, the product of the means is equal to the product of the extremes. In symbols, it looks like this:

$$\text{If } \frac{a}{b} = \frac{c}{d} \text{ then } ad = bc$$

Practice Problems

1 In the proportion $\frac{2}{3} = \frac{6}{9}$, name the four terms, the means, and the extremes.

Answer

1 See solutions section.

2 Verify the fundamental property of proportions for the following proportions.

a $\frac{5}{6} = \frac{15}{18}$

b $\frac{13}{39} = \frac{1}{3}$

EXAMPLE 2 Verify the fundamental property of proportions for the following proportions.

a $\frac{3}{4} = \frac{6}{8}$ **b** $\frac{17}{34} = \frac{1}{2}$

Solution We verify the fundamental property by finding the product of the means and the product of the extremes in each case.

Proportion	Product of the Means	Product of the Extremes
a $\frac{3}{4} = \frac{6}{8}$	$4 \cdot 6 = 24$	$3 \cdot 8 = 24$
b $\frac{17}{34} = \frac{1}{2}$	$34 \cdot 1 = 34$	$17 \cdot 2 = 34$

For each proportion the product of the means is equal to the product of the extremes.

We can use the fundamental property of proportions, along with a property we encountered in Section 4.3, to find a missing term in a proportion.

3 Find the missing term:

$\frac{3}{4} = \frac{9}{x}$

EXAMPLE 3 Find the missing term. (Solve for x.)

$$\frac{2}{3} = \frac{4}{x}$$

Solution Applying the fundamental property of proportions, we have

If $\frac{2}{3} = \frac{4}{x}$

then $2 \cdot x = 3 \cdot 4$ The product of the means equals the product of the extremes

$2x = 12$ Multiply

Note

In some of these problems you will be able to see what the solution is just by looking the problem over. In those cases it is still best to show all the work involved in solving the proportion. It is good practice for the more difficult problems.

The result is an equation. We know from Section 4.3 that we can divide both sides of an equation by the same number without changing the solution to the equation. In this case we divide both sides by 2 to solve for x:

$2x = 12$

$\frac{2x}{2} = \frac{12}{2}$ Divide both sides by 2

$x = 6$ Simplify each side

The missing term is 6. We can check our work by using the fundamental property of proportions:

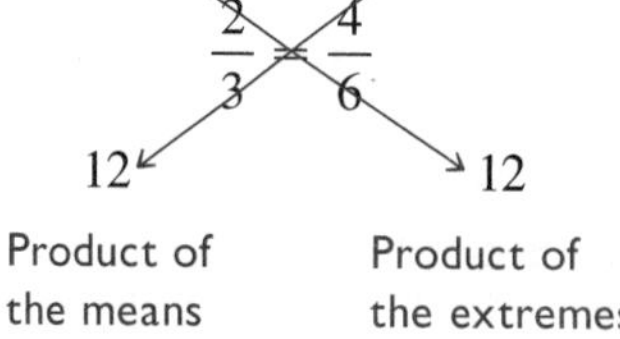

Since the product of the means and the product of the extremes are equal, our work is correct.

Answers
2 See solutions section. **3** 12

EXAMPLE 4 Solve for y: $\frac{5}{y} = \frac{10}{13}$

Solution We apply the fundamental property and solve as we did in Example 3:

If $\frac{5}{y} = \frac{10}{13}$

then $5 \cdot 13 = y \cdot 10$ — The product of the means equals the product of the extremes

$65 = 10y$ — Multiply $5 \cdot 13$

$\frac{65}{10} = \frac{\cancel{10}y}{\cancel{10}}$ — Divide both sides by 10

$6.5 = y$ — $65 \div 10 = 6.5$

The missing term is 6.5. We could check our result by substituting 6.5 for y in the original proportion and then finding the product of the means and the product of the extremes.

4 Solve for y: $\frac{2}{y} = \frac{8}{19}$

EXAMPLE 5 Find n if $\frac{n}{3} = \frac{0.4}{8}$.

Solution We proceed as we did in the previous two examples:

If $\frac{n}{3} = \frac{0.4}{8}$

then $n \cdot 8 = 3(0.4)$ — The product of the means equals the product of the extremes

$8n = 1.2$ — $3(0.4) = 1.2$

$\frac{\cancel{8}n}{\cancel{8}} = \frac{1.2}{8}$ — Divide both sides by 8

$n = 0.15$ — $1.2 \div 8 = 0.15$

The missing term is 0.15.

5 Find n if $\frac{n}{6} = \frac{0.3}{15}$.

EXAMPLE 6 Solve for x: $\frac{\frac{2}{3}}{5} = \frac{x}{6}$

Solution We begin by multiplying the means and multiplying the extremes:

If $\frac{\frac{2}{3}}{5} = \frac{x}{6}$

then $\frac{2}{3} \cdot 6 = 5 \cdot x$ — The product of the means equals the product of the extremes

$4 = 5 \cdot x$ — $\frac{2}{3} \cdot 6 = 4$

$\frac{4}{5} = \frac{\cancel{5} \cdot x}{\cancel{5}}$ — Divide both sides by 5

$\frac{4}{5} = x$

The missing term is $\frac{4}{5}$, or 0.8.

6 Solve for x: $\frac{\frac{3}{4}}{7} = \frac{x}{8}$

Answers

4 4.75 **5** 0.12 **6** $\frac{6}{7}$

The procedure for finding a missing term in a proportion is always the same. We first apply the fundamental property of proportions to find the product of the means and the product of the extremes. We then divide both sides of the resulting equation by the number being multiplied by the missing term. We then simplify the answer if possible.

Applications of Proportions

Proportions can be used to solve a variety of word problems. The examples that follow show some of these word problems. In each case we will translate the word problem into a proportion, and then solve the proportion using the method developed in this section.

7 A man drives his car 288 miles in 6 hours. If he continues at the same rate, how far will he travel in 10 hours?

EXAMPLE 7 A woman drives her car 270 miles in 6 hours. If she continues at the same rate, how far will she travel in 10 hours?

Solution We let x represent the distance traveled in 10 hours. Using x, we translate the problem into the following proportion:

$$\begin{matrix}\text{Miles} \longrightarrow \\ \text{Hours} \longrightarrow\end{matrix} \frac{x}{10} = \frac{270}{6} \begin{matrix}\longleftarrow \text{Miles} \\ \longleftarrow \text{Hours}\end{matrix}$$

Notice that the two ratios in the proportion compare the same quantities. That is, both ratios compare miles to hours. In words this proportion says:

x miles is to 10 *hours as* 270 *miles is to* 6 *hours*

$$\frac{x}{10} = \frac{270}{6}$$

Next, we solve the proportion.

$$x \cdot 6 = 10 \cdot 270$$
$$x \cdot 6 = 2{,}700$$
$$\frac{x \cdot \cancel{6}}{\cancel{6}} = \frac{2{,}700}{6}$$
$$x = 450 \text{ miles}$$

If the woman continues at the same rate, she will travel 450 miles in 10 hours.

8 A softball player gets 10 hits in the first 18 games of the season. If she continues at the same rate, how many hits will she get in 54 games?

EXAMPLE 8 A baseball player gets 8 hits in the first 18 games of the season. If he continues at the same rate, how many hits will he get in 45 games?

Solution We let x represent the number of hits he will get in 45 games. Then

x is to 45 *as* 8 *is to* 18

$$\begin{matrix}\text{Hits} \longrightarrow \\ \text{Games} \longrightarrow\end{matrix} \frac{x}{45} = \frac{8}{18} \begin{matrix}\longleftarrow \text{Hits} \\ \longleftarrow \text{Games}\end{matrix}$$

Notice again that the two ratios are comparing the same quantities, hits to games. We solve the proportion as follows:

$$18x = 360 \qquad 45 \cdot 8 = 360$$
$$\frac{\cancel{18}x}{\cancel{18}} = \frac{360}{18} \qquad \text{Divide both sides by 18}$$
$$x = 20 \qquad 360 \div 18 = 20$$

If he continues to hit at the rate of 8 hits in 18 games, he will get 20 hits in 45 games.

Answers
7 480 miles 8 30 hits

EXAMPLE 9 A solution contains 4 milliliters of alcohol and 20 milliliters of water. If another solution is to have the same ratio of milliliters of alcohol to milliliters of water and must contain 25 milliliters of water, how much alcohol should it contain?

Solution We let x represent the number of milliliters of alcohol in the second solution. The problem translates to

x milliliters is to 25 *milliliters as* 4 *milliliters is to* 20 *milliliters*

$$\text{Alcohol} \longrightarrow \frac{x}{25} = \frac{4}{20} \longleftarrow \text{Alcohol}$$
$$\text{Water} \longrightarrow \qquad \qquad \longleftarrow \text{Water}$$

$$20x = 100 \qquad 25 \cdot 4 = 100$$

$$\frac{\cancel{20}x}{\cancel{20}} = \frac{100}{20} \qquad \text{Divide both sides by 20}$$

$$x = 5 \text{ milliliters of alcohol} \qquad 100 \div 20 = 5$$

9 A solution contains 8 milliliters of alcohol and 20 milliliters of water. If another solution is to have the same ratio of milliliters of alcohol to water and must contain 35 milliliters of water, how much alcohol should it contain?

EXAMPLE 10 The scale on a map indicates that 1 inch on the map corresponds to an actual distance of 85 miles. Two cities are 3.5 inches apart on the map. What is the actual distance between the two cities?

Solution We let x represent the actual distance between the two cities. The proportion is

$$\text{Miles} \longrightarrow \frac{x}{3.5} = \frac{85}{1} \longleftarrow \text{Miles}$$
$$\text{Inches} \longrightarrow \qquad \qquad \longleftarrow \text{Inches}$$

$$x \cdot 1 = 3.5(85)$$

$$x = 297.5 \text{ miles}$$

10 The scale on a map indicates that 1 inch on the map corresponds to an actual distance of 105 miles. Two cities are 4.75 inches apart on the map. What is the actual distance between the two cities?

EXAMPLE 11 A manufacturer knows that, during an average production run, out of every 100 parts produced by a certain machine, 6 will be defective. If the machine produces 250 parts, how many can be expected to be defective?

Solution We let x represent the number of defective parts and solve the following proportion:

$$\frac{x}{250} = \frac{6}{100}$$

$$100x = 1{,}500$$

$$\frac{\cancel{100}x}{\cancel{100}} = \frac{1{,}500}{100}$$

$$x = 15$$

The manufacturer can expect to find 15 defective parts out of every 250 parts produced.

11 A manufacturer knows that, during an average production run, out of every 120 parts produced by a certain machine, 8 will be defective. If the machine produces 300 parts, how many can be expected to be defective?

Answers
9 14 mL 10 498.75 mi
11 20 parts

PROBLEM SET 6.6

For each of the following proportions, name the means, name the extremes, and show that the product of the means is equal to the product of the extremes.

1 $\frac{1}{3}=\frac{5}{15}$ **2** $\frac{6}{12}=\frac{1}{2}$ **3** $\frac{10}{25}=\frac{2}{5}$ **4** $\frac{5}{8}=\frac{10}{16}$

5 $\frac{\frac{1}{3}}{\frac{1}{2}}=\frac{4}{6}$ **6** $\frac{2}{\frac{1}{4}}=\frac{4}{\frac{1}{2}}$ **7** $\frac{0.5}{5}=\frac{1}{10}$ **8** $\frac{0.3}{1.2}=\frac{1}{4}$

Find the missing term in each of the following proportions. Set up each problem like the examples in this section. Write your answers as fractions in lowest terms.

9 $\frac{2}{5}=\frac{4}{x}$ **10** $\frac{3}{8}=\frac{9}{x}$ **11** $\frac{1}{y}=\frac{5}{12}$ **12** $\frac{2}{y}=\frac{6}{10}$ **13** $\frac{x}{4}=\frac{3}{8}$ **14** $\frac{x}{5}=\frac{7}{10}$

15 $\frac{5}{9}=\frac{x}{2}$ **16** $\frac{3}{7}=\frac{x}{3}$ **17** $\frac{3}{7}=\frac{3}{x}$ **18** $\frac{2}{9}=\frac{2}{x}$ **19** $\frac{25}{100}=\frac{x}{4}$ **20** $\frac{30}{300}=\frac{x}{10}$

21 $\frac{\frac{1}{2}}{y}=\frac{\frac{1}{3}}{12}$ **22** $\frac{\frac{2}{3}}{y}=\frac{\frac{1}{3}}{5}$ **23** $\frac{n}{12}=\frac{\frac{1}{4}}{\frac{1}{2}}$ **24** $\frac{n}{10}=\frac{\frac{3}{5}}{\frac{3}{8}}$ **25** $\frac{10}{20}=\frac{20}{n}$ **26** $\frac{8}{4}=\frac{4}{n}$

27 $\frac{x}{10}=\frac{10}{2}$ **28** $\frac{x}{12}=\frac{12}{48}$ **29** $\frac{1}{100}=\frac{y}{50}$ **30** $\frac{3}{20}=\frac{y}{10}$ **31** $\frac{0.4}{1.2}=\frac{1}{x}$ **32** $\frac{5}{0.5}=\frac{20}{x}$

33 $\frac{0.3}{0.18}=\frac{n}{0.6}$ **34** $\frac{0.01}{0.1}=\frac{n}{10}$ **35** $\frac{0.5}{x}=\frac{1.4}{0.7}$ **36** $\frac{0.3}{x}=\frac{2.4}{0.8}$

Solve each of the following word problems by translating the statement into a proportion. Be sure to show the proportion used in each case.

37 A woman drives her car 235 miles in 5 hours. At this rate how far will she travel in 7 hours?

38 An airplane flies 1,260 miles in 3 hours. How far will it fly in 5 hours?

39 A basketball player scores 162 points in 9 games. At this rate how many points will he score in 20 games?

40 In the first 4 games of the season, a football team scores a total of 68 points. At this rate how many points will the team score in 11 games?

41 A solution contains 8 pints of antifreeze and 5 pints of water. How many pints of water must be added to 24 pints of antifreeze to get a solution with the same concentration?

42 If 10 ounces of a certain breakfast cereal contains 3 ounces of sugar, how many ounces of sugar does 25 ounces of the same cereal contain?

43 The scale on a map indicates that 1 inch corresponds to an actual distance of 95 miles. Two cities are 4.5 inches apart on the map. What is the actual distance between the two cities?

44 A map is drawn so that every 2.5 inches on the map corresponds to an actual distance of 100 miles. If the actual distance between two cities is 350 miles, how far apart are they on the map?

45 A farmer knows that of every 50 eggs his chickens lay, only 45 will be marketable. If his chickens lay 1,000 eggs in a week, how many of them will be marketable?

46 Of every 17 parts manufactured by a certain machine, only 1 will be defective. How many parts were manufactured by the machine if 8 defective parts were found?

47 If a certain brand of canned fruit is on sale for 3 cans for \$1.00, how much will 7 cans cost? (Round to the nearest cent.)

48 A clothing store advertised shirts on sale at 3 for \$3.51. At this rate what would it cost for 2 shirts?

49 A traveling salesman figures it costs 21¢ for every mile he drives his car. How much does it cost him a week to drive his car if he travels 570 miles a week?

50 A family plans to drive their car during their annual vacation. The car can go 350 miles on a tank of gas, which is 18 gallons of gas. The vacation they have planned will cover 1,785 miles. How many gallons of gas will that take?

51 A 6-ounce serving of grapefruit juice contains 159 grams of water. How many grams of water are in 10 ounces of grapefruit juice?

52 If 100 grams of ice cream contains 13 grams of fat, how much fat is in 250 grams of ice cream?

Calculator Problems

Find the missing term in each proportion. Use a calculator to do the actual calculations. Write all answers in decimal form.

53 $\frac{168}{324} = \frac{56}{x}$

54 $\frac{280}{530} = \frac{112}{x}$

55 $\frac{429}{y} = \frac{858}{130}$

56 $\frac{573}{y} = \frac{2{,}292}{316}$

57 $\frac{n}{39} = \frac{533}{507}$

58 $\frac{n}{47} = \frac{1{,}003}{799}$

59 $\frac{756}{903} = \frac{x}{129}$

60 $\frac{321}{1{,}128} = \frac{x}{376}$

Use a proportion to set up each of the following problems. Then use a calculator to do the actual calculations.

61 If a car travels 378.9 miles on 50 liters of gas, how many liters of gas will it take to go 692 miles if the car travels at the same rate? (Round to the nearest tenth.)

62 If 125 grams of peas contains 26 grams of carbohydrates, how many grams of carbohydrates does 375 grams of peas contain?

63 During a recent election, 47 of every 100 registered voters in a certain city voted. If there were 127,900 registered voters in that city, how many people voted?

64 The scale on a map is drawn so that 4.5 inches corresponds to an actual distance of 250 miles. If two cities are 7.25 inches apart on the map, how many miles apart are they? (Round to the nearest tenth.)

Review Problems

The problems below are a review of some of the concepts we covered in Chapter 5.

Give the place value of the 5 in each number.

65 250.14

66 2.5014

Add or subtract as indicated.

67 $2.3 + 0.18 + 24.036$

68 $5 + 0.03 + 1.9$

69 $3.18 - 2.79$

70 $3.4 - 1.975$

Find the following products. (Multiply.)

71 2.7×0.5

72 0.7^2

73 3.18×1.2

74 0.3^4

Find the following quotients. (Divide.)

75 $2.8 \div 0.7$

76 $0.042 \div 0.21$

77 $24 \div 0.15$

78 $6.99 \div 2.33$

CHAPTER 6 SUMMARY

Ratio [6.1]

The ratio of a to b is $\frac{a}{b}$. The ratio of two numbers is a way of comparing them using fraction notation.

Rates [6.1]

A ratio that compares two different quantities, like miles and hours, gallons and seconds, etc., is called a *rate*.

Unit Pricing [6.1]

The unit price of an item is the ratio of price to quantity.

Conversion Factors [6.2, 6.3, 6.4, 6.5]

To convert from one kind of unit to another, we choose an appropriate conversion factor from one of the tables given in this chapter. For example, if we want to convert 5 feet to inches, we look for conversion factors that give the relationship between feet and inches. There are two conversion factors for feet and inches:

$$\frac{12 \text{ in.}}{1 \text{ ft}} \quad \text{and} \quad \frac{1 \text{ ft}}{12 \text{ in.}}$$

Proportion [6.6]

A proportion is an equation that indicates that two ratios are equal.

The numbers in a proportion are called *terms* and are numbered as follows:

$$\text{First term} \longrightarrow \frac{a}{b} = \frac{c}{d} \longleftarrow \text{Third term}$$

(Second term $\longrightarrow b$, $d \longleftarrow$ Fourth term)

The first and fourth terms are called the *extremes*. The second and third terms are called the *means*.

$$\text{Means} \longrightarrow \frac{a}{b} = \frac{c}{d} \longleftarrow \text{Extremes}$$

Fundamental Property of Proportions [6.6]

In any proportion the product of the means is equal to the product of the extremes. In symbols,

$$\text{If} \quad \frac{a}{b} = \frac{c}{d} \quad \text{then} \quad ad = bc$$

Examples

1 The ratio of 6 to 8 is

$$\frac{6}{8}$$

which can be reduced to

$$\frac{3}{4}$$

2 If a car travels 150 miles in 3 hours, then the ratio of miles to hours is considered a rate:

$$\frac{150 \text{ miles}}{3 \text{ hours}} = 50\frac{\text{miles}}{\text{hour}}$$
$$= 50 \text{ miles per hour}$$

3 If a 10-ounce package of frozen peas costs 69¢, then the price per ounce, or unit price, is

$$\frac{69 \text{ cents}}{10 \text{ ounces}} = 6.9\frac{\text{cents}}{\text{ounce}}$$
$$= 6.9 \text{ cents per ounce}$$

4 Convert 5 feet to inches.

$$5 \text{ ft} = 5 \cancel{\text{ft}} \times \frac{12 \text{ in.}}{1 \cancel{\text{ft}}}$$
$$= 5 \times 12 \text{ in.}$$
$$= 60 \text{ in.}$$

5 The following is a proportion:

$$\frac{6}{8} = \frac{3}{4}$$

6 Find x: $\frac{2}{5} = \frac{8}{x}$

$$2 \cdot x = 5 \cdot 8$$
$$2 \cdot x = 40$$
$$\frac{\cancel{2} \cdot x}{\cancel{2}} = \frac{40}{2}$$
$$x = 20$$

Finding a Missing Term in a Proportion [6.6]

To find the missing term in a proportion, we apply the fundamental property of proportions and solve the equation that results by dividing both sides by the number that is multiplied by the missing term. For instance, if we want to find the missing term in the proportion

$$\frac{2}{5} = \frac{8}{x}$$

we use the fundamental property of proportions to set the product of the means equal to the product of the extremes.

COMMON MISTAKES

1. A common mistake when working with ratios is to write the numbers in the wrong order when writing the ratio as a fraction. For example, the ratio 3 to 5 is equivalent to the fraction $\frac{3}{5}$. It cannot be written as $\frac{5}{3}$.
2. The most common mistake in converting from one kind of unit to another happens when we use the wrong conversion factor.

   This is a mistake:

   $$5 \text{ ft} = 5 \text{ ft} \times \frac{1 \text{ ft}}{12 \text{ in.}}$$

   Conversion factor is used incorrectly. The units, ft, don't divide out, since they are both in the numerator. We should have multiplied by $\frac{12 \text{ in.}}{1 \text{ ft}}$.

CHAPTER 6 REVIEW

Write each of the following ratios as a fraction in lowest terms. **[6.1]**

1 9 to 30

2 30 to 9

3 $\frac{3}{7}$ to $\frac{4}{7}$

4 $\frac{8}{5}$ to $\frac{8}{9}$

5 $2\frac{1}{3}$ to $1\frac{2}{3}$

6 3 to $2\frac{3}{4}$

7 0.6 to 1.2

8 0.03 to 0.24

9 2 feet to 2 yards

10 40 minutes to 1 hour

The chart shows where each dollar spent on gasoline in the United States goes.

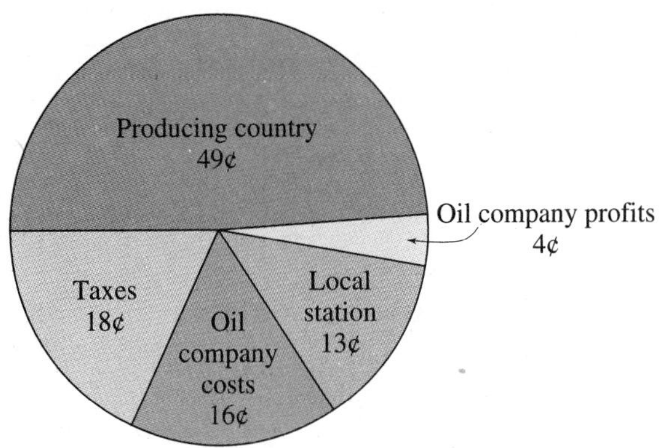

11 Find the ratio of money paid for taxes to money that goes to oil company profits. **[6.1]**

12 What is the ratio of the number of cents spent on oil company costs to the number of cents that goes to local stations? **[6.1]**

13 Give the ratio of oil company profits to oil company costs. **[6.1]**

14 Give the ratio of taxes to oil company costs and profits. **[6.1]**

15 A car travels 285 miles on 15 gallons of gas. What is the rate of gas mileage in miles per gallon? **[6.1]**

16 If it takes 2.5 seconds for sound to travel 2,750 feet, what is the speed of sound in feet per second? **[6.1]**

17 If 6 ounces of frozen orange juice costs 48¢, what is the price per ounce? **[6.1]**

18 A certain brand of frozen peas comes in two different-sized packages with the prices marked as follows:

20-ounce package 69¢
10-ounce package 43¢

Give the unit price for each package and indicate which is the better buy. **[6.1]**

Use the tables given in this chapter to make the following conversions. **[6.2–6.5]**

19 12 ft to inches

20 18 ft to yards

21 49 cm to meters

22 2 km to decimeters

23 10 acres to square feet

24 7,800 m^2 to ares

25 4 ft^2 to square inches

26 7 qt to pints

27 24 qt to gallons

28 5 L to milliliters

29 8 lb to ounces

30 2 lb 4 oz to ounces

31 5 kg to grams

32 5 t to kilograms

33 4 in. to centimeters

34 7 mi to kilometers

35 7 L to quarts

36 5 gal to liters

37 5 oz to grams

38 9 kg to pounds

39 120°C to degrees Fahrenheit

40 122°F to degrees Celsius

41 A 12-square-meter patio is to be built using bricks that measure 10 centimeters by 20 centimeters. How many bricks will be needed to cover the patio?

42 If a regular drinking glass holds 0.25 liter of liquid, how many glasses can be filled from a 6.5-liter container?

43 How many 8-fluid-ounce glasses of water will it take to fill a 5-gallon aquarium?

44 If meat costs $1.10 per pound, how much will 7 lb 4 oz cost?

Find the missing term in each of the following proportions. **[6.6]**

45 $\frac{5}{7} = \frac{35}{x}$

46 $\frac{n}{18} = \frac{18}{54}$

47 $\frac{\frac{1}{2}}{10} = \frac{y}{2}$

48 $\frac{x}{1.8} = \frac{5}{1.5}$

49 Suppose every 2,000 milliliters of a solution contains 24 milliliters of a certain drug. How many milliliters of solution are required to obtain 18 milliliters of the drug? **[6.6]**

50 If $\frac{1}{2}$ cup of breakfast cereal contains 8 milligrams of calcium, how much calcium does $1\frac{1}{2}$ cups of the cereal contain? **[6.6]**

51 A man loses 8 pounds during the first 2 weeks of a diet. If he continues losing weight at the same rate, how long will it take him to lose 20 pounds? **[6.6]**

52 If the ratio of men to women in a math class is 2 to 3, and there are 12 men in the class, how many women are in the class? **[6.6]**

CHAPTER 6 TEST

Write each ratio as a fraction in lowest terms.

1 24 to 18

2 $\frac{3}{4}$ to $\frac{5}{6}$

3 5 to $3\frac{1}{3}$

4 0.18 to 0.6

5 36 seconds to 1 minute

A family of three budgeted the following amounts for some of their monthly bills:

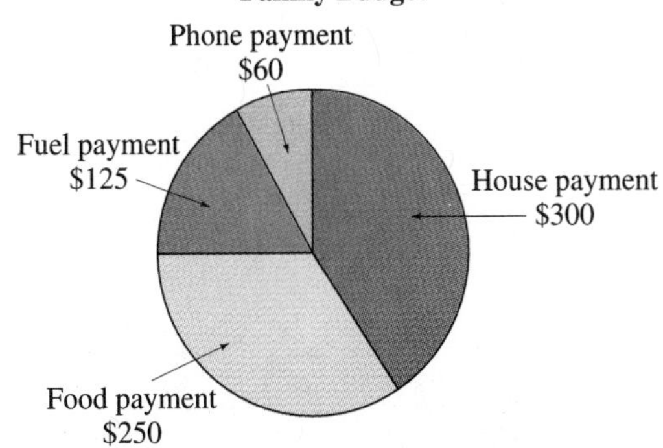

6 Find the ratio of house payment to fuel payment.

7 Find the ratio of phone payment to food payment.

8 A car travels 414 miles on 18 gallons of gas. What is the rate of gas mileage in miles per gallon?

9 A certain brand of frozen orange juice comes in two different-sized cans with prices marked as follows:

30-ounce can	$1.05
20-ounce can	$0.72

Give the unit price for each can and indicate which is the better buy.

Use the tables given in the chapter to make the following conversions.

10 7 yd to feet

11 750 m to kilometers

12 3 acres to square feet

13 432 in^2 to square feet

14 10 L to milliliters

15 5 mi to kilometers

16 10 L to quarts

17 80°F to degrees Celsius

18 A 40-square-foot pantry floor is to be tiled using tiles that measure 8 inches by 8 inches. How many tiles will be needed to cover the pantry floor?

19 How many 12-fluid-ounce glasses of water will it take to fill a 6-gallon aquarium?

20 If coffee costs $5.60 per pound, how much will 3 lb 12 oz cost?

Find the missing term in each proportion.

21 $\frac{5}{6} = \frac{30}{x}$

22 $\frac{1.8}{6} = \frac{2.4}{x}$

23 A baseball player gets 9 hits in his first 21 games of the season. If he continues at the same rate, how many hits will he get in 56 games?

24 The scale on a map indicates that 1 inch on the map corresponds to an actual distance of 60 miles. Two cities are $2\frac{1}{4}$ inches apart on the map. What is the actual distance between the two cities?

25 If the ratio of salt to sugar in a recipe is 3 to 5 and the recipe calls for 15 tablespoons of salt, how many tablespoons of sugar are needed?

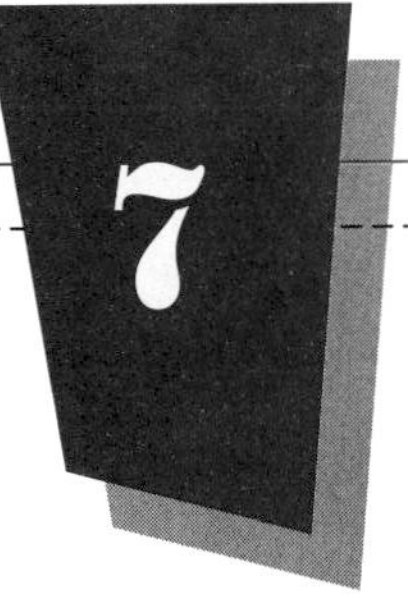

PERCENT

INTRODUCTION

In preceding chapters we have used various methods to compare quantities. For example, in Chapter 6, we used the unit price of items to find the best buy. In this chapter we will study percent. When we use percent, we are comparing everything to 100, because percent means per hundred.

The chart in Figure 1 shows the price that college bookstores paid for this book in January of each of the years shown. You can see from the chart that the largest increase in price—$4.50—occurred between January 1993 and January 1994. To see how this increase in price compares with other increases in price, we can look at the *percent increase in price.* The chart in Figure 2 shows the percent increase in the price of the book from the price the previous year.

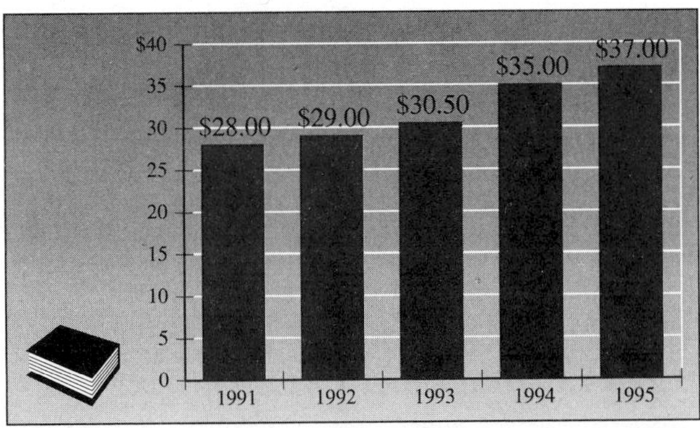

Figure 1 Price paid by college bookstores for this book

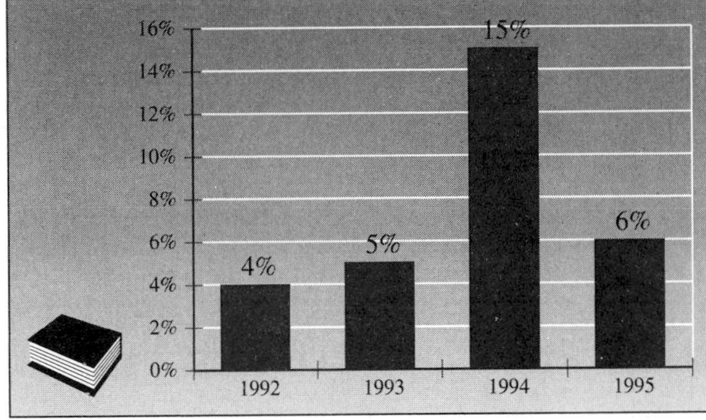

Figure 2 Percent increase in price of this book

In this chapter we will study many other applications of percent as well.

7.1 Percents, Decimals, and Fractions

In this section we will look at the meaning of percent, and we will learn to change decimals to percents and percents to decimals.

The Meaning of Percent

Percent means "per hundred." Writing a number as a percent is a way of comparing the number with 100. For example, the number 42% (the % symbol is read "percent") is the same as 42 one-hundredths. That is:

$$42\% = \frac{42}{100}$$

Percents are really fractions (or ratios) with denominator 100.

Here are some examples that show the meaning of percent.

EXAMPLE 1 $50\% = \frac{50}{100}$

EXAMPLE 2 $75\% = \frac{75}{100}$

EXAMPLE 3 $25\% = \frac{25}{100}$

EXAMPLE 4 $33\% = \frac{33}{100}$

EXAMPLE 5 $6\% = \frac{6}{100}$

EXAMPLE 6 $160\% = \frac{160}{100}$

Changing Percents to Decimals

To change a percent to a decimal number, we simply use the meaning of percent.

EXAMPLE 7 Change 35.2% to a decimal.

Solution We drop the % symbol and write 35.2 over 100.

$$35.2\% = \frac{35.2}{100} \quad \text{Use the meaning of \% to convert to a fraction with denominator 100}$$

$$= 0.352 \quad \text{Divide 35.2 by 100}$$

We see from Example 7 that 35.2% is the same as the decimal 0.352. The result is that the % symbol has been dropped and the decimal point has been moved two places to the *left*. Since % always means "per hundred," we will always end up moving the decimal point two places to the left when we change percents to decimals. Because of this, we can write the following rule.

RULE To change a percent to a decimal, drop the % symbol and move the decimal point two places to the *left*.

Practice Problems

Write each number as an equivalent fraction without the % symbol.

1 40%

2 80%

3 15%

4 37%

5 8%

6 150%

7 Change 25.2% to a decimal.

Answers

1 $\frac{40}{100}$ 2 $\frac{80}{100}$ 3 $\frac{15}{100}$

4 $\frac{37}{100}$ 5 $\frac{8}{100}$ 6 $\frac{150}{100}$

7 0.252

Change each percent to a decimal.

8 40%

9 80%

10 15%

11 5.6%

12 4.86%

13 0.6%

14 0.58%

Here are some examples to illustrate how to use this rule.

EXAMPLE 8 25% = 0.25

EXAMPLE 9 75% = 0.75

Notice that the results in Examples 8, 9, and 10 are consistent with the results in Examples 1, 2, and 3

EXAMPLE 10 50% = 0.50

EXAMPLE 11 6.8% = 0.068

Notice here that we put a 0 in front of the 6 so we can move the decimal point two places to the left

EXAMPLE 12 3.62% = 0.0362

EXAMPLE 13 0.4% = 0.004

This time we put two 0's in front of the 4 in order to be able to move the decimal point two places to the left

EXAMPLE 14 0.63% = 0.0063

Changing Decimals to Percents

Now we want to do the opposite of what we just did in Examples 7–14. We want to change decimals to percents. We know that 42% written as a decimal is 0.42, which means that in order to change 0.42 back to a percent, we must move the decimal point two places to the *right* and use the % symbol:

0.42 = 42% Notice that we don't show the new decimal point if it is at the end of the number

RULE To change a decimal to a percent, we move the decimal point two places to the *right* and use the % symbol.

Write each decimal as a percent.

15 0.35

16 5.77

17 0.4

18 0.03

19 45

20 0.69

Examples 15–20 show how we use this rule.

EXAMPLE 15 0.27 = 27%

EXAMPLE 16 4.89 = 489%

EXAMPLE 17 0.2 = 0.20 = 20%

Notice here that we put a 0 after the 2 so we can move the decimal point two places to the right

EXAMPLE 18 0.09 = 09% = 9%

Notice that we can drop the 0 at the left without changing the value of the number

EXAMPLE 19 25 = 25.00 = 2,500%

Here, we put two 0's after the 5 so we can move the decimal point two places to the right

EXAMPLE 20 0.49 = 49%

As you can see from the examples above, percent is just a way of comparing numbers to 100. To multiply decimals by 100, we move the decimal point two places to the right. To divide by 100, we move the decimal point two places to the left. Because of this, it is a fairly simple procedure to change percents to decimals and decimals to percents.

Answers

8 0.40 **9** 0.80 **10** 0.15
11 0.056 **12** 0.0486
13 0.006 **14** 0.0058
15 35% **16** 577% **17** 40%
18 3% **19** 4,500% **20** 69%

Changing Percents to Fractions

To change a percent to a fraction, drop the % symbol and write the original number over 100.

EXAMPLE 21 Change 62% to a fraction.

Solution We drop the % symbol and write 62 over 100:

$$62\% = \frac{62}{100}$$

$$= \frac{31}{50} \quad \text{Now we reduce the fraction to lowest terms}$$

EXAMPLE 22 Change 4.5% to a fraction in lowest terms.

Solution We begin by writing 4.5 over 100:

$$4.5\% = \frac{4.5}{100}$$

We now multiply the numerator and the denominator by 10 so the numerator will be a whole number:

$$\frac{4.5}{100} = \frac{4.5 \times \mathbf{10}}{100 \times \mathbf{10}} \quad \text{Multiply the numerator and the denominator by } \mathbf{10}$$

$$= \frac{45}{1{,}000}$$

$$= \frac{9}{200} \quad \text{Reduce to lowest terms}$$

EXAMPLE 23 Change $32\frac{1}{2}\%$ to a fraction in lowest terms.

Solution Writing $32\frac{1}{2}\%$ over 100 produces a complex fraction. We change $32\frac{1}{2}$ to an improper fraction and simplify:

$$32\tfrac{1}{2}\% = \frac{32\frac{1}{2}}{100}$$

$$= \frac{\frac{65}{2}}{100} \quad \text{Change } 32\tfrac{1}{2} \text{ to the improper fraction } \tfrac{65}{2}$$

$$= \frac{65}{2} \times \frac{1}{100} \quad \text{Dividing by 100 is the same as multiplying by } \tfrac{1}{100}$$

$$= \frac{65}{200} \quad \text{Multiplication}$$

$$= \frac{13}{40} \quad \text{Reduce to lowest terms}$$

EXAMPLE 24 $35\% = \frac{35}{100}$

$$= \frac{7}{20}$$

21 Change 82% to a fraction in lowest terms.

22 Change 6.5% to a fraction in lowest terms.

23 Change $42\frac{1}{2}\%$ to a fraction in lowest terms.

Change each percent to a fraction in lowest terms.

24 65%

Answers

21 $\frac{41}{50}$ **22** $\frac{13}{200}$ **23** $\frac{17}{40}$ **24** $\frac{13}{20}$

25 35.6%

EXAMPLE 25 $25.4\% = \frac{25.4}{100}$

$= \frac{254}{1{,}000}$ Multiply the numerator and the denominator by 10

$= \frac{127}{500}$ Reduce to lowest terms

26 $6\frac{1}{4}\%$

EXAMPLE 26 $8\frac{3}{4}\% = \frac{8\frac{3}{4}}{100}$

$= \frac{\frac{35}{4}}{100}$ Change $8\frac{3}{4}$ to the improper fraction $\frac{35}{4}$

$= \frac{35}{4} \times \frac{1}{100}$ Dividing by 100 is the same as multiplying by $\frac{1}{100}$

$= \frac{35}{400}$ Multiplication

$= \frac{7}{80}$ Reduce to lowest terms

Changing Fractions to Percents

To change a fraction to a percent, we can change the fraction to a decimal and then change the decimal to a percent.

27 Change $\frac{9}{10}$ to a percent.

EXAMPLE 27 Change $\frac{7}{10}$ to a percent.

Solution We can change $\frac{7}{10}$ to a decimal by dividing 7 by 10:

$$\begin{array}{r} 0.7 \\ 10\overline{)7.0} \\ \underline{7\,0} \\ 0 \end{array}$$

We then change the decimal 0.7 to a percent by moving the decimal point two places to the *right* and using the % symbol:

$0.7 = 70\%$

You may have noticed that we could have saved some time by simply writing $\frac{7}{10}$ as an equivalent fraction with denominator 100. That is:

$$\frac{7}{10} = \frac{70}{100}$$
$$= 70\%$$

This is a good way to convert fractions like $\frac{7}{10}$ to percents. It works well for fractions with denominators of 2, 4, 5, 10, 25, and 50, because they are easy to change to fractions with denominators of 100.

Answers

25 $\frac{89}{250}$ 26 $\frac{1}{16}$ 27 90%

EXAMPLE 28 Change $\frac{3}{8}$ to a percent.

Solution We write $\frac{3}{8}$ as a decimal by dividing 3 by 8. We then change the decimal to a percent by moving the decimal point two places to the right and using the % symbol.

$$\frac{3}{8} = 0.375 \qquad \begin{array}{r} .375 \\ 8\overline{)3.000} \\ \underline{2\,4} \\ 60 \\ \underline{56} \\ 40 \\ \underline{40} \\ 0 \end{array}$$

$$= 37.5\%$$

28 Change $\frac{5}{8}$ to a percent.

EXAMPLE 29 Change $\frac{5}{12}$ to a percent.

Solution We begin by dividing 5 by 12:

$$\frac{5}{12} = 0.41\overline{6} \qquad \begin{array}{r} .4166 \\ 12\overline{)5.000} \\ \underline{4\,8} \\ 20 \\ \underline{12} \\ 80 \\ \underline{72} \\ 80 \\ \underline{72} \end{array}$$

29 Change $\frac{7}{12}$ to a percent.

The 6's repeat forever. We can either write this as a decimal and a fraction, or round it off somewhere. As a decimal and a fraction, we have

$$\frac{5}{12} = 0.41\overline{6}$$

$$= 41\tfrac{2}{3}\% \qquad \text{This is the exact percent}$$

Rounding off, we have

$$\frac{5}{12} \approx 0.417$$

$$\approx 41.7\% \qquad \text{This is an approximation}$$

Note
When rounding off, let's agree to round off to the nearest thousandth and then move the decimal point. Our answers in percent form will then be accurate to the nearest tent of a percent, as in Example 29.

Answers
28 62.5% **29** $58\frac{1}{3}\% \approx 58.3\%$

30 Change $3\frac{3}{4}$ to a percent.

EXAMPLE 30 Change $2\frac{1}{2}$ to a percent.

Solution We first change to a decimal and then to a percent:

$$\begin{aligned} 2\tfrac{1}{2} &= 2.5 \\ &= 250\% \end{aligned}$$

Table 1 lists some of the most commonly used fractions and decimals and their equivalent percents.

Table 1

Fraction	Decimal	Percent
$\frac{1}{2}$	0.5	50%
$\frac{1}{4}$	0.25	25%
$\frac{3}{4}$	0.75	75%
$\frac{1}{3}$	$0.33\frac{1}{3}$	$33\frac{1}{3}\%$
$\frac{2}{3}$	$0.66\frac{2}{3}$	$66\frac{2}{3}\%$
$\frac{1}{5}$	0.2	20%
$\frac{2}{5}$	0.4	40%
$\frac{3}{5}$	0.6	60%
$\frac{4}{5}$	0.8	80%

Answer
30 375%

PROBLEM SET 7.1

Write each percent as a fraction with denominator 100.

1 20% **2** 40% **3** 60% **4** 80% **5** 24% **6** 48%

7 65% **8** 35%

Change each percent to a decimal.

9 23% **10** 34% **11** 92% **12** 87% **13** 9% **14** 7%

15 3.4% **16** 5.8% **17** 6.34% **18** 7.25% **19** 0.9% **20** 0.6%

Change each decimal to a percent.

21 0.23 **22** 0.34 **23** 0.92 **24** 0.87 **25** 0.45 **26** 0.54

27 0.03 **28** 0.04 **29** 0.6 **30** 0.9 **31** 0.8 **32** 0.5

33 0.27 **34** 0.62 **35** 1.23 **36** 2.34

Change each percent to a fraction in lowest terms.

37 60% **38** 40% **39** 75% **40** 25% **41** 4% **42** 2%

43 26.5% **44** 34.2% **45** 71.87% **46** 63.6% **47** 0.75% **48** 0.45%

49 $6\frac{1}{4}\%$ **50** $5\frac{1}{4}\%$ **51** $33\frac{1}{3}\%$ **52** $66\frac{2}{3}\%$

Change each fraction or mixed number to a percent.

53 $\frac{1}{2}$ **54** $\frac{1}{4}$ **55** $\frac{3}{4}$ **56** $\frac{2}{3}$ **57** $\frac{1}{3}$ **58** $\frac{1}{5}$

59 $\frac{4}{5}$ **60** $\frac{1}{6}$ **61** $\frac{7}{8}$ **62** $\frac{1}{8}$ **63** $\frac{7}{50}$ **64** $\frac{9}{25}$

65 $3\frac{1}{4}$ **66** $2\frac{1}{8}$ **67** $1\frac{1}{2}$ **68** $1\frac{3}{4}$

69 $\frac{21}{43}$ to the nearest tenth of a percent **70** $\frac{36}{49}$ to the nearest tenth of a percent

71 The human body is between 50% and 75% water. Write each of these percents as a decimal.

72 In the United States, 2.7% of those over 15 years of age drink more than 6.3 ounces of alcohol per day. In France, the same figure is 9%. Write each of these percents as a decimal.

73 Cancer is responsible for 18.9% of all the deaths in the United States each year. Accidents account for only 8.1% of all deaths. Write each of these percents as a decimal.

74 In the United States, 33% of the population is between 21 and 44 years of age. Write 33% as a decimal.

75 Although nutritionally breakfast is the most important meal of the day, only $\frac{1}{5}$ of the people in the United States consistently eat breakfast. What percent of the population is this?

76 In Belgium, 96% of all children between 3 and 6 years of age go to school. In Sweden, the same figure is only 25%. In the United States, the figure is 60%. Write each of these percents as a fraction in lowest terms.

77 An object weighed on the moon will be only $\frac{1}{6}$ as heavy as it is on the earth. Write this number as a percent.

78 If $\frac{1}{8}$ of the students attending college enroll in a math class, what percent of the students enroll?

Calculator Problems

Use a calculator to write each fraction as a decimal, and then change the decimal to a percent. Round all answers to the nearest tenth of a percent.

79 $\frac{29}{37}$

80 $\frac{18}{83}$

81 $\frac{6}{51}$

82 $\frac{8}{95}$

83 $\frac{236}{327}$

84 $\frac{568}{732}$

85 During World War II, $\frac{1}{12}$ of the Soviet armed forces were women, whereas today only $\frac{1}{450}$ are women. Change both fractions to percents (to the nearest tenth of a percent).

86 The ratio of the number of teachers to the number of students in secondary schools in Japan is 1 to 17. In the United States, the ratio is 1 to 19. Write each of these ratios as a fraction and then as a percent.

Review Problems

These review problems cover some material that will be needed in Section 7.2.

Multiply.

87 0.25(300)

88 0.06(2.95)

89 0.53(16)

90 0.0725(3,000)

Divide. Round the answers to the nearest thousandth, if necessary.

91 $\frac{25}{0.05}$

92 $\frac{7}{0.14}$

93 $\frac{45}{0.25}$

94 $\frac{62}{0.06}$

Solve for n.

95 $3 \cdot n = 12$

96 $16 = 9 \cdot n$

One Step Further: Extending the Concepts

97 Give a written explanation of the process you would use to change the fraction $\frac{2}{5}$ to a percent.

98 Give a written explanation of the process you would use to change 40% to a fraction in lowest terms.

99 75% is how many times as large as 0.75%?

100 60% is how many times as large as 0.30%?

101 Of the people traveling during the 4th of July weekend in 1991, 31% stayed with friends or relatives, while over $\frac{1}{3}$ stayed in commercial lodgings. Is it more likely that a person traveling during that weekend stayed with friends or stayed in commercial lodgings?

102 According to the newspaper *USA Today*, in 1991 52% of the jobs in the Army were open to women, whereas 16 of every 27 jobs in the Navy were open to women. Which of the two branches of the military had the higher percentage of jobs open to women?

7.2 Basic Percent Problems

Introduction . . . The American Dietetics Association (ADA) recommends eating foods in which the number of calories from fat is less than 30% of the total number of calories. Foods that satisfy this requirement are considered healthy foods. Is the nutrition label shown below from a food that the ADA would consider healthy? This is the type of question we will be able to answer after we have worked through the examples in this section.

Nutrition Facts
Serving Size 1/2 cup (65g)
Servings 8

Amount/Serving	
Calories 150	Calories from Fat 90
	% Daily Value*
Total Fat 10g	16%
Saturated Fat 6g	32%
Cholesterol 35mg	12%
Sodium 30mg	1%
Total Carbohydrate 14g	5%
Dietary Fiber 0g	0%
Sugars 11g	
Protein 2g	
Vitamin A 6% •	Vitamin C 0%
Calcium 6% •	Iron 0%

*Percent Daily Values are based on a 2,000 calorie diet.

Figure 1 Nutrition label from vanilla ice cream

This section is concerned with three kinds of word problems that are associated with percents. Here is an example of each type:

Type A: What number is 15% of 63?

Type B: What percent of 42 is 21?

Type C: 25 is 40% of what number?

The method we use to solve all three types of problems involves translating the sentences into equations, and then solving the equations. The following translations are used to write the sentences as equations:

English	Mathematics
is	$=$
of	$\cdot$ (multiply)
a number	n
what number	n
what percent	n

The word *is* always translates to an = sign, the word *of* almost always means multiply, and the number we are looking for can be represented with a letter, such as n or x.

Practice Problems

1 What number is 25% of 74?

EXAMPLE 1 What number is 15% of 63?

Solution We translate the sentence into an equation as follows:

What number is 15% *of* 63?

$$n = 0.15 \cdot 63$$

To do arithmetic with percents, we have to change to decimals.* That is why 15% is rewritten as 0.15. Solving the equation, we have

$$n = 0.15 \cdot 63$$
$$n = 9.45$$

15% of 63 is 9.45

2 What percent of 84 is 21?

EXAMPLE 2 What percent of 42 is 21?

Solution We translate the sentence as follows:

What percent of 42 *is* 21?

$$n \cdot 42 = 21$$

We solve for n by dividing both sides by 42.

$$\frac{n \cdot \cancel{42}}{\cancel{42}} = \frac{21}{42}$$
$$n = \frac{21}{42}$$
$$n = 0.50$$

Since the original problem asked for a percent, we change 0.50 to a percent:

$$n = 50\%$$

21 is 50% of 42

3 35 is 40% of what number?

EXAMPLE 3 25 is 40% of what number?

Solution Following the procedure from the first two examples, we have

25 *is* 40% *of what number?*

$$25 = 0.40 \cdot n$$

Again, we changed 40% to 0.40 so we can do the arithmetic involved in the prob-

*It is sometimes convenient to change percents to fractions instead of decimals. You may want to try some of the examples in this section yourself, using fractions for percents where we have used decimals.

Answers
1 18.5 **2** 25% **3** 87.5

lem. Dividing both sides of the equation by 0.40, we have

$$\frac{25}{0.40} = \frac{\cancel{0.40} \cdot n}{\cancel{0.40}}$$

$$\frac{25}{0.40} = n$$

$$62.5 = n$$

25 is 40% of 62.5

As you can see, all three types of percent problems are solved in a similar manner. We write *is* as =, *of* as ·, and *what number* as *n*. The resulting equation is then solved to obtain the answer to the original question.

EXAMPLE 4 What number is 43.5% of 25?

$$n = 0.435 \cdot 25$$

$n = 10.9$ Rounded to the nearest tenth

10.9 is 43.5% of 25

EXAMPLE 5 What percent of 78 is 31.9?

$$n \cdot 78 = 31.9$$

$$\frac{n \cdot \cancel{78}}{\cancel{78}} = \frac{31.9}{78}$$

$$n = \frac{31.9}{78}$$

$n = 0.409$ Rounded to the nearest thousandth

$$n = 40.9\%$$

31.9 is 40.9% of 78

EXAMPLE 6 34 is 29% of what number?

$$34 = 0.29 \cdot n$$

$$\frac{34}{0.29} = \frac{\cancel{0.29} \cdot n}{\cancel{0.29}}$$

$$\frac{34}{0.29} = n$$

$117.2 = n$ Rounded to the nearest tenth

34 is 29% of 117.2

4 What number is 63.5% of 45? (Round to the nearest tenth.)

5 What percent of 85 is 11.9?

6 62 is 39% of what number? (Round to the nearest tenth.)

Answers
4 28.6 **5** 14% **6** 159.0

7 The nutrition label below is from a package of vanilla frozen yogurt. What percent of the total number of calories are fat calories? Round your answer to the nearest tenth of a percent.

Nutrition Facts
Serving Size 1/2 cup (98g)
Servings 4

Amount/Serving	
Calories 160	Calories from Fat 25
	% Daily Value*
Total Fat 2.5g	4%
Saturated Fat 1.5g	7%
Cholesterol 45mg	15%
Sodium 55mg	2%
Total Carbohydrate 26g	9%
Dietary Fiber 0g	0%
Sugars 19g	
Protein 8g	
Vitamin A 0% •	Vitamin C 0%
Calcium 25% •	Iron 0%

*Percent Daily Values are based on a 2,000 calorie diet.

EXAMPLE 7 As we mentioned in the introduction to this section, the American Dietetics Association recommends eating foods in which the number of calories from fat is less than 30% of the total number of calories. According to the nutrition label below, what percent of the total number of calories are fat calories?

Nutrition Facts
Serving Size 1/2 cup (65g)
Servings 8

Amount/Serving	
Calories 150	Calories from Fat 90
	% Daily Value*
Total Fat 10g	16%
Saturated Fat 6g	32%
Cholesterol 35mg	12%
Sodium 30mg	1%
Total Carbohydrate 14g	5%
Dietary Fiber 0g	0%
Sugars 11g	
Protein 2g	
Vitamin A 6% •	Vitamin C 0%
Calcium 6% •	Iron 0%

*Percent Daily Values are based on a 2,000 calorie diet.

Figure 2 Nutrition label from vanilla ice cream

Solution To solve this problem, we must write the question in the form of one of the three basic percent problems shown in Examples 1–6. Since there are 90 calories from fat and a total of 150 calories, we can write the question this way: 90 is what percent of 150?

Now that we have written the question in the form of one of the basic percent problems, we simply translate it into an equation. Then we solve the equation.

90 *is what percent of* 150?

$$90 = n \cdot 150$$

$$\frac{90}{150} = n$$

$$n = 0.60 = 60\%$$

The number of calories from fat in this package of ice cream is 60% of the total number of calories. Thus the ADA would not consider this to be a healthy food.

Answer

7 15.6% of the calories are from fat. (So far as fat content is concerned, the frozen yogurt is a healthier choice than the ice cream.)

PROBLEM SET 7.2

Solve each of the following problems.

1 What number is 25% of 32?

2 What number is 10% of 80?

3 What number is 20% of 120?

4 What number is 15% of 75?

5 What number is 54% of 38?

6 What number is 72% of 200?

7 What number is 11% of 67?

8 What number is 2% of 49?

9 What percent of 24 is 12?

10 What percent of 80 is 20?

11 What percent of 50 is 5?

12 What percent of 20 is 4?

13 What percent of 36 is 9?

14 What percent of 70 is 14?

15 What percent of 8 is 6?

16 What percent of 16 is 9?

17 32 is 50% of what number?

18 16 is 20% of what number?

19 10 is 20% of what number?

20 11 is 25% of what number?

21 37 is 4% of what number?

22 90 is 80% of what number?

23 8 is 2% of what number?

24 6 is 3% of what number?

The following problems can be solved by the same method you used in Problems 1–24.

25 What is 6.4% of 87?

26 What is 10% of 102?

27 25% of what number is 30?

28 10% of what number is 22?

29 28% of 49 is what number?

30 97% of 28 is what number?

31 27 is 120% of what number?

32 24 is 150% of what number?

33 65 is what percent of 130?

34 26 is what percent of 78?

35 What is 0.4% of 235,671?

36 What is 0.8% of 721,423?

37 4.89% of 2,000 is what number?

38 3.75% of 4,000 is what number?

For each nutrition label in Problems 39–42, find what percent of the total number of calories comes from fat calories. Then indicate whether the label is from a food considered healthy by the American Dietetics Association. Round to the nearest tenth of a percent if necessary.

39 Spaghetti noodles

Nutrition Facts
Serving Size 2 oz. (56g/ 1/8 of pkg) dry
Servings Per Container: 8

Amount Per Serving	
Calories 210	Calories from Fat 10
	% Daily Value*
Total Fat 1g	2%
Saturated Fat 0g	0%
Polyunsaturated Fat 0.5g	
Monounsaturated Fat 0g	
Cholesterol 0mg	0%
Sodium 0mg**	0%
Total Carbohydrate 42g	14%
Dietary Fiber 2g	7%
Sugars 3g	
Protein 7g	
Vitamin A 0% •	Vitamin C 0%
Calcium 0% •	Iron 10%
Thiamin 30% •	Riboflavin 10%
Niacin 15%	

*Percent Daily Values are based on a 2,000 calorie diet.

40 Canned Italian tomatoes

Nutrition Facts
Serving Size 1/2 cup (121g)
Servings Per Container: about 3 1/2

Amount Per Serving	
Calories 25	Calories from Fat 0
	% Daily Value*
Total Fat 0g	0%
Saturated Fat 0g	0%
Cholesterol 0mg	0%
Sodium 300mg	12%
Potassium 145mg	4%
Total Carbohydrate 4g	2%
Dietary Fiber 1g	4%
Sugars 4g	
Protein 1g	
Vitamin A 20% •	Vitamin C 15%
Calcium 4% •	Iron 15%

*Percent Daily Values are based on a 2,000 calorie diet.

41 Shredded romano cheese

Nutrition Facts
Serving Size 2 tsp (5g)
Servings Per Container: 34

Amount Per Serving	
Calories 20	Calories from Fat 10
	% Daily Value*
Total Fat 1.5g	2%
Saturated Fat 1g	5%
Cholesterol 5mg	2%
Sodium 70mg	3%
Total Carbohydrate 0g	0%
Fiber 0g	0%
Sugars 0g	
Protein 2g	
Vitamin A 0% •	Vitamin C 0%
Calcium 4% •	Iron 0%

*Percent Daily Values are based on a 2,000 calorie diet.

42 Tortilla chips

Nutrition Facts
Serving Size 1 oz. (28g/About 12 chips)
Servings Per Container: about 2

Amount Per Serving	
Calories 140	Calories from Fat 60
	% Daily Value*
Total Fat 7g	1%
Saturated Fat 1g	6%
Cholesterol 0mg	0%
Sodium 170mg	7%
Total Carbohydrate 18g	6%
Dietary Fiber 1g	4%
Sugars less than 1g	
Protein 2g	
Vitamin A 0% •	Vitamin C 0%
Calcium 4% •	Iron 2%

*Percent Daily Values are based on a 2,000 calorie diet.

Calculator Problems

Set up each of the following problems using the method developed in this section. Then do all the calculations on a calculator.

43 What number is 62.5% of 398?

44 What number is 4.25% of 907?

45 What percent of 789 is 204? (Round to the nearest tenth of a percent.)

46 What percent of 123 is 119? (Round to the nearest tenth of a percent.)

47 19 is 38.4% of what number? (Round to the nearest tenth.)

48 249 is 16.9% of what number? (Round to the nearest tenth.)

Review Problems

The problems below are review from Section 3.2.

Factor each of the following into a product of prime factors:

49 45

50 12

51 105

52 75

53 36

54 72

55 210

56 150

One Step Further: Writing Mathematics

57 Write a basic percent problem, the solution to which can be found by solving the equation $n = 0.25(350)$.

58 Write a basic percent problem, the solution to which can be found by solving the equation $n = 0.35(250)$.

59 Write a basic percent problem, the solution to which can be found by solving the equation $n \cdot 24 = 16$.

60 Write a basic percent problem, the solution to which can be found by solving the equation $n \cdot 16 = 24$.

61 Write a basic percent problem, the solution to which can be found by solving the equation $46 = 0.75 \cdot n$.

62 Write a basic percent problem, the solution to which can be found by solving the equation $75 = 0.46 \cdot n$.

7.3 General Applications of Percent

Introduction . . . As you know from watching television and reading the newspaper, we encounter percent in many situations in everyday life. A 1995 newspaper article discussing the effects of a cholesterol-lowering drug stated that the drug in question "lowered levels of LDL cholesterol by an average of 35%." As we progress through this chapter, we will become more and more familiar with percent, and as a result, we will be better equipped to understand statements like the one above concerning cholesterol.

In this section we continue our study of percent by doing more of the translations that were introduced at the end of Section 7.2. The better you are at working the problems in Section 7.2, the easier it will be for you to get started on the problems in this section.

EXAMPLE 1 On a 120-question test a student answered 96 correctly. What percent of the problems did the student work correctly?

Solution We have 96 correct answers out of a possible 120. The problem can be restated as

96 *is what percent of* 120?

$96 = n \cdot 120$

$\frac{96}{120} = \frac{n \cdot \cancel{120}}{\cancel{120}}$ Divide both sides by 120

$n = \frac{96}{120}$ Switch the left and right sides of the equation

$n = 0.80$ Divide 96 by 120

$= 80\%$ Rewrite as a percent

When we write a test score as a percent, we are comparing the original score to an equivalent score on a 100-question test. That is, 96 correct out of 120 is the same as 80 correct out of 100.

EXAMPLE 2 How much HCl (hydrochloric acid) is in a 60-milliliter bottle that is marked 80% HCl?

Solution If the bottle is marked 80% HCl, that means 80% of the solution is HCl and the rest is water. Since the bottle contains 60 milliliters, we can restate the question as:

What is 80% *of* 60?

$n = 0.80 \cdot 60$

$n = 48$

There are 48 milliliters of HCl in 60 milliliters of 80% HCl solution.

Practice Problems

1 On a 150-question test a student answered 114 correctly. What percent of the problems did the student work correctly?

2 How much HCl is in a 40-milliliter bottle that is marked 75% HCl?

Answers

1 76% 2 30 milliliters

3 If 42% of the students in a certain college are female and there are 3,360 female students, what is the total number of students in the college?

EXAMPLE 3 If 48% of the students in a certain college are female and there are 2,400 female students, what is the total number of students in the college?

Solution We restate the problem as:

2,400 *is* 48% *of what number?*

$$2{,}400 = 0.48 \cdot n$$

$$\frac{2{,}400}{0.48} = \frac{0.48 \cdot n}{0.48} \quad \text{Divide both sides by 0.48}$$

$$n = \frac{2{,}400}{0.48} \quad \text{Switch the left and right sides of the equation}$$

$$n = 5{,}000$$

There are 5,000 students.

4 Suppose in Example 4 that 35% of the students receive a grade of A. How many of the 300 students is that?

EXAMPLE 4 If 25% of the students in elementary algebra courses receive a grade of A, and there are 300 students enrolled in elementary algebra this year, how many students will receive A's?

Solution After reading the question a few times, we find that it is the same as this question:

What number is 25% *of* 300?

$$n = 0.25 \cdot 300$$

$$n = 75$$

Thus, 75 students will receive A's in elementary algebra.

Almost all application problems involving percents can be restated as one of the three basic percent problems we listed in Section 7.2. It takes some practice before the restating of application problems becomes automatic. You may have to review Section 7.2 and Examples 1–4 above several times before you can translate word problems into mathematical expressions yourself.

Answers
3 8,000 **4** 105

PROBLEM SET 7.3

Solve each of the following problems by first restating it as one of the three basic percent problems of Section 7.2. In each case, be sure to show the equation.

1 On a 120-question test a student answered 84 correctly. What percent of the problems did the student work correctly?

2 An engineering student answered 81 questions correctly on a 90-question trigonometry test. What percent of the questions did she answer correctly? What percent were answered incorrectly?

3 A basketball player made 63 out of 75 free throws. What percent is this?

4 A family spends \$450 every month on food. If the family's income each month is \$1,800, what percent of the family's income is spent on food?

5 How much HCl (hydrochloric acid) is in a 60-milliliter bottle that is marked 75% HCl?

6 How much acetic acid is in a 5-liter container of acetic acid and water that is marked 80% acetic acid? How much is water?

7 A farmer owns 28 acres of land. Of the 28 acres, only 65% can be farmed. How many acres are available for farming? How many are not available for farming?

8 Of the 420 students enrolled in a basic math class, only 30% are first-year students. How many are first-year students? How many are not?

9 If 48% of the students in a certain college are female and there are 1,440 female students, what is the total number of students in the college?

10 A solution of alcohol and water is 80% alcohol. The solution is found to contain 32 milliliters of alcohol. How many milliliters total (both alcohol and water) were in the original solution?

11 Suppose 60% of the graduating class in a certain high school goes on to college. If 240 students from this graduating class are going on to college, how many students are there in the graduating class?

12 In a shipment of airplane parts, 3% are known to be defective. If 15 parts are found to be defective, how many parts are in the shipment?

13 There are 3,200 students at our school. If 52% of them are women, how many women students are there at our school?

14 In a certain school, 75% of the students in first-year chemistry have had algebra. If there are 300 students in first-year chemistry, how many of them have had algebra?

15 In a city of 32,000 people, there are 10,000 people under 25 years of age. What percent of the population is under 25 years of age?

16 If 45 people enrolled in a psychology course but only 35 completed it, what percent of the students completed the course? (Round to the nearest tenth of a percent.)

Calculator Problems

The following problems are similar to Problems 1–16. They should be set up in the same way. Then the actual calculations should be done on a calculator.

17 Of 7,892 people attending an outdoor concert in Los Angeles, 3,972 are over 18 years of age. What percent is this? (Round to the nearest whole-number percent.)

18 A car manufacturer estimates that 25% of the new cars sold in one city have defective engine mounts. If 2,136 new cars are sold in that city, how many will have defective engine mounts?

19 Suppose a solution of alcohol and water contains 31.8 milliliters of alcohol. If the solution is 75.6% alcohol, how many milliliters of solution are there? (Round to the nearest whole number.)

20 A married couple makes $43,698 a year. If 23.4% of their income is used for house payments, how much did they spend on house payments for the year?

Review Problems

The problems below are review from Sections 3.3 and 3.7.

Multiply.

21 $\frac{1}{2} \cdot \frac{2}{5}$

22 $\frac{3}{4} \cdot \frac{1}{3}$

23 $\frac{3}{4} \cdot \frac{5}{9}$

24 $\frac{5}{6} \cdot \frac{12}{13}$

25 $2 \cdot \frac{3}{8}$

26 $3 \cdot \frac{5}{12}$

27 $1\frac{1}{4} \cdot \frac{8}{15}$

28 $2\frac{1}{3} \cdot \frac{9}{10}$

One Step Further: Batting Averages

Batting averages in baseball are given as decimal numbers, rounded to the nearest thousandth. For example, on July 1, 1991, Cal Ripken had the highest batting average in the American League. At that time, he had 103 hits in 290 times at bat. His batting average was .355, which is found by dividing the number of hits by the number of times he was at bat, and then rounding to the nearest thousandth.

$$\text{Batting average} = \frac{\text{number of hits}}{\text{number of times at bat}} = \frac{103}{290} \approx .355$$

Since we can write any decimal number as a percent, we can convert batting averages to percents and use our knowledge of percent to solve problems. Looking at Cal Ripken's batting average as a percent, we can say that he will get a hit 35.5% of the times he is at bat.

Each of the following problems can be solved by converting batting averages to percents and translating the problem into one of our three basic percent problems.

29 On July 1, 1991, Tony Gwynn had the highest batting average in the National League with 110 hits in 303 times at bat. What percent of the time Tony Gwynn is at bat can we expect him to get a hit?

30 Third behind Tony Gwynn in batting averages was Terry Pendleton with 72 hits in 222 times at bat. What percent of the time can we expect Pendleton to get a hit?

31 On July 1, 1991, Wally Joyner was batting .326. If he had been at bat 279 times, how many hits did he have? (Remember, his batting average has been rounded to the nearest thousandth.)

32 On July 1, 1991, Ozzie Smith was batting .320. If he had been at bat 244 times, how many hits did he have? (Remember, his batting average has been rounded to the nearest thousandth.)

33 How many hits must Cal Ripken have in his next 50 times at bat to maintain a batting average of at least .355?

34 How many hits must Tony Gwynn have in his next 50 times at bat to maintain a batting average of at least .363?

7.4 Sales Tax and Commission

Introduction . . . In February 1995, *USA Today* reported that Delta Air Lines was limiting the amount of commission it paid travel agents for selling airline tickets. According to the article, in the past Delta paid 10% commission on all airline tickets. In February Delta put a cap on its commission: \$25 on a one-way ticket and \$50 on a round-trip ticket. The mathematics necessary to understand this newspaper article is what we will study in this section.

To solve the problems in this section, we will first restate them in terms of the problems we have already learned how to solve. Then we will solve the restated problems just like we solved the problems in Sections 7.2 and 7.3.

Sales Tax

EXAMPLE 1 Suppose the sales tax rate in Mississippi is 6% of the purchase price. If the price of a used refrigerator is \$550, how much sales tax must be paid?

Solution Since the sales tax is 6% of the purchase price, and the purchase price is \$550, the problem can be restated as:

What is 6% of \$550?

We solve this problem, as we did in Section 7.2, by translating it into an equation:

What is 6% *of* \$550?

$$n = 0.06 \cdot 550$$

$$n = 33$$

The sales tax is \$33. The total price of the refrigerator would be

Purchase price		*Sales tax*		*Total price*
↓		↓		↓
\$550	+	\$33	=	\$583

> Total price = Purchase price + Sales tax

EXAMPLE 2 Suppose the sales tax rate is 4%. If the sales tax on a 10-speed bicycle is \$5.44, what is the purchase price, and what is the total price of the bicycle?

Solution We know that 4% of the purchase price is \$5.44. We find the purchase price first by restating the problem as:

\$5.44 *is* 4% *of what number?*

$$5.44 = 0.04 \cdot n$$

We solve the equation by dividing both sides by 0.04:

$$\frac{5.44}{0.04} = \frac{0.04 \cdot n}{0.04} \qquad \text{Divide both sides by 0.04}$$

$$n = \frac{5.44}{0.04} \qquad \text{Switch the left and right sides of the equation}$$

$$n = 136 \qquad \text{Divide}$$

Practice Problems

1 What is the sales tax on a new washing machine if the machine is purchased for \$625 and the sales tax rate is 6%?

2 Suppose the sales tax rate is 3%. If the sales tax on a 10-speed bicycle is \$4.35, what is the purchase price, and what is the total price of the bicycle?

Note

In Example 1, the *sales tax rate* is 6%, and the *sales tax* is \$33. In most everyday communications, people say "The sales tax is 6%," which is incorrect. The 6% is the tax *rate*, and the \$33 is the actual tax.

Answers

1 \$37.50 2 \$145; \$149.35

The purchase price is \$136. The total price is the sum of the purchase price and the sales tax.

Purchase price	=	\$136.00
Sales tax	=	5.44
Total price	=	\$141.44

3 Suppose the purchase price of two speakers is \$197 and the sales tax is \$11.82. What is the sales tax rate?

EXAMPLE 3 Suppose the purchase price of a stereo system is \$396 and the sales tax is \$19.80. What is the sales tax rate?

Solution We restate the problem as:

\$19.80 *is what percent of* \$396?

$$19.80 = n \cdot 396$$

To solve this equation, we divide both sides by 396:

$$\frac{19.80}{396} = \frac{n \cdot \cancel{396}}{\cancel{396}} \quad \text{Divide both sides by 396}$$

$$n = \frac{19.80}{396} \quad \text{Switch the left and right sides of the equation}$$

$$n = 0.05 \quad \text{Divide}$$

$$n = 5\% \quad 0.05 = 5\%$$

The sales tax rate is 5%.

Commission

Many salespeople work on a *commission* basis. That is, their earnings are a percentage of the amount they sell. The *commission rate* is a percent, and the actual commission they receive is a dollar amount.

4 A real estate agent gets 3% of the price of each house she sells. If she sells a house for \$115,000, how much money does she earn?

EXAMPLE 4 A real estate agent gets 6% of the price of each house she sells. If she sells a house for \$89,500, how much money does she earn?

Solution The commission is 6% of the price of the house, which is \$89,500. We restate the problem as:

What is 6% *of* \$89,500?

$$n = 0.06 \cdot 89{,}500$$

$$n = 5{,}370$$

The commission is \$5,370. (Actually, most real estate salespeople work for a real estate broker. In most cases, the broker gets half of the commission. If that were true in this case, the agent would get \$2,685 and the broker would get \$2,685.)

Answers
3 6% **4** \$3,450

EXAMPLE 5 Suppose a car salesperson's commission rate is 12%. If the commission on one of the cars is $1,836, what is the purchase price of the car?

Solution 12% of the sales price is $1,836. The problem can be restated as:

12% *of what number is* $1,836?

$$0.12 \cdot n = 1{,}836$$

$$\frac{0.12 \cdot n}{0.12} = \frac{1{,}836}{0.12} \quad \text{Divide both sides by 0.12}$$

$$n = 15{,}300$$

The car sells for $15,300.

EXAMPLE 6 If the commission on a $600 dining room set is $90, what is the commission rate?

Solution The commission rate is a percentage of the selling price. What we want to know is:

$90 *is what percent of* $600?

$$90 = n \cdot 600$$

$$\frac{90}{600} = \frac{n \cdot 600}{600} \quad \text{Divide both sides by 600}$$

$$n = \frac{90}{600} \quad \text{Switch the left and right sides of the equation}$$

$$n = 0.15 \quad \text{Divide}$$

$$n = 15\% \quad \text{Change to a percent}$$

The commission rate is 15%.

5 An appliance salesperson's commission rate is 10%. If the commission on one of the ovens is $115, what is the purchase price of the oven?

6 If the commission on a $750 sofa is $105, what is the commission rate?

Answers
5 $1,150 **6** 14%

PROBLEM SET 7.4

These problems should be solved by the method shown in this section. In each case show the equation needed to solve the problem. Write neatly and show your work.

1 Suppose the sales tax rate in Mississippi is 6% of the purchase price. If a new food processor sells for $750, how much is the sales tax?

2 If the sales tax rate is 5% of the purchase price, how much sales tax is paid on a television that sells for $980?

3 Suppose the sales tax rate in Michigan is 5%. How much is the sales tax on a $15 concert ticket? What is the total price?

4 Suppose the sales tax rate in Hawaii is 4%. How much sales tax is charged on a new car if the purchase price is $6,400? What is the total price?

5 The sales tax rate is 4%. If the sales tax on a 10-speed bicycle is $6, what is the purchase price? What is the total price?

6 The sales tax on a new microwave oven is $30. If the sales tax rate is 5%, what is the purchase price? What is the total price?

7 Suppose the purchase price of a dining room set is $450. If the sales tax is $22.50, what is the sales tax rate?

8 If the purchase price of a case of California wine is $24 and the sales tax is $1.50, what is the sales tax rate?

9 A real estate agent has a commission rate of 6%. If a piece of property sells for $24,000, what is his commission?

10 A tire salesperson has a 12% commission rate. If he sells a set of radial tires for $400, what is his commission?

11 Suppose a salesperson gets a commission rate of 12% on the lawnmowers she sells. If the commission on one of the mowers is $24, what is the purchase price of the lawnmower?

12 If an appliance salesperson gets 9% on all the appliances she sells, what is the price of a refrigerator if her commission is $67.50?

13 If the commission on an $800 washer is $112, what is the commission rate?

14 A salesperson makes a commission of $3,600 on a $90,000 house he sells. What is his commission rate?

Calculator Problems

The following problems are similar to Problems 1–14. Set them up in the same way, but use a calculator for the calculations.

15 The sales tax rate on a certain item is 5.5%. If the purchase price is \$216.95, how much is the sales tax? (Round to the nearest cent.)

16 If the sales tax rate is 4.75% and the sales tax is \$18.95, what is the purchase price? What is the total price? (Both answers should be rounded to the nearest cent.)

17 The purchase price for a new suit is \$129.50. If the sales tax is \$5.83, what is the tax rate? (Round to the nearest tenth of a percent.)

18 If the commission rate for a mobile home salesperson is 11%, what is the commission on the sale of a \$15,794 mobile home?

19 Suppose the commission rate on the sale of used cars is 13%. If the commission on one of the cars is \$519.35, what did the car sell for?

20 If the commission on the sale of \$79.40 worth of clothes is \$14.29, what is the commission rate? (Round to the nearest percent.)

Review Problems

The problems below review some basic concepts of division with fractions and mixed numbers from Sections 3.4 and 3.7.

Divide.

21 $\frac{1}{3} \div \frac{2}{3}$

22 $\frac{2}{3} \div \frac{1}{3}$

23 $2 \div \frac{3}{4}$

24 $3 \div \frac{1}{2}$

25 $\frac{3}{8} \div \frac{1}{4}$

26 $\frac{5}{9} \div \frac{2}{3}$

27 $2\frac{1}{4} \div \frac{1}{2}$

28 $1\frac{1}{4} \div 2\frac{1}{2}$

One Step Further: Luxury Taxes

In 1990, Congress passed a law, which took effect on January 1, 1991, requiring an additional tax of 10% on a portion of the purchase price of certain luxury items. For expensive cars, it was paid on the part of the purchase price that exceeded $30,000. For example, if you purchased a Jaguar XJ-S for $53,000, you would pay a luxury tax of 10% of $23,000, because the purchase price, $53,000, is $23,000 above $30,000.

29 If you purchased a Jaguar XJ-S for $53,000 on February 1, 1991, in California, where the sales tax rate was 6%, how much would you pay in luxury tax and how much would you pay in sales tax?

30 If you purchased a Mercedes 300E for $43,500 on January 20, 1991, in California, where the sales tax rate was 6%, how much more would you pay in sales tax than luxury tax?

31 How much would you have saved if you had purchased the Jaguar mentioned in Problem 29 on December 31, 1990?

32 How much would you have saved if you bought a car with a purchase price of $45,000 on December 31, 1990, instead of January 1, 1991?

33 Suppose you are buying a car in 1991. How much will you save on a car with a sticker price of $31,500, if you can persuade the car dealer to reduce the price to $29,900?

34 Suppose you are buying a car in 1991. One of the cars you are interested in has a sticker price of $35,500, while another has a sticker price of $28,500. If you expect to pay full price for either car, how much will you save if you buy the less expensive car?

7.5 Percent Increase or Decrease, and Discount

Increases and decreases are often expressed in terms of percents. For example, salary raises are usually given as percent increases. If you get a 5% salary increase, your new salary is 5% more than your old salary.

EXAMPLE 1 If a person earns $22,000 a year and gets a 5% increase in salary, what is the new salary?

Solution We can find the dollar amount of the salary increase by finding 5% of $22,000:

$$0.05 \times 22{,}000 = 1{,}100$$

The increase in salary is $1,100. The new salary is the old salary plus the raise:

$22,000	Old salary
+ 1,100	Raise (5% of $22,000)
$23,100	New salary

EXAMPLE 2 In 1976, there were approximately 553,000 arrests in the United States for drug-related offenses. In 1977, there was a 3% decrease in the number of these arrests. Approximately how many people were arrested in 1977 for drug-related offenses? Round the answer to the nearest thousand.

Solution The decrease in the number of arrests is 3% of 553,000, or

$$0.03 \times 553{,}000 = 16{,}590$$

There were 553,000 arrests in 1976. Since the number of arrests in 1977 was 16,590 fewer than in 1976, we have

553,000	Number of arrests in 1976
− 16,590	Decreased by 3% of 553,000
536,410	Number of arrests in 1977

To the nearest thousand, there were approximately 536,000 arrests in 1977 in the United States for drug-related offenses.

EXAMPLE 3 Shoes that usually sell for $25 are on sale for $21. What is the percent decrease in price?

Solution We must first find the decrease in price. Subtracting the sale price from the original price, we have

$$\$25 - \$21 = \$4$$

The decrease is $4. To find the percent decrease (from the original price), we have

$4 is *what percent of* $25?

$$4 = n \cdot 25$$

$$\frac{4}{25} = \frac{n \cdot \cancel{25}}{\cancel{25}} \quad \text{Divide both sides by 25}$$

$$n = \frac{4}{25} \quad \text{Switch the left and right sides of the equation}$$

Practice Problems

1 A person earning $18,000 a year gets a 7% increase in salary. What is the new salary?

2 In 1976, there were approximately 659,000 arrests for disorderly conduct in the United States. In 1977, there was a 9% decrease in the number of these arrests. Approximately how many people were arrested in 1977 for disorderly conduct?

3 Shoes that usually sell for $35 are on sale for $28. What is the percent decrease in price?

Answers
1 $19,260 2 599,690 3 20%

$n = 0.16$ Divide

$n = 16\%$ Change to a percent

The shoes that sold for $25 have been reduced by 16% to $21. In a problem like this, $25 is the *original* (or *marked*) *price*, $21 is the *sale price*, $4 is the *discount*, and 16% is the *rate of discount.*

4 During a sale, a microwave oven that usually sells for $550 is marked "15% off." What is the discount? What is the sale price?

EXAMPLE 4 During a clearance sale, a suit that usually sells for $300 is marked "25% off." What is the discount? What is the sale price?

Solution To find the discount, we restate the problem as:

What is 25% *of* 300?

↓ ↓ ↓ ↓ ↓

$n = 0.25 \cdot 300$

$n = 75$

The discount is $75. The sale price is the original price less the discount:

$300	Original price
− 75	Less the discount (25% of $300)
$225	Sale price

5 A woman buys a new coat on sale. The coat usually sells for $45, but it is on sale at 15% off. If the sales tax rate is 5%, how much is the total bill for the coat?

EXAMPLE 5 A man buys a washing machine that is on sale. The machine usually sells for $450, but it is on sale at 12% off. If the sales tax rate is 5%, how much is the total bill for the washer?

Solution First we have to find the sale price of the washing machine, and we begin by finding the discount:

What is 12% *of* $450?

↓ ↓ ↓ ↓ ↙

$n = 0.12 \cdot 450$

$n = 54$

The washing machine is marked down $54. The sale price is

$450	Original price
− 54	Discount (12% of $450)
$396	Sale price

Since the sales tax rate is 5%, we find the sales tax as follows:

What is 5% *of* $396?

↓ ↓ ↓ ↓ ↓

$n = 0.05 \cdot 396$

$n = 19.80$

The sales tax is $19.80. The total price the man pays for the washing machine is

$396.00	Sale price
+ 19.80	Sales tax
$415.80	Total price

Answers

4 $82.50; $467.50 **5** $40.16

PROBLEM SET 7.5

Solve each of these problems using the method developed in this section.

1 If a person earns \$23,000 a year and gets a 7% increase in salary, what is the new salary?

2 A computer programmer's yearly income of \$27,000 is increased by 8%. What is the dollar amount of the increase and what is her new salary?

3 The yearly tuition at a college is presently \$3,000. Next year it is expected to increase by 17%. What will the tuition at this school be next year?

4 A supermarket increased the price of cheese that sold for \$1.98 per pound by 3%. What is the new price for a pound of this cheese? (Round to the nearest cent.)

5 In one year a new car decreased in value by 20%. If it sold for \$16,500 when it was new, what was it worth after one year?

6 A certain light beer has 20% fewer calories than the regular beer. If the regular beer has 120 calories per bottle, how many calories are in the same-sized bottle of the light beer?

7 A person earning \$1,500 a month gets a raise of \$120 per month. What is the percent increase in salary?

8 A student reader is making \$4.50 per hour and gets a \$0.70 raise. What is the percent increase? (Round to the nearest tenth of a percent.)

9 Shoes that usually sell for $25 are on sale for $20. What is the percent decrease in price?

10 The enrollment in a certain elementary school was 410 in 1990. In 1991, the enrollment in the same school was 328. Find the percent decrease in enrollment from 1990 to 1991.

11 Hamburger that sold for $0.69 a pound in 1976 now sells for $2.07 a pound. Find the percent increase in price per pound from 1976 to the present.

12 At the beginning of a 2-month crash diet, Jo Ann weighed 140 pounds. At the end of 2 months, she weighed 119 pounds. What is the percent decrease in weight from the beginning of the diet to the end of the diet?

13 During a clearance sale, a three-piece suit that usually sells for $300 is marked "15% off." What is the discount? What is the sale price?

14 On opening day a new music store offers a 12% discount on all electric guitars. If the regular price on a guitar is $550, what is the sale price?

15 A man buys a washing machine that is on sale. The washing machine usually sells for $450, but is on sale at 20% off. If the sales tax rate in his state is 6%, how much is the total bill for the washer?

16 A bedroom set that normally sells for $1,450 is on sale for 10% off. If the sales tax rate is 5%, what is the total price of the bedroom set if it is bought while on sale?

Calculator Problems

Set up the following problems the same way you set up Problems 1–16. Then use a calculator to do the calculations.

17 A teacher making $23,752 per year gets a 6.5% raise. What is the new salary?

18 A homeowner had a $15.90 electric bill in December. In January the bill was $17.81. Find the percent increase in the electric bill from December to January. (Round to the nearest whole number.)

19 House plants that usually sell for $4.95 are marked "25% off" for an annual spring sale. What is the sale price?

20 A used car that normally sells for $3,522 is on sale for 17% off. If the sales tax rate is 5.5%, what is the total price paid for the car if it was bought on sale?

Review Problems

The problems below review some basic concepts of addition of fractions and mixed numbers from Sections 3.5 and 3.8.

Add each of the following and reduce all answers to lowest terms.

21 $\frac{1}{3} + \frac{2}{3}$

22 $\frac{3}{8} + \frac{1}{8}$

23 $\frac{1}{2} + \frac{1}{4}$

24 $\frac{1}{5} + \frac{3}{10}$

25 $\frac{3}{4} + \frac{2}{3}$

26 $\frac{3}{8} + \frac{1}{6}$

27 $2\frac{1}{2} + 3\frac{1}{2}$

28 $3\frac{1}{4} + 2\frac{1}{8}$

One Step Further: Batting Averages

As you know from the One Step Further problems in Problem Set 7.3, on July 1, 1991, Cal Ripken had 103 hits in 290 times at bat for a batting average of .355. On the same date, Tony Gwynn had accumulated 110 hits in 303 times at bat. Suppose that on July 2, Cal Ripken had 2 hits in 5 times at bat, and Tony Gwynn had only 1 hit in 5 times at bat.

29 What was Cal Ripken's batting average for the game on July 2?

30 What was Cal Ripken's overall batting average after the game on July 2?

31 By how much did Cal Ripken's batting average increase from July 1 to July 2?

32 What is the percent increase in Cal Ripken's batting average from July 1 to July 2?

33 What was Tony Gwynn's batting average for the game on July 2?

34 What was Tony Gwynn's overall batting average after the game on July 2?

35 By how much did Tony Gwynn's batting average decrease from July 1 to July 2?

36 What is the percent decrease in Tony Gwynn's batting average from July 1 to July 2?

7.6 Interest

Anyone who has borrowed money from a bank or other lending institution, or who has invested money in a savings account, is aware of *interest*. Interest is the amount of money paid for the use of money. If we put \$500 in a savings account that pays 6% annually, the interest will be 6% of \$500, or 0.06(500) = \$30. The amount we invest (\$500) is called the *principal*, the percent (6%) is the *interest rate*, and the money earned (\$30) is the *interest*.

EXAMPLE 1 A man invests \$2,000 in a savings plan that pays 7% per year. How much money will be in the account at the end of 1 year?

Solution We first find the interest by taking 7% of the principal, \$2,000:

$$\begin{aligned} \text{Interest} &= 0.07(\$2{,}000) \\ &= \$140 \end{aligned}$$

The interest earned in 1 year is \$140. The total amount of money in the account at the end of a year is the original amount plus the \$140 interest:

\$2,000	Original investment (principal)
+ 140	Interest (7% of \$2,000)
\$2,140	Amount after 1 year

The amount in the account after 1 year is \$2,140.

EXAMPLE 2 A farmer borrows \$8,000 from his local bank at 12%. How much does he pay back to the bank at the end of the year to pay off the loan?

Solution The interest he pays on the \$8,000 is

$$\begin{aligned} \text{Interest} &= 0.12(\$8{,}000) \\ &= \$960 \end{aligned}$$

At the end of the year, he must pay back the original amount he borrowed (\$8,000) plus the interest at 12%:

\$8,000	Amount borrowed (principal)
+ 960	Interest at 12%
\$8,960	Total amount to pay back

The total amount that the farmer pays back is \$8,960.

There are many situations in which interest on a loan is figured on other than a yearly basis. Many short-term loans are for only 30 or 60 days. In these cases we can use a formula to calculate the interest that has accumulated. This type of interest is called *simple interest*. The formula is

$$I = P \cdot R \cdot T$$

where

I = Interest

P = Principal

R = Interest rate (this is the percent)

T = Time (in years, 1 year = 360 days)

Practice Problems

1 A man invests \$3,000 in a savings plan that pays 8% per year. How much money will be in the account at the end of 1 year?

2 If a woman borrows \$7,500 from her local bank at 12% interest, how much does she pay back to the bank if she pays off the loan in 1 year? If she pays the loan back in 12 equal monthly payments, how much is each payment?

Answers

1 \$3,240

2 \$8,400; \$700 a month

We could have used this formula to find the interest in Examples 1 and 2. In those two cases, T is 1. When the length of time is in days rather than years, it is common practice to use 360 days for 1 year, and we write T as a fraction. Examples 3 and 4 illustrate this procedure.

3 Another student takes out a loan like the one in Example 3. This loan is for \$700 at 4%. How much interest does this student pay if the loan is paid back in 90 days?

EXAMPLE 3 A student takes out an emergency loan for tuition, books, and supplies. The loan is for \$600 at an interest rate of 4%. How much interest does the student pay if the loan is paid back in 60 days?

Solution The principal P is \$600, the rate R is $4\% = 0.04$, and the time T is $\frac{60}{360}$. Notice that T must be given in years, and 60 days $= \frac{60}{360}$ year. Applying the formula, we have

$$I = P \cdot R \cdot T$$

$$I = 600 \times 0.04 \times \frac{60}{360}$$

$$I = 600 \times 0.04 \times \frac{1}{6} \qquad \frac{60}{360} = \frac{1}{6}$$

$$I = 4 \qquad \text{Multiplication}$$

The interest is \$4.

4 Suppose \$1,200 is deposited in an account that pays 9.5% interest per year. If all the money is withdrawn after 120 days, how much money is withdrawn?

EXAMPLE 4 A woman deposits \$900 in an account that pays 6% annually. If she withdraws all the money in the account after 90 days, how much does she withdraw?

Solution We have $P = \$900$, $R = 0.06$, and $T = 90$ days $= \frac{90}{360}$ year. Using these numbers in the formula, we have

$$I = P \cdot R \cdot T$$

$$I = 900 \times 0.06 \times \frac{90}{360}$$

$$I = 900 \times 0.06 \times \frac{1}{4} \qquad \frac{90}{360} = \frac{1}{4}$$

$$I = 13.5 \qquad \text{Multiplication}$$

The interest earned in 90 days is \$13.50. If the woman withdraws all the money in her account, she will withdraw

\$900.00	Original amount (principal)
+ 13.50	Interest for 90 days
\$913.50	Total amount withdrawn

The woman has withdrawn \$913.50.

A second common kind of interest is *compound interest.* Compound interest includes interest paid on interest. We can use what we know about simple interest to help us solve problems involving compound interest.

Answers
3 \$7 **4** \$1,238

EXAMPLE 5 A homemaker puts $3,000 into a savings account that pays 7% compounded annually. How much money is in the account at the end of 2 years?

Solution Since the account pays 7% annually, the simple interest at the end of 1 year is 7% of $3,000:

$$\text{Interest after 1 year} = 0.07(\$3{,}000) = \$210$$

Since the interest is paid annually, at the end of 1 year the total amount of money in the account is

$3,000	Original amount
+ 210	Interest for 1 year
$3,210	Total in account after 1 year

The interest paid for the second year is 7% of this new total, or

$$\text{Interest paid the second year} = 0.07(\$3{,}210) = \$224.70$$

At the end of 2 years, the total in the account is

$3,210.00	Amount at the beginning of year 2
+ 224.70	Interest paid for year 2
$3,434.70	Account after 2 years

At the end of 2 years, the account totals $3,434.70. The total interest earned during this 2-year period is $210 (first year) + $224.70 (second year) = $434.70.

5 If $5,000 is put into an account that pays 6% annually, how much money is in the account at the end of 2 years?

Note
If the interest earned in Example 5 were calculated using the formula for simple interest, $I = P \cdot R \cdot T$, the amount of money in the account at the end of two years would be $3,420.00.

You may have heard of savings and loan companies that offer interest rates that are compounded quarterly. If the interest rate is, say, 6% and it is compounded quarterly, then after every 90 days $\left(\frac{1}{4}\text{ of a year}\right)$ the interest is added to the account. If it is compounded semiannually, then the interest is added to the account every 6 months. Most accounts have interest rates that are compounded daily, which means the simple interest is computed daily and added to the account.

EXAMPLE 6 If $10,000 is invested in a savings account that pays 6% compounded quarterly, how much is in the account at the end of a year?

Solution The interest for the first quarter $\left(\frac{1}{4}\text{ of a year}\right)$ is calculated using the formula for simple interest:

$$I = P \cdot R \cdot T$$

$$I = \$10{,}000 \times 0.06 \times \frac{1}{4} \quad \text{First quarter}$$

$$I = \$150$$

At the end of the first quarter, this interest is added to the original principal. The new principal is $10,000 + $150 = $10,150. Again we apply the formula to calculate the interest for the second quarter:

$$I = \$10{,}150 \times 0.06 \times \frac{1}{4} \quad \text{Second quarter}$$

$$I = \$152.25$$

6 If $20,000 is invested in an account that pays 8% compounded quarterly, how much is in the account at the end of a year?

Answers
5 $5,618 6 $21,648.64

The principal at the end of the second quarter is \$10,150 + \$152.25 = \$10,302.25. The interest earned during the third quarter is

$$I = \$10{,}302.25 \times 0.06 \times \frac{1}{4} \qquad \text{Third quarter}$$

$$I = \$154.53 \qquad \text{To the nearest cent}$$

The new principal is \$10,302.25 + \$154.53 = \$10,456.78. Interest for the fourth quarter is

$$I = \$10{,}456.78 \times 0.06 \times \frac{1}{4} \qquad \text{Fourth quarter}$$

$$I = \$156.85 \qquad \text{To the nearest cent}$$

The total amount of money in this account at the end of 1 year is

$$\$10{,}456.78 + \$156.85 = \$10{,}613.63$$

For comparison, let's compute the total amount of money that would be in the account in Example 6 at the end of 1 year using simple interest. The interest earned during the year would be

$$I = \$10{,}000 \times 0.06 \times 1$$
$$= \$600$$

The total amount in the account at the end of the year would be

$$\$10{,}000 + \$600 = \$10{,}600$$

which is \$13.63 less than what we calculated using compound interest. Compounding interest means more money to the saver.

PROBLEM SET 7.6

These problems are similar to the examples found in this section. They should be set up and solved in the same way. (Problems 1–12 involve simple interest.)

1 A man invests $2,000 in a savings plan that pays 8% per year. How much money will be in the account at the end of 1 year?

2 How much simple interest is earned on $5,000 if it is invested for 1 year at 5%?

3 A savings account pays 7% per year. How much interest will $9,500 invested in such an account earn in a year?

4 A local bank pays 5.5% annual interest on all savings accounts. If $600 is invested in this account, how much will be in the account at the end of a year?

5 A farmer borrows $8,000 from his local bank at 7%. How much does he pay back to the bank at the end of the year when he pays off the loan?

6 If $400 is borrowed at a rate of 12% for 1 year, how much is the interest?

7 A bank lends one of its customers $2,000 at 8% for 1 year. If the customer pays the loan back in 12 equal monthly payments, how much is each payment?

8 If a loan of $2,000 at 20% for 1 year is to be paid back in 12 equal monthly payments, how much is each payment?

9 A student takes out an emergency loan for tuition, books, and supplies. The loan is for \$600 with an interest rate of 5%. How much interest does the student pay if the loan is paid back in 60 days?

10 If a loan of \$1,200 at 9% is paid off in 90 days, what is the interest?

11 A woman deposits \$800 in a savings account that pays 5%. If she withdraws all the money in the account after 120 days, how much does she withdraw?

12 \$1,800 is deposited in a savings account that pays 6%. If the money is withdrawn at the end of 30 days, how much interest is earned?

The problems that follow involve compound interest.

13 A woman puts \$5,000 into a savings account that pays 6% compounded annually. How much money is in the account at the end of 2 years?

14 A savings account pays 5% compounded annually. If \$10,000 is deposited in the account, how much is in the account at the end of 2 years?

15 If \$8,000 is invested in a savings account that pays 5% compounded quarterly, how much is in the account at the end of a year?

16 Suppose \$1,200 is invested in a savings account that pays 6% compounded semiannually. How much is in the account at the end of $1\frac{1}{2}$ years?

Calculator Problems

The following problems should be set up in the same way in which Problems 1–16 have been set up. Then the calculations should be done on a calculator.

17 A woman invests \$917.26 in a savings account that pays 6.25% annually. How much is in the account at the end of a year?

18 The owner of a clothing store borrows \$6,210 for 1 year at 11.5% interest. If he pays the loan back in 12 equal monthly payments, how much is each payment?

19 Suppose \$617 is invested in an account that pays 6.75% compounded annually. How much is in the account at the end of 2 years?

20 A savings and loan company pays 6% interest compounded quarterly on all its savings accounts. If \$425 is invested in one of those accounts, how much interest is earned in 1 year?

Review Problems

The problems below will allow you to review subtraction of fractions and mixed numbers from Sections 3.5 and 3.8.

Subtract.

21 $\frac{3}{4} - \frac{1}{4}$

22 $\frac{9}{10} - \frac{7}{10}$

23 $\frac{5}{8} - \frac{1}{4}$

24 $\frac{7}{10} - \frac{1}{5}$

25 $\frac{1}{3} - \frac{1}{4}$

26 $\frac{9}{12} - \frac{1}{5}$

27 $3\frac{1}{4} - 2$

28 $5\frac{1}{6} - 3\frac{1}{4}$

One Step Further: Extending the Concepts

The following problems are a random assortment of percent problems. Use any of the methods developed in this chapter to solve them.

29 During the month of June 1991, G. Heilemann Brewing Company announced plans to market a new malt liquor, which had 50% more alcohol per can than regular beer. If regular beer is 3.8% alcohol, what is the percent alcohol of the new malt liquor? (In July 1991, the company canceled its plans for the new malt liquor, after the Federal Bureau of Alcohol, Tobacco and Firearms withdrew its approval for the beer's label because it was to be called *PowerMaster*. Federal law prohibits companies from marketing beer on the basis of strength.)

30 In 1991, the Pentagon began cutting its budget by closing military bases. The newspaper *USA Today* reported that the Pentagon would save $1.7 billion annually by cutting 25% from its budget. If that is the case, what was the annual budget for the Pentagon before they began closing bases?

The 1984 movie *Terminator 1* was produced for $6 million. The sequel, *Terminator 2: Judgment Day*, was released in 1991. Production costs for *Terminator 2* were estimated to be $85 million.

31 *Terminator 2* was how many times more expensive to make than *Terminator 1*?

32 What was the percent increase in production costs from *Terminator 1* to *Terminator 2*?

Holiday travel over the 4th of July weekend in 1991 was up 12.5% from the previous year. The American Petroleum Institute reported that demand for gasoline averaged 320 million gallons per day during the 4th of July weekend, 1991. The price of regular unleaded gasoline averaged $1.19 per gallon during the 4th of July weekend in 1991, which was up $0.06 a gallon from the year before.

33 Estimate the total number of gallons used in the United States on July 4, 1990.

34 Estimate the total amount of money spent on gasoline in the United States on July 4, 1990.

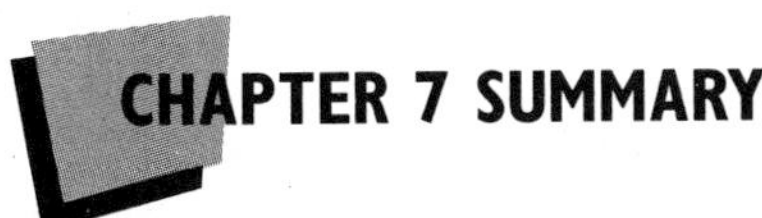

CHAPTER 7 SUMMARY

Examples

The Meaning of Percent [7.1]
Percent means "per hundred." It is a way of comparing numbers to the number 100.

1 42% means 42 per hundred or $\frac{42}{100}$

Changing Percents to Decimals [7.1]
To change a percent to a decimal, drop the percent symbol (%) and move the decimal point two places to the *left*.

2 75% = 0.75

Changing Decimals to Percents [7.1]
To change a decimal to a percent, move the decimal point two places to the *right* and use the % symbol.

3 0.25 = 25%

Changing Percents to Fractions [7.1]
To change a percent to a fraction, drop the % symbol and use a denominator of 100. Reduce the resulting fraction to lowest terms if necessary.

4 $6\% = \frac{6}{100} = \frac{3}{50}$

Changing Fractions to Percents [7.1]
To change a fraction to a percent, either write the fraction as a decimal and then change the decimal to a percent, or write the fraction as an equivalent fraction with denominator 100, drop the 100, and use the % symbol.

5 $\frac{3}{4} = 0.75 = 75\%$

or

$\frac{9}{10} = \frac{90}{100} = 90\%$

Basic Word Problems Involving Percents [7.2]
There are three basic types of word problems:

Type A: What number is 14% of 68?

Type B: What percent of 75 is 25?

Type C: 25 is 40% of what number?

To solve them, we write *is* as =, *of* as · (multiply), and *what number* or *what percent* as n. We then solve the resulting equation to find the answer to the original question.

6 Translating to equations, we have:

Type A: $n = 0.14(68)$

Type B: $75n = 25$

Type C: $25 = 0.40n$

Applications of Percent [7.3, 7.4, 7.5, 7.6]
There are many different kinds of application problems involving percent. They include problems on income tax, sales tax, commission, discount, percent increase and decrease, and interest. Generally, to solve these problems, we restate them as an equivalent problem of Type A, B, or C above. Problems involving simple interest can be solved using the formula

$$I = P \cdot R \cdot T$$

where I = interest, P = principal, R = interest rate, and T = time (in years). It is standard procedure with simple interest problems to use 360 days = 1 year.

COMMON MISTAKES

1 A common mistake is forgetting to change a percent to a decimal when working problems that involve percents in the calculations. We always change percents to decimals before doing any calculations.

2 Moving the decimal point in the wrong direction when converting percents to decimals or decimals to percents is another common mistake. Remember, *percent* means "per hundred." Rewriting a number expressed as a percent as a decimal will make the numerical part smaller.

$$25\% = 0.25$$

CHAPTER 7 REVIEW

Write each percent as a decimal. **[7.1]**

1 35%
2 17.8%
3 5%
4 0.2%

Write each decimal as a percent. **[7.1]**

5 0.95
6 0.8
7 0.495
8 1.65

Write each percent as a fraction or mixed number in lowest terms. **[7.1]**

9 75%
10 4%
11 145%
12 2.5%

Write each fraction or mixed number as a percent. **[7.1]**

13 $\frac{3}{10}$
14 $\frac{5}{8}$
15 $\frac{2}{3}$
16 $4\frac{3}{4}$

Solve the following problems. **[7.2, 7.3, 7.4, 7.5, 7.6]**

17 What number is 60% of 28?

18 What percent of 38 is 19?

19 24 is 30% of what number?

20 Suppose 45 out of 60 people surveyed believe a college education will increase a person's earning potential. What percent believe this?

21 A salesperson gets 12% commission rate on all appliances she sells. If she sells $600 in appliances in 1 day, what is her commission?

22 A lawnmower that usually sells for $175 is marked down to $140. What is the discount? What is the discount rate?

23 A sewing machine that normally sells for \$600 is on sale for 25% off. If the sales tax rate is 6%, what is the total price of the sewing machine if it is purchased during the sale?

24 If the interest rate on a home mortgage is 9%, then each year you pay 9% of the unpaid balance in interest. If the unpaid balance on one such loan is \$60,000 at the beginning of the year, how much interest must be paid that year?

25 In 1978, the United States government claimed that an urban family of four had to gross \$19,000 per year in order to maintain a moderate standard of living. This was 9% more than the previous year. How much did an urban family of four have to gross in 1977 in order to maintain a moderate standard of living? (Use a calculator.)

26 In 1978, the United States government estimated that a family of four living in Denver had to gross \$18,565 per year in order to maintain a "middle level" standard of living. In San Francisco, the same figure was \$19,427. Write the difference in the two amounts as a percent of the smaller amount. (Use a calculator, and round to the nearest tenth of a percent.)

27 A solution of sodium bicarbonate and water is 2% sodium bicarbonate by weight. If the whole solution is 40 grams, how many grams of sodium bicarbonate are there?

28 A real estate agent gets a commission of 6% on all houses he sells. If his total sales for December are \$420,000, how much money does he make?

29 If \$1,800 is invested at 7% simple interest for 120 days, how much interest is earned?

30 How much interest will be earned on a savings account that pays 8% compounded semiannually if \$1,000 is invested for 2 years?

CHAPTER 7 TEST

Write each percent as a decimal.

1 18% **2** 4% **3** 0.5%

Write each decimal as a percent.

4 0.45 **5** 0.7 **6** 1.35

Write each percent as a fraction or a mixed number in lowest terms.

7 65% **8** 146% **9** 3.5%

Write each number as a percent.

10 $\frac{7}{20}$ **11** $\frac{3}{8}$ **12** $1\frac{3}{4}$

13 What number is 75% of 60?

14 What percent of 40 is 18?

15 16 is 20% of what number?

16 On a driver's test, a student answered 23 questions correctly on a 25-question test. What percent of the questions did the student answer correctly?

17 A salesperson gets an 8% commission rate on all computers she sells. If she sells $12,000 in computers in 1 day, what is her commission?

18 A washing machine that usually sells for $250 is marked down to $210. What is the discount? What is the discount rate?

19 A tennis racket that normally sells for $280 is on sale for 25% off. If the sales tax rate is 5%, what is the total price of the tennis racket if it is purchased during the sale?

20 If $5000 is invested at 8% simple interest for 3 months, how much interest is earned?

21 How much interest will be earned on a savings account that pays 10% compounded annually, if $12,000 is invested for 2 years?

APPENDIX I

ONE HUNDRED ADDITION FACTS

The following 100 problems should be done mentally. You should be able to find these sums quickly and accurately. Do all 100 problems and then check your answers. Make a list of each problem you missed and then go over the list as many times as it takes to memorize the correct answers. Once this has been done, go back and work all 100 problems again. Repeat this process until you get all 100 problems correct.

Add

1 $0 + 4$	**2** $1 + 9$	**3** $2 + 7$	**4** $0 + 3$	**5** $4 + 1$	**6** $2 + 5$	**7** $5 + 6$	**8** $2 + 6$
9 $9 + 8$	**10** $4 + 9$	**11** $3 + 7$	**12** $9 + 4$	**13** $4 + 5$	**14** $1 + 4$	**15** $0 + 1$	**16** $6 + 4$
17 $5 + 3$	**18** $9 + 5$	**19** $5 + 8$	**20** $0 + 7$	**21** $5 + 1$	**22** $1 + 0$	**23** $6 + 5$	**24** $5 + 2$
25 $5 + 7$	**26** $3 + 6$	**27** $7 + 0$	**28** $4 + 6$	**29** $1 + 5$	**30** $3 + 1$	**31** $1 + 7$	**32** $7 + 6$
33 $3 + 3$	**34** $8 + 0$	**35** $1 + 8$	**36** $5 + 4$	**37** $2 + 8$	**38** $0 + 5$	**39** $5 + 9$	**40** $9 + 9$
41 $0 + 0$	**42** $6 + 9$	**43** $7 + 4$	**44** $8 + 2$	**45** $7 + 3$	**46** $8 + 5$	**47** $8 + 4$	**48** $6 + 7$
49 $8 + 1$	**50** $1 + 3$	**51** $7 + 5$	**52** $0 + 1$	**53** $8 + 3$	**54** $6 + 6$	**55** $9 + 6$	**56** $8 + 9$
57 $3 + 0$	**58** $6 + 2$	**59** $2 + 1$	**60** $6 + 0$	**61** $4 + 8$	**62** $6 + 1$	**63** $2 + 0$	**64** $7 + 9$
65 $2 + 9$	**66** $9 + 3$	**67** $3 + 8$	**68** $7 + 2$	**69** $8 + 8$	**70** $4 + 4$	**71** $8 + 7$	**72** $2 + 4$
73 $3 + 2$	**74** $4 + 7$	**75** $9 + 7$	**76** $1 + 2$	**77** $6 + 3$	**78** $2 + 2$	**79** $9 + 2$	**80** $9 + 1$
81 $3 + 5$	**82** $1 + 1$	**83** $5 + 5$	**84** $7 + 7$	**85** $0 + 8$	**86** $7 + 8$	**87** $2 + 3$	**88** $3 + 9$
89 $6 + 8$	**90** $4 + 0$	**91** $0 + 6$	**92** $4 + 3$	**93** $7 + 1$	**94** $0 + 9$	**95** $9 + 0$	**96** $8 + 6$
97 $3 + 4$	**98** $5 + 0$	**99** $1 + 6$	**100** $4 + 2$				

APPENDIX 2

ONE HUNDRED MULTIPLICATION FACTS

The following 100 problems should be done mentally. You should be able to find these products quickly and accurately. Do all 100 problems and then check your answers. Make a list of each problem you missed and then go over the list as many times as it takes to memorize the correct answers. Once this has been done, go back and work all 100 problems again. Repeat this process until you get all 100 problems correct.

Multiply

1 3×4	**2** 5×4	**3** 0×5	**4** 0×6	**5** 2×5	**6** 1×7	**7** 4×4	**8** 7×2
9 7×1	**10** 5×9	**11** 2×6	**12** 5×3	**13** 3×5	**14** 4×5	**15** 8×1	**16** 3×7
17 5×8	**18** 0×3	**19** 6×0	**20** 0×4	**21** 4×6	**22** 8×2	**23** 5×7	**24** 0×9
25 2×7	**26** 6×1	**27** 1×8	**28** 4×7	**29** 3×8	**30** 6×8	**31** 5×4	**32** 7×3
33 0×8	**34** 4×8	**35** 2×9	**36** 5×6	**37** 1×9	**38** 2×8	**39** 0×7	**40** 6×2
41 4×9	**42** 8×0	**43** 3×0	**44** 9×6	**45** 3×4	**46** 8×3	**47** 7×9	**48** 1×9
49 1×3	**50** 9×9	**51** 2×4	**52** 8×7	**53** 5×2	**54** 7×8	**55** 1×4	**56** 0×2
57 9×8	**58** 2×3	**59** 9×5	**60** 7×7	**61** 6×5	**62** 0×1	**63** 9×3	**64** 8×9
65 9×7	**66** 6×4	**67** 3×1	**68** 1×5	**69** 9×6	**70** 1×2	**71** 4×1	**72** 2×2
73 6×3	**74** 5×1	**75** 9×2	**76** 6×9	**77** 7×0	**78** 4×0	**79** 7×6	**80** 5×0
81 0×0	**82** 3×9	**83** 1×1	**84** 8×6	**85** 9×0	**86** 7×4	**87** 8×4	**88** 2×0
89 9×1	**90** 8×5	**91** 4×2	**92** 8×3	**93** 7×5	**94** 2×1	**95** 6×6	**96** 3×2
97 4×3	**98** 1×6	**99** 6×7	**100** 1×0				

SOLUTIONS TO SELECTED PRACTICE PROBLEMS

Solutions to all practice problems that require more than one step are shown here. Before you look back here to see where you have made a mistake, you should try the problem you are working on twice. If you do not get the correct answer the second time you work the problem, then the solution shown here should show you where you went wrong.

CHAPTER 1

Section 1.2

4 $(x + 5) + 9 = x + (5 + 9)$
$= x + 14$

5 $6 + (8 + y) = (6 + 8) + y$
$= 14 + y$

6 $(1 + x) + 4 = (x + 1) + 4$
$= x + (1 + 4)$
$= x + 5$

7 $8 + (a + 4) = 8 + (4 + a)$
$= (8 + 4) + a$
$= 12 + a$

8 a $n = 8$, since $8 + 9 = 17$
b $n = 8$, since $8 + 2 = 10$
c $n = 1$, since $8 + 1 = 9$
d $n = 6$, since $16 = 6 + 10$

9 $(x + 5) + 4 = 10$
$x + (5 + 4) = 10$
$x + 9 = 10$
$x = 1$

10 $(a + 4) + 2 = 7 + 9$
$a + (4 + 2) = 16$
$a + 6 = 16$
$a = 10$

Section 1.3

1

63	=	6 tens + 3 ones
+ 25	=	2 tens + 5 ones
		8 tens + 8 ones

Answer: 88

2

342	=	3 hundreds + 4 tens + 2 ones
+ 605	=	6 hundreds + 0 tens + 5 ones
		9 hundreds + 4 tens + 7 ones

Answer: 947

3

375	=	3 hundreds + 7 tens + 5 ones
121	=	1 hundred + 2 tens + 1 one
+ 473	=	4 hundreds + 7 tens + 3 ones
		8 hundreds + 16 tens + 9 ones = 9 hundreds + 6 tens + 9 ones

Answer: 969

4

```
  11 11
  57,904
   7,193
     655
  ------
  65,752
```

5 a $7 + 7 + 7 + 7 = 28$ ft **b** $88 + 88 + 33 + 33 = 242$ in. **c** $44 + 66 + 77 = 187$ yd

Section 1.4

5 a

Food	$ 5,296
Car	4,847
Total	$10,143 = $10,140 to the nearest ten dollars

b

Savings	$2,149
Taxes	6,137
Total	$8,286 = $8,300 to the nearest hundred dollars

c

House	$10,200
Taxes	6,137
Misc.	6,142
Car	4,847
Savings	2,149
Total	$29,475 = $29,000 to the nearest thousand dollars

6 We round each of the four numbers in the sum to the nearest thousand, and then we add the rounded numbers.

5,287	rounds to	5,000
2,561	rounds to	3,000
888	rounds to	1,000
+ 4,898	rounds to	+ 5,000
		14,000

We estimate the answer to this problem to be approximately 14,000. The actual answer, found by adding the original, unrounded numbers, is 13,634.

7

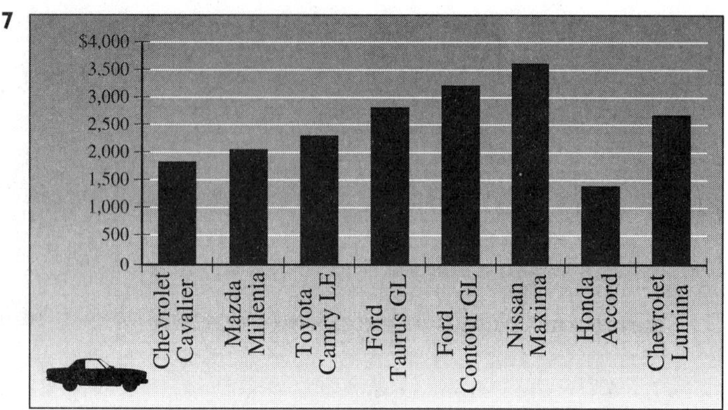

Section 1.5

1

684	= 6 hundreds	+ 8 tens	+ 4 ones	
− 431	= 4 hundreds	3 tens	1 one	
	2 hundreds	+ 5 tens	+ 3 ones	

Answer: 253

2

6,857
− 405
6,452

3

63	= 6 tens + 3 ones	= 5 tens	+ 13 ones
− 47	= 4 tens 7 ones	= 4 tens	7 ones
		1 ten	+ 6 ones

Answer: 16

4

5	15	
~~6~~	~~5~~	6
− 2	8	3
3	7	3

5

$127
− 52
$ 75

Section 1.6

5 $4(7x) = (4 \cdot 7)x$
$= 28x$

6 $7(9y) = (7 \cdot 9)y$
$= 63y$

7 **a** $n = 7$, since $5 \cdot 7 = 35$ **b** $n = 9$, since $8 \cdot 9 = 72$ **c** $n = 7$, since $49 = 7 \cdot 7$ **d** $n = 3$, since $27 = 9 \cdot 3$

8 $5 \cdot 100 = 100 + 100 + 100 + 100 + 100$
$= 500$

9 $4 \cdot 7{,}000 = 7{,}000 + 7{,}000 + 7{,}000 + 7{,}000$
$= 28{,}000$

10 $9 \cdot 10 = 90$

11 $5 \cdot 10 = 50$ **12** $14 \cdot 10 = 140$ **13** $8 \cdot 100 = 800$ **14** $8 \cdot 40 = 320$ **15** $6 \cdot 700 = 4{,}200$

16 $3 \cdot 6{,}000 = 18{,}000$ **17** $6 \cdot 9{,}000 = 54{,}000$

Section 1.7

1 $3(5 + 7) = 3(5) + 3(7)$
$= 15 + 21$
$= 36$

2 $5(6 + 8) = 5(6) + 5(8)$
$= 30 + 40$
$= 70$

3 $4(x + 2) = 4 \cdot x + 4 \cdot 2$
$= 4x + 8$

4 $7(a + 9) = 7 \cdot a + 7 \cdot 9$
$= 7a + 63$

5 $4(7x + 2) = 4(7x) + 4 \cdot 2$
$= (4 \cdot 7)x + 4 \cdot 2$
$= 28x + 8$

6 $7(9y + 6) = 7(9y) + 7 \cdot 6$
$= (7 \cdot 9)y + 7 \cdot 6$
$= 63y + 42$

	Long Method	Shortcut Method		Long Method	Shortcut Method

7 Long Method:
$$\begin{array}{r} 50 + 7 \\ \times \quad 8 \\ \hline 56 \\ +400 \\ \hline 456 \end{array}$$
Shortcut Method:
$$\begin{array}{r} {}^{5} \\ 57 \\ \times\ 8 \\ \hline 456 \end{array}$$

8 Long Method:
$$\begin{array}{r} 500 + 70 + 2 \\ \times \qquad 6 \\ \hline 12 \\ 420 \\ +3{,}000 \\ \hline 3{,}432 \end{array}$$
Shortcut Method:
$$\begin{array}{r} {}^{41} \\ 572 \\ \times\ 6 \\ \hline 3{,}432 \end{array}$$

9
$$\begin{array}{rl} 45 & \\ \times 62 & \\ \hline 90 & \leftarrow 2(45) = 90 \\ +2{,}700 & \leftarrow 60(45) = 2{,}700 \\ \hline 2{,}790 & \end{array}$$

10
$$\begin{array}{rl} 356 & \\ \times 641 & \\ \hline 356 & \leftarrow 1(356) = 356 \\ 14{,}240 & \leftarrow 40(356) = 14{,}240 \\ 213{,}600 & \leftarrow 600(356) = 213{,}600 \\ \hline 228{,}196 & \end{array}$$

11
$$\begin{array}{r} 365 \\ \times 550 \\ \hline 18{,}250 \\ 182{,}500 \\ \hline 200{,}750 \text{ mg} \end{array}$$

12 36($12) = $432 Total weekly earnings
$432 − $109 = $323 Take-home pay

13 Fat: 3(10) = 30 grams of fat; sodium: 3(160) = 480 milligrams of sodium

14 Bowling for 3 hours burns 3(265) = 795 calories. Eating two bags of chips means you are consuming 2(3)(160) = 960 calories. No; bowling won't burn all the calories.

Section 1.8

1
$$\begin{array}{r} 72 \\ 4\overline{)288} \\ \underline{28}\downarrow \\ 08 \\ \underline{8} \\ 0 \end{array}$$

2
$$\begin{array}{r} 283 \\ 24\overline{)6{,}792} \\ \underline{48}\downarrow \\ 199 \\ \underline{192}\downarrow \\ 72 \\ \underline{72} \\ 0 \end{array}$$

3
$$\begin{array}{r} 69 \text{ R } 20, \text{ or } 69\tfrac{20}{27} \\ 27\overline{)1{,}883} \\ \underline{162}\downarrow \\ 263 \\ \underline{243} \\ 20 \end{array}$$

4
$$\begin{array}{r} 156 \\ 12\overline{)1{,}872} \\ \underline{12}\downarrow \\ 67 \\ \underline{60}\downarrow \\ 72 \\ \underline{72} \\ 0 \end{array}$$
The family spent $156 per day

Section 1.9

1 Base 5, exponent 2; 5 to the second power, or 5 squared

2 Base 2, exponent 3; 2 to the third power, or 2 cubed

3 Base 1, exponent 4; 1 to the fourth power

4 $5^2 = 5 \cdot 5 = 25$

5 $9^2 = 9 \cdot 9 = 81$

6 $2^3 = 2 \cdot 2 \cdot 2 = 8$

7 $1^4 = 1 \cdot 1 \cdot 1 \cdot 1 = 1$

8 $2^5 = 2 \cdot 2 \cdot 2 \cdot 2 \cdot 2 = 8 \cdot 4 = 32$

9 $7^1 = 7$

10 $4^1 = 4$

11 $9^0 = 1$

12 $1^0 = 1$

13 $5 \cdot 7 - 3 \cdot 6 = 35 - 18$
$= 17$

14 $7 + 3(6 + 4) = 7 + 3(10)$
$= 7 + 30$
$= 37$

15 $6 \cdot 9 + 64 \div 16 - 2 = 54 + 4 - 2$
$= 58 - 2$
$= 56$

16 $5 + 3[24 - 5(6 - 2)] = 5 + 3[24 - 5(4)]$
$= 5 + 3[24 - 20]$
$= 5 + 3[4]$
$= 5 + 12$
$= 17$

17
$$\begin{array}{r} 187 \\ 273 \\ 173 \\ \underline{227} \\ 860 \end{array}$$
$\frac{860}{4} = 215$ miles (average)

Section 1.10

1 $A = 10 \cdot 2 + 4 \cdot 4$
$= 20 + 16$
$= 36 \text{ ft}^2$

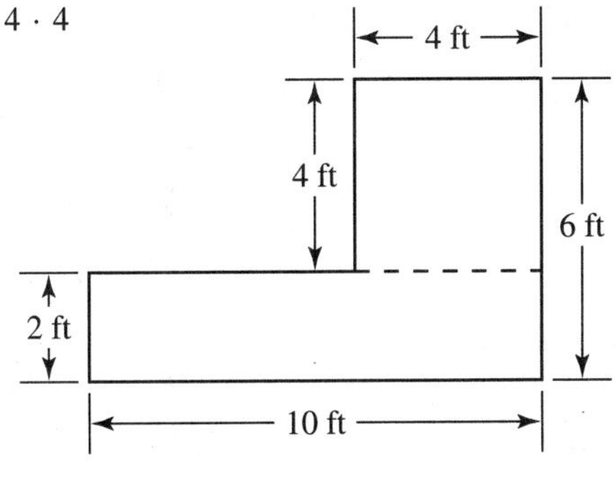

4 ft
4 ft
6 ft
2 ft
10 ft

2 $A = (1 \text{ yd}) \cdot (1 \text{ yd})$
$= (3 \text{ ft}) \cdot (3 \text{ ft})$
$= (3 \cdot 3) \text{ ft}^2$
$= 9 \text{ ft}^2$

3 Area of small square $= 4 \cdot 4 = 16 \text{ ft}^2$
Area of large square $= 12 \cdot 12 = 144 \text{ ft}^2$
Since $9 \cdot 16 = 144$, the area of the larger square is 9 times the area of the small square.

4 $V = 15 \cdot 12 \cdot 8 = 1{,}440 \text{ ft}^3$

5 **a** Surface area $= 2(15 \cdot 8) + 2(8 \cdot 12) + (15 \cdot 12) = 612 \text{ ft}^2$

b 2 gallons will cover it, with some paint left over

CHAPTER 2

Section 2.2

1 $2 + (-5) = -3$ **2** $-2 + 5 = 3$ **3** $-2 + (-5) = -7$ **4** $2 + 6 = 8$ **5** $2 + (-6) = -4$ **6** $-2 + 6 = 4$

7 $-2 + (-6) = -8$

8 $15 + 12 = 27$
$15 + (-12) = 3$
$-15 + 12 = -3$
$-15 + (-12) = -27$

9 $12 + (-3) + (-7) + 5 = 9 + (-7) + 5$
$= 2 + 5$
$= 7$

10 $[-2 + (-12)] + [7 + (-5)] = [-14] + [2]$
$= -12$

Section 2.3

1 $7 - 3 = 7 + (-3)$
$= 4$

2 $-7 - 3 = -7 + (-3)$
$= -10$

3 $-8 - 6 = -8 + (-6)$
$= -14$

4 $10 - (-6) = 10 + 6$
$= 16$

5 $-10 - (-15) = -10 + 15$
$= 5$

6 **a** $8 - 5 = 8 + (-5)$
$= 3$

b $-8 - 5 = -8 + (-5)$
$= -13$

c $8 - (-5) = 8 + 5$
$= 13$

d $-8 - (-5) = -8 + 5$
$= -3$

e $12 - 10 = 12 + (-10)$
$= 2$

f $-12 - 10 = -12 + (-10)$
$= -22$

g $12 - (-10) = 12 + 10$
$= 22$

h $-12 - (-10) = -12 + 10$
$= -2$

7 $-4 + 6 - 7 = -4 + 6 + (-7)$
$= 2 + (-7)$
$= -5$

8 $15 - (-5) - 8 = 15 + 5 + (-8)$
$= 20 + (-8)$
$= 12$

9 $-8 - 2 = -8 + (-2)$
$= -10$

10 $7 - (-5) = 7 + 5$
$= 12$

11 $7 - (-3) = 7 + 3$
$= 10$

12 $-8 - (-6) = -8 + 6$
$= -2$

13 $42 - (-42) = 42 + 42 = 84°F$

Section 2.4

1 $2(-6) = (-6) + (-6)$
$= -12$

2 $-2(6) = 6(-2) = (-2) + (-2) + (-2) + (-2) + (-2) + (-2)$
$= -12$

3 $-2(-6) = 12$

10 $-5(2)(-4) = -10(-4)$
$= 40$

11 $(-5)^2 = (-5)(-5)$
$= 25$

12 $-2[5 + (-8)] = -2[-3]$
$= 6$

13 $-3 + 4(-7 + 3) = -3 + 4(-4)$
$= -3 + (-16)$
$= -19$

14 $-3(5) + 4(-4) = -15 + (-16)$
$= -31$

15 $-2(3 - 5) - 7(-2 - 4) = -2(-2) - 7(-6)$
$= 4 - (-42)$
$= 4 + 42$
$= 46$

16 $(-6 - 1)(4 - 9) = (-7)(-5)$
$= 35$

Section 2.5

6 $\dfrac{8(-5)}{-4} = \dfrac{-40}{-4}$
$= 10$

7 $\dfrac{-20 + 6(-2)}{7 - 11} = \dfrac{-20 + (-12)}{-4}$
$= \dfrac{-32}{-4}$
$= 8$

8 $-3(4^2) + 10 \div (-5) = -3(16) + 10 \div (-5)$
$= -48 + (-2)$
$= -50$

9 $-80 \div 2 \div 10 = -40 \div 10$
$= -4$

Section 2.6

1 $5(7a) = (5 \cdot 7)a$
$= 35a$

2 $-3(9x) = (-3 \cdot 9)x$
$= -27x$

3 $5(-8y) = [5(-8)]y$
$= -40y$

4 $6 + (9 + x) = (6 + 9) + x$
$= 15 + x$

5 $(3x + 7) + 4 = 3x + (7 + 4)$
$= 3x + 11$

6 $6(x + 4) = 6(x) + 6(4)$
$= 6x + 24$

7 $7(a - 5) = 7(a) - 7(5)$
$= 7a - 35$

8 $6(4x + 5) = 6(4x) + 6(5)$
$= (6 \cdot 4)x + 6(5)$
$= 24x + 30$

9 $3(8a - 4) = 3(8a) - 3(4)$
$= 24a - 12$

10 $8(3x + 4y) = 8(3x) + 8(4y)$
$= 24x + 32y$

11 $(2a)^3 = (2a)(2a)(2a)$
$= (2 \cdot 2 \cdot 2)(a \cdot a \cdot a)$
$= 8a^3$

12 $(7xy)^2 = (7xy)(7xy)$
$= (7 \cdot 7)(x \cdot x)(y \cdot y)$
$= 49x^2y^2$

13 $(3x)^2(7xy)^2 = (3x)(3x)(7xy)(7xy)$
$= (3 \cdot 3 \cdot 7 \cdot 7)(x \cdot x \cdot x \cdot x)(y \cdot y)$
$= 441x^4y^2$

14 $(5x)^3(2x)^2 = (5x)(5x)(5x)(2x)(2x)$
$= (5 \cdot 5 \cdot 5 \cdot 2 \cdot 2)(x \cdot x \cdot x \cdot x \cdot x)$
$= 500x^5$

Section 2.7

1 $x^3 \cdot x^5 = (x \cdot x \cdot x)(x \cdot x \cdot x \cdot x \cdot x)$
$= x \cdot x \cdot x \cdot x \cdot x \cdot x \cdot x \cdot x$
$= x^8$

2 $4x^2 \cdot 6x^5 = (4 \cdot 6)(x^2 \cdot x^5)$
$= (4 \cdot 6)(x^{2+5})$
$= 24x^7$

3 $2x^5 \cdot 3x^4 \cdot 4x^3 = (2 \cdot 3 \cdot 4)(x^5 \cdot x^4 \cdot x^3)$
$= (2 \cdot 3 \cdot 4)(x^{5+4+3})$
$= 24x^{12}$

4 $(10xy^2)(9x^3y^5) = (10 \cdot 9)(x \cdot x^3)(y^2 \cdot y^5)$
$= (10 \cdot 9)(x^{1+3})(y^{2+5})$
$= 90x^4y^7$

5 $(2^2)^3 = 2^{2 \cdot 3}$
$= 2^6$
$= 64$

6 $(x^3)^4 \cdot (x^5)^2 = x^{3 \cdot 4} \cdot x^{5 \cdot 2}$
$= x^{12} \cdot x^{10}$
$= x^{12+10}$
$= x^{22}$

7 $(5xy)^3 = 5^3 \cdot x^3 \cdot y^3$
$= 125x^3y^3$

8 $(4x^5y^2)^3 = 4^3(x^5)^3(y^2)^3$
$= 64x^{15}y^6$

9 $(2x^3y^4)^3(3xy^5)^2 = 2^3(x^3)^3(y^4)^3 \cdot 3^2x^2(y^5)^2$
$= 8x^9y^{12} \cdot 9x^2y^{10}$
$= (8 \cdot 9)(x^9x^2)(y^{12}y^{10})$
$= 72x^{11}y^{22}$

Section 2.8

1 $(5x^2 - 3x + 2) + (2x^2 + 10x - 9) = (5x^2 + 2x^2) + (-3x + 10x) + (2 - 9)$
$= (5 + 2)x^2 + (-3 + 10)x + (2 - 9)$
$= 7x^2 + 7x - 7$

$$\begin{array}{r} 5x^2 - 3x + 2 \\ \underline{2x^2 + 10x - 9} \\ 7x^2 + 7x - 7 \end{array}$$

2 $(3y^2 + 9y - 5) + (6y^2 - 4) = (3y^2 + 6y^2) + 9y + (-5 - 4)$
$= (3 + 6)y^2 + 9y + (-5 - 4)$
$= 9y^2 + 9y - 9$

$$\begin{array}{r} 3y^2 + 9y - 5 \\ \underline{6y^2 \qquad - 4} \\ 9y^2 + 9y - 9 \end{array}$$

3 $(8x^3 + 4x^2 + 3x + 2) + (4x^2 + 5x + 6) = 8x^3 + (4x^2 + 4x^2) + (3x + 5x) + (2 + 6)$
$= 8x^3 + (4 + 4)x^2 + (3 + 5)x + (2 + 6)$
$= 8x^3 + 8x^2 + 8x + 8$

$$\begin{array}{r} 8x^3 + 4x^2 + 3x + 2 \\ \underline{4x^2 + 5x + 6} \\ 8x^3 + 8x^2 + 8x + 8 \end{array}$$

4 $(5x^2 - 2x + 7) - (4x^2 + 8x - 4) = 5x^2 - 2x + 7 - 4x^2 - 8x + 4$
$= (5x^2 - 4x^2) + (-2x - 8x) + (7 + 4)$
$= (5 - 4)x^2 + (-2 - 8)x + (7 + 4)$
$= 1x^2 - 10x + 11$
$= x^2 - 10x + 11$

5 $(3y^3 - 2y^2 + 7y - 6) - (8y^3 - 6y^2 + 4y - 8) = 3y^3 - 2y^2 + 7y - 6 - 8y^3 + 6y^2 - 4y + 8$
$= (3y^3 - 8y^3) + (-2y^2 + 6y^2) + (7y - 4y) + (-6 + 8)$
$= (3 - 8)y^3 + (-2 + 6)y^2 + (7 - 4)y + (-6 + 8)$
$= -5y^3 + 4y^2 + 3y + 2$

6 $(-2x^2 + 5x - 1) - (6x^2 - 2x + 5) = -2x^2 + 5x - 1 - 6x^2 + 2x - 5$
$= (-2x^2 - 6x^2) + (5x + 2x) + (-1 - 5)$
$= -8x^2 + 7x - 6$

7 When $x = -3$
the polynomial $5x^2 - 3x + 8$
becomes $5(-3)^2 - 3(-3) + 8 = 5(9) + 9 + 8$
$= 45 + 9 + 8$
$= 62$

Section 2.9

1 $x^3(x^5 + x^7) = x^3 \cdot x^5 + x^3 \cdot x^7$
$= x^8 + x^{10}$

2 $(x^5 + x^7)x^3 = x^5 \cdot x^3 + x^7 \cdot x^3$
$= x^8 + x^{10}$

3 $5x^2(6x^3 - 4) = 5x^2 \cdot 6x^3 - 5x^2 \cdot 4$
$= (5 \cdot 6)(x^2 \cdot x^3) - (5 \cdot 4)x^2$
$= 30x^5 - 20x^2$

4 $5a^3b^5(2a^2 + 7b^2) = 5a^3b^5 \cdot 2a^2 + 5a^3b^5 \cdot 7b^2$
$= (5 \cdot 2)(a^3 \cdot a^2)b^5 + (5 \cdot 7)(a^3)(b^5 \cdot b^2)$
$= 10a^5b^5 + 35a^3b^7$

5 $(x + 2)(x + 6) = (x + 2)x + (x + 2)6$
$= x \cdot x + 2 \cdot x + x \cdot 6 + 2 \cdot 6$
$= x^2 + 2x + 6x + 12$
$= x^2 + 8x + 12$

6 $(x - 2)(x + 6) = (x - 2)x + (x - 2)6$
$= x \cdot x - 2 \cdot x + x \cdot 6 - 2 \cdot 6$
$= x^2 - 2x + 6x - 12$
$= x^2 + 4x - 12$

7 $(3x - 2)(5x + 4) = (3x - 2) \cdot 5x + (3x - 2) \cdot 4$
$= 3x \cdot 5x - 2 \cdot 5x + 3x \cdot 4 - 2 \cdot 4$
$= 15x^2 - 10x + 12x - 8$
$= 15x^2 + 2x - 8$

8 $(x + 3)^2 = (x + 3)(x + 3)$
$= (x + 3) \cdot x + (x + 3) \cdot 3$
$= x \cdot x + 3 \cdot x + x \cdot 3 + 3 \cdot 3$
$= x^2 + 3x + 3x + 9$
$= x^2 + 6x + 9$

9 $(3x - 5)^2 = (3x - 5)(3x - 5)$
$= (3x - 5)[3x + (-5)]$
$= (3x - 5) \cdot 3x + (3x - 5)(-5)$
$= 3x \cdot 3x - 5 \cdot 3x + 3x(-5) - 5(-5)$
$= 9x^2 - 15x - 15x + 25$
$= 9x^2 - 30x + 25$

CHAPTER 3

Section 3.1

6 $\frac{2}{3} = \frac{2 \cdot 4}{3 \cdot 4} = \frac{8}{12}$ **7** $\frac{2}{3} = \frac{2 \cdot 4x}{3 \cdot 4x} = \frac{8x}{12x}$ **8** $\frac{15}{20} = \frac{15 \div 5}{20 \div 5} = \frac{3}{4}$

Section 3.2

1 37 and 59 are prime numbers; 39 is divisible by 3 and 13; 51 is divisible by 3 and 17

2 $90 = 9 \cdot 10$
$= 3 \cdot 3 \cdot 2 \cdot 5$
$= 2 \cdot 3^2 \cdot 5$

4 $\frac{12}{18} = \frac{12 \div 6}{18 \div 6} = \frac{2}{3}$ **5** $\frac{15}{20} = \frac{3 \cdot \cancel{5}}{2 \cdot 2 \cdot \cancel{5}} = \frac{3}{4}$ **6** $\frac{30}{35} = \frac{2 \cdot 3 \cdot \cancel{5}}{\cancel{5} \cdot 7} = \frac{6}{7}$ **7** $\frac{8}{72} = \frac{\cancel{8} \cdot 1}{\cancel{8} \cdot 9} = \frac{1}{9}$ **8** $\frac{5}{50} = \frac{\cancel{5} \cdot 1}{\cancel{5} \cdot 10} = \frac{1}{10}$

9 $\frac{120}{25} = \frac{2 \cdot 2 \cdot 2 \cdot 3 \cdot \cancel{5}}{5 \cdot \cancel{5}} = \frac{24}{5}$ **10** $\frac{54x}{90xy} = \frac{\cancel{2} \cdot \cancel{3} \cdot \cancel{3} \cdot 3 \cdot \cancel{x}}{\cancel{2} \cdot \cancel{3} \cdot \cancel{3} \cdot 5 \cdot \cancel{x} \cdot y} = \frac{3}{5y}$ **11** $\frac{306a^2}{228a} = \frac{\cancel{2} \cdot \cancel{3} \cdot 3 \cdot 17 \cdot \cancel{a} \cdot a}{\cancel{2} \cdot 2 \cdot \cancel{3} \cdot 19 \cdot \cancel{a}} = \frac{51a}{38}$

Section 3.3

1 $\frac{2}{3} \cdot \frac{5}{9} = \frac{10}{27}$

2 $-\frac{2}{5} \cdot 7 = -\frac{2}{5} \cdot \frac{7}{1}$
$= -\frac{14}{5}$

3 $\frac{1}{3}\left(\frac{4}{5} \cdot \frac{1}{3}\right) = \frac{1}{3}\left(\frac{4}{15}\right)$
$= \frac{4}{45}$

4 $\frac{1}{4}(4y) = \left(\frac{1}{4} \cdot 4\right)y$
$= 1 \cdot y$
$= y$

5 $\frac{12}{25} \cdot \frac{5}{6} = \frac{12 \cdot 5}{25 \cdot 6}$
$= \frac{(2 \cdot \cancel{2} \cdot \cancel{3}) \cdot \cancel{5}}{(\cancel{5} \cdot 5) \cdot (\cancel{2} \cdot \cancel{3})}$
$= \frac{2}{5}$

6 $\frac{8}{3} \cdot \frac{9}{24} = \frac{8 \cdot 9}{3 \cdot 24}$
$= \frac{(\cancel{2} \cdot \cancel{2} \cdot \cancel{2}) \cdot (\cancel{3} \cdot \cancel{3})}{\cancel{3} \cdot (\cancel{2} \cdot \cancel{2} \cdot \cancel{2} \cdot \cancel{3})}$
$= \frac{1}{1}$
$= 1$

7 $\frac{yz^2}{x} \cdot \frac{x^3}{yz} = \frac{\cancel{y} \cdot \cancel{z} \cdot z \cdot \cancel{x} \cdot x \cdot x}{\cancel{x} \cdot \cancel{y} \cdot \cancel{z}}$
$= \frac{z \cdot x \cdot x}{1}$
$= x^2z$

8 $\frac{3}{4} \cdot \frac{8}{3} \cdot \frac{1}{6} = \frac{3 \cdot 8 \cdot 1}{4 \cdot 3 \cdot 6}$
$= \frac{\cancel{3} \cdot (\cancel{2} \cdot \cancel{2} \cdot \cancel{2}) \cdot 1}{(\cancel{2} \cdot \cancel{2}) \cdot \cancel{3} \cdot (\cancel{2} \cdot 3)}$
$= \frac{1}{3}$

9 $\left(\frac{2}{3}\right)^2 = \frac{2}{3} \cdot \frac{2}{3}$
$= \frac{4}{9}$

10 $\left(\frac{3}{4}\right)^2 \cdot \frac{1}{2} = \frac{3}{4} \cdot \frac{3}{4} \cdot \frac{1}{2}$
$= \frac{9}{32}$

11 $\frac{2}{3} \cdot \frac{1}{2} = \frac{\cancel{2} \cdot 1}{3 \cdot \cancel{2}}$
$= \frac{1}{3}$

12 $\frac{2}{3}(-12) = \frac{2}{3}\left(-\frac{12}{1}\right)$

$= -\frac{2 \cdot 2 \cdot 2 \cdot \not{3}}{\not{3} \cdot 1}$

$= -\frac{8}{1}$

$= -8$

13 $A = \frac{1}{2}(7)(10)$

$= 35 \text{ in}^2$

14

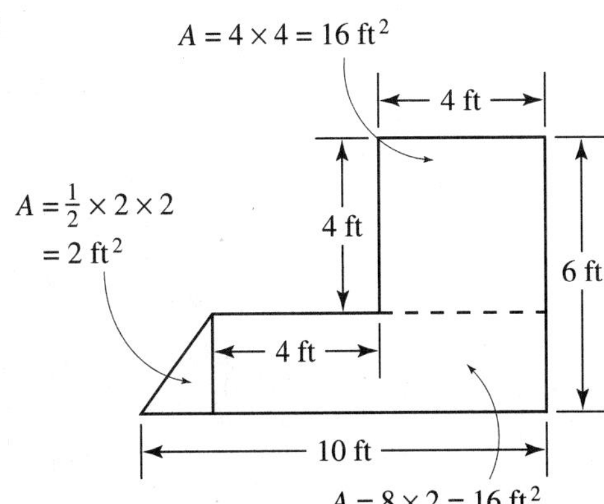

Total area = 2 + 16 + 16 = 34 ft^2

Section 3.4

1 $\frac{1}{3} \div \frac{1}{6} = \frac{1}{3} \cdot \frac{6}{1}$

$= \frac{6}{3}$

$= 2$

2 $\frac{5}{9} \div \frac{10}{3} = \frac{5}{9} \cdot \frac{3}{10}$

$= \frac{\not{5} \cdot \not{3}}{\not{3} \cdot 3 \cdot 2 \cdot \not{5}}$

$= \frac{1}{6}$

3 $-\frac{3}{4} \div 3 = -\frac{3}{4} \cdot \frac{1}{3}$

$= -\frac{1}{4}$

4 $-4 \div \left(-\frac{1}{5}\right) = -4(-5)$

$= 20$

5 $\frac{5}{32} \div \frac{10}{42} = \frac{5}{32} \cdot \frac{42}{10}$

$= \frac{\not{5} \cdot \not{2} \cdot 3 \cdot 7}{\not{2} \cdot 2 \cdot 2 \cdot 2 \cdot 2 \cdot 2 \cdot \not{5}}$

$= \frac{21}{32}$

6 $\frac{12}{25} \div 6 = \frac{12}{25} \cdot \frac{1}{6}$

$= \frac{2}{25}$

7 $-12 \div \left(-\frac{4}{3}\right) = -12\left(-\frac{3}{4}\right)$

$= 9$

8 $\frac{x^3}{y} \div \frac{x^2}{y^2} = \frac{x^3}{y} \cdot \frac{y^2}{x^2}$

$= \frac{\not{x} \cdot \not{x} \cdot x \cdot y \cdot \not{y}}{\not{y} \cdot \not{x} \cdot \not{x}}$

$= \frac{x \cdot y}{1}$

$= xy$

9 $\frac{5}{4} \div \frac{1}{8} + 8 = \frac{5}{4} \cdot \frac{8}{1} + 8$

$= 10 + 8$

$= 18$

10 $18 \div \left(\frac{3}{5}\right)^2 + 48 \div \left(\frac{2}{5}\right)^2 = 18 \div \frac{9}{25} + 48 \div \frac{4}{25}$

$= 18 \cdot \frac{25}{9} + 48 \cdot \frac{25}{4}$

$= 50 + 300$

$= 350$

11 $12 \div \frac{3}{4} = 12 \cdot \frac{4}{3}$

$= 4 \cdot 4$

$= 16$ blankets

Section 3.5

1 $\frac{3}{10} + \frac{1}{10} = \frac{3 + 1}{10}$

$= \frac{4}{10}$

$= \frac{2}{5}$

2 $\frac{a - 5}{12} + \frac{3}{12} = \frac{a - 5 + 3}{12}$

$= \frac{a - 2}{12}$

3 $\frac{8}{x} - \frac{5}{x} = \frac{8 - 5}{x}$

$= \frac{3}{x}$

4 $\frac{5}{9} - \frac{8}{9} + \frac{5}{9} = \frac{5 - 8 + 5}{9}$

$= \frac{2}{9}$

5 $\left.\begin{array}{l}18 = 2 \cdot 3 \cdot 3\\ 14 = 2 \cdot 7\end{array}\right\}$ $\begin{aligned}\text{LCD} &= 2 \cdot 3 \cdot 3 \cdot 7\\ &= 126\end{aligned}$

6 $\begin{aligned}\frac{5}{18} + \frac{3}{14} &= \frac{5 \cdot 7}{18 \cdot 7} + \frac{3 \cdot 9}{14 \cdot 9}\\ &= \frac{35}{126} + \frac{27}{126}\\ &= \frac{62}{126}\\ &= \frac{31}{63}\end{aligned}$

7 $\left.\begin{array}{l}9 = 3 \cdot 3\\ 15 = 3 \cdot 5\end{array}\right\}$ $\begin{aligned}\text{LCD} &= 3 \cdot 3 \cdot 5\\ &= 45\end{aligned}$

8 $\begin{aligned}\frac{2}{9} + \frac{4}{15} &= \frac{2 \cdot 5}{9 \cdot 5} + \frac{4 \cdot 3}{15 \cdot 3}\\ &= \frac{10}{45} + \frac{12}{45}\\ &= \frac{22}{45}\end{aligned}$

9 LCD = 100; $\begin{aligned}\frac{8}{25} - \frac{3}{20} &= \frac{8 \cdot 4}{25 \cdot 4} - \frac{3 \cdot 5}{20 \cdot 5}\\ &= \frac{32}{100} - \frac{15}{100}\\ &= \frac{17}{100}\end{aligned}$

10 LCD = 20; $\begin{aligned}\frac{x}{4} - \frac{1}{5} &= \frac{x \cdot 5}{4 \cdot 5} - \frac{1 \cdot 4}{5 \cdot 4}\\ &= \frac{5x}{20} - \frac{4}{20}\\ &= \frac{5x - 4}{20}\end{aligned}$

11 LCD = 36; $\begin{aligned}\frac{1}{9} + \frac{1}{4} + \frac{1}{6} &= \frac{1 \cdot 4}{9 \cdot 4} + \frac{1 \cdot 9}{4 \cdot 9} + \frac{1 \cdot 6}{6 \cdot 6}\\ &= \frac{4}{36} + \frac{9}{36} + \frac{6}{36}\\ &= \frac{19}{36}\end{aligned}$

12 $\begin{aligned}2 - \frac{3}{4} &= \frac{2}{1} - \frac{3}{4}\\ &= \frac{2 \cdot 4}{1 \cdot 4} - \frac{3}{4}\\ &= \frac{8}{4} - \frac{3}{4}\\ &= \frac{5}{4}\end{aligned}$

13 $\begin{aligned}\frac{5}{x} + \frac{2}{3} &= \frac{5 \cdot 3}{x \cdot 3} + \frac{2 \cdot x}{3 \cdot x}\\ &= \frac{15}{3x} + \frac{2x}{3x}\\ &= \frac{15 + 2x}{3x}\end{aligned}$

Section 3.6

1 $\begin{aligned}5\tfrac{2}{3} &= 5 + \frac{2}{3}\\ &= \frac{5}{1} + \frac{2}{3}\\ &= \frac{5 \cdot 3}{1 \cdot 3} + \frac{2}{3}\\ &= \frac{15}{3} + \frac{2}{3}\\ &= \frac{17}{3}\end{aligned}$

2 $\begin{aligned}3\tfrac{1}{6} &= 3 + \frac{1}{6}\\ &= \frac{18}{6} + \frac{1}{6}\\ &= \frac{19}{6}\end{aligned}$

3 $\begin{aligned}5\tfrac{2}{3} &= \frac{(3 \cdot 5) + 2}{3}\\ &= \frac{17}{3}\end{aligned}$

4 $\begin{aligned}6\tfrac{4}{9} &= \frac{(9 \cdot 6) + 4}{9}\\ &= \frac{58}{9}\end{aligned}$

5 $\begin{array}{r}3\\ 3\overline{)11}\\ \underline{9}\\ 2\end{array}$ so $\frac{11}{3} = 3\tfrac{2}{3}$

6 $\begin{array}{r}2\\ 5\overline{)14}\\ \underline{10}\\ 4\end{array}$ so $\frac{14}{5} = 2\tfrac{4}{5}$

7 $\begin{array}{r}7\\ 26\overline{)207}\\ \underline{182}\\ 25\end{array}$ so $\frac{207}{26} = 7\tfrac{25}{26}$

8 $\begin{aligned}3 + \frac{2}{x} &= \frac{3}{1} + \frac{2}{x}\\ &= \frac{3 \cdot x}{1 \cdot x} + \frac{2}{x}\\ &= \frac{3x}{x} + \frac{2}{x}\\ &= \frac{3x + 2}{x}\end{aligned}$

9 $\begin{aligned}x - \frac{2}{3} &= \frac{x}{1} - \frac{2}{3}\\ &= \frac{x \cdot 3}{1 \cdot 3} - \frac{2}{3}\\ &= \frac{3x}{3} - \frac{2}{3}\\ &= \frac{3x - 2}{3}\end{aligned}$

Section 3.7

1 $\begin{aligned}2\tfrac{3}{4} \cdot 4\tfrac{1}{3} &= \frac{11}{4} \cdot \frac{13}{3}\\ &= \frac{143}{12}\\ &= 11\tfrac{11}{12}\end{aligned}$

2 $\begin{aligned}2 \cdot 3\tfrac{5}{8} &= \frac{2}{1} \cdot \frac{29}{8}\\ &= \frac{58}{8}\\ &= 7\tfrac{2}{8}\\ &= 7\tfrac{1}{4}\end{aligned}$

3 $\begin{aligned}1\tfrac{3}{5} \div 3\tfrac{2}{5} &= \frac{8}{5} \div \frac{17}{5}\\ &= \frac{8}{5} \cdot \frac{5}{17}\\ &= \frac{8}{17}\end{aligned}$

4 $\begin{aligned}4\tfrac{5}{8} \div 2 &= \frac{37}{8} \div \frac{2}{1}\\ &= \frac{37}{8} \cdot \frac{1}{2}\\ &= \frac{37}{16}\\ &= 2\tfrac{5}{16}\end{aligned}$

Section 3.8

1
$$\begin{aligned} 3\tfrac{2}{3} + 2\tfrac{1}{4} &= 3 + \frac{2}{3} + 2 + \frac{1}{4} \\ &= (3 + 2) + \left(\frac{2}{3} + \frac{1}{4}\right) \\ &= 5 + \left(\frac{2 \cdot 4}{3 \cdot 4} + \frac{1 \cdot 3}{4 \cdot 3}\right) \\ &= 5 + \left(\frac{8}{12} + \frac{3}{12}\right) \\ &= 5 + \frac{11}{12} = 5\tfrac{11}{12} \end{aligned}$$

2
$$\begin{array}{rcrcr} 5\frac{3}{4} & = & 5\frac{3\cdot 5}{4\cdot 5} & = & 5\frac{15}{20} \\ +\,6\frac{4}{5} & = & 6\frac{4\cdot 4}{5\cdot 4} & = & 6\frac{16}{20} \\ \hline & & & & 11\frac{31}{20} = 12\frac{11}{20} \end{array}$$

3
$$\begin{array}{rcrcr} 6\frac{3}{4} & = & 6\frac{3\cdot 2}{4\cdot 2} & = & 6\frac{6}{8} \\ +\,2\frac{7}{8} & = & 2\frac{7}{8} & = & 2\frac{7}{8} \\ \hline & & & & 8\frac{13}{8} = 9\frac{5}{8} \end{array}$$

4
$$\begin{array}{rcrcr} 2\frac{1}{3} & = & 2\frac{1\cdot 4}{3\cdot 4} & = & 2\frac{4}{12} \\ 1\frac{1}{4} & = & 1\frac{1\cdot 3}{4\cdot 3} & = & 1\frac{3}{12} \\ +\,3\frac{11}{12} & = & 3\frac{11}{12} & = & 3\frac{11}{12} \\ \hline & & & & 6\frac{18}{12} = 7\frac{6}{12} = 7\frac{1}{2} \end{array}$$

5
$$\begin{array}{r} 4\frac{7}{8} \\ -\,1\frac{5}{8} \\ \hline 3\frac{2}{8} = 3\frac{1}{4} \end{array}$$

6
$$\begin{array}{rcrcr} 12\frac{7}{10} & = & 12\frac{7}{10} & = & 12\frac{7}{10} \\ -\,7\frac{2}{5} & = & -\,7\frac{2\cdot 2}{5\cdot 2} & = & -\,7\frac{4}{10} \\ \hline & & & & 5\frac{3}{10} \end{array}$$

7
$$\begin{array}{rcr} 10 & = & 9\frac{7}{7} \\ -\,5\frac{4}{7} & = & -\,5\frac{4}{7} \\ \hline & & 4\frac{3}{7} \end{array}$$

8
$$\begin{array}{rcr} 6\frac{1}{3} & = & 5\frac{4}{3} \\ -\,2\frac{2}{3} & = & -\,2\frac{2}{3} \\ \hline & & 3\frac{2}{3} \end{array}$$

9
$$\begin{array}{rcrcrcr} 6\frac{3}{4} & = & 6\frac{3\cdot 3}{4\cdot 3} & = & 6\frac{9}{12} & = & 5\frac{21}{12} \\ -\,2\frac{5}{6} & = & -\,2\frac{5\cdot 2}{6\cdot 2} & = & -\,2\frac{10}{12} & = & -\,2\frac{10}{12} \\ \hline & & & & & & 3\frac{11}{12} \end{array}$$

10
$$\begin{array}{rcrcrcr} 17\frac{1}{8} & = & 17\frac{1\cdot 5}{8\cdot 5} & = & 17\frac{5}{40} & = & 16\frac{45}{40} \\ -\,12\frac{4}{5} & = & -\,12\frac{4\cdot 8}{5\cdot 8} & = & -\,12\frac{32}{40} & = & -\,12\frac{32}{40} \\ \hline & & & & & & 4\frac{13}{40} \end{array}$$

11 $10{,}000\left(2\tfrac{1}{8} - 1\tfrac{15}{16}\right) = 10{,}000\left(\dfrac{3}{16}\right) = \dfrac{30{,}000}{16} = \$1{,}875$

Section 3.9

1
$$\begin{aligned} 4 + \left(1\tfrac{1}{2}\right)\left(2\tfrac{3}{4}\right) &= 4 + \left(\frac{3}{2}\right)\left(\frac{11}{4}\right) \\ &= 4 + \frac{33}{8} \\ &= \frac{32}{8} + \frac{33}{8} \\ &= \frac{65}{8} \\ &= 8\tfrac{1}{8} \end{aligned}$$

2
$$\begin{aligned} \left(\frac{2}{3} + \frac{1}{6}\right)\left(2\tfrac{5}{6} + 1\tfrac{1}{3}\right) &= \left(\frac{5}{6}\right)\left(4\tfrac{1}{6}\right) \\ &= \frac{5}{6}\left(\frac{25}{6}\right) \\ &= \frac{125}{36} \\ &= 3\tfrac{17}{36} \end{aligned}$$

3
$$\begin{aligned} \frac{3}{7} + \frac{1}{3}\left(1\tfrac{1}{2} + 4\tfrac{1}{2}\right)^2 &= \frac{3}{7} + \frac{1}{3}(6)^2 \\ &= \frac{3}{7} + \frac{1}{3}(36) \\ &= \frac{3}{7} + 12 \\ &= 12\tfrac{3}{7} \end{aligned}$$

4
$$\begin{aligned} \frac{\frac{2}{3}}{\frac{5}{9}} &= \frac{2}{3} \div \frac{5}{9} \\ &= \frac{2}{3} \cdot \frac{9}{5} \\ &= \frac{18}{15} \\ &= \frac{6}{5} = 1\tfrac{1}{5} \end{aligned}$$

5
$$\begin{aligned} \frac{\frac{1}{2} + \frac{3}{4}}{\frac{2}{3} - \frac{1}{4}} &= \frac{12\left(\frac{1}{2} + \frac{3}{4}\right)}{12\left(\frac{2}{3} - \frac{1}{4}\right)} \\ &= \frac{12 \cdot \frac{1}{2} + 12 \cdot \frac{3}{4}}{12 \cdot \frac{2}{3} - 12 \cdot \frac{1}{4}} \\ &= \frac{6 + 9}{8 - 3} \\ &= \frac{15}{5} = 3 \end{aligned}$$

6
$$\begin{aligned} \frac{4 + \frac{2}{3}}{3 - \frac{1}{4}} &= \frac{12\left(4 + \frac{2}{3}\right)}{12\left(3 - \frac{1}{4}\right)} \\ &= \frac{12 \cdot 4 + 12 \cdot \frac{2}{3}}{12 \cdot 3 - 12 \cdot \frac{1}{4}} \\ &= \frac{48 + 8}{36 - 3} \\ &= \frac{56}{33} = 1\tfrac{23}{33} \end{aligned}$$

7
$$\begin{aligned} \frac{12\frac{1}{3}}{6\frac{2}{3}} &= 12\tfrac{1}{3} \div 6\tfrac{2}{3} \\ &= \frac{37}{3} \div \frac{20}{3} \\ &= \frac{37}{3} \cdot \frac{3}{20} \\ &= \frac{37}{20} \\ &= 1\tfrac{17}{20} \end{aligned}$$

CHAPTER 4

Section 4.1

1 $6(x+4) = 6(x) + 6(4)$
$= 6x + 24$

2 $-3(2x+4) = -3(2x) + (-3)(4)$
$= -6x + (-12)$
$= -6x - 12$

3 $6x - 2 + 3x + 8 = 6x + 3x + (-2) + 8$
$= 9x + 6$

4 $2(4x+3) + 7 = 2(4x) + 2(3) + 7$
$= 6x + 3 + 20x - 15$
$= 26x - 12$

5 $3(2x+1) + 5(4x-3) = 3(2x) + 3(1) + 5(4x) - 5(3)$
$= 8x + 6 + 7$
$= 8x + 13$

Section 4.2

1 When $x = 3$
the equation $5x - 4 = 11$
becomes $5(3) - 4 = 11$
or $15 - 4 = 11$
$11 = 11$

2 When $a = -3$
the equation $6a - 3 = 2a + 4$
becomes $6(-3) - 3 = 2(-3) + 4$
$-18 - 3 = -6 + 4$
$-21 = -2$
This is a false statement, so $a = -3$ is not a solution.

3
$$\begin{aligned} x + 5 &= -2 \\ x + 5 + (-5) &= -2 + (-5) \\ x + 0 &= -7 \\ x &= -7 \end{aligned}$$

4
$$\begin{aligned} a - 2 &= 7 \\ a - 2 + 2 &= 7 + 2 \\ a + 0 &= 9 \\ a &= 9 \end{aligned}$$

5
$$\begin{aligned} y + 6 - 2 &= 8 - 9 \\ y + 4 &= -1 \\ y + 4 + (-4) &= -1 + (-4) \\ y + 0 &= -5 \\ y &= -5 \end{aligned}$$

6
$$\begin{aligned} 5x - 3 - 4x &= 4 - 7 \\ x - 3 &= -3 \\ x - 3 + 3 &= -3 + 3 \\ x + 0 &= 0 \\ x &= 0 \end{aligned}$$

7
$$\begin{aligned} -5 - 7 &= x + 2 \\ -12 &= x + 2 \\ -12 + (-2) &= x + 2 + (-2) \\ -14 &= x + 0 \\ -14 &= x \end{aligned}$$

8
$$\begin{aligned} a - \frac{2}{3} &= \frac{5}{6} \\ a - \frac{2}{3} + \frac{2}{3} &= \frac{5}{6} + \frac{2}{3} \\ a &= \frac{9}{6} = \frac{3}{2} \end{aligned}$$

Section 4.3

1
$$\begin{aligned} \frac{1}{3}x &= 5 \\ 3 \cdot \frac{1}{3}x &= 3 \cdot 5 \\ x &= 15 \end{aligned}$$

2
$$\begin{aligned} \frac{1}{5}a + 3 &= 7 \\ \frac{1}{5}a + 3 + (-3) &= 7 + (-3) \\ \frac{1}{5}a &= 4 \\ 5 \cdot \frac{1}{5}a &= 5 \cdot 4 \\ a &= 20 \end{aligned}$$

3
$$\begin{aligned} \frac{3}{5}y &= 6 \\ \frac{5}{3} \cdot \frac{3}{5}y &= \frac{5}{3} \cdot 6 \\ y &= 10 \end{aligned}$$

4
$$\begin{aligned} -\frac{3}{4}x &= \frac{6}{5} \\ -\frac{4}{3}\left(-\frac{3}{4}x\right) &= -\frac{4}{3} \cdot \frac{6}{5} \\ x &= -\frac{8}{5} \end{aligned}$$

5
$$\begin{aligned} 6x &= -42 \\ \frac{6x}{6} &= \frac{-42}{6} \\ x &= -7 \end{aligned}$$

6
$$\begin{aligned} -5x + 6 &= -14 \\ -5x + 6 + (-6) &= -14 + (-6) \\ -5x &= -20 \\ \frac{-5x}{-5} &= \frac{-20}{-5} \\ x &= 4 \end{aligned}$$

7
$$\begin{aligned} 3x - 7x + 5 &= 3 - 18 \\ -4x + 5 &= -15 \\ -4x + 5 + (-5) &= -15 + (-5) \\ -4x &= -20 \\ \frac{-4x}{-4} &= \frac{-20}{-4} \\ x &= 5 \end{aligned}$$

8
$$\begin{aligned} -5 + 4 &= 2x - 11 + 3x \\ -1 &= 5x - 11 \\ -1 + 11 &= 5x - 11 + 11 \\ 10 &= 5x \\ \frac{10}{5} &= \frac{5x}{5} \\ 2 &= x \end{aligned}$$

Section 4.4

1
$$\begin{aligned} 4(x+3) &= -8 \\ 4x+12 &= -8 \\ 4x+12+(-12) &= -8+(-12) \\ 4x &= -20 \\ \frac{4x}{4} &= \frac{-20}{4} \\ x &= -5 \end{aligned}$$

2
$$\begin{aligned} 6a+7 &= 4a-3 \\ 6a+(-4a)+7 &= 4a+(-4a)-3 \\ 2a+7 &= -3 \\ 2a+7+(-7) &= -3+(-7) \\ 2a &= -10 \\ \frac{2a}{2} &= \frac{-10}{2} \\ a &= -5 \end{aligned}$$

3
$$\begin{aligned} 5(x-2)+3 &= -12 \\ 5x-10+3 &= -12 \\ 5x-7 &= -12 \\ 5x-7+7 &= -12+7 \\ 5x &= -5 \\ \frac{5x}{5} &= \frac{-5}{5} \\ x &= -1 \end{aligned}$$

4
$$\begin{aligned} 3(4x-5)+6 &= 3x+9 \\ 12x-15+6 &= 3x+9 \\ 12x-9 &= 3x+9 \\ 12x+(-3x)-9 &= 3x+(-3x)+9 \\ 9x-9 &= 9 \\ 9x-9+9 &= 9+9 \\ 9x &= 18 \\ \frac{9x}{9} &= \frac{18}{9} \\ x &= 2 \end{aligned}$$

5
$$\begin{aligned} \frac{x}{3}+\frac{x}{6} &= 9 \\ 6\left(\frac{x}{3}+\frac{x}{6}\right) &= 6(9) \\ 6\left(\frac{x}{3}\right)+6\left(\frac{x}{6}\right) &= 6(9) \\ 2x+x &= 54 \\ 3x &= 54 \\ x &= 18 \end{aligned}$$

6
$$\begin{aligned} 3x+\frac{1}{4} &= \frac{5}{8} \\ 8\left(3x+\frac{1}{4}\right) &= 8\left(\frac{5}{8}\right) \\ 8(3x)+8\left(\frac{1}{4}\right) &= 8\left(\frac{5}{8}\right) \\ 24x+2 &= 5 \\ 24x &= 3 \\ x &= \frac{1}{8} \end{aligned}$$

7
$$\begin{aligned} \frac{4}{x}+3 &= \frac{11}{5} \\ 5x\left(\frac{4}{x}+3\right) &= 5x\left(\frac{11}{5}\right) \\ 5x\left(\frac{4}{x}\right)+5x(3) &= 5x\left(\frac{11}{5}\right) \\ 20+15x &= 11x \\ 20 &= -4x \\ -5 &= x \end{aligned}$$

Section 4.5

1 **Step 1** ***Read and list.***

Known items: The numbers 3 and 10

Unknown item: The number in question

Step 2 ***Assign a variable and translate the information.***

Let x = the number asked for in the problem.

Then "The sum of a number and 3" translates to $x + 3$.

Step 3 ***Reread and write an equation.***

The sum of x and 3 | is | 10.
↓ ↓
$x + 3 = 10$

Step 4 ***Solve the equation.***

$$\begin{aligned} x+3 &= 10 \\ x &= 7 \end{aligned}$$

Step 5 ***Write your answer.***

The number is 7.

Step 6 ***Reread and check.***

The sum of **7** and 3 is 10.

2 **Step 1** ***Read and list.***

Known items: The numbers 4 and 34, twice a number, and three times a number

Unknown item: The number in question

Step 2 ***Assign a variable and translate the information.***

Let x = the number asked for in the problem.

Then "The sum of twice a number and three times the number" translates to $2x + 3x$.

Step 3 ***Reread and write an equation.***

4 added to | the sum of twice a number and three times the number | is | 34
↓ ↓ ↓ ↓
$4 + \quad 2x + 3x \quad = \quad 34$

Step 4 ***Solve the equation.***

$$\begin{aligned} 4+2x+3x &= 34 \\ 5x+4 &= 34 \\ 5x &= 30 \\ x &= 6 \end{aligned}$$

Step 5 ***Write your answer.***

The number is 6.

Step 6 ***Reread and check.***

Twice **6** is 12 and three times **6** is 18. Their sum is $12 + 18 = 30$. Four added to this is 34. Therefore, 4 added to the sum of twice **6** and three times **6** is 34.

3 Step 1 ***Read and list.***

Known items: Length is twice width; perimeter 42 cm

Unknown items: The length and the width

Step 2 ***Assign a variable and translate the information.***

Let $x =$ the width. Since the length is twice the width, the length must be $2x$. Here is a picture.

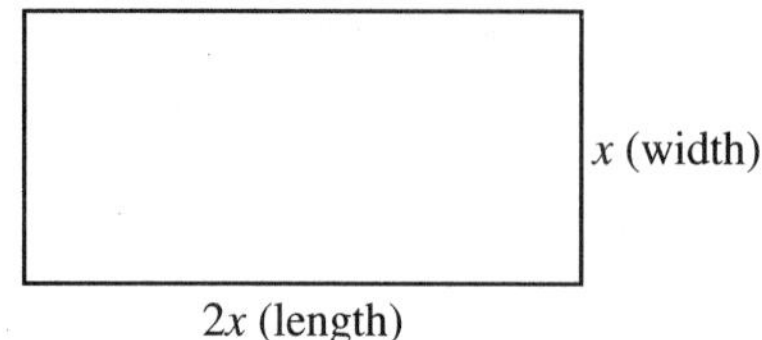

Step 3 ***Reread and write an equation.***

The perimeter is the sum of the sides, and is also given as 42; therefore,

$$x + x + 2x + 2x = 42$$

Step 4 ***Solve the equation.***

$$\begin{aligned} x + x + 2x + 2x &= 42 \\ 6x &= 42 \\ x &= 7 \end{aligned}$$

Step 5 ***Write your answer.***

The width is 7 centimeters and the length is $2(7) = 14$ centimeters.

Step 6 ***Reread and check.***

The length, 14, is twice the width, 7. The perimeter is $7 + 7 + 14 + 14 = 42$ centimeters.

4 Step 1 ***Read and list.***

Known items: Joyce is 21 years older than Travis. Six years from now their ages will add to 49.

Unknown items: Their ages now

Step 2 ***Assign a variable and translate the information.***

Let $x =$ Travis's age now; then since Joyce is 21 years older than that, she is $x + 21$ years old.

Step 3 ***Reread and write an equation.***

	Now	In 6 years
Joyce	$x + 21$	$x + 27$
Travis	x	$x + 6$

$$x + 27 + x + 6 = 49$$

Step 4 ***Solve the equation.***

$$\begin{aligned} x + 27 + x + 6 &= 49 \\ 2x + 33 &= 49 \\ 2x &= 16 \\ x &= 8 \end{aligned}$$

Travis is now 8 years old, and Joyce is $8 + 21 = 29$ years old.

Step 5 ***Write your answer.***

Travis is 8 years old and Joyce is 29 years old.

Step 6 ***Reread and check.***

Joyce is 21 years older than Travis. In six years, Joyce will be 35 years old and Travis will be 14 years old. At that time, the sum of their ages will be $35 + 14 = 49$.

Section 4.6

1
$$\begin{aligned} P &= 2w + 2l \\ &= 2 \cdot 10 + 2 \cdot 15 \\ &= 20 + 30 \\ &= 50 \text{ feet} \end{aligned}$$

2 When $F = 77$

the formula $C = \frac{5}{9}(F - 32)$

becomes
$$\begin{aligned} C &= \frac{5}{9}(77 - 32) \\ &= \frac{5}{9}(45) \\ &= \frac{5}{9} \cdot \frac{45}{1} \\ &= \frac{225}{9} \\ &= 25 \text{ degrees Celsius} \end{aligned}$$

3 When $x = 0$

the formula $y = 2x + 6$

becomes
$$\begin{aligned} y &= 2 \cdot 0 + 6 \\ &= 0 + 6 \\ &= 6 \end{aligned}$$

4 When $x = -3$

the formula $2x + 3y = 4$

becomes
$$\begin{aligned} 2(-3) + 3y &= 4 \\ -6 + 3y &= 4 \\ 3y &= 10 \\ y &= \frac{10}{3} \end{aligned}$$

Section 4.7

1 When $x = 0$ the equation $3x + 5y = 15$ becomes

$$\begin{aligned} 3 \cdot 0 + 5y &= 15 \\ 5y &= 15 \\ y &= 3 \end{aligned}$$

which gives (0, 3) as one solution

When $y = 0$ the equation $3x + 5y = 15$ becomes

$$\begin{aligned} 3x + 5 \cdot 0 &= 15 \\ 3x &= 15 \\ x &= 5 \end{aligned}$$

which means (5, 0) is a second solution

When $x = -5$ the equation $3x + 5y = 15$ becomes

$$\begin{aligned} 3(-5) + 5y &= 15 \\ -15 + 5y &= 15 \\ 5y &= 30 \\ y &= 6 \end{aligned}$$

which gives (−5, 6) as a third solution

2 When $x = 2$, we have

$$\begin{aligned} 5 \cdot 2 + 2y &= 20 \\ 10 + 2y &= 20 \\ 2y &= 10 \\ y &= 5 \end{aligned}$$

When $x = 0$, we have

$$\begin{aligned} 5 \cdot 0 + 2y &= 20 \\ 2y &= 20 \\ y &= 10 \end{aligned}$$

When $y = 5$, we have

$$\begin{aligned} 5x + 2 \cdot 5 &= 20 \\ 5x + 10 &= 20 \\ 5x &= 10 \\ x &= 2 \end{aligned}$$

When $y = 0$, we have

$$\begin{aligned} 5x + 2 \cdot 0 &= 20 \\ 5x &= 20 \\ x &= 4 \end{aligned}$$

3 When $x = 0$, we have

$$\begin{aligned} y &= \frac{1}{2} \cdot 0 + 1 \\ y &= 1 \end{aligned}$$

When $x = 4$, we have

$$\begin{aligned} y &= \frac{1}{2} \cdot 4 + 1 \\ y &= 2 + 1 \\ y &= 3 \end{aligned}$$

When $y = 7$, we have

$$\begin{aligned} 7 &= \frac{1}{2}x + 1 \\ 6 &= \frac{1}{2}x \\ 12 &= x \end{aligned}$$

When $y = -3$, we have

$$\begin{aligned} -3 &= \frac{1}{2}x + 1 \\ -4 &= \frac{1}{2}x \\ -8 &= x \end{aligned}$$

4 (1, 5) is not a solution. When we substitute 1 for x and 5 for y into $y = 5x - 6$, we get a false statement.

$$\begin{aligned} 5 &= 5 \cdot 1 - 6 \\ 5 &= 5 - 6 \\ 5 &= -1 \quad \text{a false statement} \end{aligned}$$

(2, 4) is a solution. Substituting 2 for x and 4 for y in the equation $y = 5x - 6$, yields a true statement.

$$\begin{aligned} 4 &= 5 \cdot 2 - 6 \\ 4 &= 10 - 6 \\ 4 &= 4 \quad \text{a true statement} \end{aligned}$$

Section 4.8

1 **2**

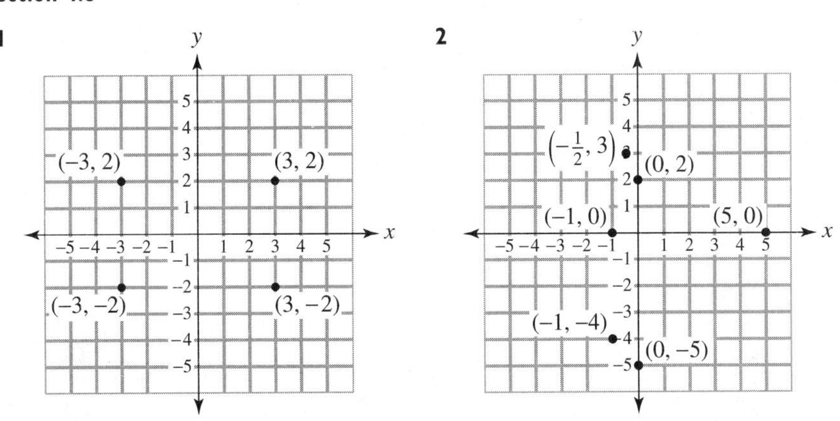

Section 4.9

1

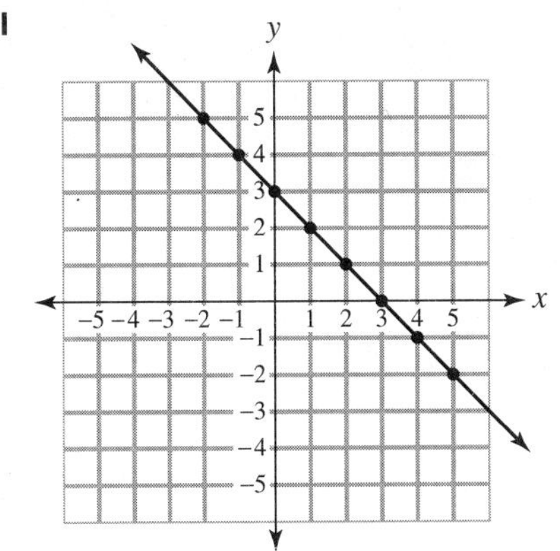

All the points shown on the graph have coordinates that add to 3.

2 When $x = 1$, $y = -2(1) + 1$

$$y = -2 + 1$$
$$y = -1$$

$(1, -1)$ is one solution

When $x = 0$, $y = -2(0) + 1$

$$y = 0 + 1$$
$$y = 1$$

$(0, 1)$ is a second solution

When $x = -1$, $y = -2(-1) + 1$

$$y = 2 + 1$$
$$y = 3$$

$(-1, 3)$ is our third solution

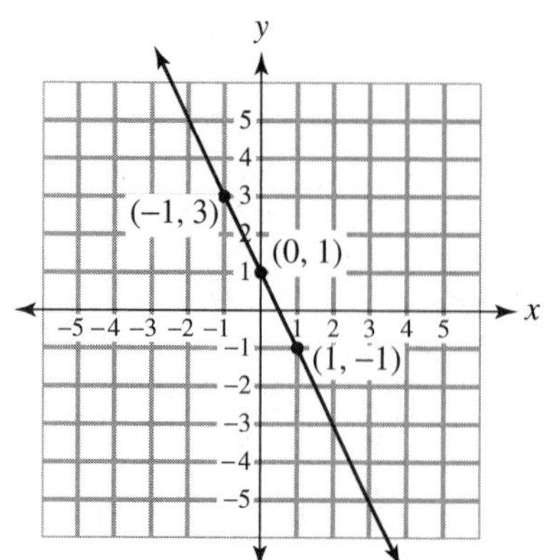

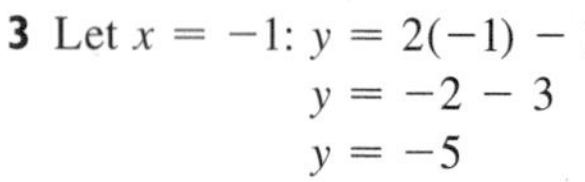

3 Let $x = -1$: $y = 2(-1) - 3$

$$y = -2 - 3$$
$$y = -5$$

$(-1, -5)$ is one solution

Let $x = 0$: $y = 2(0) - 3$

$$y = 0 - 3$$
$$y = -3$$

$(0, -3)$ is another solution

Let $x = 2$: $y = 2(2) - 3$

$$y = 4 - 3$$
$$y = 1$$

$(2, 1)$ is a third solution

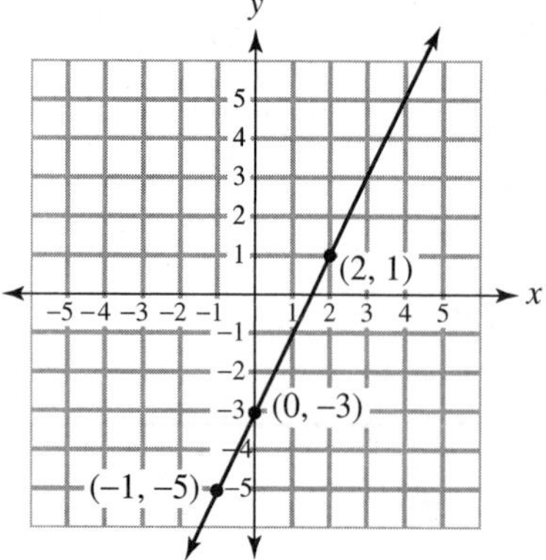

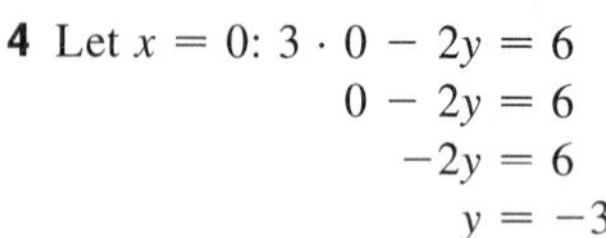

4 Let $x = 0$: $3 \cdot 0 - 2y = 6$

$$0 - 2y = 6$$
$$-2y = 6$$
$$y = -3$$

$(0, -3)$ is one solution

Let $y = 0$: $3x - 2 \cdot 0 = 6$

$$3x - 0 = 6$$
$$3x = 6$$
$$x = 2$$

$(2, 0)$ is a second solution

The third point is up to you to find. Substituting -2, 4, or 6 for x will make your work easier. Do you know why?

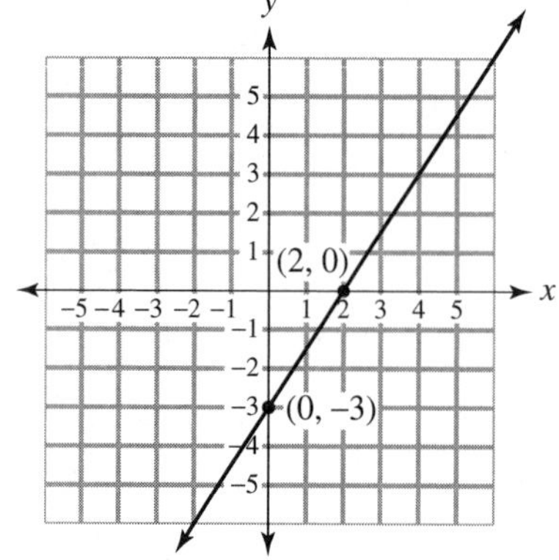

5 Any ordered pair with an x-coordinate of -3 is a solution to the equation $x = -3$. Here are a few of them: $(-3, -4)$, $(-3, -2)$, $(-3, 0)$, $(-3, 2)$, and $(-3, 4)$.

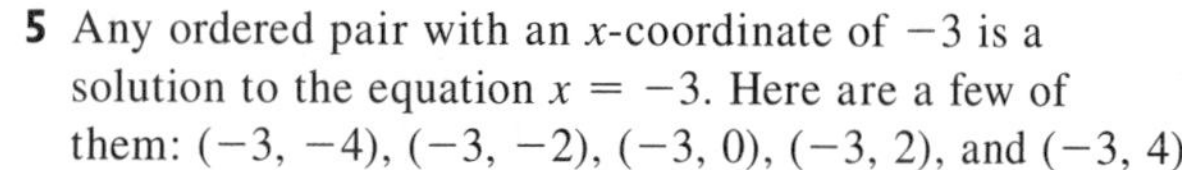

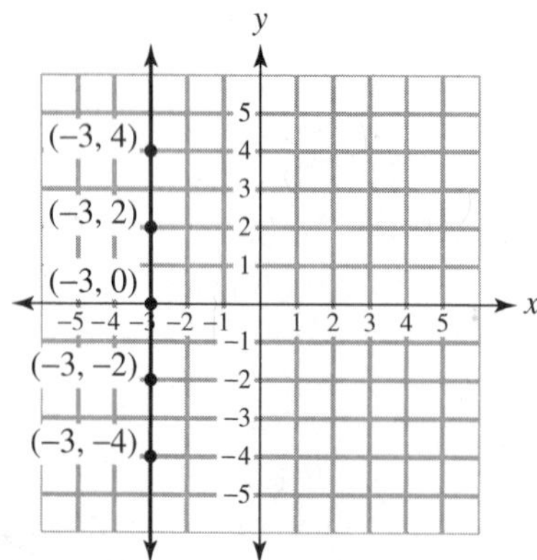

CHAPTER 5

Section 5.1

1 $700 + 80 + 5 + \frac{4}{10} + \frac{6}{100} + \frac{2}{1,000}$ **2 a** Six hundredths **b** Seven tenths **c** Eight thousandths

3 a Five and six hundredths **b** Four and seven tenths **c** Three and eight thousandths

4 Five and ninety-eight hundredths **5** Three hundred five and four hundred six thousandths

Section 5.2

1 $\begin{array}{r} 38.45 \\ +456.073 \\ \hline \end{array} = \begin{array}{r} 38\frac{45}{100} \\ 456\frac{73}{1,000} \\ \hline \end{array} = \begin{array}{r} 38\frac{450}{1,000} \\ 456\frac{73}{1,000} \\ \hline 494\frac{523}{1,000} \end{array} = 494.523$

2 $\begin{array}{r} 78.674 \\ -23.431 \\ \hline 55.243 \end{array}$ **3** $\begin{array}{r} 16.000 \\ 0.033 \\ 4.600 \\ +\ 0.080 \\ \hline 20.713 \end{array}$ **4** $\begin{array}{r} 6.7000 \\ -2.0563 \\ \hline 4.6437 \end{array}$ **5** $\begin{array}{r} 7.000 \\ +\ 3.567 \\ \hline 10.567 \end{array}$ and $\begin{array}{r} 10.567 \\ -\ 5.890 \\ \hline 4.677 \end{array}$

8 $\begin{array}{r} \$10.00 \\ -\ \ 9.56 \\ \hline \$\ \ .44 \end{array}$ 1 quarter + 2 dimes + 4 pennies = 0.25 + 0.20 + 0.04 = 0.49, which is too much change. One of the dimes should be a nickel.

9

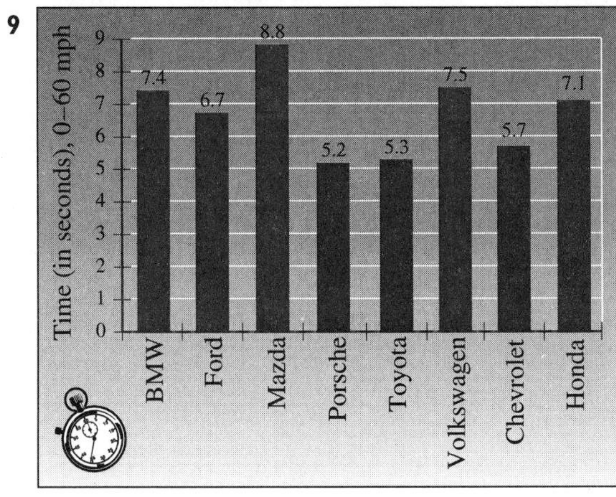

Section 5.3

1 $0.4 \times 0.6 = \frac{4}{10} \times \frac{6}{10} = \frac{24}{100} = 0.24$

2 $0.5 \times 0.007 = \frac{5}{10} \times \frac{7}{1,000} = \frac{35}{10,000} = 0.0035$

3 $3.5 \times 0.04 = 3\frac{5}{10} \times \frac{4}{100} = \frac{35}{10} \times \frac{4}{100} = \frac{140}{1,000} = \frac{14}{100} = 0.14$

4 3 + 2 = 5 digits to the right

5 $\begin{array}{r} 4.03 \\ \times 5.22 \\ \hline 806 \\ 8060 \\ 201500 \\ \hline 21.0366 \end{array}$

8 a $80 \times 6 = 480$ **b** $40 \times 180 = 7,200$ **c** $8^2 = 64$

9 $0.03(5.5 + 0.02) = 0.03(5.52) = 0.1656$

10 $5.7 + 14(2.4)^2 = 5.7 + 14(5.76) = 5.7 + 80.64 = 86.34$

11 $0.43 + 19(0.32) = 0.43 + 6.08 = \$6.51$

12 $6.32(36) + 9.48(14) = 227.52 + 132.72 = \$360.24$

13 $C = 3.14(3) = 9.42$ cm

14 $C = 2(3.14)(5) = 31.4$ ft

15 Radius $= \frac{1}{2}(20) = 10$ ft

$A = \pi r^2 = (3.14)(10)^2$
$= 314 \text{ ft}^2$

16 $V = \pi r^2 h$
$= (3.14)(0.250)^2(6)$
$= 1.178 \text{ in}^3$

Section 5.4

1
$$\begin{array}{r} 154.2 \\ 30\overline{)4{,}626.0} \\ \underline{3\,0} \\ 1\,62 \\ \underline{1\,50} \\ 126 \\ \underline{120} \\ 6\,0 \\ \underline{6\,0} \\ 0 \end{array}$$

2
$$\begin{array}{r} 6.7 \\ 5\overline{)33.5} \\ \underline{30} \\ 3\,5 \\ \underline{3\,5} \\ 0 \end{array}$$

3
$$\begin{array}{r} 2.636 \\ 18\overline{)47.448} \\ \underline{36} \\ 11\,4 \\ \underline{10\,8} \\ 64 \\ \underline{54} \\ 108 \\ \underline{108} \\ 0 \end{array}$$

4
$$\begin{array}{r} 45.54 \\ 25\overline{)1{,}138.50} \\ \underline{1\,00} \\ 138 \\ \underline{125} \\ 13\,5 \\ \underline{12\,5} \\ 1\,00 \\ \underline{1\,00} \\ 0 \end{array}$$

5
$$\begin{array}{r} 3.15 \\ 42\overline{)132.30} \\ \underline{126} \\ 6\,3 \\ \underline{4\,2} \\ 2\,10 \\ \underline{2\,10} \\ 0 \end{array}$$

6
$$\begin{array}{r} 1.422 \\ 0.32.\overline{)0.45.530} \\ \underline{32} \\ 13\,5 \\ \underline{12\,8} \\ 73 \\ \underline{64} \\ 90 \\ \underline{64} \\ 26 \end{array}$$

Answer to nearest hundredth is 1.42

7
$$\begin{array}{r} 3\,16.66 \\ 0.06.\overline{)19.00.00} \\ \underline{18} \\ 1\,0 \\ \underline{6} \\ 40 \\ \underline{36} \\ 4\,0 \\ \underline{3\,6} \\ 40 \\ \underline{36} \\ 4 \end{array}$$

Answer to nearest tenth is 316.7

8
$$\begin{array}{r} 28.5 \text{ hours} \\ 6.54.\overline{)186.39.0} \\ \underline{130\,8} \\ 55\,59 \\ \underline{52\,32} \\ 3\,27\,0 \\ \underline{3\,27\,0} \\ 0 \end{array}$$

9 $\dfrac{4.39 - 0.43}{0.33} = \dfrac{3.96}{0.33}$
$= 12$ additional minutes

The call was 13 minutes long.

10

Class	Units	Grade	Value	Grade Points
Algebra	5	B	3	$5 \times 3 = 15$
Chemistry	4	B	3	$4 \times 3 = 12$
English	3	A	4	$3 \times 4 = 12$
History	3	B	3	$3 \times 3 = 9$
Total Units:	15		Total Grade Points:	48

$\text{GPA} = \dfrac{48}{15} = 3.20$

Section 5.5

1
$$\begin{array}{r} 0.6 \\ 5\overline{)3.0} \\ \underline{3\,0} \\ 0 \end{array}$$
so $\dfrac{3}{5} = 0.6$

2
$$\begin{array}{r} 0.9166 \\ 12\overline{)11.0000} \\ \underline{10\,8} \\ 20 \\ \underline{12} \\ 80 \\ \underline{72} \\ 80 \\ \underline{72} \\ 8 \end{array}$$
so $\dfrac{11}{12} = 0.917$ to the nearest thousandth

3
$$\begin{array}{r} 0.4545 \\ 11\overline{)5.0000} \\ \underline{4\,4} \\ 60 \\ \underline{55} \\ 50 \\ \underline{44} \\ 60 \\ \underline{55} \\ 5 \end{array}$$
so $\dfrac{5}{11} = 0.\overline{45}$

4 $0.48 = \dfrac{48}{100} = \dfrac{12}{25}$

5 $0.025 = \dfrac{25}{1{,}000} = \dfrac{1}{40}$

6 $12.8 = 12\frac{8}{10} = 12\frac{4}{5}$

7 $\dfrac{14}{25}(2.43 + 0.27) = 0.56(2.43 + 0.27)$
$= 0.56(2.70)$
$= 1.512$

8 $\frac{1}{4} + 0.25\left(\frac{3}{5}\right) = \frac{1}{4} + \frac{1}{4}\left(\frac{3}{5}\right)$

$= \frac{1}{4} + \frac{3}{20}$

$= \frac{5}{20} + \frac{3}{20}$

$= \frac{8}{20}$

$= \frac{2}{5}$ or 0.4

9 $\left(\frac{1}{3}\right)^3(5.4) + \left(\frac{1}{5}\right)^2(2.5) = \frac{1}{27}(5.4) + \frac{1}{25}(2.5)$

$= 0.2 + 0.1$

$= 0.3$

10 $\frac{3}{4}(35.50) = 26.625$

$= \$26.63$ to the nearest cent

11 $V = \pi r^2 h + \frac{1}{2} \cdot \frac{4}{3}\pi r^3$

$= (3.14)(10)^2(10) + \frac{1}{2} \cdot \frac{4}{3}(3.14)(10)^3$

$= 3{,}140 + \frac{2}{3}(3{,}140)$

$= 3{,}140 + 2{,}093.3$

$= 5{,}233.3 \text{ in}^3$

Section 5.6

1
$$x - 3.4 = 6.7$$
$$x - 3.4 + \mathbf{3.4} = 6.7 + \mathbf{3.4}$$
$$x + 0 = 10.1$$
$$x = 10.1$$

2
$$4y = 3.48$$
$$\frac{4y}{\mathbf{4}} = \frac{3.48}{\mathbf{4}}$$
$$y = 0.87$$

3
$$\frac{1}{5}x - 2.4 = 8.3$$
$$\frac{1}{5}x - 2.4 + \mathbf{2.4} = 8.3 + \mathbf{2.4}$$
$$\frac{1}{5}x = 10.7$$
$$\mathbf{5}\left(\frac{1}{5}x\right) = \mathbf{5}(10.7)$$
$$x = 53.5$$

4
$$7a - 0.18 = 2a + 0.77$$
$$7a + (\mathbf{-2a}) - 0.18 = 2a + (\mathbf{-2a}) + 0.77$$
$$5a - 0.18 = 0.77$$
$$5a - 0.18 + \mathbf{0.18} = 0.77 + \mathbf{0.18}$$
$$5a = 0.95$$
$$\frac{5a}{\mathbf{5}} = \frac{0.95}{\mathbf{5}}$$
$$a = 0.19$$

5 Let x = the number of miles driven
$$2(11) + 0.16x = 41$$
$$22 + 0.16x = 41$$
$$22 + (\mathbf{-22}) + 0.16x = 41 + (\mathbf{-22})$$
$$0.16x = 19$$
$$\frac{0.16x}{\mathbf{0.16}} = \frac{19}{\mathbf{0.16}}$$
$$x = 118.75 \text{ miles}$$

6 Let x = the number of quarters.

	Quarters	Dimes
Number of	x	$x + 7$
Value of	$0.25x$	$0.10(x + 7)$

$$0.25x + 0.10(x + 7) = 1.75$$
$$0.25x + 0.10x + 0.70 = 1.75$$
$$0.35x + 0.70 = 1.75$$
$$0.35x = 1.05$$
$$x = 3$$

She has 3 quarters and 10 dimes.

Section 5.7

1 $4\sqrt{25} = 4 \cdot 5 = 20$

2 $\sqrt{36} + \sqrt{4} = 6 + 2 = 8$

3 $\sqrt{\frac{36}{100}} = \frac{6}{10} = \frac{3}{5}$

4 $14\sqrt{36} = 14 \cdot 6 = 84$

5 $\sqrt{81} - \sqrt{25} = 9 - 5 = 4$

6 $\sqrt{\frac{64}{121}} = \frac{8}{11}$

7 $5\sqrt{14} \approx 5(3.7416574) = 18.708287 = 18.7083$ to the nearest ten thousandth

8 $\sqrt{405} + \sqrt{147} \approx 20.124612 + 12.124356 = 32.248968 = 32.25$ to the nearest hundredth

9 $\sqrt{\frac{7}{12}} \approx \sqrt{0.5833333}, \approx 0.7637626 = 0.764$ to the nearest thousandth

10 a $c = \sqrt{16^2 + 12^2} = \sqrt{256 + 144} = \sqrt{400}$, $c = 20$ cm

b $c = \sqrt{5^2 + 5^2} = \sqrt{25 + 25} = \sqrt{50}$, $c = 7.07$ ft (to the nearest hundredth)

11 $c = \sqrt{12^2 + 5^2} = \sqrt{144 + 25} = \sqrt{169} = 13$ ft

Section 5.8

1 $\sqrt{63} = \sqrt{3 \cdot 3 \cdot 7} = \sqrt{3 \cdot 3} \cdot \sqrt{7} = 3\sqrt{7}$

2 $\sqrt{45x^2} = \sqrt{3 \cdot 3 \cdot 5 \cdot x \cdot x} = 3 \cdot x \cdot \sqrt{5} = 3x\sqrt{5}$

3 $\sqrt{300} = \sqrt{10 \cdot 10 \cdot 3} = 10\sqrt{3}$

4 $\sqrt{50x^3} = \sqrt{5 \cdot 5 \cdot 2 \cdot x \cdot x \cdot x} = 5 \cdot x \cdot \sqrt{2x} = 5x\sqrt{2x}$

Section 5.9

1 $5\sqrt{3} - 2\sqrt{3} = (5 - 2)\sqrt{3} = 3\sqrt{3}$

2 $7\sqrt{y} + 3\sqrt{y} = (7 + 3)\sqrt{y} = 10\sqrt{y}$

3 $8\sqrt{5} - 2\sqrt{5} + 9\sqrt{5} = (8 - 2 + 9)\sqrt{5} = 15\sqrt{5}$

4 $\sqrt{27} + \sqrt{75} - \sqrt{12} = \sqrt{3 \cdot 3 \cdot 3} + \sqrt{5 \cdot 5 \cdot 3} - \sqrt{2 \cdot 2 \cdot 3} = 3\sqrt{3} + 5\sqrt{3} - 2\sqrt{3} = (3 + 5 - 2)\sqrt{3} = 6\sqrt{3}$

5 $5\sqrt{45} - 3\sqrt{20} = 5\sqrt{3 \cdot 3 \cdot 5} - 3\sqrt{2 \cdot 2 \cdot 5} = 5 \cdot 3\sqrt{5} - 3 \cdot 2\sqrt{5} = 15\sqrt{5} - 6\sqrt{5} = (15 - 6)\sqrt{5} = 9\sqrt{5}$

6 $7\sqrt{18x^3} - 3\sqrt{50x^3} = 7\sqrt{3 \cdot 3 \cdot 2 \cdot x \cdot x \cdot x} - 3\sqrt{5 \cdot 5 \cdot 2 \cdot x \cdot x \cdot x} = 7 \cdot 3 \cdot x\sqrt{2x} - 3 \cdot 5 \cdot x\sqrt{2x} = 21x\sqrt{2x} - 15x\sqrt{2x} = 6x\sqrt{2x}$

CHAPTER 6

Section 6.1

2 $\dfrac{\frac{3}{5}}{\frac{9}{10}} = \frac{3}{5} \cdot \frac{10}{9} = \frac{2}{3}$

3 $\dfrac{0.06}{0.12} = \dfrac{6}{12} = \dfrac{1}{2}$

5 Alcohol to water: $\frac{4}{12} = \frac{1}{3}$; water to alcohol: $\frac{12}{4} = \frac{3}{1}$; water to total solution: $\frac{12}{16} = \frac{3}{4}$

6 $\dfrac{107 \text{ miles}}{2 \text{ hours}} = 53.5$ miles/hour

7 $\dfrac{192 \text{ miles}}{6 \text{ gallons}} = 32$ miles/gallon

8 $\dfrac{18¢}{6 \text{ ounces}} = 3¢$/ounce; $\dfrac{60¢}{12 \text{ ounces}} = 5¢$/ounce; $\dfrac{128¢}{32 \text{ ounces}} = 4¢$/ounce

Section 6.2

1 8 ft = 8 × 12 in. = 96 in.

2 $26 \text{ ft} = 26 \cancel{\text{ft}} \times \dfrac{1 \text{ yd}}{3 \cancel{\text{ft}}} = \dfrac{26}{3} \text{ yd} = 8\frac{2}{3}$ yd, or 8.67 yd

3 $220 \text{ yd} = 220 \cancel{\text{yd}} \times \dfrac{3 \cancel{\text{ft}}}{1 \cancel{\text{yd}}} \times \dfrac{12 \text{ in.}}{1 \cancel{\text{ft}}} = 220 \times 3 \times 12 \text{ in.} = 7{,}920$ in.

4 $67 \text{ cm} = 67 \cancel{\text{cm}} \times \dfrac{1 \text{ m}}{100 \cancel{\text{cm}}} = \dfrac{67 \text{ m}}{100} = 0.67 \text{ m}$

5 $78.4 \text{ mm} = 78.4 \cancel{\text{mm}} \times \dfrac{1 \cancel{\text{m}}}{1{,}000 \cancel{\text{mm}}} \times \dfrac{10 \text{ dm}}{1 \cancel{\text{m}}} = \dfrac{78.4 \times 10}{1{,}000} \text{ dm} = 0.784 \text{ dm}$

6 $6 \text{ pens} = 6 \cancel{\text{pens}} \times \dfrac{28 \cancel{\text{feet of fencing}}}{1 \cancel{\text{pen}}} \times \dfrac{1.72 \text{ dollars}}{1 \cancel{\text{foot of fencing}}} = 6 \times 28 \times 1.72 \text{ dollars} = \$288.96$

7 $50 \text{ miles/hour} = \dfrac{50 \cancel{\text{miles}}}{1 \cancel{\text{hour}}} \times \dfrac{5{,}280 \text{ feet}}{1 \cancel{\text{mile}}} \times \dfrac{1 \cancel{\text{hour}}}{60 \cancel{\text{minutes}}} \times \dfrac{1 \cancel{\text{minute}}}{60 \text{ seconds}} = \dfrac{50 \times 5{,}280 \text{ feet}}{60 \times 60 \text{ seconds}} = 73.3 \text{ feet/second to the nearest tenth}$

Section 6.3

1 $1 \text{ yd}^2 = 1 \cancel{\text{yd}} \times \cancel{\text{yd}} \times \dfrac{3 \text{ ft}}{1 \cancel{\text{yd}}} \times \dfrac{3 \text{ ft}}{1 \cancel{\text{yd}}} = 3 \times 3 \text{ ft} \times \text{ft} = 9 \text{ ft}^2$

2 Length = 36 in. + 12 in. = 48 in.; Width = 24 in. + 12 in. = 36 in.;
Area = 48 in. × 36 in. = $1{,}728 \text{ in}^2$

Area in square feet $= 1{,}728 \cancel{\text{in}^2} \times \dfrac{1 \text{ ft}^2}{144 \cancel{\text{in}^2}} = \dfrac{1{,}728}{144} \text{ ft}^2 = 12 \text{ ft}^2$

3 $A = 1.5 \times 45 = 67.5 \text{ yd}^2$

$67.5 \text{ yd}^2 = 67.5 \cancel{\text{yd}^2} \times \dfrac{9 \text{ ft}^2}{1 \cancel{\text{yd}^2}} = 607.5 \text{ ft}^2$

4 $55 \text{ acres} = 55 \cancel{\text{acres}} \times \dfrac{43{,}560 \text{ ft}^2}{1 \cancel{\text{acre}}} = 55 \times 43{,}560 \text{ ft}^2 = 2{,}395{,}800 \text{ ft}^2$

5 $960 \text{ acres} = 960 \cancel{\text{acres}} \times \dfrac{1 \text{ mi}^2}{640 \cancel{\text{acres}}} = \dfrac{960}{640} \text{ mi}^2 = 1.5 \text{ mi}^2$

6 $1 \text{ m}^2 = 1 \cancel{\text{m}^2} \times \dfrac{100 \cancel{\text{dm}^2}}{1 \cancel{\text{m}^2}} \times \dfrac{100 \text{ cm}^2}{1 \cancel{\text{dm}^2}} = 10{,}000 \text{ cm}^2$

7 $5 \text{ gal} = 5 \cancel{\text{gal}} \times \dfrac{4 \cancel{\text{qt}}}{1 \cancel{\text{gal}}} \times \dfrac{2 \text{ pt}}{1 \cancel{\text{qt}}} = 5 \times 4 \times 2 \text{ pt} = 40 \text{ pt}$

8 $2{,}000 \text{ qt} = 2{,}000 \cancel{\text{qt}} \times \dfrac{1 \text{ gal}}{4 \cancel{\text{qt}}} = \dfrac{2{,}000}{4} \text{ gal} = 500 \text{ gal}$

The number of 10-gal containers in 500 gal is $\frac{500}{10} = 50$ containers.

9 $3.5 \text{ liters} = 3.5 \cancel{\text{liters}} \times \dfrac{1{,}000 \text{ mL}}{1 \cancel{\text{liter}}} = 3.5 \times 1{,}000 \text{ mL} = 3{,}500 \text{ mL}$

Section 6.4

1 $15 \text{ lb} = 15 \cancel{\text{lb}} \times \dfrac{16 \text{ oz}}{1 \cancel{\text{lb}}} = 15 \times 16 \text{ oz} = 240 \text{ oz}$

2 $2 \text{ lb } 4 \text{ oz} = 2 \text{ lb} + \dfrac{4}{16} \text{ lb} = 2 \text{ lb} + 0.25 \text{ lb} = 2.25 \text{ lb}$

Total price for 2.25 lb is $2.09 × 2.25 = $4.70 (rounded to nearest cent)

3 $5 \text{ kg} = 5 \cancel{\text{kg}} \times \dfrac{1{,}000 \cancel{\text{g}}}{1 \cancel{\text{kg}}} \times \dfrac{1{,}000 \text{ mg}}{1 \cancel{\text{g}}} = 5 \times 1{,}000 \times 1{,}000 \text{ mg} = 5{,}000{,}000 \text{ mg}$

4 Total number of milligrams in bottle = 75 × 200 = 15,000 mg

$15{,}000 \text{ mg} = 15{,}000 \cancel{\text{mg}} \times \dfrac{1 \text{ g}}{1{,}000 \cancel{\text{mg}}} = \dfrac{15{,}000}{1{,}000} \text{ g} = 15 \text{ g}$

Section 6.5

1 $10 \text{ in.} = 10\,\cancel{\text{in.}} \times \dfrac{2.54 \text{ cm}}{1\,\cancel{\text{in.}}} = 10 \times 2.54 \text{ cm} = 25.4 \text{ cm}$

2 $9 \text{ m} = 9\,\cancel{\text{m}} \times \dfrac{3.28 \text{ ft}}{1\,\cancel{\text{m}}} = 9 \times 3.28 \text{ ft} = 29.52 \text{ ft}$

3 $15 \text{ gal} = 15\,\cancel{\text{gal}} \times \dfrac{3.79 \text{ liters}}{1\,\cancel{\text{gal}}} = 15 \times 3.79 \text{ liters} = 56.85 \text{ liters}$

4 $2.2 \text{ liters} = 2.2\,\cancel{\text{liters}} \times \dfrac{1{,}000\,\cancel{\text{mL}}}{1\,\cancel{\text{liter}}} \times \dfrac{1 \text{ in}^3}{16.39\,\cancel{\text{mL}}} = \dfrac{2.2 \times 1{,}000}{16.39} \text{ in}^3$
$= 134 \text{ in}^3$ (rounded to the nearest cubic inch)

5 $165 \text{ lb} = 165\,\cancel{\text{lb}} \times \dfrac{1 \text{ kg}}{2.20\,\cancel{\text{lb}}} = \dfrac{165}{2.20} \text{ kg} = 75 \text{ kg}$

6 $F = \dfrac{9}{5}(40) + 32 = 72 + 32 = 104°F$

7 $C = \dfrac{5(101.6 - 32)}{9}$
$= 38.7°C$ (rounded to the nearest tenth)

Section 6.6

1 First term = 2, second term = 3, third term = 6, fourth term = 9; means: 3 and 6; extremes: 2 and 9

2 a $5 \cdot 18 = 90$, $6 \cdot 15 = 90$ **b** $13 \cdot 3 = 39$, $39 \cdot 1 = 39$

3 $3 \cdot x = 4 \cdot 9$; $3 \cdot x = 36$; $\dfrac{\cancel{3} \cdot x}{\cancel{3}} = \dfrac{36}{3}$; $x = 12$

4 $2 \cdot 19 = 8 \cdot y$; $38 = 8 \cdot y$; $\dfrac{38}{8} = \dfrac{\cancel{8} \cdot y}{\cancel{8}}$; $4.75 = y$

5 $15 \cdot n = 6(0.3)$; $15 \cdot n = 1.8$; $\dfrac{\cancel{15} \cdot n}{\cancel{15}} = \dfrac{1.8}{15}$; $n = 0.12$

6 $\dfrac{3}{4} \cdot 8 = 7 \cdot x$; $6 = 7 \cdot x$; $\dfrac{6}{7} = \dfrac{\cancel{7} \cdot x}{\cancel{7}}$; $\dfrac{6}{7} = x$

7 $\dfrac{x}{10} = \dfrac{288}{6}$; $x \cdot 6 = 10 \cdot 288$; $x \cdot 6 = 2{,}880$; $\dfrac{x \cdot \cancel{6}}{\cancel{6}} = \dfrac{2{,}880}{6}$; $x = 480$ miles

8 $\dfrac{x}{54} = \dfrac{10}{18}$; $x \cdot 18 = 54 \cdot 10$; $x \cdot 18 = 540$; $\dfrac{x \cdot \cancel{18}}{\cancel{18}} = \dfrac{540}{18}$; $x = 30$ hits

9 $\dfrac{x}{35} = \dfrac{8}{20}$; $x \cdot 20 = 35 \cdot 8$; $x \cdot 20 = 280$; $\dfrac{x \cdot \cancel{20}}{\cancel{20}} = \dfrac{280}{20}$; $x = 14$ milliliters of alcohol

10 $\dfrac{x}{4.75} = \dfrac{105}{1}$; $x \cdot 1 = 4.75(105)$; $x = 498.75$ miles

11 $\dfrac{x}{300} = \dfrac{8}{120}$; $x \cdot 120 = 300 \cdot 8$; $x \cdot 120 = 2{,}400$; $\dfrac{x \cdot \cancel{120}}{\cancel{120}} = \dfrac{2{,}400}{120}$; $x = 20$ parts

CHAPTER 7

Section 7.2

1 $n = 0.25(74) = 18.5$

2 $n \cdot 84 = 21$; $\dfrac{n \cdot \cancel{84}}{\cancel{84}} = \dfrac{21}{84}$; $n = 0.25$; $n = 25\%$

3 $35 = 0.40 \cdot n$; $\dfrac{35}{0.40} = \dfrac{\cancel{0.40} \cdot n}{\cancel{0.40}}$; $87.5 = n$

4 $n = 0.635(45)$; $n \approx 28.6$

5 $n \cdot 85 = 11.9$; $\dfrac{n \cdot \cancel{85}}{\cancel{85}} = \dfrac{11.9}{85}$; $n = 0.14$; $n = 14\%$

6 $62 = 0.39 \cdot n$; $\dfrac{62}{0.39} = \dfrac{\cancel{0.39} \cdot n}{\cancel{0.39}}$; $159.0 \approx n$

7 25 is what percent of 160?
$25 = n \cdot 160$; $\dfrac{25}{160} = n$
$n = 0.156 = 15.6\%$ to the nearest tenth of a percent

Section 7.3

1 114 is what percent of 150?

$$114 = n \cdot 150$$
$$\frac{114}{150} = \frac{n \cdot \cancel{150}}{\cancel{150}}$$
$$n = 0.76$$
$$n = 76\%$$

2 What is 75% of 40?

$n = 0.75(40)$
$n = 30$ milliliters HCl

3 3,360 is 42% of what number?

$$3{,}360 = 0.42 \cdot n$$
$$\frac{3{,}360}{0.42} = \frac{\cancel{0.42} \cdot n}{\cancel{0.42}}$$
$n = 8{,}000$ students

4 $n = 0.35(300)$
$= 105$

Section 7.4

1 What is 6% of \$625?

$n = 0.06(625)$
$n = \$37.50$

2 \$4.35 is 3% of what number?

$4.35 = 0.03 \cdot n$
$n = \$145$ Purchase price
Total price $= \$145 + \$4.35$
$= \$149.35$

3 \$11.82 is what percent of \$197?

$11.82 = n \cdot 197$
$n = 6\%$

4 What is 3% of \$115,000?

$n = 0.03(115{,}000)$
$n = \$3{,}450$

5 10% of what number is \$115?

$0.10 \cdot n = 115$
$n = \$1{,}150$

6 \$105 is what percent of \$750?

$105 = n \cdot 750$
$n = 14\%$

Section 7.5

1 $0.07(18{,}000) = 1{,}260$

\$18,000 Old salary
\+ 1,260 Raise
\$19,260 New salary

2 $0.09(659{,}000) = 59{,}310$

659,000
− 59,310
599,690 Arrests in 1977

3 \$35 − \$28 = \$7
\$7 is what percent of \$35?
$7 = n \cdot 35$
$n = 20\%$ Decrease

4 What is 15% of \$550?

$n = 0.15(550)$
$n = \$82.50$ Discount
\$550.00 Original price
− 82.50 Less discount
\$467.50 Sale price

5 What is 15% of \$45?

$n = 0.15(45)$
$n = \$6.75$
\$45.00 Original price
− 6.75 Less discount
\$38.25 Sale price

What is 5% of \$38.25?

$n = 0.05(38.25)$
$n = \$1.91$ to the nearest cent
\$38.25 Sale price
\+ 1.91 Sales tax
\$40.16 Total price

Section 7.6

1 Interest = 0.08(\$3,000)
= \$240
\$3,000 Principal
\+ 240 Interest
\$3,240 Amount after 1 year

2 Interest = 0.12(\$7,500)
= \$900
\$7,500 Principal
\+ 900 Interest
\$8,400 Total amount to pay back

$\frac{\$8{,}400}{12} = \$700$ a month

3 $I = P \cdot R \cdot T$

$$I = 700 \times 0.04 \times \frac{90}{360}$$
$$I = 700 \times 0.04 \times \frac{1}{4}$$
$I = \$7$ Interest

4 $I = P \cdot R \cdot T$

$$I = 1{,}200 \times 0.095 \times \frac{120}{360}$$
$$I = 1{,}200 \times 0.095 \times \frac{1}{3}$$
$I = \$38$ Interest
\$1,200 Principal
\+ 38 Interest
\$1,238 Total amount withdrawn

5 Interest after 1 year is
0.06(\$5,000) = \$300
Total in account after 1 year is
\$5,000
\+ 300
\$5,300

Interest paid the second year is
0.06(\$5,300) = \$318
Total in account after 2 years is
\$5,300
\+ 318
\$5,618

6 Interest at the end of first quarter

$$I = \$20{,}000 \times 0.08 \times \frac{1}{4} = \$400$$

Total in account at end of first quarter

$\$20{,}000 + \$400 = \$20{,}400$

Interest for the second quarter

$$I = \$20{,}400 \times 0.08 \times \frac{1}{4} = \$408$$

Total in account at end of second quarter

$\$20{,}400 + \$408 = \$20{,}808$

Interest for the third quarter

$$I = \$20{,}808 \times 0.08 \times \frac{1}{4} = \$416.16$$

Total in account at the end of third quarter

$\$20{,}808 + \$416.16 = \$21{,}224.16$

Interest for the fourth quarter

$$I = \$21{,}224.16 \times 0.08 \times \frac{1}{4} = \$424.48 \text{ to the nearest cent}$$

Total in account at end of 1 year

$$\begin{array}{r} \$21{,}224.16 \\ +\quad 424.48 \\ \hline \$21{,}648.64 \end{array}$$

ANSWERS TO ODD-NUMBERED PROBLEMS

CHAPTER 1

Problem Set 1.1

1 8 ones, 7 tens **3** 5 ones, 4 tens **5** 8 ones, 4 tens, 3 hundreds **7** 8 ones, 0 tens, 6 hundreds
9 8 ones, 7 tens, 3 hundreds, 2 thousands **11** 9 ones, 6 tens, 5 hundreds, 3 thousands, 7 ten thousands, 2 hundred thousands
13 Ten thousands **15** Hundred millions **17** Ones **19** Hundred thousands **21** 600 + 50 + 8 **23** 60 + 8
25 4,000 + 500 + 80 + 7 **27** 30,000 + 2,000 + 600 + 70 + 4
29 3,000,000 + 400,000 + 60,000 + 2,000 + 500 + 70 + 7 **31** 400 + 7 **33** 30,000 + 60 + 8
35 3,000,000 + 4,000 + 8 **37** Twenty-nine **39** Forty **41** Five hundred seventy-three **43** Seven hundred seven
45 Seven hundred seventy **47** Twenty-three thousand, five hundred forty **49** Three thousand, four
51 Three thousand, forty **53** One hundred four million, sixty-five thousand, seven hundred eighty
55 Five billion, three million, forty thousand, eight
57 Two million, five hundred forty-six thousand, seven hundred thirty-one **59** 325 **61** 5,432 **63** 86,762
65 2,000,200 **67** 2,002,200 **69** Thousands **71** Millions
73 90,000,000 + 2,000,000 + 800,000 + 90,000 + 7,000 + 400 + 10 + 6 **75** 249,000,000
77 One hundred twenty-four million **79** 30,000,000 **81** Eight million, eight hundred thousand **83** 3 in. **85** 5 in.
87 6 **89** 3 **91** 12

Problem Set 1.2

1 12 **3** 14 **5** 11 **7** 13 **9** 15 **11** 16 **13** 13 **15** 18 **17** 14 **19** 12 **21** 9 + 5 **23** 8 + 3
25 4 + 6 **27** $5 + n$ **29** $1 + x$ **31** 1 + (2 + 3) **33** 2 + (1 + 6) **35** (1 + 9) + 1 **37** $4 + (n + 1)$
39 $a + (b + c)$ **41** $x + 11$ **43** $a + 16$ **45** $5 + y$ **47** $16 + t$ **49** $n = 5$ **51** $n = 4$ **53** $n = 5$
55 $n = 8$ **57** $n = 9$ **59** $n = 8$ **61** $n = 8$ **63** c **65** b **67** c **69** b **71** a **73** c **75** b
77 $x = 5$ **79** $a = 7$ **81** $y = 10$ **83** $x = 0$ **85** $a = 3$ **87** $t = 4$ **89** The sum of 4 and 9
91 The sum of 8 and 1 **93** The sum of 2 and 3 is 5. **95** The sum of a and b is c. **97** 5 + 2 **99** $5 + a$
101 6 + 8 = 14 **103** 2 + 8 **105** $x + 4$ **107** $t + 6$
109 Your written answer should include the idea that the commutative property allows us to change the order of numbers in a sum without changing the answer.
111 Subtraction is a commutative operation if changing the order of the numbers in a difference does not change the result. What do you think? Is 5 − 3 the same as 3 − 5?

Problem Set 1.3

1 15 **3** 14 **5** 24 **7** 15 **9** 20 **11** 68 **13** 98 **15** 7,297 **17** 6,487 **19** 96 **21** 7,449 **23** 65
25 102 **27** 875 **29** 829 **31** 10,391 **33** 16,204 **35** 155,554 **37** 111,110 **39** 17,391 **41** 14,892
43 180 **45** 2,220 **47** 18,285 **49** 7,730 **51** 154,833 **53** 34 gallons **55** \$349 **57** \$1,148 **59** \$384

61 12 in. **63** 16 ft **65** 26 yd **67** 18 in. **69** 76 in. **71** 168 ft **73** 11,619,201 **75** 2,593,614
77 47,883,755 **79** 64,943,560 **81** $w = 1$ and $l = 5$, $w = 2$ and $l = 4$, $w = 3$ and $l = 3$.
83 Yes, when the width and length are both 5 ft. **85** 5 **87** 10
89 14; the sequence is an arithmetic sequence, starting with 5, in which each term comes from adding 3 to the term before it.
91 85; the sequence starts with 10 and every number after that is 25 more than the number just before it.

Problem Set 1.4

1 40 **3** 50 **5** 50 **7** 80 **9** 460 **11** 470 **13** 56,780 **15** 4,500 **17** 500 **19** 800 **21** 900
23 1,100 **25** 5,000 **27** 39,600 **29** 5,000 **31** 10,000 **33** 1,000 **35** 658,000 **37** 510,000 **39** 3,789,000
41 7,820; 7,800; 8,000 **43** 6,000; 6,000; 6,000 **45** 10,990; 11,000; 11,000 **47** 100,000; 100,000; 100,000
49 110,000 **51** 15,500,000 **53** $4,680,000,000 **55** $15,200 **57** $31,000 **59** 1,200 **61** 1,900 **63** 64,000
65 160 miles per hour **67** Answers will vary, but 75 miles per hour is a good estimate.

69

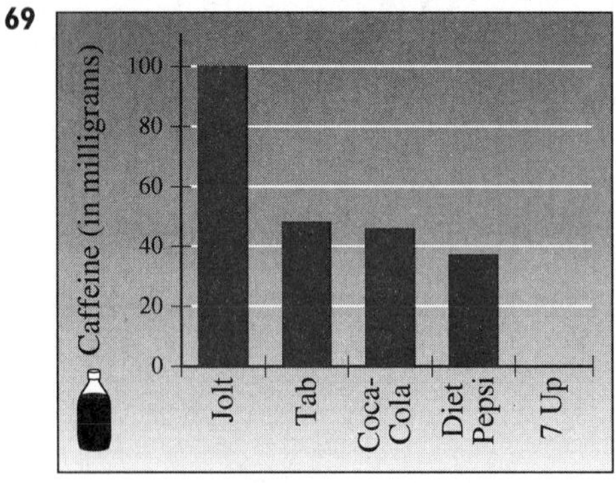

71

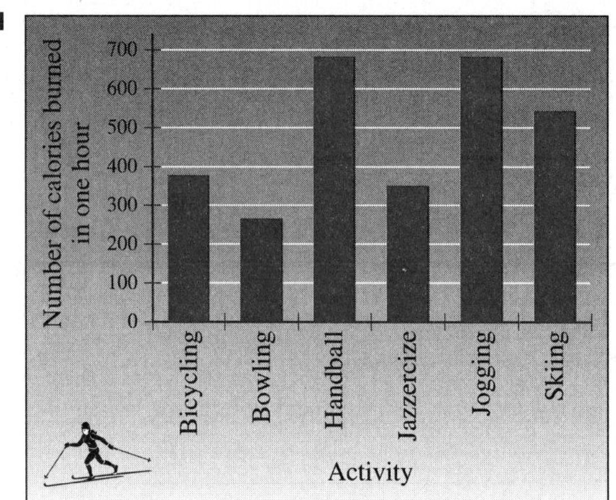

Problem Set 1.5

1 32 **3** 22 **5** 10 **7** 111 **9** 312 **11** 403 **13** 1,111 **15** 4,544 **17** 15 **19** 33 **21** 5 **23** 33
25 95 **27** 152 **29** 274 **31** 488 **33** 538 **35** 163 **37** 1,610 **39** 2,143 **41** 23,790 **43** 46,083
45 $255 **47** 179 **49** 168 **51** $574 **53** $350 **55** 219,312 **57** 1,708,029 **59** 990,000 **61** 700
63 3,000 **65** 20,000 **67** 50 miles per hour **69** Answers will vary, but 50 miles per hour is a good estimate.
71 1,772,098 **73** 31,420,426 **75** 1,062,000 **77** 184,000 **79** 667,525 **81** 1,147,421

Problem Set 1.6

1 The product of 6 and 7 **3** The product of 4 and 9 **5** The product of 2 and n **7** The product of 9 and 7 is 63
9 The product of 2 and 9 is 18 **11** $8 \cdot 3$ **13** $6 \cdot 6$ **15** $7 \cdot n$ **17** $6 \cdot 7 = 42$ **19** $0 \cdot 6 = 0$
21 Products: $9 \cdot 7$ and 63 **23** Products: 4(4) and 16 **25** Factors: 2, 3, and 4 **27** Factors: 2, 2, and 3 **29** 9(5)
31 $7 \cdot 6$ **33** 0(1) **35** $n \cdot 3$ **37** $(2 \cdot 7) \cdot 6$ **39** $(3 \times 9) \times 1$ **41** $(2 \cdot 6) \cdot 5$ **43** $(3 \cdot 4) \cdot n$ **45** d **47** a
49 c **51** c **53** c **55** $42x$ **57** $27x$ **59** $28a$ **61** $18y$ **63** $n = 3$ **65** $n = 9$ **67** $n = 0$ **69** $n = 8$
71 $n = 3$ **73** $n = 7$ **75** 60 **77** 150 **79** 900 **81** 300 **83** 600 **85** 3,000 **87** 5,000 **89** 21,000
91 81,000 **93** 2,400,000 **95** 300 **97** 30,000 **99** 720 **101** 720,000 **103** 400 **105** $270 **107** 225
109 45 **111** 380 **113** $1,800 **115** $312 **117** 1 and 12; 2 and 6; 3 and 4
119 The product of 8 and 5 is 27 more than the sum of 8 and 5.
121 Check your answer to see that you have written in complete sentences and that you have indicated that the commutative property of multiplication tells us we can change the order of two numbers being multiplied together without changing the result.
123 A factor of a number is a whole number that divides the original number evenly, without remainder. For example, 5 is a factor of 15, while 6 is not a factor of 15.

Problem Set 1.7

1 7(2) + 7(3) = 35 **3** 9(4) + 9(7) = 99 **5** $3x + 3$ **7** $4a + 36$ **9** $15x + 20$ **11** $48y + 12$ **13** 100
15 228 **17** 36 **19** 1,440 **21** 950 **23** 1,725 **25** 121 **27** 1,552 **29** 4,200 **31** 66,248 **33** 279,200
35 12,321 **37** 106,400 **39** 198,592 **41** 612,928 **43** 333,180 **45** 18,053,805 **47** 263,646,976
49 81,647,808 **51** 31,648,320 **53** 1,234,321 **55** 1,800 **57** 5,060 miles **59** 42 **61** $6,600 **63** $250
65 8,000 **67** 1,500,000 **69** 1,400,000 **71** 2,081 calories **73** 280 calories **75** Yes **77** 16,978,130
79 43,605,972 **81** 462,672
83 40; the sequence starts with 5. Then each number comes from the number before it by multiplying by 2 each time.
85 54; the sequence is a geometric sequence, starting with 2, in which each term is found by multiplying the term before it by 3.
87 Arithmetic **89** Geometric **91** Neither

Problem Set 1.8

1 6 ÷ 3 **3** 45 ÷ 9 **5** $r \div s$ **7** 20 ÷ 4 = 5 **9** 2 · 3 = 6 **11** 9 · 4 = 36 **13** 6 · 8 = 48 **15** 7 · 4 = 28
17 5 **19** 8 **21** Undefined **23** 45 **25** 23 **27** 1,530 **29** 1,350 **31** 18,000 **33** 16,680 **35** a **37** b
39 1 **41** 2 **43** 4 **45** 6 **47** 45 **49** 49 **51** 432 **53** 1,438 **55** 345 **57** 61 R 4 **59** 90 R 1
61 13 R 7 **63** 234 R 6 **65** 452 R 4 **67** 35 R 35 **69** $1,850 **71** 16¢ **73** 8 glasses
75 6 glasses with 2 oz left **77** 3 bottles **79** $12 **81** 665 mg **83** 437 **85** 3,247 **87** 869 **89** 5,684
91 169 gal

Problem Set 1.9

	Base	*Exponent*
1	4	5
3	3	6
5	8	2
7	9	1
9	4	0

11 36 **13** 8 **15** 1 **17** 1 **19** 81 **21** 10 **23** 12 **25** 1 **27** 43 **29** 16 **31** 84 **33** 416 **35** 66
37 21 **39** 7 **41** 124 **43** 11 **45** 91 **47** 8(4 + 2) = 48 **49** 2(10 + 3) = 26 **51** 3(3 + 4) + 4 = 25
53 (20 ÷ 2) − 9 = 1 **55** (8 · 5) + (5 · 4) = 60 **57** 255 calories **59** 465 calories **61** 30 calories
63 Big Mac has twice the calories. **65** 3 **67** 6 **69** 5 **71** 54 **73** 37 **75** $16,000 **77** 77 **79** 5,932
81 775,029 **83** 63,259 **85** 4 **87** 16 **89** Answers will vary. **91** 13, 21, 34

Problem Set 1.10

1 25 in^2 **3** 84 m^2 **5** 60 ft^2 **7** 192 cm^2 **9** 2,200 ft^2 **11** 945 cm^2 **13** 64 cm^3 **15** 420 ft^3 **17** 162 in^3
19 96 cm^2 **21** 148 ft^2 **23** 7 ft **25** 9 ft **27** 124 tiles **29** 288 ft^2
31 The area increases from 25 ft^2 to 49 ft^2, which is an increase of 24 ft^2.
33 a

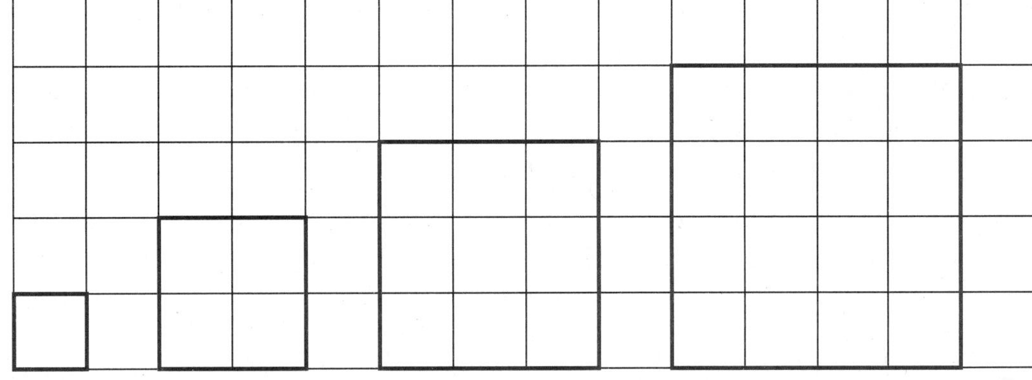

b Perimeters of Squares

Length of Each Side (in centimeters)	Perimeter (in centimeters)
1	4
2	8
3	12
4	16

Areas of Squares

Length of Each Side (in centimeters)	Area (in square centimeters)
1	1
2	4
3	9
4	16

c

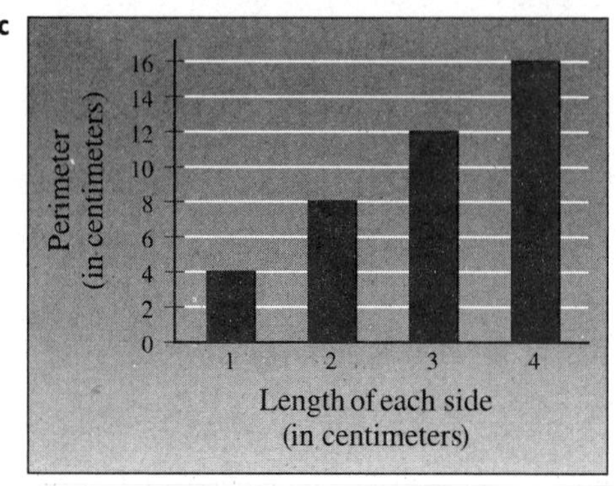

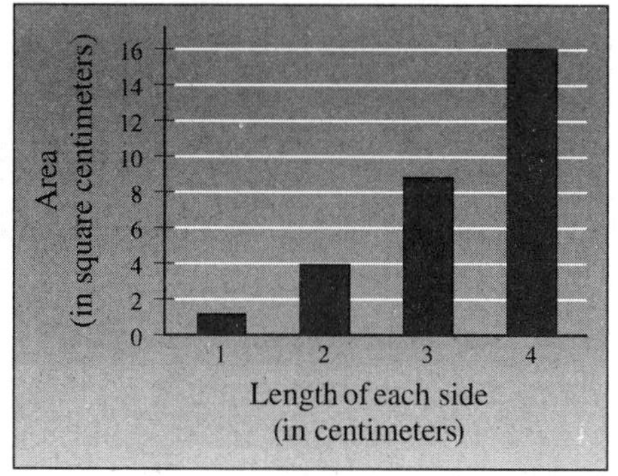

35 12,321 ft^2 **37** Volume = 1,331 ft^3; surface area = 726 ft^2 **39** 1,356,531 ft^3

Chapter 1 Review

1 One thousand, one hundred fifty-three **3** 5,245,652 **5** 500,000 + 20,000 + 7,000 + 7 **7** d **9** c **11** b **13** g **15** 749 **17** 8,272 **19** 314 **21** 3,149 **23** 584 **25** 3,717 **27** 173 **29** 428 **31** 3,781,090 **33** 3,800,000 **35** 79 **37** 222 **39** 79 **41** $3(4 + 6) = 30$ **43** $2(17 - 5) = 24$ **45** $488 **47** 2,100 parts **49** $2,032 **51** $1,938 **53** $183 **55** Perimeter = 40 ft; area = 84 ft^2 **57** 1,470 calories **59** 250 more calories **61** No **63** Answers will vary.

Chapter 1 Test

1 Twenty thousand, three hundred forty-seven **2** 2,045,006 **3** 100,000 + 20,000 + 3,000 + 400 + 7 **4** f **5** c **6** a **7** e **8** 876 **9** 16,383 **10** 524 **11** 3,085 **12** 1,674 **13** 22,258 **14** 85 **15** 21 **16** 520,000 **17** 11 **18** 4 **19** 164 **20** $2(11 + 7) = 36$ **21** $(20 \div 5) + 9 = 13$ **22** $1,527 **23** $235 **24** $P = 14$ ft; $A = 12$ ft^2 **25** $V = 96$ in^3; $S = 136$ in^2

CHAPTER 2

Problem Set 2.1

1 4 is less than 7 **3** 5 is greater than −2 **5** −10 is less than −3 **7** 0 is greater than −4 **9** $30 > -30$ **11** $-10 < 0$ **13** $-3 > -15$ **15** −3 **17** 2 **19** −75 **21** 0 **23** 5 **25** −100 **27** $<$ **29** $>$ **31** $<$ **33** $<$ **35** $>$ **37** $<$ **39** $<$ **41** $<$ **43** $>$ **45** $<$ **47** 2 **49** 100 **51** 8 **53** 231 **55** 6 **57** 200 **59** 8 **61** 231 **63** 2 **65** 8 **67** −2 **69** −8 **71** 0 **73** −20° **75** −25 ft **77** Positive **79** −100 **81** −20 **83** −360 **85** −7°F **87** 10°F and 25-mph wind

89

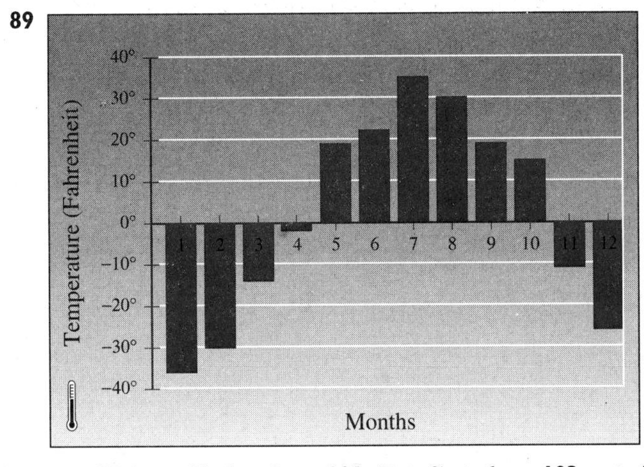

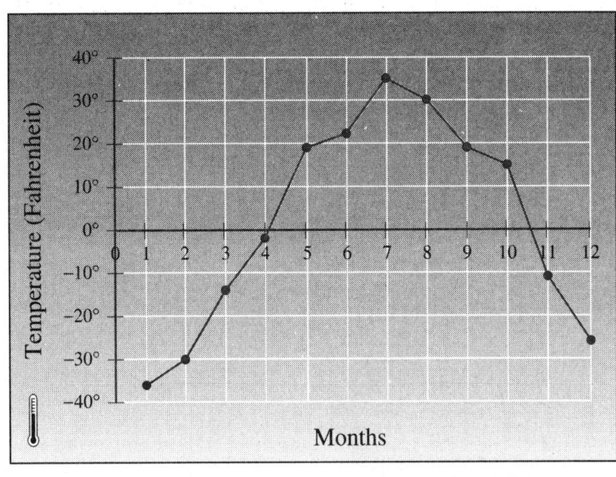

95 0 **97** 0 **99** 5 + 3 **101** (7 + 2) + 6 **103** $x + 4$ **105** $y + 5$ **107** 4,313 **109** 131,221 **111** −3
113 Answers will vary. **115** The opposite of the absolute value of −3. It simplifies to −3.

Problem Set 2.2

1 5 **3** 1 **5** −2 **7** −6 **9** 4 **11** 4 **13** −9 **15** 15 **17** −3 **19** −11 **21** −7 **23** −3
25 −16 **27** −8 **29** −127 **31** 49 **33** 34 **35** 398 **37** −4 **39** −4 **41** 10 **43** −445 **45** 107
47 −1 **49** −20 **51** −17 **53** −50 **55** −7 **57** 3 **59** 50 **61** −73 **63** −11 **65** 17 **67** −21
69 −5 **71** −4 **73** 7 **75** 10 **77** a **79** b **81** d **83** c **85** −1,000 **87** \$74 + (−\$141) = −\$67
89 3 + (−5) = −2 **91** −7 and 13 **93** $10 - x$ **95** $y - 17$ **97** 474 **99** 8 **101** 14 **103** 730

Problem Set 2.3

1 2 **3** 2 **5** −8 **7** −5 **9** 7 **11** 12 **13** 3 **15** −7 **17** −3 **19** −13 **21** −50 **23** −100
25 399 **27** −21 **29** −154 **31** −7 **33** −9 **35** −14 **37** −1 **39** 5 **41** 11 **43** −4 **45** 8
47 6 **49** b **51** a **53** −100 **55** b **57** a **59** 44° **61** −\$22 **63** 11°F **65** 27°F **67** 86°F
69 12°F **71** $3 \cdot 5$ **73** $7x$ **75** 5(3) **77** $(5 \cdot 7) \cdot 8$ **79** 2(3) + 2(4) = 6 + 8 = 14 **81** 2,352
83 $3 - 5 \neq 5 - 3$ **85** Answers will vary. **87** −5, −10 **89** 2, 6 **91** −36 **93** −57 **95** 8 **97** 3

Problem Set 2.4

1 −56 **3** −60 **5** 56 **7** 81 **9** −24 **11** 30 **13** −50 **15** −24 **17** 24 **19** −6 **21** 9 **23** 25
25 −8 **27** −125 **29** 16 **31** −12 **33** −4 **35** 50 **37** 1 **39** −35 **41** −22 **43** −30 **45** −25
47 9 **49** −13 **51** 19 **53** 6 **55** −6 **57** −4 **59** −17 **61** 2,000 **63** a **65** d **67** a **69** b
71 $12 \div 6$, or $\frac{12}{6}$ **73** $6 \div 3 = 2$ **75** $5 \cdot 2 = 10$ **77** 89 **79** a, c **81** Answers will vary. **83** −54, 162
85 54, −162 **87** 44 **89** 19 **91** −17

Problem Set 2.5

1 −3 **3** −5 **5** 3 **7** 2 **9** −4 **11** −2 **13** 0 **15** 5 **17** −2 **19** 10 **21** 1 **23** −6 **25** −2
27 −1 **29** −1 **31** 2 **33** −3 **35** −7 **37** 30 **39** 4 **41** −5 **43** −20 **45** −5 **47** −5 **49** 35
51 6 **53** −1 **55** c **57** a **59** d **61** 0

63

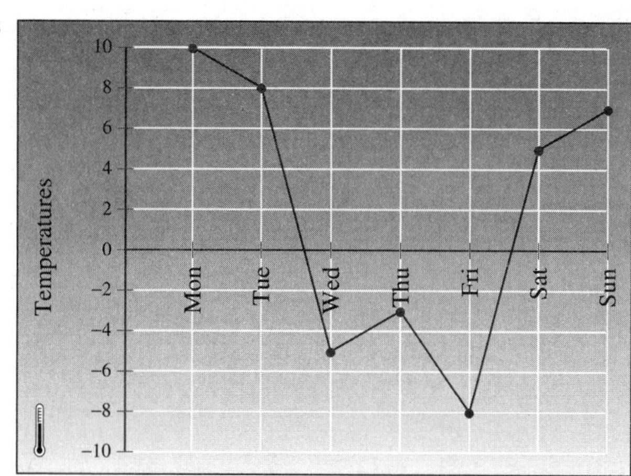

65 2°F **67** $x + 3$ **69** $y \cdot 4$ **71** $(5 + 7) + a$

73 $(3 \cdot 4)y$ **75** $5(3) + 5(7)$ **77** a **79** x **81** $2 \div 6 \neq 6 \div 2$ **83** Answers will vary. **85** -4 **87** 4 **89** 4 **91** -1 **93** -1

Problem Set 2.6

1 $20a$ **3** $48a$ **5** $-18x$ **7** $-27x$ **9** $-10y$ **11** $-60y$ **13** $5 + x$ **15** $13 + x$ **17** $10 + y$ **19** $8 + y$ **21** $5x + 6$ **23** $6y + 7$ **25** $12a + 21$ **27** $7x + 28$ **29** $7x + 35$ **31** $6a - 42$ **33** $2x - 2y$ **35** $20 + 4x$ **37** $6x + 15$ **39** $18a + 6$ **41** $12x - 6y$ **43** $35 - 20y$ **45** $8x$ **47** $4a$ **49** $4x$ **51** $5y$ **53** $-10a$ **55** $-5x$ **57** $36x^2$ **59** $16x^2$ **61** $27a^3$ **63** $8a^3b^3$ **65** $81x^2y^2$ **67** $25x^2y^2z^2$ **69** $1{,}296x^4y^2$ **71** $576x^6$ **73** $200x^5$ **75** $7{,}200a^7$ **77** 4 **79** -12 **81** 12 **83** -32 **85** 32 **87** -2

Problem Set 2.7

1 x^8 **3** a^8 **5** y^{15} **7** x^5 **9** x^9 **11** a^{12} **13** y^6 **15** $14x^9$ **17** $24y^8$ **19** $21x^2$ **21** $192x^{15}$ **23** $24a^3$ **25** $125y^{12}$ **27** $6x^5y^8$ **29** $56a^4b^5$ **31** $24a^6b^6$ **33** $28a^{12}b^{12}$ **35** x^{10} **37** a^{21} **39** $125y^{12}$ **41** $81x^8$ **43** $8a^3b^3$ **45** $625x^8y^{12}$ **47** x^{14} **49** a^{29} **51** $72x^{12}y^{22}$ **53** $175a^{12}b^9$ **55** $a^{18}b^{34}$ **57** 52 ft **59** 84 ft **61** 9 in.

Problem Set 2.8

1 $6x^2 + 6x + 8$ **3** $5a^2 - 9a + 8$ **5** $6x^3 + 4x^2 - 3x + 2$ **7** $13t^2 - 2$ **9** $33x^3 - 10x^2 - 4x - 4$ **11** $2x^2 - 2x + 1$ **13** $4y^3 - 8y^2 + 3y - 12$ **15** $-4x - 4$ **17** $x^2 - 30x + 64$ **19** $7y^2 + 3y + 25$ **21** 9 **23** 0 **25** 1 **27** $V = 175$ in^3; $S = 190$ in^2 **29** $V = 2{,}340$ ft^3; $S = 1{,}062$ ft^2

Problem Set 2.9

1 $x^6 + x^5$ **3** $x^6 + x^5$ **5** $6x^6 - 14x^4$ **7** $12x^6y^3 + 8x^2y^6$ **9** $6x^6 - 14x^4$ **11** $12x^2y^2 - 27xy$ **13** $12x^2y^2 - 27xy$ **15** $6x^5y^7 + 4x^6y^6 + 10x^4y^5$ **17** $x^2 + 5x + 6$ **19** $x^2 + x - 6$ **21** $6x^2 + 19x + 10$ **23** $3x^2 + 6x - 24$ **25** $16x^2 - 9$ **27** $12x^2 - x - 1$ **29** $6x^2 - 23x + 20$ **31** $x^2 + 6x + 9$ **33** $x^2 + 12x + 36$ **35** $9x^2 + 12x + 4$ **37** $4x^2 + 12x + 9$ **39** $49x^2 + 84x + 36$ **41** $16x^2 + 8x + 1$ **43** $x^2 - 10x + 25$ **45** $4x^2 - 16x + 16$ **47** 169 ft^2 **49** 210 in^2 **51** 480 yd^2

Chapter 2 Review

1 -17 **3** $>$ **5** $>$ **7** 4 **9** -2 **11** -971 **13** -17 **15** -20 **17** -3 **19** 2 **21** -8 **23** 2 **25** -42 **27** -7 **29** -11 **31** -129 **33** -2 **35** 39 **37** -19 **39** True **41** False **43** True **45** -1 and -15 **47** 8 or 2 **49** $24x$ **51** $-30y$ **53** $2x - 10$ **55** $6a + 15b$ **57** $125x^3$ **59** x^8 **61** $81x^4y^4$ **63** $125x^{18}y^6$ **65** $12x^2 - 3x - 3$ **67** $3x - 12$ **69** $x^5 + x^7$ **71** $x^2 + 9x + 14$ **73** $x^2 + 10x + 25$

Chapter 2 Test

1 −14 **2** 5 **3** > **4** > **5** 7 **6** −2 **7** −9 **8** −6 **9** −21 **10** −36 **11** 42 **12** 54 **13** −5 **14** 5 **15** 9 **16** −8 **17** −11 **18** −15 **19** 7 **20** −4 **21** −61 **22** −7 **23** 24 **24** −5 **25** 28 **26** −\$35 **27** 25° **28** $30x^7$ **29** $8a^5 - 12a^2$ **30** $6x^2 + 2x - 20$ **31** $x^2 + 6x + 9$ **32** $5x^2 - 3x - 5$ **33** $3x - 8$

CHAPTER 3

Problem Set 3.1

1 1 **3** 2 **5** x **7** a **9** 5 **11** 1 **13** 12 **15** y **17** $\frac{3}{4}, \frac{1}{2}, \frac{9}{10}$ **19** True **21** False **23** False **25** $\frac{3}{4}$ **27** $\frac{43}{47}$ **29** $\frac{4}{3}$ **31** $\frac{13}{17}$ **33** $\frac{4}{6}$ **35** $\frac{5}{6}$ **37** $\frac{8}{12}$ **39** $\frac{8}{12}$ **41** $\frac{2x}{12x}$ **43** $\frac{9x}{12x}$ **45** $\frac{9a}{24a}$ **47** $\frac{20a}{24a}$ **49** 3 **51** 2 **53** 37 **55** 4 **57** $\frac{4}{5}$ **59** $\frac{2}{15}$ **61** $\frac{29}{43}$ **63** $\frac{1{,}121}{1{,}791}$ **65** $\frac{238}{527}$ **67** **a** $\frac{1}{2}$ **b** $\frac{1}{2}$ **c** $\frac{1}{4}$ **d** $\frac{1}{4}$ **69–77**

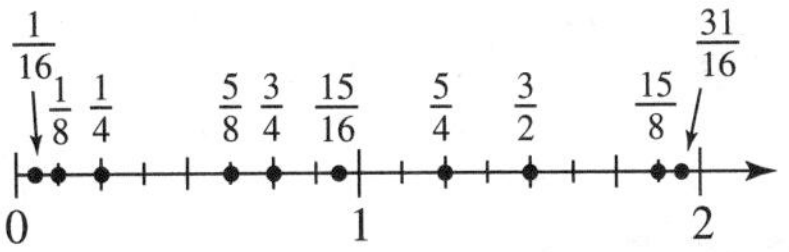

79 d **81** a **83** Yes **85** $\frac{735}{2{,}205}$ **87** $\frac{96}{384}$ **89** $\frac{1{,}580}{1{,}896}$ **91** 23 **93** 32 **95** 16 **97** 18 **99** Correct

Problem Set 3.2

1 Prime **3** Composite; 3, 5, and 7 are factors **5** Composite; 3 is a factor **7** Prime **9** $2^2 \cdot 3$ **11** 3^4 **13** $5 \cdot 43$ **15** $3 \cdot 5$ **17** $\frac{1}{2}$ **19** $\frac{2}{3}$ **21** $\frac{4}{5}$ **23** $\frac{9}{5}$ **25** $\frac{7}{11}$ **27** $\frac{3x}{5}$ **29** $\frac{1}{7}$ **31** $\frac{7}{9}$ **33** $\frac{7x}{5}$ **35** $\frac{a}{5}$ **37** $\frac{11}{7}$ **39** $\frac{5z}{3}$ **41** $\frac{8x}{9y}$ **43** $\frac{42}{55}$ **45** $\frac{17ac}{19b}$ **47** $\frac{14}{33}$ **49** $\frac{2}{17}, \frac{3}{26}, \frac{1}{9}, \frac{3}{28}, \frac{2}{19}$ **51** $\frac{1}{45}, \frac{1}{30}, \frac{1}{18}, \frac{1}{15}, \frac{1}{10}$ **53** **a** $\frac{1}{3}$ **b** $\frac{5}{6}$ **c** $\frac{1}{5}$ **55** $\frac{9}{16}$ **57** $\frac{1}{4}$ **59** $\frac{1}{6}$ **61** $\frac{2}{3}$ **63–65**

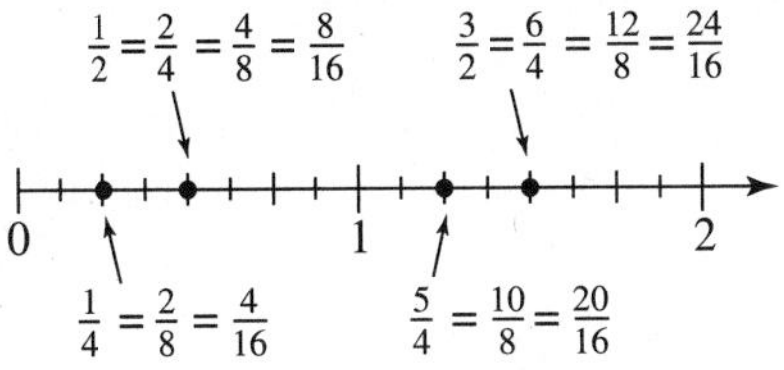

67 b **69** c **71** $\frac{1}{3}$ **73** $\frac{1}{8}$ **75** $\frac{3}{8}$ **77** 410 **79** 819 **81** 7 **83** 2 **85** 72 **87** 45 **89** $5x + 20$ **91** $12x + 15$

93 Your answer must include the fact that the numerator and the denominator must be factored completely and then any factors common to them can be divided out.

95 Six of them: 3, 6, 9, 12, 15, and 18

97 No, because the square of a prime number will always be divisible by the original prime number.

Problem Set 3.3

1 $\frac{8}{15}$ **3** $\frac{7}{8}$ **5** −1 **7** $\frac{27}{4}$ **9** 1 **11** $\frac{1}{24}$ **13** $\frac{24}{125}$ **15** 1 **17** $\frac{3}{5}$ **19** $-\frac{27}{25}$ **21** $\frac{1}{6}$ **23** 9 **25** 1 **27** 8 **29** $\frac{1}{15}$ **31** $\frac{ac^2}{b}$ **33** 9 **35** 4 **37** x **39** y **41** a **43** $\frac{1}{4}$ **45** $-\frac{8}{27}$ **47** $\frac{1}{2}$ **49** $\frac{9}{100}$ **51** 3

53 24 **55** 4 **57** 9 **59** Numerator should be 3, not 4 **61** Denominator should be 9, not 5 **63** 133 in^2

65 $\frac{4}{9}$ ft^2 **67** 3 yd^2 **69** 138 in^2 **71** 215 **73** 30 **75** $\frac{3}{8}$ cup

77 a

b Perimeters of squares

Length of Each Side (in inches)	Perimeter (in inches)
$\frac{1}{4}$	1
$\frac{2}{4} = \frac{1}{2}$	2
$\frac{3}{4}$	3
$\frac{4}{4} = 1$	4

Areas of squares

Length of Each Side (in inches)	Area (in square inches)
$\frac{1}{4}$	$\frac{1}{16}$
$\frac{1}{2}$	$\frac{1}{4}$
$\frac{3}{4}$	$\frac{9}{16}$
1	1

c

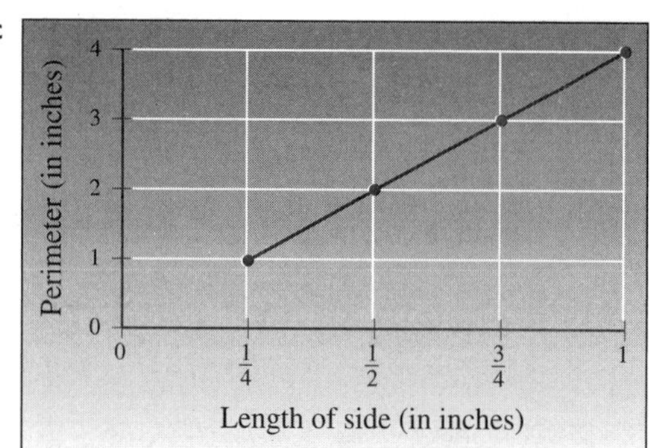

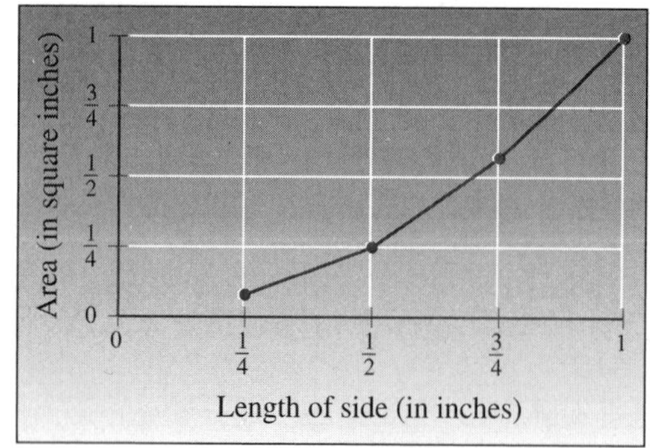

79 2 **81** 3 **83** 0 **85** $\frac{54,653}{59,136}$ **87** $\frac{77,004}{5,141}$ **89** 3 **91** 2 **93** 5 **95** 3 **97** $\frac{1}{27}$ **99** $\frac{8}{27}$

Problem Set 3.4

1 $\frac{15}{4}$ **3** $-\frac{4}{3}$ **5** -9 **7** 200 **9** $-\frac{3}{8}$ **11** 1 **13** $\frac{49}{64}$ **15** $-\frac{3}{4}$ **17** $\frac{15}{16}$ **19** $\frac{1}{6}$ **21** 6 **23** $\frac{x}{y}$

25 ab **27** $4a$ **29** 40 **31** $\frac{7}{10}$ **33** 13 **35** 12 **37** 186 **39** 646 **41** $\frac{3}{5}$ **43** 40

45 $3 \div \frac{1}{5} = 3 \cdot \frac{5}{1} = 3 \cdot 5$ **47** 14 blankets **49** 48 **51** 6 **53** $\frac{14}{32} = \frac{7}{16}$ **55** $\frac{7}{16}(4{,}064) = 7(254) = 1{,}778$

57 28 **59** 2 **61** 2 **63** 0 **65** $\frac{3{,}479}{3{,}589}$ **67** $\frac{493}{1{,}245}$ **69** $\frac{21{,}113}{19{,}341}$ **71** $\frac{3}{6}$ **73** $\frac{9}{6}$ **75** $\frac{4}{12}$ **77** $\frac{8}{12}$

79 $\frac{10x}{12x}$ **81** $\frac{8x}{12x}$

Problem Set 3.5

1 $\frac{2}{3}$ **3** $-\frac{1}{4}$ **5** $\frac{1}{2}$ **7** $\frac{x-1}{3}$ **9** $\frac{3}{2}$ **11** $\frac{x+6}{2}$ **13** $-\frac{3}{5}$ **15** $\frac{10}{a}$ **17** $\frac{7}{8}$ **19** $\frac{1}{10}$ **21** $\frac{7}{9}$ **23** $\frac{7}{3}$

25 $\frac{1}{4}$ **27** $\frac{7}{6}$ **29** $\frac{5x+4}{20}$ **31** $\frac{10+3x}{5x}$ **33** $\frac{19}{24}$ **35** $\frac{13}{60}$ **37** $\frac{10a+1}{100}$ **39** $\frac{3x+28}{7x}$ **41** $\frac{29}{35}$ **43** $\frac{949}{1{,}260}$

45 $\frac{13}{420}$ **47** $\frac{41}{24}$ **49** $\frac{7y+24}{6y}$ **51** $\frac{5}{4}$ **53** $\frac{160}{63}$ **55** $\frac{5}{8}$ **57** $\frac{9}{2}$ pints **59** $\frac{5}{8}(2{,}120) = 5(265) = \$1{,}325$

61 $\frac{7}{10}$ **63** $\frac{7}{40}(40) = 7$ **65** 10 lots **67** $\frac{3}{2}$ in. **69** $\frac{9}{5}$ ft **71** $\frac{66}{497}$ **73** $\frac{610}{437}$ **75** $\frac{9}{10}$ **77** -8 **79** 3

81 2 **83** xy^2 **85** $\frac{2}{7}$ **87** $\frac{15}{22}$ **89** $\frac{7}{3}$ **91** 3

Problem Set 3.6

1 $\frac{14}{3}$ **3** $\frac{21}{4}$ **5** $\frac{13}{8}$ **7** $\frac{47}{3}$ **9** $\frac{104}{21}$ **11** $\frac{427}{33}$ **13** $1\frac{1}{8}$ **15** $4\frac{3}{4}$ **17** $4\frac{5}{6}$ **19** $3\frac{1}{4}$ **21** $4\frac{1}{27}$ **23** $28\frac{8}{15}$

25 $\frac{8x+3}{x}$ **27** $\frac{2y-5}{y}$ **29** $\frac{6x+5}{6}$ **31** $\frac{3a-1}{3}$ **33** $\frac{17}{8}$ **35** \$6 **37** $\frac{71}{12}$ **39** 6 ft **41** $\frac{1{,}359}{10}$ **43** $\frac{824}{23}$

45 $\frac{2{,}512}{17}$ **47** $\frac{826}{139}$ **49** $\frac{1{,}018}{51}$ **51** $\frac{9}{40}$ **53** $\frac{3}{8}$ **55** $-\frac{32}{35}$ **57** $\frac{3}{2} = 1\frac{1}{2}$ **59** $\frac{4}{5}$ **61** $\frac{ab}{c}$

Problem Set 3.7

1 $5\frac{1}{10}$ **3** $13\frac{2}{3}$ **5** $6\frac{93}{100}$ **7** $5\frac{5}{6}$ **9** $9\frac{3}{4}$ **11** $3\frac{1}{5}$ **13** $12\frac{1}{2}$ **15** $9\frac{9}{20}$ **17** $\frac{32}{45}$ **19** $1\frac{2}{3}$ **21** 4 **23** $4\frac{3}{10}$

25 $\frac{1}{10}$ **27** $3\frac{1}{5}$ **29** $2\frac{1}{8}$ **31** $7\frac{1}{2}$ **33** $\frac{11}{13}$ **35** $5\frac{1}{2}$ cups **37** $\frac{1}{3} \cdot 2\frac{1}{2} = \frac{5}{6}$ cup **39** $1\frac{1}{3}$ **41** $1{,}087\frac{1}{5}$ cents

43 $163\frac{3}{4}$ mi **45** $4\frac{1}{2}$ yd **47** $2\frac{1}{4}$ yd^2 **49** $9\frac{3}{8}$ in^2 **51** Can 1 has 70 calories more than Can 2.

53 Can 1 has 910 milligrams more sodium than Can 2. **55** $\frac{3{,}638}{9}$ **57** $\frac{2{,}516}{437}$ **59** $\frac{4{,}117}{629}$ **61** 2 **63** $\frac{25}{168}$

65 $\frac{10}{9} = 1\frac{1}{9}$ **67** $2\frac{1}{4}$ **69** $3\frac{1}{16}$ **71** $1\frac{1}{2}$ ft **73** $1\frac{1}{2}$ ft

Problem Set 3.8

1 $5\frac{4}{5}$ **3** $12\frac{2}{5}$ **5** $3\frac{4}{9}$ **7** 12 **9** $1\frac{3}{8}$ **11** $14\frac{1}{6}$ **13** $4\frac{1}{12}$ **15** $2\frac{1}{12}$ **17** $26\frac{7}{12}$ **19** 12 **21** $2\frac{1}{2}$ **23** $8\frac{6}{7}$

25 $3\frac{3}{8}$ **27** $10\frac{4}{15}$ **29** $2\frac{1}{15}$ **31** 9 **33** $18\frac{1}{10}$ **35** 14 **37** 17 **39** $24\frac{1}{24}$ **41** $27\frac{6}{7}$ **43** $37\frac{3}{20}$ **45** $6\frac{1}{4}$ **47** $9\frac{7}{10}$

49 $5\frac{1}{2}$ **51** $\frac{2}{3}$ **53** $1\frac{11}{12}$ **55** $3\frac{11}{12}$ **57** $5\frac{19}{20}$ **59** $5\frac{1}{2}$ **61** $\frac{13}{24}$ **63** $3\frac{1}{2}$ **65** $5\frac{29}{40}$ **67** $12\frac{1}{4}$ in. **69** $\frac{3}{4}$ in.

71 $31\frac{1}{6}$ in. **73** $98\frac{1}{8}$ yd **75** $5\frac{7}{8}$¢ **77** **a** $\$2\frac{1}{2}$ **b** \$250 **79** \$300 **81** $7\frac{1}{2}$ yd **83** $15\frac{3}{4}$ in. **85** $1{,}154\frac{790}{2{,}173}$

87 $279\frac{3{,}691}{5{,}777}$ **89** 17 **91** 14 **93** 104 **95** 96 **97** 40

Problem Set 3.9

1 7 **3** 7 **5** 2 **7** 35 **9** $\frac{7}{8}$ **11** $8\frac{1}{3}$ **13** $\frac{11}{36}$ **15** $3\frac{2}{3}$ **17** $6\frac{3}{8}$ **19** $4\frac{5}{12}$ **21** $\frac{8}{9}$ **23** $\frac{1}{2}$ **25** $1\frac{1}{10}$

27 5 **29** $\frac{3}{5}$ **31** $\frac{7}{11}$ **33** 5 **35** $\frac{17}{28}$ **37** $1\frac{7}{16}$ **39** $\frac{13}{22}$ **41** $\frac{5}{22}$ **43** $\frac{15}{16}$ **45** $1\frac{5}{17}$ **47** $\frac{3}{29}$ **49** $1\frac{34}{67}$

51 $\frac{346}{441}$ **53** $5\frac{2}{5}$ **55** 8 **57** $115\frac{2}{3}$ yd **59** \$690 **61** 114 yd **63** $\$212\frac{1}{2}$ **65** $\frac{2}{3}$ **67** $-\frac{1}{6}$ **69** $-\frac{1}{7}$

71 $9\frac{7}{9}$ **73** $\frac{5x + 12}{15}$ **75** $\frac{4x - 2}{x}$ **77** $\frac{21 + 2a}{3a}$ **79** $3\frac{1}{2}$ **81** $1\frac{1}{2}$

Chapter 3 Review

1 $\frac{3}{4}$ **3** $\frac{11a^2}{7}$ **5** x **7** $\frac{8}{21}$ **9** $\frac{2}{3}$ **11** a^2b **13** $\frac{1}{2}$ **15** $-\frac{7}{2}$ **17** $\frac{1}{36}$ **19** $\frac{29}{8}$ **21** $\frac{4x - 3}{4}$ **23** $\frac{8}{13}$

25 12 **27** $17\frac{11}{12}$ **29** $11\frac{2}{3}$ **31** 5 **33** $\frac{1}{2}$ **35** 20 **37** 9 **39** $1\frac{7}{8}$ cups **41** $10\frac{1}{2}$ tablespoons

43 $A = 22$ ft^2; $P = 26\frac{1}{2}$ ft

Chapter 3 Test

1 $\frac{2}{3}$ **2** $\frac{13y}{5}$ **3** $18x$ **4** $\frac{8}{35}$ **5** $\frac{8}{27}$ **6** $-\frac{1}{10}$ **7** $\frac{2}{5}$ **8** $\frac{3}{x}$ **9** $-\frac{23}{5}$ **10** $\frac{15 + 2x}{5x}$ **11** $\frac{47}{36}$ **12** $\frac{37}{7}$

13 $8\frac{3}{5}$ **14** $\frac{5x + 4}{x}$ **15** $\frac{9}{2} = 4\frac{1}{2}$ **16** $9\frac{17}{24}$ **17** $3\frac{2}{3}$ **18** $16\frac{3}{4}$ **19** $9\frac{11}{12}$ **20** $\frac{1}{2}$ **21** 40 grapefruit

22 $27\frac{1}{2}$ in. **23** $9\frac{1}{3}$ cups **24** $3\frac{2}{15}$ ft **25** $P = 23\frac{1}{6}$ ft; $A = 25$ ft^2

CHAPTER 4

Problem Set 4.1

1 $10x$ **3** $4a$ **5** y **7** $-8x$ **9** $3a$ **11** $-5x$ **13** $6x + 11$ **15** $2x + 2$ **17** $-a + 12$ **19** $4y - 4$

21 $-2x + 4$ **23** $8x - 6$ **25** $5a + 9$ **27** $-x + 3$ **29** $17y + 3$ **31** $a - 3$ **33** $6x + 16$ **35** $10x - 11$

37 $19y + 32$ **39** $30y - 18$ **41** $6x + 14$ **43** $12x - 1$ **45** $27a + 5$ **47** 14 **49** 27 **51** -19 **53** 14

55 7 **57** 1 **59** -9 **61** 18 **63** 12 **65** -10 **67** 28 **69** 40 **71** 26 **73** 4

75 $5x + 3$ cannot be simplified. **77** $4x - x = 3x$ **79** $5(y + 4) = 5y + 20$ **81** -9 **83** 6

85 a Distributive property **87 c** Commutative property **89 c** Commutative property **91** 2, 4, 6, 8, . . .

93 1, 3, 5, 7, . . . **95** 1, 4, 9, 16, . . .

Problem Set 4.2

1 Yes **3** Yes **5** No **7** Yes **9** No **11** 6 **13** 11 **15** -15 **17** 1 **19** -3 **21** -1 **23** $\frac{7}{5}$

25 $\frac{11}{12}$ **27** $\frac{23}{26}$ **29** -4 **31** -3 **33** 2 **35** -6 **37** -6 **39** -1 **41** 1 **43** 3 **45** $20x$ **47** $-6y$

49 $-12a$ **51** x **53** y **55** a

Problem Set 4.3

1 8 **3** -6 **5** -6 **7** 6 **9** 16 **11** 16 **13** $-\frac{3}{2}$ **15** -7 **17** -8 **19** 6 **21** 2 **23** 3 **25** 4

27 -24 **29** 12 **31** -8 **33** 15 **35** 3 **37** -1 **39** 1 **41** 3 **43** 1 **45** -1 **47** $-\frac{1}{3}$ **49** -14

51 $3x + 12$ **53** $6a - 16$ **55** $-15x + 3$ **57** $3y - 9$ **59** $16x - 6$

Problem Set 4.4

1 3 **3** 2 **5** −3 **7** −4 **9** 1 **11** 0 **13** 2 **15** −2 **17** 3 **19** −1 **21** 7 **23** −3 **25** 1

27 −2 **29** 4 **31** 10 **33** −5 **35** $-\frac{1}{12}$ **37** −3 **39** 20 **41** −1 **43** 5 **45** 4 **47** $x + 2$ **49** $2x$

51 $2(x + 6)$ **53** $x - 4$ **55** $2x + 5$

Problem Set 4.5

1 $x + 3$ **3** $2x + 1$ **5** $5x - 6$ **7** $3(x + 1)$ **9** $5(3x + 4)$ **11** 2 **13** −2 **15** 3 **17** 5 **19** −2

21 Length = 10 m; width = 5 m **23** 8 cm **25** Sue is 35; Dale is 39 **27** $\frac{17}{12}$, or $1\frac{5}{12}$ **29** $\frac{1}{2}$ **31** $2\frac{4}{5}$

33 $\frac{8}{5}$, or $1\frac{3}{5}$ **35** $5\frac{1}{2}$ **37** $\frac{65}{9}$, or $7\frac{2}{9}$

Problem Set 4.6

1 704 ft^2 **3** $\frac{9}{8}$ in^2 **5** \$240 **7** \$285 **9** 12 ft **11** 8 ft **13** \$140 **15** 58 in. **17** $3\frac{1}{4} = \frac{13}{4}$ ft

19 $C = 100°$C; yes **21** $C = 20°$C; yes **23** 0°C **25** 5°F **27** 44 in. **29** 22 in. **31** 154 ft^2 **33** $\frac{11}{14}$ ft^2

35 360 in^3 **37** 1 yd^3 **39** $y = 7$ **41** $y = -3$ **43** $y = -2$ **45** $x = 3$ **47** $x = 5$ **49** $x = 8$ **51** $y = 1$

53 $y = 3$ **55** $y = \frac{5}{2}$ **57** $y = 0$ **59** $y = \frac{13}{3}$ **61** $y = 4$ **63** $x = 0$ **65** $x = \frac{13}{4}$ **67** $x = 3$ **69** $\frac{3}{4}$

71 3 **73** 2 **75** 6

Problem Set 4.7

1 (0, 4), (3, 1), (−2, 6) **3** (0, 3), (2, 2), (18, −6) **5** (0, 4), (3, 0), (−3, 8) **7** (1, 1), (0, −3), (5, 17)

9 (0, 3), (2, 7), (−2, −1) **11** (2, 14), $\left(\frac{6}{7}, 6\right)$, (0, 0) **13** (0, 0), (−2, 4), (2, −4) **15** (0, 0), (2, 1), (4, 2)

17 (−2, 3), (0, 2), (2, 1)

19

x	y
2	3
3	2
1	4
0	5

21

x	y
0	0
−1	4
−2	8
1	−4

23

x	y
−2	−6
2	0
4	3
0	−3

25

x	y
−1	−7
1	5
−2	−13
0	−1

27 (0, −2) **29** (0, 3) **31** (0, 0), (5, −5), (−3, 3) **33** (1, 5) **35** $A = 6$ ft^2; $P = 12$ ft **37** $A = 12$ in^2; $P = 18$ in.

Problem Set 4.8

1–17 Odd **19** (2, 2) **21** (−3, 2) **23** (3, −3) **25** (−4, 0)

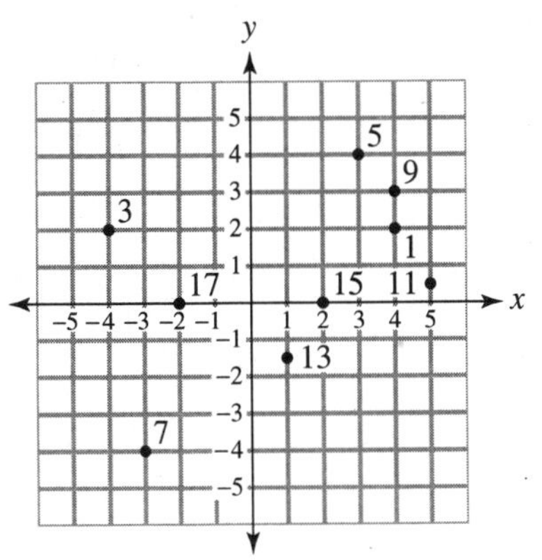

27 Yes **29** No

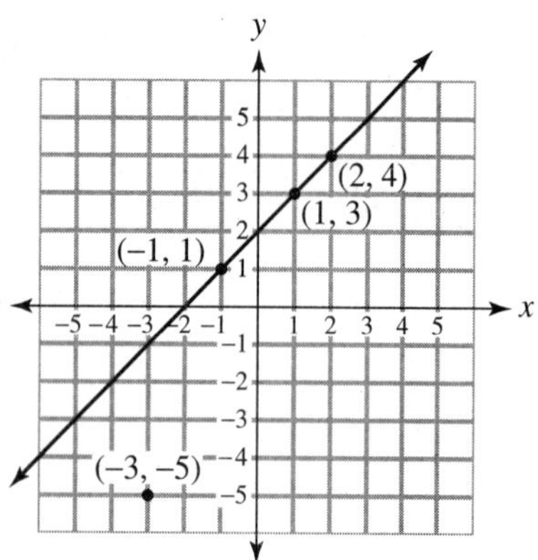

31 Yes **33** No

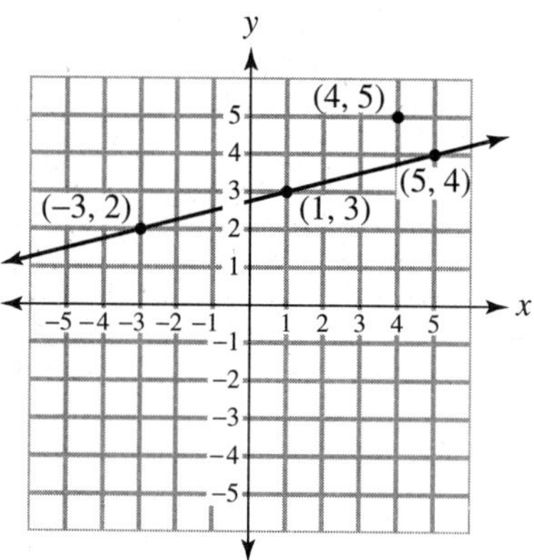

35 xy^2 **37** $\frac{1}{xy}$ **39** $\frac{4x + 15}{20}$ **41** $\frac{3x - 1}{x}$

Problem Set 4.9

1

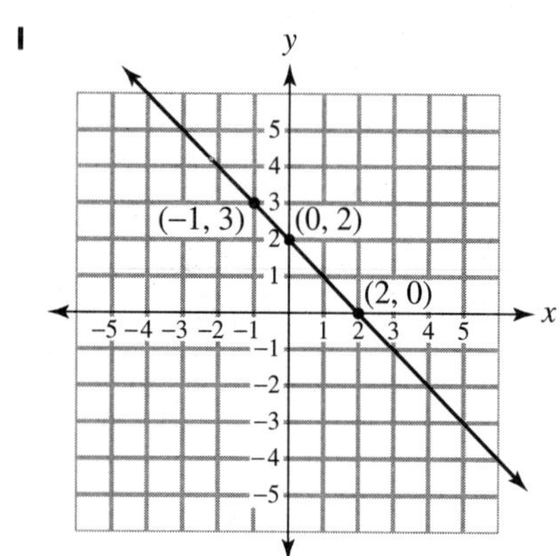

3

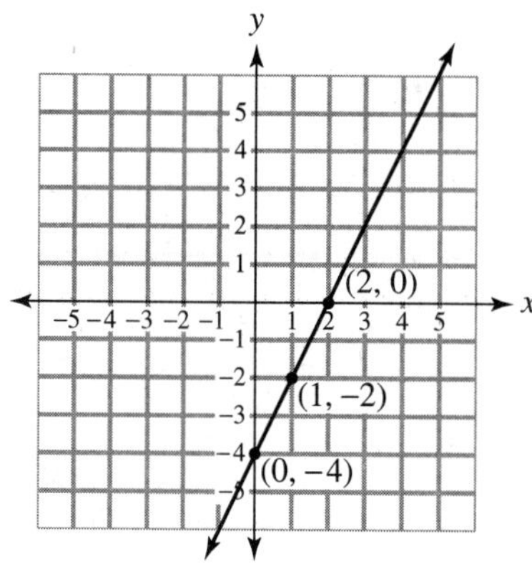

5

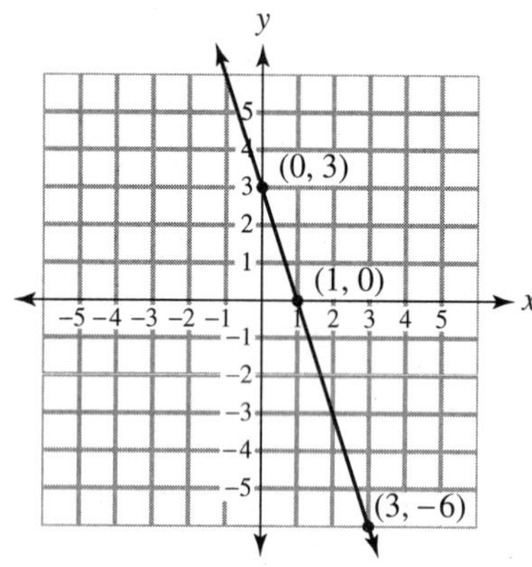

7

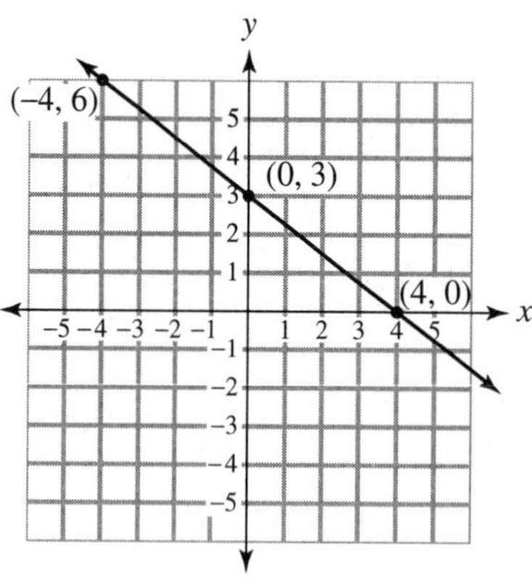

9

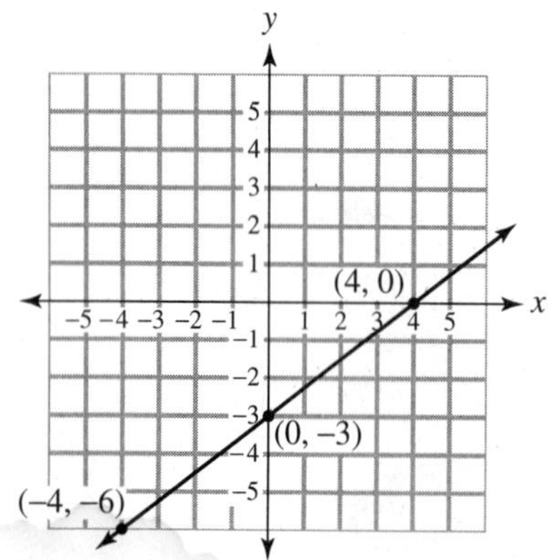

11

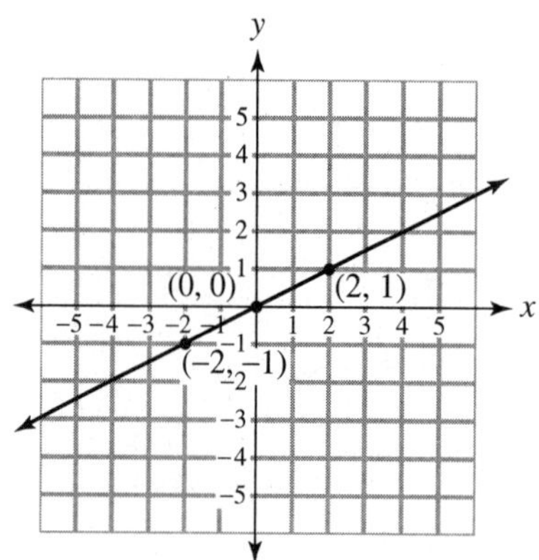

13

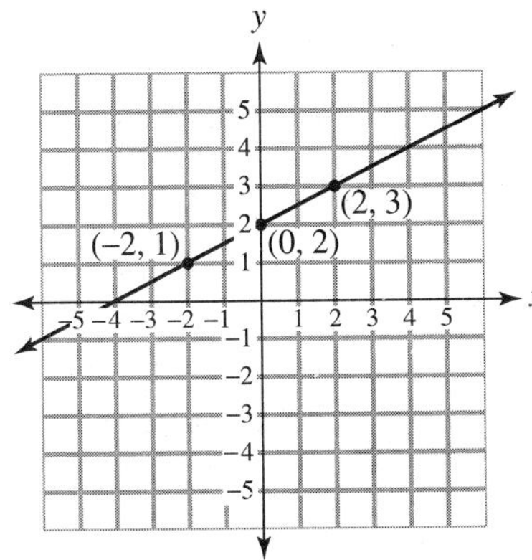

15

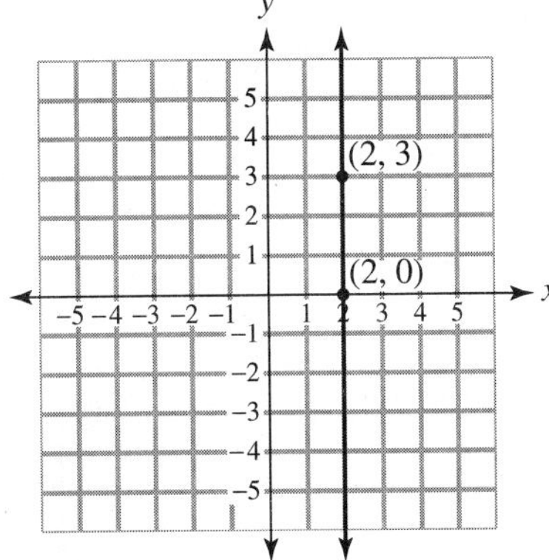

17

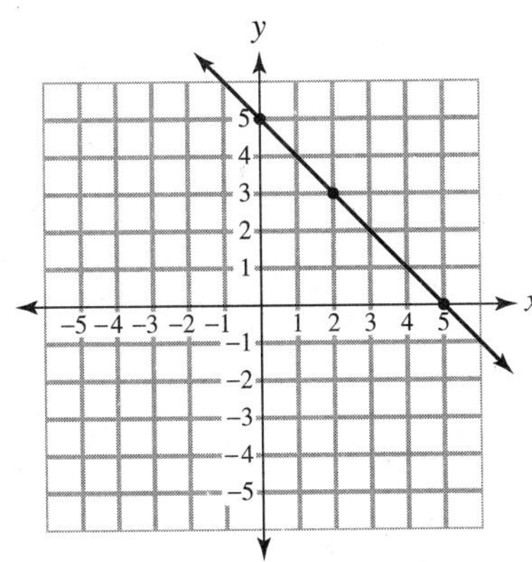

19

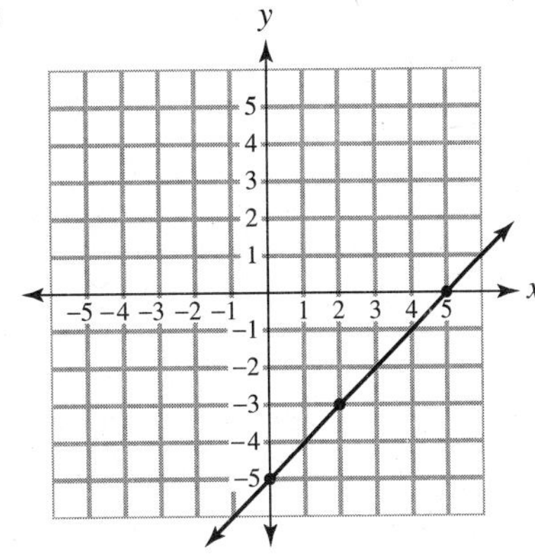

21

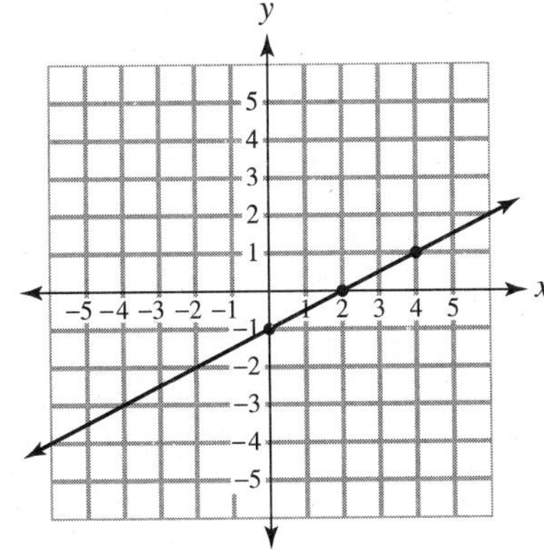

23

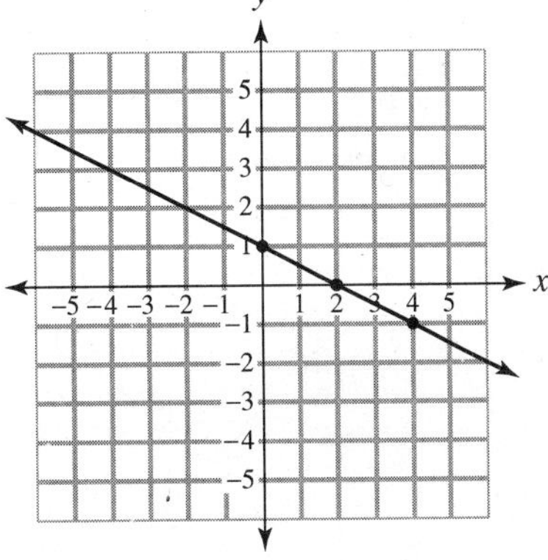

25

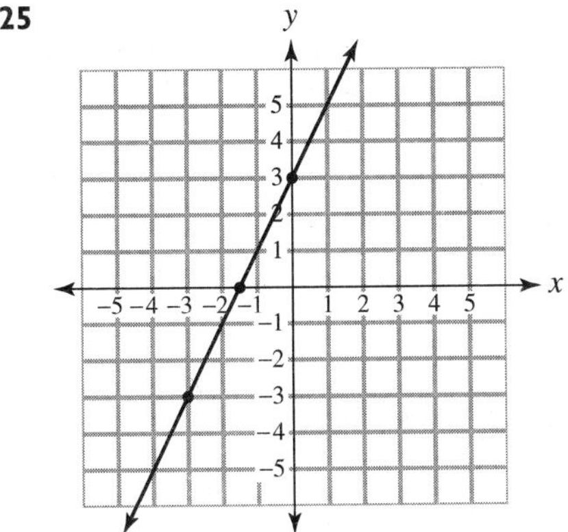

27

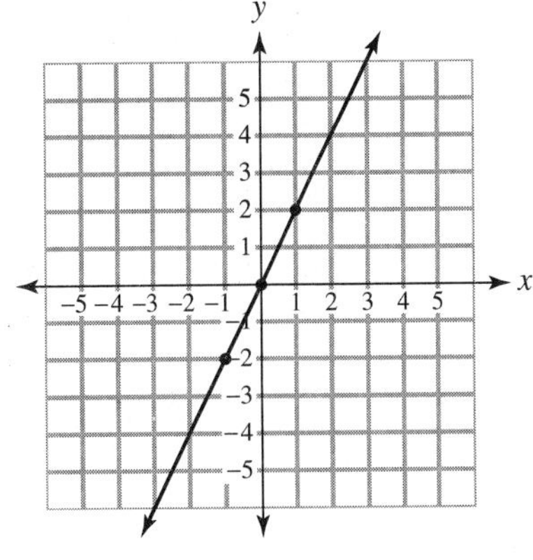

29

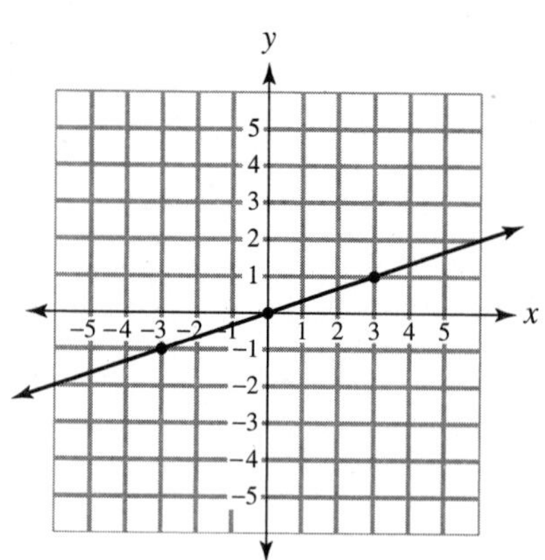

31

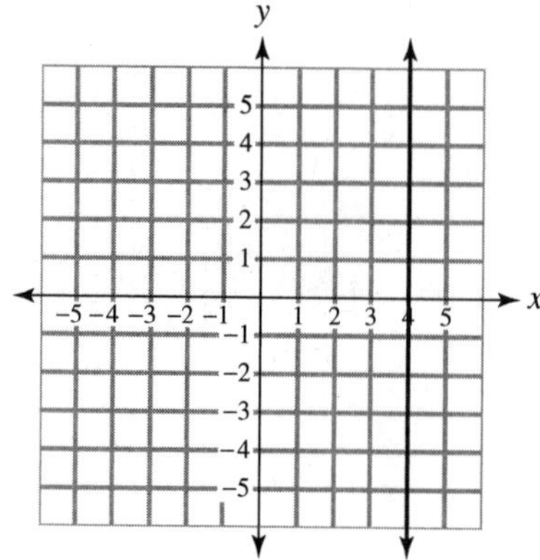

33

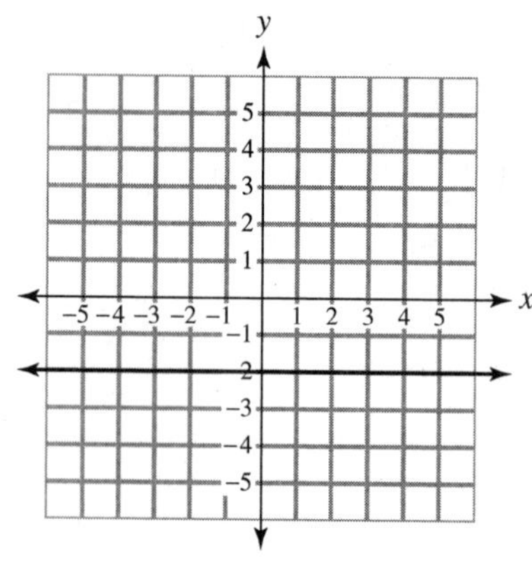

35 $\frac{33}{50}$ **37** $\frac{56}{225}$ **39** $\frac{1}{100}$

Chapter 4 Review

1 $17x$ **3** $11a - 3$ **5** $5x + 4$ **7** $10a - 4$ **9** 42 **11** 1 **13** Yes **15** 9 **17** 3 **19** 9 **21** 5

23 -1 **25** $-\frac{1}{8}$ **27** -2 **29** -4 **31** Length = 14 m; width = 7 m **33** $y = 6$ **35** $y = 3$ **37** $x = 0$

39 $x = 2$

41–47

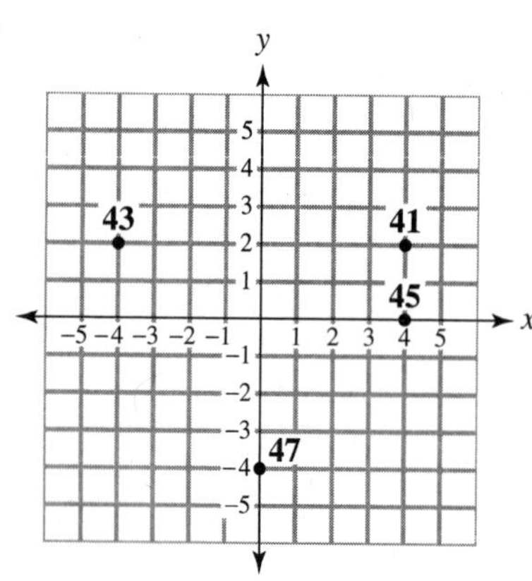

49

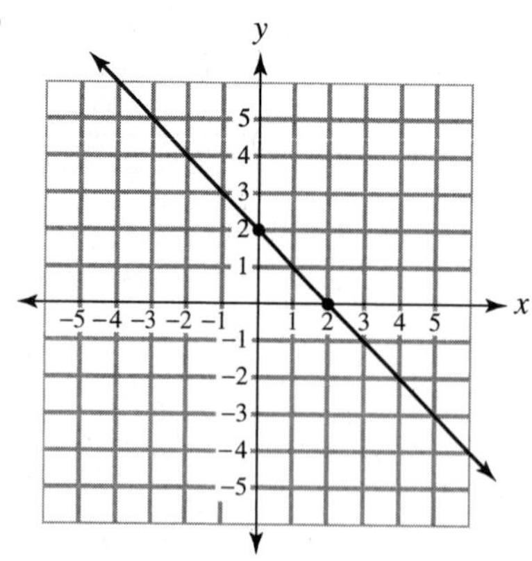

51

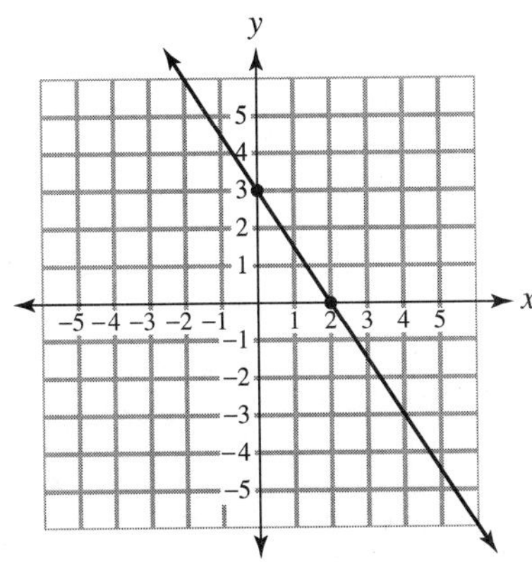

53

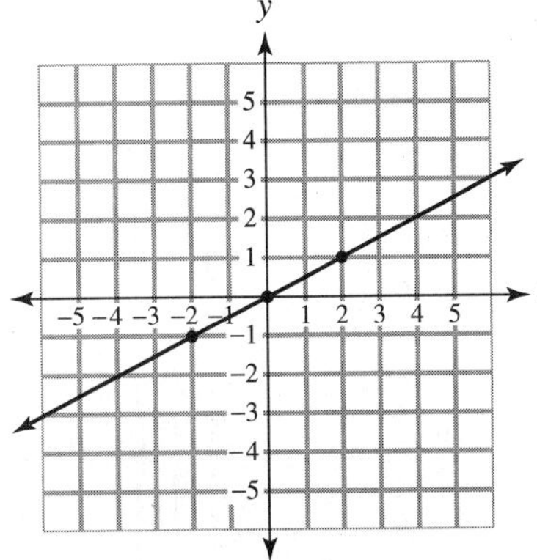

55

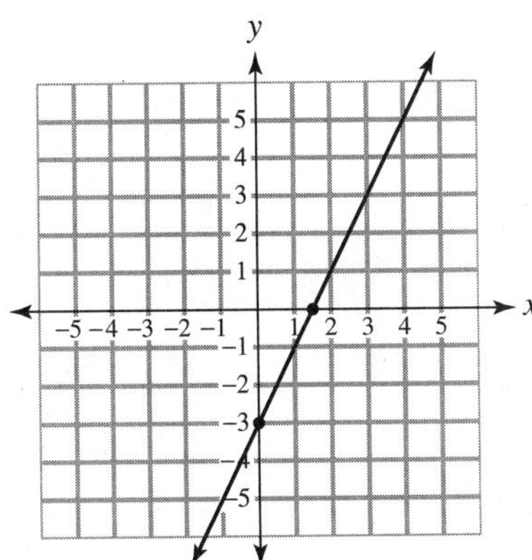

57

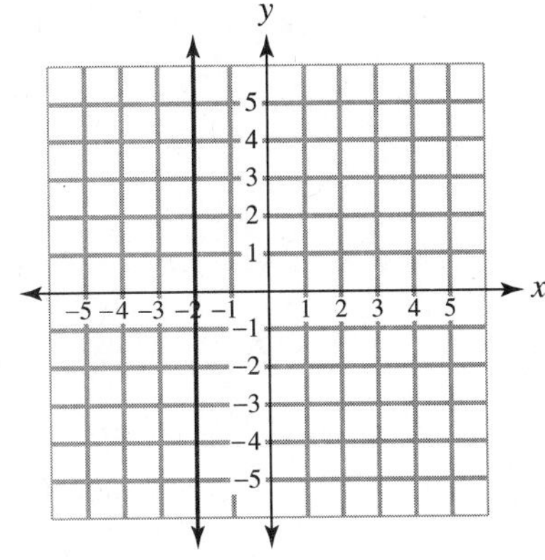

Chapter 4 Test

1 $6x - 5$ **2** $3b - 4$ **3** -3 **4** 9 **5** Yes **6** 4 **7** 27 **8** -4 **9** 1 **10** $\frac{2}{3}$ **11** 1 **12** -8 **13** 3

14 Length = 9 cm; width = 5 cm **15** Susan is 11; Karen is 6 **16** 8

17

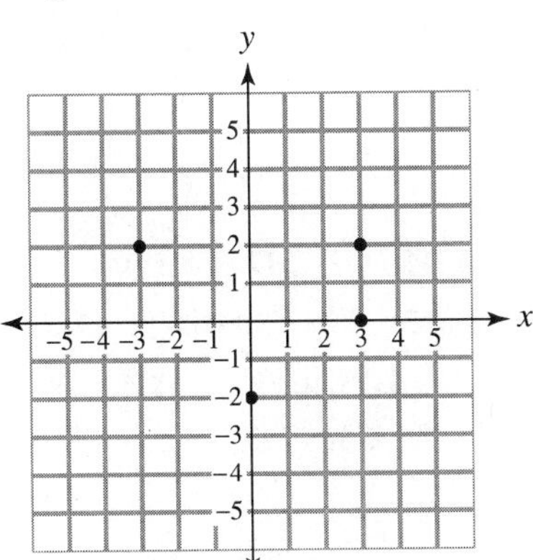

18

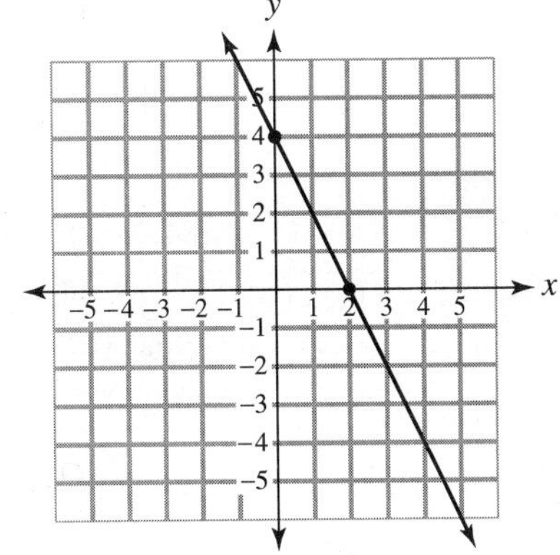

19

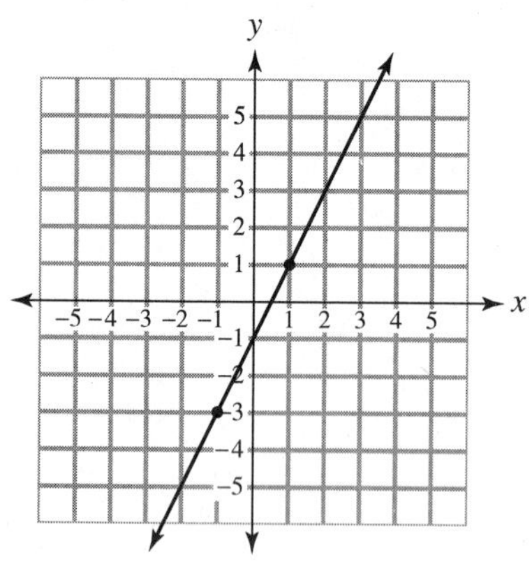

20

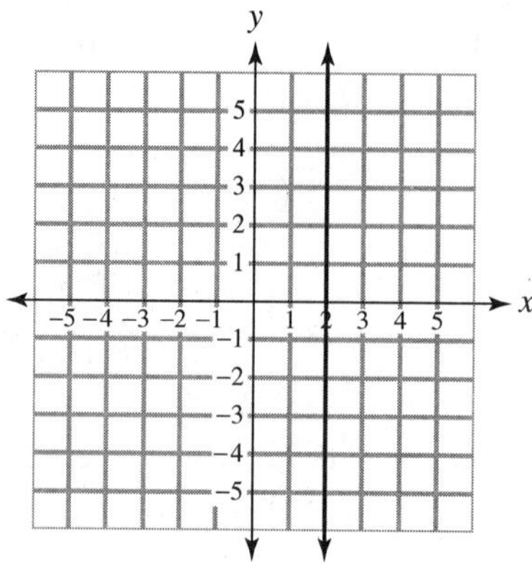

CHAPTER 5

Problem Set 5.1

1 Three tenths **3** Fifteen thousandths **5** Three and four tenths **7** Fifty-two and seven tenths **9** $405\frac{36}{100}$

11 $9\frac{9}{1{,}000}$ **13** $1\frac{234}{1{,}000}$ **15** $\frac{305}{100{,}000}$ **17** Tens **19** Tenths **21** Hundred thousandths **23** Ones **25** Hundreds

27 0.55 **29** 6.9 **31** 11.11 **33** 100.02 **35** 3,000.003 **37** 48; 47.5; 47.55; 47.548 **39** 1; 0.8; 0.82; 0.818

41 0; 0.2; 0.16; 0.156 **43** 2,789; 2,789.3; 2,789.32; 2,789.324 **45** 100; 100.0; 100.00; 100.000

47 Three and eleven hundredths; two and five tenths **49** 186,282.40 **51** Fifteen hundredths

53 0.002 0.005 0.02 0.025 0.05 0.052 **55** 7.451 7.54 **57** $\frac{1}{4}$ **59** $\frac{1}{2}$ **61** $\frac{7}{20}$ **63** $\frac{1}{8}$ **65** $\frac{5}{8}$ **67** $\frac{7}{8}$

69 9.99 **71** 10.05 **73** 0.05 **75** 0.01 **77** $6\frac{31}{100}$ **79** $6\frac{23}{50}$ **81** $18\frac{123}{1{,}000}$ **83** $\frac{2x+3}{x}$ **85** $\frac{4x-3}{4}$

87 0.285 0.286 0.287 0.288 0.289 0.290 0.291 0.292 0.293 0.294

89 2.5 2.6 2.7 2.8 2.9 3.0 3.1 3.2 3.3 3.4

Problem Set 5.2

1 6.19 **3** 1.13 **5** 1.49 **7** 9.042 **9** −1.979 **11** 11.7843 **13** 24.343 **15** 24.111 **17** 258.5414

19 666.66 **21** 11.11 **23** −3.57 **25** 4.22 **27** −120.41 **29** 44.933 **31** −8.327 **33** 530.865 **35** 27.89

37 35.64 **39** 411.438 **41** 6 **43** 1 **45** 3.1 **47** 5.9 **49** 3.272 **51** 4.001 **53** \$116.82 **55** \$0.29

57 \$1,571.10 **59** \$5.43 **61** \$154.12 **63** \$4.83

65

5.6 seconds **67** b **69** 219.0842

71 69.34537 **73** 1,035.7943 **75** $\frac{3}{100}$ **77** $\frac{51}{10,000}$ **79** $1\frac{1}{2}$ **81** 1,400 **83** 2 in.

85 \$3.25; three \$1 bills and a quarter **87** 3.25

Problem Set 5.3

1 0.28 **3** 0.028 **5** 0.0027 **7** −0.78 **9** 0.792 **11** 0.0156 **13** 24.29821 **15** −0.03 **17** 187.85 **19** 0.002 **21** −27.96 **23** 0.43 **25** 49,940 **27** 9,876,540 **29** 1.89 **31** −0.0025 **33** −4.9894 **35** 7.3485 **37** 4.4 **39** 2.074 **41** 3.58 **43** 187.4 **45** 116.64 **47** 20.75 **49** 0.126 **51** Moves it two places to the right **53** \$6.93 **55** 1,319¢ or \$13.19 **57** \$415.38 **59** \$7.10 **61** \$44.40 **63** \$293.04 **65** $C = 25.12$ in.; $A = 50.24$ in^2 **67** $C = 37.68$ in.; $A = 113.04$ in^2 **69** 24,492 mi **71** 100.48 ft^3 **73** 50.24 ft^3 **75** 105,957.65 **77** 202.60296 **79** 0.421875 **81** $A = 77.56$ ft^2; $C = 31.21$ ft **83** 1,879 **85** 1,516 R 4 **87** $\frac{3}{4}$ **89** $\frac{1}{5}$ **91** 90 **93** x^2y^3 **95** b^2c^3 **97** $4x^2y^2$

99 The next number is 0.4. The sequence is an arithmetic sequence, starting with 0.1, in which each term after that is found by adding 0.1 to the term before it.

101 The next number is 0.0001. The sequence is a geometric sequence, starting with 0.1, in which each term after that is found by multiplying the term before it by 0.1.

103 0.04, 0.008, 0.0016; a decreasing sequence **105** 1.1, 1.21, 1.331; an increasing sequence

Problem Set 5.4

1 19.7 **3** 6.2 **5** 5.2 **7** 11.04 **9** −4.8 **11** 9.7 **13** 2.63 **15** 4.24 **17** 2.55 **19** −1.35 **21** 6.5 **23** 9.9 **25** 0.05 **27** 89 **29** 2.2 **31** 1.35 **33** 16.97 **35** 0.25 **37** 2.71 **39** 11.69 **41** False **43** True **45** 5 **47** \$4.15 **49** 25.5 hr **51** 22.4 mi **53** 5 hr **55** 7 min **57** 2.73 **59** 0.13 **61** 0.77778 **63** 307.20607 **65** 0.70945 **67** $\frac{3}{4}$ **69** $\frac{2}{3y}$ **71** $\frac{3xy}{4}$ **73** $\frac{6}{10}$ **75** $\frac{60}{100}$ **77** $\frac{12x}{15x}$ **79** $\frac{60}{15x}$ **81** $\frac{18}{15x}$ **83** 2.35 ft **85** Diameter = 2.5 m; radius = 1.25 m **87** Radius = 4 m; diameter = 8 m

Problem Set 5.5

1 0.5 **3** 0.375 **5** 0.875 **7** 0.48 **9** 0.4375 **11** 0.6 **13** 0.92 **15** 0.27 **17** 0.09 **19** 0.28 **21** $\frac{16}{25}$ **23** $\frac{3}{20}$ **25** $\frac{2}{25}$ **27** $\frac{3}{8}$ **29** $5\frac{3}{5}$ **31** $5\frac{3}{50}$ **33** $17\frac{13}{50}$ **35** $1\frac{11}{50}$ **37** 2.4 **39** 1.7625 **41** 3.02 **43** 0.3 **45** 0.072 **47** 0.8 **49** 1 **51** 0.2 or $\frac{1}{5}$ **53** $\frac{1}{4}$ or 0.25 **55** \$8.42 **57** 4 glasses **59** \$38.66 **61** \$561.63 to the nearest cent **63** 104.625 calories **65** 33.49 mi^3 **67** 248.35 in^3 **69** 22.28 in^2 **71** 226.1 ft^3 **73** $0.\overline{7}$ **75** $0.\overline{123}$ **77** 53.11894 **79** −9 **81** −7 **83** −49 **85** −6 **87** −3 **89** 2 **91** 1 **93** 10.125, $11\frac{3}{8}$ **95** $\frac{1}{10,000}$, 0.00001

Problem Set 5.6

1 −1.5 **3** 0.77 **5** 0.15 **7** −0.35 **9** −0.8 **11** 2.05 **13** −0.064 **15** 12.03 **17** 0.44 **19** −0.175 **21** 2 **23** 3,000 **25** 87 mi **27** 147 mi **29** 8 nickels, 18 dimes **31** 16 dimes, 32 quarters **33** 125 **35** 9 **37** $\frac{1}{81}$ **39** $\frac{25}{36}$ **41** 0.25 **43** 1.44 **45** x^{13} **47** $45a^7$ **49** $225x^{10}y^6$

Problem Set 5.7

1 8 **3** 9 **5** 6 **7** 5 **9** 15 **11** 48 **13** 45 **15** 48 **17** 15 **19** 1 **21** 78 **23** 9 **25** $\frac{4}{7}$

27 $\frac{3}{4}$ **29** False **31** True **33** 10 in. **35** 13 ft **37** 6.40 in. **39** 17.49 m **41** 25 ft **43** 1.4142

45 3.4641 **47** 11.1803 **49** 18.0000 **51** 3.46 **53** 11.18 **55** 0.58 **57** 0.58 **59** 12.124 **61** 9.327

63 12.124 **65** 12.124 **67** 7.54 in. **69** $\frac{2}{5}$ **71** $9\frac{8}{25}$ **73** 8 **75** $2\frac{4}{5}$ **77** 6 in., 24 in^2 **79** 5 ft

Problem Set 5.8

1 $2\sqrt{3}$ **3** $2\sqrt{5}$ **5** $6\sqrt{2}$ **7** $7\sqrt{2}$ **9** $2\sqrt{7}$ **11** $10\sqrt{2}$ **13** $2x\sqrt{3}$ **15** $5x\sqrt{2}$ **17** $5x\sqrt{3x}$ **19** $5x\sqrt{2x}$
21 $4xy\sqrt{2y}$ **23** $9x^2\sqrt{3}$ **25** $6xy^2\sqrt{2}$ **27** $2xy\sqrt{3xy}$ **29** $3\sqrt{2}$ in. **31** $4\sqrt{2}$ in. **33** $x\sqrt{2}$ in. **35** 8.4852814
37 4.8989795 **39** 11.18034 **41** 55.901699 **43** -3 **45** -5 **47** -2 **49** 2 **51** 0 **53** 5 **55** 10

Problem Set 5.9

1 $10\sqrt{3}$ **3** $4\sqrt{5}$ **5** $7\sqrt{x}$ **7** $9\sqrt{7}$ **9** $6\sqrt{y}$ **11** $7\sqrt{2}$ **13** $8\sqrt{3}$ **15** $-2\sqrt{3}$ **17** $8\sqrt{10}$ **19** $x\sqrt{2}$
21 $17x\sqrt{5x}$ **23** $7\sqrt{2}$ ft **25** $\sqrt{2} + \sqrt{3} \approx 3.1463$; $\sqrt{5} \approx 2.2361$
27 $\sqrt{16} - \sqrt{10} \approx 0.8377$; $\sqrt{6} \approx 2.4495$ **29** $\sqrt{8} + \sqrt{2} \approx 4.2426$; $\sqrt{18} \approx 4.2426$; $\sqrt{8} + \sqrt{2} = 2\sqrt{2} + \sqrt{2} = 3\sqrt{2}$; $\sqrt{18} = \sqrt{3 \cdot 3 \cdot 2} = 3\sqrt{2}$

31

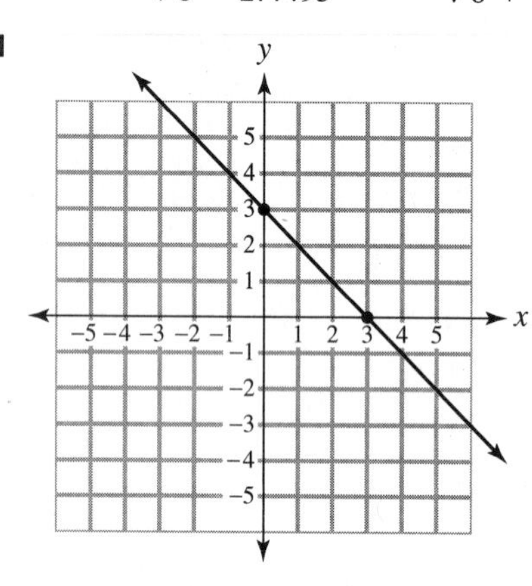

33

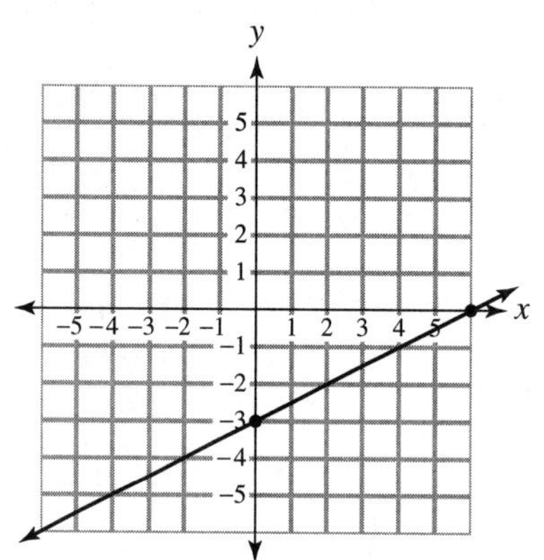

35

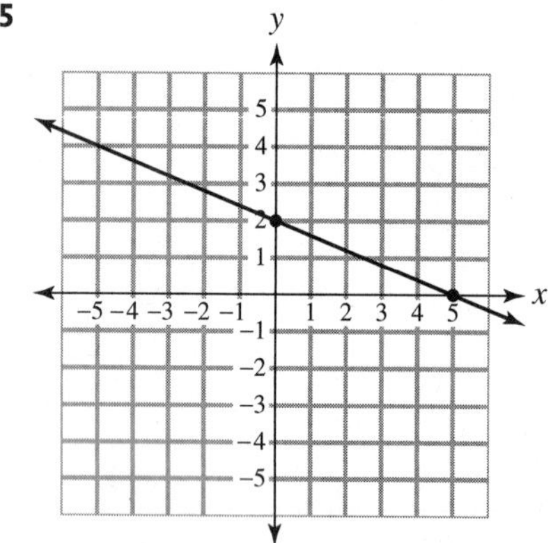

37

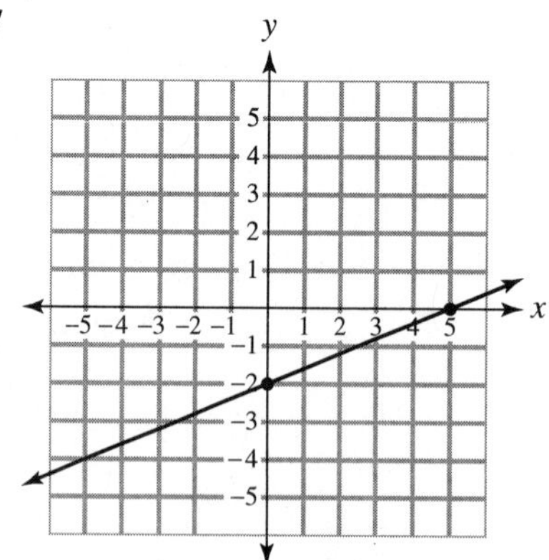

39

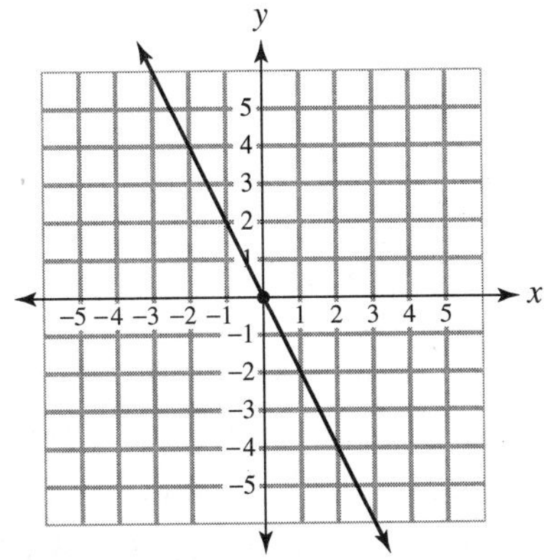

Chapter 5 Review

1 Thousandths **3** 37.0042 **5** 98.77 **7** 7.779 **9** −21.5736 **11** 1.23 **13** $\frac{141}{200}$ **15** 2.42 **17** 79

19 −5.9 **21** 6.4 **23** 15 **25** 47 **27** \$267.70 **29** 4.875 **31** 2.80 **33** 5.4 ft **35** $P = 12.4$ ft, $A = 5$ ft^2

37 $2\sqrt{3}$ **39** $4\sqrt{2}$ **41** $2x\sqrt{5}$ **43** $3xy\sqrt{2y}$ **45** $10\sqrt{3}$ **47** $5\sqrt{3}$ **49** $5\sqrt{6}$ **51** $-9\sqrt{3}$

Chapter 5 Test

1 Five and fifty-three thousandths **2** Thousandths **3** 17.0406 **4** 46.75 **5** 8.18 **6** 6.056 **7** 35.568

8 8.72 **9** 0.92 **10** $\frac{14}{25}$ **11** 14.664 **12** 4.69 **13** 17.129 **14** 0.26 **15** 0.19 **16** 36 **17** $\frac{5}{9}$

18 \$11.53 **19** \$19.04 **20** \$6.55 **21** 5 in. **22** $6\sqrt{2}$ **23** $4x\sqrt{3y}$ **24** $8\sqrt{7}$ **25** $-8\sqrt{3}$

CHAPTER 6

Problem Set 6.1

1 $\frac{4}{3}$ **3** $\frac{16}{3}$ **5** $\frac{2}{5}$ **7** $\frac{1}{2}$ **9** $\frac{3}{1}$ **11** $\frac{7}{6}$ **13** $\frac{7}{5}$ **15** $\frac{5}{7}$ **17** $\frac{8}{5}$ **19** $\frac{1}{3}$ **21** $\frac{1}{10}$ **23** $\frac{3}{25}$ **25** $\frac{1}{3}$

27 $\frac{5}{3}$ **29 a** $\frac{5}{8}$ **b** $\frac{1}{4}$ **c** $\frac{3}{8}$ **d** $\frac{5}{3}$ **31** 55 miles per hour **33** 84 kilometers per hour

35 0.2 gallon per second **37** 14 liters per minute **39** 19 miles per gallon **41** $4\frac{1}{3}$ miles per liter **43** 8¢ per ounce

45 3.45¢ per ounce **47 a** $\frac{2{,}314}{2{,}408} = 0.96$ **b** $\frac{2{,}408}{2{,}314} = 1.04$ **c** $\frac{2{,}314}{4{,}722} = 0.49$ **d** $\frac{4{,}722}{2{,}408} = 1.96$

49 54.03 miles per hour **51** 9.3 miles per gallon **53** $\frac{8}{3}$, or $2\frac{2}{3}$ **55** $\frac{1}{40}$ **57** $\frac{73}{20}$, or $3\frac{13}{20}$ **59** $\frac{62}{25}$, or $2\frac{12}{25}$

61 $\frac{4}{3}$, or $1\frac{1}{3}$

Problem Set 6.2

1 60 in. **3** 120 in. **5** 6 ft **7** 162 in. **9** $2\frac{1}{4}$ ft **11** $6\frac{1}{3}$ yd **13** $1\frac{1}{3}$ yd **15** 1,800 cm **17** 4,800 m

19 50 cm **21** 0.248 km **23** 670 mm **25** 34.98 m **27** 6.34 dm **29** 20 yd **31** 80 in. **33** 244 cm

35 65 mm **37** 2,960 chains **39** 120,000 μm **41** 7,920 ft **43** 80.7 ft/sec **45** 19.5 mi/hr **47** 1,023 mi/hr

49 \$18,216 **51** \$78.75 **53** 3,965,280 ft **55** 179,352 in. **57** 2.7 mi **59** 18,094,560 ft **61** $\frac{1}{3}$ **63** 6

65 $3\frac{1}{2}$ **67** 6 **69** 6 **71** $\frac{7}{10}$ **73** 125 calories **75** 28 days **77** 375 Cal/hour

Problem Set 6.3

1 432 in^2 **3** 2 ft^2 **5** 1,306,800 ft^2 **7** 1,280 acres **9** 3 mi^2 **11** 108 ft^2 **13** 1,700 mm^2 **15** 28,000 cm^2 **17** 0.0012 m^2 **19** 500 m^2 **21** 700 a **23** 3.42 ha **25** 18,000 ft^2 **27** 30 a **29** 5,500 bricks **31** 135 ft^3 **33** 48 fl oz **35** 8 qt **37** 20 pt **39** 480 fl oz **41** 8 gal **43** 6 qt **45** 9 yd^3 **47** 5,000 mL **49** 0.127 liter **51** 4,000,000 mL **53** 14,920 liters **55** 16 cups **57** 34,560 in^3 **59** 48 glasses **61** 20,288,000 acres **63** 3,230.93 mi^2 **65** 23.35 gal **67** 21,492 ft^3 **69** 285,795,000 ft^3 **71** $\frac{5}{8}$ **73** 9 **75** $\frac{1}{3}$ **77** $\frac{17}{144}$ **79** $3\frac{1}{18}$

Problem Set 6.4

1 128 oz **3** 4,000 lb **5** 12 lb **7** 0.9 T **9** 32,000 oz **11** 56 oz **13** 13,000 lb **15** 2,000 g **17** 40 mg **19** 200,000 cg **21** 508 cg **23** 4.5 g **25** 47.895 cg **27** 1.578 g **29** 0.42 kg **31** $9\frac{1}{4}$ lb, or 9.25 lb **33** \$9.14 **35** 15 g **37** 63.68 oz **39** 12.5625 lb **41** \$76.42 **43** $\frac{9}{50}$ **45** $\frac{9}{100}$ **47** $\frac{4}{5}$ **49** 0.9 **51** 0.125

Problem Set 6.5

1 15.24 cm **3** 13.12 ft **5** 6.56 yd **7** 32,200 m **9** 5.98 yd^2 **11** 24.7 acres **13** 8,195 mL **15** 2.12 qt **17** 75.8 liters **19** 339.6 g **21** 33 lb **23** 365°F **25** 30°C **27** 3.94 in. **29** 7.62 m **31** 46.23 liters **33** 17.67 oz **37** 91.46 m **39** 20.90 m^2 **41** 88.55 km/hr **43** 2.03 m **45** 38.3°C **47** 0.75 **49** 5.5 **51** 0.03 **53** $\frac{17}{50}$ **55** $2\frac{2}{5}$ **57** $1\frac{3}{4}$

Problem Set 6.6

1 Means: 3, 5; extremes: 1, 15; products: 15 **3** Means: 25, 2; extremes: 10, 5; products: 50 **5** Means: $\frac{1}{2}$, 4; extremes: $\frac{1}{3}$, 6; products: 2 **7** Means: 5, 1; extremes: 0.5, 10; products: 5 **9** 10 **11** $\frac{12}{5}$ **13** $\frac{3}{2}$ **15** $\frac{10}{9}$ **17** 7 **19** 1 **21** 18 **23** 6 **25** 40 **27** 50 **29** $\frac{1}{2}$ **31** 3 **33** 1 **35** $\frac{1}{4}$ **37** 329 mi **39** 360 points **41** 15 pt **43** 427.5 mi **45** 900 eggs **47** \$2.33 **49** \$119.70 **51** 265 g **53** 108 **55** 65 **57** 41 **59** 108 **61** 91.3 liters **63** 60,113 **65** Tens **67** 26.516 **69** 0.39 **71** 1.35 **73** 3.816 **75** 4 **77** 160

Chapter 6 Review

1 $\frac{3}{10}$ **3** $\frac{3}{4}$ **5** $\frac{7}{5}$ **7** $\frac{1}{2}$ **9** $\frac{1}{3}$ **11** $\frac{9}{2}$ **13** $\frac{1}{4}$ **15** 19 mi/gal **17** 8¢ per oz **19** 144 in. **21** 0.49 m **23** 435,600 ft^2 **25** 576 in^2 **27** 6 gal **29** 128 oz **31** 5,000 g **33** 10.16 cm **35** 7.42 qt **37** 141.5 g **39** 248°F **41** 600 bricks **43** 80 glasses **45** 49 **47** $\frac{1}{10}$ **49** 1,500 mL **51** 5 weeks

Chapter 6 Test

1 $\frac{4}{3}$ **2** $\frac{9}{10}$ **3** $\frac{3}{2}$ **4** $\frac{3}{10}$ **5** $\frac{3}{5}$ **6** $\frac{12}{5}$ **7** $\frac{6}{25}$ **8** 23 mpg **9** 30-ounce: \$0.035/ounce; 20-ounce: \$0.036/ounce; 30-ounce can is the better buy. **10** 21 ft **11** 0.75 km **12** 130,680 ft^2 **13** 3 ft^2 **14** 10,000 mL **15** 8 km **16** 10.6 qt **17** 26.7°C **18** 90 tiles **19** 64 glasses **20** \$21.00 **21** 36 **22** 8 **23** 24 **24** 135 miles **25** 25 tbsp

CHAPTER 7

Problem Set 7.1

1 $\frac{20}{100}$ **3** $\frac{60}{100}$ **5** $\frac{24}{100}$ **7** $\frac{65}{100}$ **9** 0.23 **11** 0.92 **13** 0.09 **15** 0.034 **17** 0.0634 **19** 0.009

21 23% **23** 92% **25** 45% **27** 3% **29** 60% **31** 80% **33** 27% **35** 123% **37** $\frac{3}{5}$ **39** $\frac{3}{4}$ **41** $\frac{1}{25}$

43 $\frac{53}{200}$ **45** $\frac{7{,}187}{10{,}000}$ **47** $\frac{3}{400}$ **49** $\frac{1}{16}$ **51** $\frac{1}{3}$ **53** 50% **55** 75% **57** $33\frac{1}{3}$% **59** 80% **61** 87.5%

63 14% **65** 325% **67** 150% **69** 48.8% **71** 0.50, 0.75 **73** 0.189, 0.081 **75** 20% **77** $16\frac{2}{3}$% **79** 78.4%

81 11.8% **83** 72.2% **85** 8.3%, 0.2% **87** 75 **89** 8.48 **91** 500 **93** 180 **95** 4

97 Divide 2 by 5 to obtain 0.40. Then move the decimal point two places to the right and attach the % symbol. The result is 40%.

99 75% is 100 times as large as 0.75%.

101 The only numbers we have to work with are 1/3 and 31%. Since 1/3 converts to $33\frac{1}{3}$%, which is larger than 31%, it is more likely that they will stay in a commercial lodging.

Problem Set 7.2

1 8 **3** 24 **5** 20.52 **7** 7.37 **9** 50% **11** 10% **13** 25% **15** 75% **17** 64 **19** 50 **21** 925

23 400 **25** 5.568 **27** 120 **29** 13.72 **31** 22.5 **33** 50% **35** 942.684 **37** 97.8

39 4.8% calories from fat; healthy **41** 50% calories from fat; not healthy **43** 248.75 **45** 25.9% **47** 49.5

49 $3 \cdot 3 \cdot 5$ **51** $3 \cdot 5 \cdot 7$ **53** $2 \cdot 2 \cdot 3 \cdot 3$ **55** $2 \cdot 3 \cdot 5 \cdot 7$ **57** What number is 25% of 350?

59 What percent of 24 is 16? **61** 46 is 75% of what number?

Problem Set 7.3

1 70% **3** 84% **5** 45 mL **7** 18.2 acres for farming; 9.8 acres not available for farming **9** 3,000 **11** 400

13 1,664 **15** 31.25% **17** 50% **19** 42 mL **21** $\frac{1}{5}$ **23** $\frac{5}{12}$ **25** $\frac{3}{4}$ **27** $\frac{2}{3}$

29 36.3% to the nearest tenth of a percent **31** 91 hits **33** At least 18 hits

Problem Set 7.4

1 \$45 **3** \$0.75, \$15.75 **5** \$150, \$156 **7** 5% **9** \$1,440 **11** \$200 **13** 14% **15** \$11.93 **17** 4.5%

19 \$3,995 **21** $\frac{1}{2}$ **23** $2\frac{2}{3}$ **25** $1\frac{1}{2}$ **27** $4\frac{1}{2}$ **29** Sales tax = \$3,180; luxury tax = \$2,300 **31** \$2,300

33 You will save \$1,600 on the sticker price and \$150 in luxury tax. If you live in a state with a 6% sales tax rate, you will save an additional 0.06(\$1,600) = \$96.

Problem Set 7.5

1 \$24,610 **3** \$3,510 **5** \$13,200 **7** 8% **9** 20% **11** 200% **13** \$45, \$255 **15** \$381.60 **17** \$25,295.88

19 \$3.71 **21** 1 **23** $\frac{3}{4}$ **25** $1\frac{5}{12}$ **27** 6 **29** .400 **31** .001 **33** .200 **35** .003

Problem Set 7.6

1 \$2,160 **3** \$665 **5** \$8,560 **7** \$180 **9** \$5 **11** \$813.33 **13** \$5,618 **15** \$8,407.56 **17** \$974.59

19 \$703.11 **21** $\frac{1}{2}$ **23** $\frac{3}{8}$ **25** $\frac{1}{12}$ **27** $1\frac{1}{4}$ **29** 5.7% alcohol

31 To the nearest tenth, it was 14.2 times as expensive. **33** 284 million gallons (to the nearest million gallons)

Chapter 7 Review

1 0.35 **3** 0.05 **5** 95% **7** 49.5% **9** $\frac{3}{4}$ **11** $1\frac{9}{20}$ **13** 30% **15** $66\frac{2}{3}\%$ **17** 16.8 **19** 80 **21** \$72
23 \$477 **25** \$17,431.19 **27** 0.8 g **29** \$42

Chapter 7 Test

1 0.18 **2** 0.04 **3** 0.005 **4** 45% **5** 70% **6** 135% **7** $\frac{13}{20}$ **8** $1\frac{23}{50}$ **9** $\frac{7}{200}$ **10** 35% **11** 37.5%
12 175% **13** 45 **14** 45% **15** 80 **16** 92% **17** \$960 **18** 16% off; \$40 **19** \$220.50 **20** \$100
21 \$2,520

APPENDIX 1

1 4 **2** 10 **3** 9 **4** 3 **5** 5 **6** 7 **7** 11 **8** 8 **9** 17 **10** 13 **11** 10 **12** 13 **13** 9 **14** 5
15 1 **16** 10 **17** 8 **18** 14 **19** 13 **20** 7 **21** 6 **22** 1 **23** 11 **24** 7 **25** 12 **26** 9 **27** 7
28 10 **29** 6 **30** 4 **31** 8 **32** 13 **33** 6 **34** 8 **35** 9 **36** 9 **37** 10 **38** 5 **39** 14 **40** 18
41 0 **42** 15 **43** 11 **44** 10 **45** 10 **46** 13 **47** 12 **48** 13 **49** 9 **50** 4 **51** 12 **52** 1 **53** 11
54 12 **55** 15 **56** 17 **57** 3 **58** 8 **59** 3 **60** 6 **61** 12 **62** 7 **63** 2 **64** 16 **65** 11 **66** 12
67 11 **68** 9 **69** 16 **70** 8 **71** 15 **72** 6 **73** 5 **74** 11 **75** 16 **76** 3 **77** 9 **78** 4 **79** 11
80 10 **81** 8 **82** 2 **83** 10 **84** 14 **85** 8 **86** 15 **87** 5 **88** 12 **89** 14 **90** 4 **91** 6 **92** 7
93 8 **94** 9 **95** 9 **96** 14 **97** 7 **98** 5 **99** 7 **100** 6

APPENDIX 2

1 12 **2** 20 **3** 0 **4** 0 **5** 10 **6** 7 **7** 16 **8** 14 **9** 7 **10** 45 **11** 12 **12** 15 **13** 15 **14** 20
15 8 **16** 21 **17** 40 **18** 0 **19** 0 **20** 0 **21** 24 **22** 16 **23** 35 **24** 0 **25** 14 **26** 6 **27** 8
28 28 **29** 24 **30** 48 **31** 20 **32** 21 **33** 0 **34** 32 **35** 18 **36** 30 **37** 9 **38** 16 **39** 0
40 12 **41** 36 **42** 0 **43** 0 **44** 54 **45** 12 **46** 24 **47** 63 **48** 9 **49** 3 **50** 81 **51** 8 **52** 56
53 10 **54** 56 **55** 4 **56** 0 **57** 72 **58** 6 **59** 45 **60** 49 **61** 30 **62** 0 **63** 27 **64** 72 **65** 63
66 24 **67** 3 **68** 5 **69** 54 **70** 2 **71** 4 **72** 4 **73** 18 **74** 5 **75** 18 **76** 54 **77** 0 **78** 0
79 42 **80** 0 **81** 0 **82** 27 **83** 1 **84** 48 **85** 0 **86** 28 **87** 32 **88** 0 **89** 9 **90** 40 **91** 8
92 24 **93** 35 **94** 2 **95** 36 **96** 6 **97** 12 **98** 6 **99** 42 **100** 0

INDEX

W

X

Y

Z

STUDENT QUESTIONNAIRE

Your chance to rate ***Prealgebra, Third Edition*** (McKeague):

In order to keep this text responsive to your needs, it would help us to know what you, the student, thought of it. We would appreciate it if you would answer the following questions. Then cut out the page, fold, seal, and mail it; no postage required. Thank you for your help.

What chapters did you cover? (circle) 1 2 3 4 5 6 7 All

Does the book have enough worked-out examples? Yes ______ No ______

enough exercises? Yes ______ No ______

Which helped most?

Explanations ______ Examples ______ Exercises ______ All three ______ Other ____________________ *(fill in)*

Were the answers at the back of the book helpful? Yes ______ No ______

Did the answers have any typos or misprints? If so, where?

For you, was the course elective? ______ Required? ______

Do you plan to take more mathematics courses? Yes ______ No ______

If yes, which ones?

How much mathematics did you have before this course? Terms in high school (circle) 1 2 3 4

Courses in college (circle) 1 2 3

If you had mathematics before, how long ago?

Last 2 years ______ 3-5 years ago ______ 5 years of longer ______

What is your major or your career goal? ______________________________ Your age? ______

What did you like the most about ***Prealgebra, Third Edition***?

Can we quote you? Yes ______ No ______

FOLD HERE

What did you like the least about the book?

College ________________________________ State ______________

FOLD HERE

NO POSTAGE NECESSARY IF MAILED IN THE UNITED STATES

BUSINESS REPLY MAIL

FIRST-CLASS MAIL PERMIT NO.5438 BOSTON, MA

POSTAGE WILL BE PAID BY ADDRESSEE

Developmental Math Editor

PWS PUBLISHING COMPANY

20 Park Plaza 13th floor

BOSTON MA 02116-9900